THIRD EDITION

BIOPHYSICAL FOUNDATIONS OF HUMAN MOVEMENT

BRUCE ABERNETHY, PhD
University of Queensland

VAUGHAN KIPPERS, PhD
University of Queensland

STEPHANIE J. HANRAHAN, PhD
University of Queensland

MARCUS G. PANDY, PhD
University of Melbourne

ALISON M. MCMANUS, PhD
University of Hong Kong

LAUREL MACKINNON, PhD
University of Queensland

HUMAN KINETICS

Library of Congress Cataloging-in-Publication Data

Biophysical foundations of human movement / Bruce Abernethy . . . [et. al.]. -- 3rd ed.
   p. ; cm.
   Includes bibliographical references and index.
   ISBN 978-1-4504-3165-1 (print) -- ISBN 1-4504-3165-8 (print)
   1. Abernethy, Bruce, 1958-
   [DNLM: 1. Movement--physiology. 2. Biomechanics. 3. Biophysics. 4. Exercise--physiology. 5. Sports--psychology. WE 103]

   612.7'6--dc23

2012040566

ISBN-10: 1-4504-3165-8 (print)
ISBN-13: 978-1-4504-3165-1 (print)

The web addresses cited in this text were current as of August 2012, unless otherwise noted.

Acquisitions Editor: Myles Schrag; **Developmental Editor:** Melissa J. Zavala; **Assistant Editors:** Elizabeth Evans, Amy Akin, and Red Inc.; **Copyeditor:** Amanda Eastin-Allen; **Indexer:** Nancy Ball; **Permissions Manager:** Dalene Reeder; **Graphic Designer:** Joe Buck; **Graphic Artist:** Yvonne Griffith; **Cover Designer:** Keith Blomberg; **Photograph (cover):** © Dannie Walls/Icon SMI; **Photographs (interior):** © Human Kinetics, unless otherwise noted; © Jana Lumley/fotolia.com (p. 1, left; p. 3); © From the Esquiline Hill; former collection Massimo-Lancellotti (p. 1, right; p. 13); © 2001 Brand X Pictures (p. 25, far left; p. 31); © iStockphoto/Eduardo Jose Bernardino (p. 25, third from left; p. 59); © Comstock (p. 25, far right; p. 73); © Stockdisc Royalty Free Photos (p. 83, third from left; p. 121); © Photodisc (p. 155, far left; p. 159); © Wojciech Gajda/fotolia.com (p. 155, third from left; p. 181); © Anton/fotolia.com (p. 155, far right; p. 193); © Eyewire (p. 213, third from left; p. 257); © Suprijono Suharjoto/fotolia.com (p. 293, third from left; p. 321); © Brand X Pictures (p. 337, far left; p. 341); **Photo Asset Manager:** Laura Fitch; **Photo Production Manager:** Jason Allen; **Art Manager:** Kelly Hendren; **Associate Art Manager:** Alan L. Wilborn; **Illustrations:** © Human Kinetics, unless otherwise noted; **Printer:** Total Printing Systems

Printed in the United States of America   10 9 8

The paper in this book is certified under a sustainable forestry program.

Human Kinetics
1607 N. Market Street
Champaign, IL 61820
USA

United States and International
Website: **US.HumanKinetics.com**
Email: info@hkusa.com
Phone: 1-800-747-4457

Canada
Website: **Canada.HumanKinetics.com**
Email: info@hkcanada.com

E5730

Tell us what you think!
Human Kinetics would love to hear what we can do to improve the customer experience. Use this QR code to take our brief survey.

To all those scholars, past and present, who have contributed to the academic credibility of the study of human movement.

CONTENTS

Part IV Physiological Bases of Human Movement: Exercise Physiology **155**

| **Part V** | **Neural Bases of Human Movement: Motor Control** | **213** |

PREFACE

This third edition of *Biophysical Foundations of Human Movement*, like the first two, was written with three main purposes in mind. The first purpose of this text is to provide an introduction to key concepts concerning the anatomical, mechanical, physiological, neural, and psychological bases of human movement. Fulfilling this purpose involves both considering the biophysical dimensions of the field of study known in different parts of the world as human movement studies, human movement science, kinesiology, or sport and exercise science and examining the discipline–profession links in this field. Gaining an overview of the field provides students of kinesiology with an entrée to more detailed study in one or more of the subdisciplines of human movement studies and helps lay the foundations for integrative, multidisciplinary, and cross-disciplinary studies.

The second purpose is to provide an overview of the multidimensional changes in movement and movement potential that occur with the processes of growth, development, maturation, and ageing throughout the life span. The third purpose is to provide a comparable overview of the changes that occur in movement and movement potential as an adaptation to training, practice, and other lifestyle factors. Fulfilling the second and third purposes is important as a means of exposing readers to fundamental issues in biology and in positioning the study of movement as a major topic in human biology. We deliberately selected life span changes and adaptation as key organising themes for this book because of their centrality to biology.

Gaining knowledge about the processes of growth, development, maturation, and ageing aids understanding of key changes in movement potential throughout the life span that, because of their inevitability, directly impact all of us. Although the processes of maturation and ageing are inevitable, adaptation (through training, practice, and lifestyle decisions) offers humans some degree of control of their own destiny and capabilities. It is our sincere hope that the clear message about the important role physical activity plays in the maintenance of health, which arises from consideration of adaptation in each of the biophysical subdisciplines of human movement, will be one readers of the text will both take to heart themselves in making personal lifestyle decisions and communicate to others.

The text is structured in seven parts to broadly reflect its main purposes. Part I provides a general introduction to human movement studies as a field of study. It does this by examining the disciplinary and professional structure of contemporary human movement studies (in chapter 1) and providing a brief overview of the historical origins of the current academic field (in chapter 2).

Part II provides an introduction to basic concepts, adaptations that arise in response to training, and life span changes in each of the five major biophysical subdisciplines of human movement: functional anatomy, biomechanics, exercise physiology, motor control, and sport and exercise psychology. These subdisciplines represent the anatomical, mechanical, physiological, neural, and psychological bases of human movement, respectively.

To explain these subdisciplines, one can make a crude analogy to automotive engineering. To understand and optimise the performance of a motor vehicle, the automotive engineer needs specific knowledge about the vehicle's material structure (its anatomical basis); mechanical design characteristics (its mechanical basis); motor, capacity, and fuel consumption (its physiological basis); and electrical wiring, steering, and control mechanisms (its neural basis) and the characteristics and capabilities of its driver (its psychological basis). In the case of the human vehicle, this information is provided by the subdisciplines of functional anatomy, biomechanics, exercise physiology, motor control, and sport and exercise psychology and the interactions between them.

The structure of parts II through VI is broadly the same. Each part begins with a brief introduction that defines the subdiscipline and provides information on its historical development, the typical issues and problems it addresses, the level(s) of analysis it uses, and relevant professional training and organisations. This introduction is then followed by chapters that provide an overview of the basic concepts in each of the subdisciplines and then by chapters on life span changes and adaptation.

Part VII of the book is devoted to multidisciplinary and cross-disciplinary approaches to human movement and provides some examples of contemporary issues in which the application and integration of knowledge from a number of the biophysical subdisciplines is fundamental to understanding. The material presented in this part of the text demonstrates that integrating information from different subdisciplines is the essential strength and prospective direction of the academic discipline of human movement studies and of practice in those professions grounded in its knowledge base. Separate chapters are devoted to applications to chronic-disease prevention and management, injury prevention and management, and performance enhancement.

Throughout the book, special "In Focus" sidebars features highlight key organisations, individuals, and studies that have contributed to our understanding of human movement. Although the body of the text does not, as a general rule, focus on specific research studies or methods, the "In Focus" sidebars give the reader some feel for the research methods in this field and provide more detailed exposition of some of the pivotal studies that underpin current knowledge. In this way these featured sections present a snapshot (albeit a limited one) of the kind of research methods and information students can expect to encounter in more detailed investigations in each subdiscipline of human movement studies. Such investigations are typically undertaken in the second and subsequent years of formal courses in kinesiology or human movement studies. The "In Focus" sections contain research examples from around the world, highlighting the importance of positioning research and knowledge in a relevant social and cultural context and illustrating the genuinely international nature of research activity in this field.

A glossary at the end of the text defines key terms introduced throughout the book and is presented as an aid to student comprehension and review. The glossary presents definitions of all terms that appear in bold in the text.

As the title of the book indicates, our intention in writing and revising this book has been to provide an introduction to the biophysical fundamentals of human movement. Therefore, by necessity, our coverage of the topic is broad and illustrative rather than detailed and exhaustive. Students with an interest in further study may refer to the many excellent specialised texts now available for each of the subdisciplines, a number of which are listed in the Further Reading section of each chapter. Details of relevant websites are also added to assist those interested in seeking out further information.

In making revisions for this third edition, we included new and updated information that has become available since the previous edition was published in 2005. We also made some changes in formatting and presentation style in response to the helpful comments made by readers of the earlier editions. The major change has been made to the final part of book, where the treatment of integrative perspectives has been expanded from one chapter to three to provide more examples of the type of interdisciplinary research that is increasingly at the forefront of new knowledge and understanding about human movement. These revisions place increased emphasis on the application of knowledge from across the various biophysical subdisciplines to health and to performance enhancement and include examples drawn from fields including public health, sport, and the workplace. A new chapter on the historical origins of the academic study of human movement (chapter 2) has also been added to part I to help students understand where the field of study has come from and better predict where it might be going in the future. The material on exercise physiology (part IV) has been reorganised in this edition to provide more discrete coverage of key concepts in nutrition. In keeping with the changes introduced in the previous edition, we continue in this edition to strengthen the use of supporting web materials. Ancillary products supporting this text are free to adopting instructors. Contact your Sales Manager for details about how to access HK*Propel*, our ancillary delivery and learning platform. The instructor guide offers suggestions for class activities. The test package includes questions that instructors may use to create customized tests. The image bank includes most of the art, tables, and content photos, sorted by chapter, and can be used to create customized presentations based on specific course requirements or used in creating other learning aids for students. We trust that these many changes will enhance both the readability and pedagogical value of the text.

The study and understanding of human movement presents an exciting challenge for students, scientists, and practitioners alike. Given how central an understanding of human movement and its enhancement is to a wide range of human endeavours, it is our hope that this third edition of *Biophysical Foundations of Human Movement* will serve as a readable introduction for both students and professionals involved in the many professions—sport and exercise science, physical education, ergonomics, music and performing arts, physiotherapy, occupational therapy, nursing, medicine, health education and health promotion, and other rehabilitation and health sciences to name but a few—that are grounded in an understanding of human movement or kinesiology. We trust that the text will help convey to our readers some of the fascination that this subject matter holds for us.

ACKNOWLEDGMENTS

In addition to those people acknowledged for their contribution to the previous editions of this book, we also express our sincere appreciation to those instructors who adopted the earlier editions for their courses and those students who read the text as part of their courses and provided valuable feedback;

- to the staff at Human Kinetics (Chris Halbert, Myles Schrag, and Melissa Zavala in particular) for their encouragement to produce a third edition of the text and their support, patience, and contributions in ensuring this came to fruition; and

- to our families, friends, and colleagues for their support, understanding, and tolerance while this book was being written (again!).

Without the collective encouragement of each and every one of these people, there would not have been a third edition.

The introductory chapters (chapters 1-2) and chapters 15 through 18 were written by Bruce Abernethy; Vaughan Kippers wrote chapters 3 through 6; Marcus Pandy wrote chapters 7 through 10; and Stephanie Hanrahan wrote chapters 19 through 22. Ali McManus, joining the author team for the first time, wrote chapters 11 through 14 and 23, basing a significant part of these chapters on material prepared for the previous editions by Laurel Mackinnon. Chapters 24 and 25, befitting the section to which they belong, were a collaborative, integrative effort.

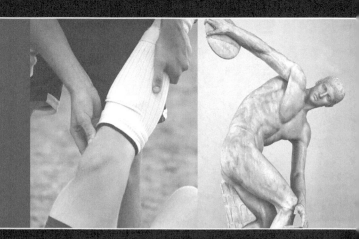

PART I

INTRODUCTION TO HUMAN MOVEMENT STUDIES

The field of human movement studies is not necessarily well known to the general public and to students who have not had specialist training in the field. Because of this, the first part of this text provides a broad introduction to the field and its historical origins, contemporary structure, and likely future directions. Chapter 1 describes the focus, significance, and structure of both the academic discipline (or knowledge base) of human movement studies and the numerous professions that use this knowledge in day-to-day practice. Chapter 2 provides a brief overview of the historical development of human movement studies as an academic discipline. This background knowledge is fundamental to better understanding the present structure of the field and to predicting future directions, opportunities, and challenges.

CHAPTER 1

HUMAN MOVEMENT STUDIES AS A DISCIPLINE AND A PROFESSION

The major learning concepts in this chapter relate to

- ▶ the focus and importance of human movement studies,
- ▶ the characteristics of disciplines and professions,
- ▶ the disciplinary and professional elements of human movement studies,
- ▶ naming the discipline of human movement studies, and
- ▶ the relationships between human movement studies and its key professions.

This chapter has a number of purposes. The first is to define and describe the field of human movement studies and outline some of the reasons it is significant. The second is to explore the distinction and relationship between the discipline of human movement studies and a range of professions that are informed by knowledge about human movement. A third purpose is to look at how scholarly knowledge about human movement is organised and structured and how finding an appropriate, universally accepted name to describe this body of knowledge has proven problematic. Finally, through the use of sidebars, this chapter also introduces some of the key professional associations relevant to human movement studies. Understanding how knowledge about human movement is organised, and how human movement studies relates to other disciplines and professions, is an important stepping stone to scholarly and professional development in this field.

WHAT IS HUMAN MOVEMENT STUDIES AND WHY IS IT IMPORTANT?

Human movement studies is the field of academic inquiry concerned with systematically understanding *how* and *why* people move and the factors that limit and enhance our capacity to move.

Two key points need to be emphasised in this very simple definition of the field of human movement studies. The first key point is that the unique focus of the field is on human movement. This is true regardless of whether the movement to be ultimately understood is one performed in the context of undertaking a fundamental daily skill (such as walking, speaking, or reaching and grasping), executing a highly practiced sport or musical skill, exercising for health, or regaining the function of an injured limb. The study of human movement is important in and of itself because movement is a central biological and social phenomenon.

The study of movement is central to the understanding of human biology because movement is a fundamental property—indeed, indicator—of life (remembering that biology is the study of life). Human movement, as we note in the preface, offers a valuable medium for the study of biological phenomena fundamental to developmental changes across the life span (changes that occur with ageing as a consequence of internal body processes), to **adaptation** (changes that occur as an accommodation or adjustment to environmental processes), and to the interactions of genetic and environmental factors (nature and nurture) that dictate human phenotypic expression.

Human movement, especially that which occurs in collective settings such as organised sport, exercise and rehabilitation settings, and health and physical education classes, also clearly has an essential social and cultural component that warrants intensive study. Understanding individual and group motives and opportunities and barriers to involvement in different types of human movement, for instance, provides an important window into the nature of human society, just as understanding the mechanisms of individual human movement provides an important window into the understanding of human biology. Movement, in short, plays a fundamental

role in human existence and what it means to be human and for these reasons warrants our very best efforts to understand it. Knowledge about human movement is fundamental to optimising health and performance and preventing injury and illness.

The second key point that needs to be highlighted in our definition of human movement studies is the importance of a systematic, research-based approach to the generation of knowledge. Because many aspects of the current practice of human movement involve practices based as much on fads, folklore, tradition, and intuition as on sound, logical theory substantiated by systematically collected, reproducible data, it is imperative that the knowledge base for human movement studies be one based upon research conducted with a methodological rigor equivalent to that of other established biological, physical, and social sciences. Only through such an approach can fact be separated from fiction and can a sound basis for best practice in the profession, based on the knowledge base of human movement studies, be established.

In the tradition of the scientific method, the field of human movement studies aims not only to describe key phenomena in its domain but also to move beyond *description* to *understanding* through explanation and prediction. Human movement studies therefore carries the twin goals of all fields of science:

1. generating knowledge through understanding of basic phenomena, and
2. applying knowledge for the benefit of society.

The basic understanding of human movement, for which the field strives, has applicability to all of the many areas and professions that deal with the enhancement of our capacity to move. Obvious areas of application of the knowledge base of human movement studies are sport, exercise, health and physical education, the workplace, and rehabilitation. Professionals who rely on such information include those working in the health and fitness industry, sport scientists, physical educators, health promoters, doctors, nurses, and rehabilitation therapists.

How is this vital knowledge about human movement organised and how is it translated into the practices of relevant professions? To answer this question we must consider the discipline and professions of human movement studies.

DISCIPLINES AND PROFESSIONS

According to the pioneering American physical educator Franklin Henry (1964):

> *An academic discipline is an organised body of knowledge collectively embraced in a formal course of learning. The acquisition of such knowledge is assumed to be an adequate and worthy objective as such, without any demonstration or requirement of practical application. The content is theoretical and scholarly as distinguished from technical and professional.*

The principal function of a discipline is therefore to develop a coherent body of knowledge that describes, explains, and predicts key phenomena from the domain of interest (or subject matter).

In contrast, professions, as a general rule, attempt to improve the conditions of society by providing a regulated service in which practices and educational or training programs are developed that are in accordance with knowledge available from one or more relevant disciplines. Practice in the profession of engineering, for example, is based on application of knowledge from disciplines such as physics, mathematics, chemistry, and computer science, whereas practice in the medical profession is based on knowledge from disciplines such as anatomy, physiology, pharmacology, biochemistry, and psychology. Established professions share a number of characteristics, including

1. an identified set of jobs or service tasks over which they have jurisdiction or monopoly;
2. organisation under the framework of a publicly recognised association;
3. identified educational competencies and formalised training and education criteria (this generally includes the mastery of complex skills and the presence of a theory and evidence base for their practice);
4. political recognition, usually through acts of government legislation (including, in some cases, establishment of licensing or registration boards); and
5. a code of ethics defining minimal standards of acceptable practice.

Disciplines therefore seek to understand subject matter and professions seek to implement change based on this understanding. The emphases in disciplines and professions are often characterised as theory and research versus application and practice, but such a distinction is overly simplistic and potentially misleading. Applied research (including research on aspects of professional practice) is now an accepted part of the business of the discipline just as the profession may frequently be the site for original, discovery-type research.

IS HUMAN MOVEMENT STUDIES A DISCIPLINE?

Remembering our earlier definition of human movement studies and Henry's definition of a discipline, an important practical and philosophical question for our field is whether there is a unique, organised body of knowledge on how and why people move (to satisfy the criteria for a discipline of human movement studies) or whether human movement studies is simply the application of knowledge from other disciplines such as anatomy, physiology, and so on (thereby making it, by definition, more a profession than a discipline). This question has been the source of much debate both in and beyond the field for at least 50 yr. The extent of the uniqueness and collective coherence of the knowledge in the field has been the source of most contention. The establishment of university departments of human movement studies, kinesiology, and the like, independent of traditional professions such as physical education teacher training, is predicated on the assumption that human movement studies possesses an organised body of knowledge in much the same way as traditional disciplines such as physics, chemistry, and psychology and that human movement studies is more than simply a loose collection of the applications of knowledge from other fields.

The forebears of the modern field of human movement studies were, as discussed in the next chapter, primarily physical educators. A number of these, most particularly the physiologist and psychologist Franklin Henry [see "Franklin Henry (1904-1993)" in chapter 2] and the motor developmentalist Lawrence Rarick, created strong arguments in the 1960s for both the importance and the existence of a discipline of human movement studies, claiming that such a field (or its precursor, physical education) asked questions that would not have arisen from **cognate disciplines.**

Rarick (1967) argued that

> Most certainly human movement is a legitimate field of study and research. We have only just begun to explore it. There is need for a well-organised body of knowledge about how and why the human body moves, how simple and complex motor skills are acquired and executed, and how the effects (physical, psychological, and emotional) of physical activity may be immediate or lasting. (p. 51)

In a similar vein, Henry, in a much-quoted 1964 paper titled "Physical Education: An Academic Discipline," claimed the pre-existence of a discipline base for the study of human movement:

> There is indeed a scholarly field of knowledge basic to physical education. It is constituted of certain portions of such diverse fields as anatomy, physics and physiology, cultural anthropology, history and sociology, as well as psychology. The focus of attention is on the study of man as an individual, engaging in the motor performances required by his daily life and in other motor performances yielding aesthetic values or serving as expressions of his physical and competitive nature, accepting challenges of his capability in putting himself against a hostile environment, and participating in the leisure time activities that have become of increasing importance in our culture.

Henry's definition of the discipline base of our field has generally stood well the test of time, save some relatively minor changes, most obviously

▸ the substitution of the term *human movement studies* for *physical education* (now typically defined in a narrower, professional sense),

▸ the extension of the focus of the field beyond solely the study of the person as an individual to also incorporate the study of the person as an element of a social system, and

▸ the alteration of the language to acknowledge the equal involvement of females and males.

STRUCTURE OF A DISCIPLINE OF HUMAN MOVEMENT STUDIES

Because movement potential and performance are known to be influenced by many things, including biological factors (such as maturation, ageing, train-

ing, and lifestyle), health factors (such as disease, disuse, and injury), and social factors (such as motivation, incentive, and opportunity), it is clear that a discipline of human movement studies must draw heavily, but not exclusively, on the methods, theories, and knowledge of a wide range of other disciplines and provide them with an integrative focus on human movement. Information of relevance for a discipline of human movement studies may be gleaned from biological science disciplines such as anatomical science, physiology, and biochemistry; physical science disciplines such as physics, chemistry, mathematics, and computer science; social science disciplines such as psychology, sociology, and education; and disciplines in the humanities such as history and philosophy.

Figure 1.1 presents one possible way of conceptualising the organisation of knowledge in a discipline of human movement studies. In this conceptualisation the discipline of human movement studies consists of the collective knowledge contained in and between each of the subdisciplines of functional anatomy, exercise physiology, biomechanics, motor control, sport and exercise psychology, and the pedagogy, sociology, history, and philosophy of sport and physical activity (as illustrated by the shaded box in figure 1.1). The subdisciplines of functional anatomy, exercise physiology, biomechanics, motor control, and sport and exercise psychology (in so much as it focuses on the individual rather than group or societal behaviour) constitute the biophysical foundations of human movement and are afforded coverage in this text. The social psychology of sport and exercise along with the pedagogical, sociological, historical, philosophical, political, and cultural aspects of physical activity and sport constitute the sociocultural foundations of human movement.

Figure 1.1 clearly indicates that each of the subdisciplines draws theories, methods, and knowledge from one or more cognate disciplines. It is also important to recognise, however, that each of the specialist subdisciplines draws on only a subset of the knowledge contained in its cognate discipline(s) and that the subdisciplines can, and frequently do, generate theories, methods, and approaches of their own not acquired from the cognate discipline(s).

Figure 1.1 also suggests a number of other points with respect to the conceptualisation of the discipline of human movement studies.

▸ The clustering of disciplines into discipline groups and the selection of the subdiscipline

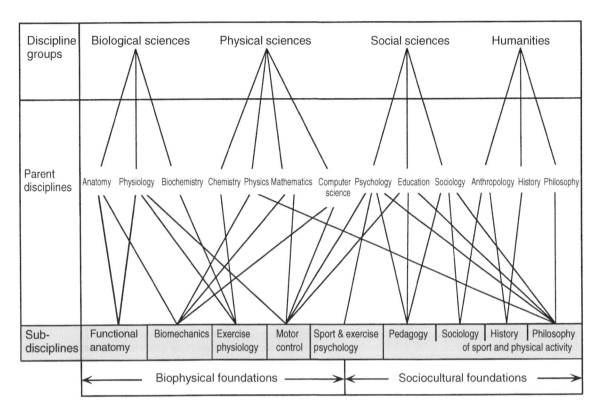

Figure 1.1 One possible conceptualisation of the structure of knowledge about human movement. The discipline of human movement studies is represented by the shaded area.

groups is necessarily somewhat arbitrary, and no two conceptualisations of the field, its interrelations, and the naming of its component parts are likely to be identical.

▶ The interrelationships of cognitive disciplines and human movement studies subdisciplines are frequently reflected in the scheduling and prerequisite course structuring of many tertiary programs in human movement studies. Basic exposure to the cognate disciplines generally precedes exposure to each of the subdisciplines of human movement studies.

▶ The broad general education in the sciences and the arts that students of human movement studies or kinesiology must first undertake in order to understand the foundations of their own discipline positions them well for careers in professions linked to human movement studies and for further study in many cognate fields.

▶ The disciplines and subdisciplines are organised in figure 1.1 in order to present a generally progressive shift from a focus on the microphenomena to the macrophenomena of human movement.

The subdisciplines are presented in figure 1.1 as essentially insular components, having as much (or more) contact with the cognate discipline as with the other subdisciplines of human movement studies. To some extent this representation is an accurate reflection of the current state of the field, and there is increasing concern in many quarters that the growing differentiation and specialisation in human movement studies may produce fragmentation and an inevitable loss of integrity in the discipline base. This fragmentation is reinforced by the dominant tendency in tertiary programs to teach each of the subdisciplinary fields in a separate, independent course. To this end the field of human movement studies is probably most accurately depicted as currently being **multidisciplinary**, whereas the desirable direction is to make it more **cross-disciplinary** and ultimately **interdisciplinary** or **transdisciplinary** (figure 1.2). A good indicator of the maturation of a discipline of human movement studies will be the extent to which it becomes more interdisciplinary and advances in knowledge are made by crossing the traditional (but arbitrary) boundaries between the subdisciplines and by synthesising material from the subdisciplines rather than importing ideas from the mainstream disciplines.

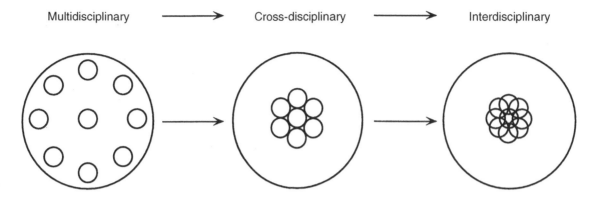

Figure 1.2 A desirable evolutionary progression for a discipline of human movement studies. The components shown by circles represent the subdisciplines identified in figure 1.1.

Adapted, by permission, from E.F. Zeigler, 1990, Don't forget the profession when choosing a name! In *The evolving undergraduate major*, edited by C.B. Corbin and H.M. Eckert (Champaign, IL: Human Kinetics), 67-77.

WHAT SHOULD THE DISCIPLINE OF HUMAN MOVEMENT STUDIES BE CALLED?

It is clear from the previous sections that we favour the use of the term *human movement studies* to describe the discipline; however, such a term is far from universally accepted. There is essentially no uniformity in the names selected by the numerous university departments offering courses in human movement studies either between countries or in the same country. The fact that finding a widely accepted name for the discipline always has been, and still remains, a persistent source of debate perhaps indicates the relative immaturity and diversity of our field of study compared with some others.

The term *human movement studies* was first coined by the British physical educator and psychologist H.T.A. (John) Whiting in the early 1970s [see "H.T.A. (John) Whiting (1929-2001)" in chapter 2], and it has subsequently been adopted in parts of Europe, the United States, and Australia and at one point became the title for one of the field's journals. The term is useful because it clearly encapsulates the unique and unifying theme in the body of knowledge in our field, but it is considered by some as too cumbersome, especially in a context of public marketing. The alternative term *human movement science* has had some following, especially in Europe, and is probably appropriate for the biophysical aspects of human movement.

Where this term is problematic is with aspects of the sociocultural foundations of human movement, where knowledge acquisition occurs through methods in addition to those of traditional science.

In North America, the term *kinesiology*, literally meaning the study of movement (from the Greek *kinein*, to move, and *logos*, a branch of learning), is widely used. The National Academy of Kinesiology, which has championed the cause for universal use of the term in degree titles and department names, defines kinesiology as ". . . a multifaceted field of study in which movement or physical activity is the intellectual focus" (http://www.nationalacademyofkinesiology.org/what-is-kinesiology). Knowledge drawn from experience of movement, as well as knowledge drawn from scholarly research, is seen as an integral part of the study of kinesiology. The term *kinesiology* has not yet been widely adopted internationally, presumably because, in many countries, it is poorly understood in general use, is in general use already in relation to some forms of alternative therapies practices that are by and large not evidence based, or is used in a scholarly but much narrower context to refer simply to the mechanics of human movement.

In contrast to the terms *human movement studies* and *kinesiology*, the terms *exercise and sport science* and *physical education* are well understood by the general public but are much narrower in focus. The current use of the term *physical education* is more closely tied to notions of a profession than to notions of an all-encompassing discipline.

PROFESSIONS BASED ON HUMAN MOVEMENT STUDIES

As noted earlier, professions, unlike disciplines, have a specific interest in the application of knowledge as a means of solving specific problems, enhancing the quality of life, and providing a service to society. Standards of practice in specific professions are typically controlled by professional bodies that impose minimal training and education requirements, set membership and accreditation criteria, and establish codes of professional ethics. A wide range of professions draw on knowledge from the discipline of human movement studies, and each of these is represented by one or more professional bodies. Like other disciplines, human movement studies informs multiple professions, but in a nonexclusive way. For example, human movement studies informs professions such as sports medicine, physical therapy, and health promotion, but all of these professions are also informed by knowledge generated in other disciplines such as anatomical sciences, physiology, and pharmacology.

Traditionally, the profession most linked to the knowledge base of human movement studies has been physical education, which, as discussed in chapter 2, is the principal historical forebear of human movement studies. Over time, new professions (such as those related to exercise prescription, sport and exercise science, exercise rehabilitation and physical therapy, ergonomics, sports medicine, athletic training, coaching, sport management, and sport and exercise psychology) have emerged, all of which draw to some degree on the discipline of human movement studies.

PROFESSIONAL ORGANISATIONS

There are currently a number of organisations, both nationally and internationally, that represent key professions whose practices are grounded in human movement studies. "Examples of Some Major Professional Organisations Relevant to the Discipline of Human Movement Studies" lists some of the major organisations (along with their acronyms) in sports medicine, exercise science, and physical education, recreation, and dance. Additional details on the functions and goals of three of the major professional organisations follow.

EXAMPLES OF SOME MAJOR PROFESSIONAL ORGANISATIONS RELEVANT TO THE DISCIPLINE OF HUMAN MOVEMENT STUDIES

International

International Council of Sport Science and Physical Education (ICSSPE): www.icsspe.org

Fédération Internationale de Médecine du Sport (FIMS): http://www.fims.org/

International Council for Health, Physical Education, Recreation, Sport, and Dance (ICHPER·SD): www.ichpersd.org

Association Internationale des Ecoles Superieures d'Education Physique (AISEP): www.aiesep.ulg.ac.be

North American

American College of Sports Medicine (ACSM): www.acsm.org

American Alliance for Health, Physical Education, Recreation, and Dance (AAHPERD): www.aahperd.org

National Academy of Kinesiology: www.nationalacademyofkinesiology.org

American Kinesiology Association: www.americankinesiology.org

Canadian Kinesiology Alliance: www.cka.ca

Physical and Health Education Canada: www.phecanada.ca

Canadian Academy of Sport and Exercise Medicine (CASEM): www.casm-acms.org

European

European College of Sport Science (ECSS): www.ecss.mobi

European Federation of Sports Medicine Associations (EFSMA): www.efsma.net

L'association des Chercheurs en Activités Physiques et Sportives (ACAPS) (France)

(continued)

Examples of Some Major Professional Organisations Relevant to the Disciplines of Human Movement Studies *(continued)*

European *(continued)*

Deutschen Vereinigung für Sportwissenschaft (DVS) (Germany): www.sportwissenschaft.de

Netherlands Association of Sports Medicine: www.sportgeneeskunde.com

The British Association of Sport and Exercise Sciences (BASES): www.bases.org.uk

British Association of Sport and Exercise Medicine (BASEM): www.basem.co.uk

Association for Physical Education (AfPE): www.afpe.org.uk

Asian and South Pacific

Asian Federation of Sports Medicine (AFSM): www.afsmonline.com

Hong Kong Association of Sports Medicine and Sports Science: www.hkasmss.org.hk

Chinese Association of Sports Medicine

The Society of Chinese Scholars on Exercise Physiology and Fitness (SCSEPF): www.scsepf.org

Japanese Federation of Physical Fitness and Sports Medicine

Korean Alliance for Health, Physical Education, Recreation, and Dance (KAHPERD): www.kahperd.or.kr

Indian Association of Sports Medicine: www.iasm.co.in

National Association of Physical Education and Sports Science (NAPESS): www.napess.org

Sports Medicine Australia (SMA): www.sma.org.au

Exercise and Sports Science Australia (ESSA): www.essa.org.au

The Australian Council for Health, Physical Education, and Recreation (ACHPER): www.achper.org.au

Sports Medicine New Zealand (SMNZ): www.sportsmedicine.co.nz

Africa

Biokinetics Association of South Africa: www.biokinetics.org.za

South African Sports Medicine Association (SASMA): www.sasma.org.za

South America

Confederación Sudamericana de Medicina del Deporte

Sociedade Brasileira de Medicina do Esporte

This listing is by no means exhaustive. More specialised bodies are listed in part II.

International Council of Sport Science and Physical Education

The International Council of Sport Science and Physical Education (ICSSPE) was founded in 1958 in Paris, France, under the name International Council of Sport and Physical Education. ICSSPE's main purpose is to serve as an international umbrella organisation concerned with the promotion and dissemination of results and findings in the field of sport science and the practical application of these findings in culture and educational contexts. Its aims are to contribute to the awareness of human values inherent in sport, to improve health and physical efficiency, and to develop physical education and sport in all countries to a high level. As a world organisation, ICSSPE endeavours to bridge the gap between developed and developing countries and to promote cooperation between scientists and organisations from countries with different political systems.

ICSSPE's five stated fundamental objectives are to

▸ encourage international cooperation in the field of sport science;

▸ promote, stimulate, and coordinate scientific research in the field of physical education and sport throughout the world and to support the application of its results in the many practical areas of sport;

▸ make scientific knowledge of sport and practical experiences available to all interested national and international organisations and institutions, especially to those in developing countries;

▸ facilitate differentiation in sport science while promoting the integration of the various branches; and

▸ support and implement initiatives with aims similar to those pursued by itself and initiated or developed by any other appropriate agency or organisation in the field.

ICSSPE conducts its scientific work in three main areas: sport science; physical activity, physical education, and sport; and scientific services (information dissemination). ICSSPE serves as a permanent advisory body to the United Nations Economic, Social, and Cultural Organisation (UNESCO) and regularly conducts research projects on behalf of UNESCO and the International Olympic Committee. It has eight regional bureaus throughout the world, plus close links to a number of other international organisations of physical education and sport science. Membership of ICSSPE is open to organisations and institutions rather than individual subscribers, and there are currently some 300 organisations and institutions affiliated with ICSSPE. More information about ICSSPE can be obtained at www.icsspe.org.

INTERNATIONAL FEDERATION OF SPORTS MEDICINE

The Fédération Internationale de Médecine Sportive (FIMS) is an organisation made up of the national sports medicine associations of more than 100 countries. It was founded in 1928 during a meeting in St. Moritz, Switzerland, of Olympic medical doctors with the principal purpose of promoting the study and development of sports medicine throughout the world.

The specific objectives of FIMS are to

- promote the study and development of sports medicine throughout the world;
- preserve and improve the health of mankind through participation in physical fitness and sport;
- study scientifically the natural and pathological implications of physical training and sport participation;
- organise or sponsor scientific meetings, courses, congresses, and exhibits on an international basis in the field of sports medicine;
- cooperate with national and international organisations in sports medicine and related fields; and
- publish scientific information in the field of sports medicine and in related fields.

FIMS hosts a major international conference in sports medicine every 3 yr. In addition, FIMS is active in the production of position statements on aspects of health, physical activity, and sports medicine. The most influential of these position statements is probably its 1995 joint statement with the World Health Organisation (WHO) titled *Physical Activity for Health*. FIMS now also publishes a bimonthly electronic journal titled *International SportMed Journal*. In addition to its member national associations, FIMS makes available individual membership for holders of doctoral degrees in medicine and related sciences who are affiliated with one of the member national bodies in sports medicine. More information about FIMS is available at www.fims.org.

AMERICAN COLLEGE OF SPORTS MEDICINE

The mission of the American College of Sports Medicine (ACSM) is to promote and integrate scientific research, education, and practical applications of sports medicine and exercise science to maintain and enhance physical performance, fitness, health, and quality of life. ACSM was founded in 1954 by a group of 11 physicians, physiologists, and educators. The organisation has grown rapidly since its formation, to the point where ACSM is presently the largest sports medicine and exercise science organisation in the world, comprising more than 45,000 members and certified professionals from North America and more than 90 other countries.

ACSM's research and educational programs are broad ranging. Its annual 4 d meeting held in May of each year is one of the major international conferences for the presentation and discussion of new research in sports medicine and exercise science. ACSM's official journal, *Medicine and Science in Sport and Exercise* (first published in 1969), is one of the principal international journals in the field for the publication of original research. Its other publications such as *Exercise and Sport Science Reviews*, *Current Sports Medicine Reports*, and *Health and Fitness Journal* are valuable sources of state-of-the-art reviews on key topics in sports medicine and exercise science. In addition to its role in advancing the discipline through disseminating basic and applied scientific research on physical activity, ACSM is the peak professional body in North America for the certification of individuals seeking to work in clinical sports medicine, exercise science, and the health and fitness industry. This role has been facilitated by the development of key guidelines and policy documents, such as its *Guidelines for Exercise Testing and Prescription*, which have internationally become the gold standard for professional practice. ACSM's recent global Exercise Is Medicine initiative (www.ExerciseisMedicine.org) aims to make physical activity and exercise a standard part of the prevention and medical treatment of chronic disease. ACSM offers a variety of membership types, details about which can be found at www.acsm.org.

RELATIONSHIPS BETWEEN THE DISCIPLINE AND THE PROFESSIONS

Thus far, our discussions of discipline–profession relations may have created the impression that the flow of information between the two is unidirectional, with the role of the discipline being to generate knowledge that can then be used by the profession(s) as a basis for practice. While this information flow is important, it is essential to recognise that the ideal relationship between discipline and profession should be one of mutual benefit (i.e., symbiosis) such that information flows as much from the professions to the discipline as it does in the reverse direction. In particular, the professions are well positioned to provide questions, problems, observations, and issues that can function as a valuable guide and focus for the knowledge-seeking, research-based activities of the discipline. Observations made in practice frequently form the initial basis of hypothesis testing in the discipline.

SUMMARY

Human movement studies is a discipline (an organised body of knowledge) concerned with understanding how and why people move and the factors that limit and enhance that capacity to move. The discipline goes by different names throughout the world, being known most frequently in North America as kinesiology. Human movement studies is important because movement is a fundamental biological and social phenomenon that has profound implications for human health and existence. Knowledge about human movement comes from a range of subdisciplines that span the focus from microlevel biophysical aspects of movement to macrolevel social and cultural phenomena related to organised forms of physical activity. As the discipline of human movement studies matures, it is expected that these subdisciplines will become less discrete and that cross-disciplinary and inter-disciplinary studies will become more prevalent. Knowledge about human movement forms the basis for the activities of an increasingly wide range of professions.

FURTHER READING AND REFERENCES

Bouchard, C., McPherson, B.D., & Taylor, A.W. (Eds.). (1992). *Physical activity sciences*. Champaign, IL: Human Kinetics.

Brooke, J.D., & Whiting, H.T.A. (Eds.). (1973). *Human movement: A field of study*. London: Henry Kimpton.

Brooks, G.A. (Ed.). (1981). *Perspectives on the academic discipline of physical education*. Champaign, IL: Human Kinetics.

Curtis, J.E., & Russell, S.J. (Eds.). (1997). *Physical activity in human experience: Interdisciplinary perspectives*. Champaign, IL: Human Kinetics.

Henry, F.M. (1964). Physical education: An academic discipline. *Proceedings of the 67th Annual Meeting of the National College Physical Education Association for Men* (pp. 6-9). Washington, D.C.: AAHPERD.

Hoffman, S.J. (Ed.). (2013). *Introduction to kinesiology: Studying physical activity* (4th ed.). Champaign, IL: Human Kinetics.

Newell, K.M. (1990). Kinesiology: The label for the study of physical activity in higher education. *Quest, 42*, 269-278.

Rarick, G.L. (1967). The domain of physical education. *Quest, 9*, 49-52.

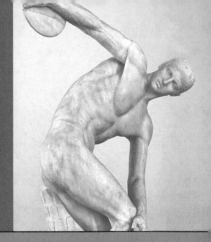

CHAPTER 2

HISTORICAL ORIGINS OF THE ACADEMIC STUDY OF HUMAN MOVEMENT

The major learning concepts in this chapter relate to

- ▶ the ancient origins of scholarly work on human movement;

- ▶ the ways in which the body and physical activity were viewed in the Middle Ages and in the Renaissance and Reformation periods;

- ▶ the professionalisation of physical education beginning in the late 1800s and, from this, the emergence of organised research efforts to understand and improve physical performance;

- ▶ the beginnings of a discipline of human movement studies in the 1960s and 1970s and, soon after, the emergence of subdisciplines and specialisations in the field;

- ▶ noting how key individuals influenced the development of the field internationally; and

- ▶ using a knowledge of the field's origins to understand the present and attempt to predict future developments and challenges.

To understand the current structure and status of the discipline of human movement studies and to attempt to predict, from a factual base, the future direction of the field, it is necessary to have some appreciation of the historical antecedents to the contemporary field. This chapter

- ▶ provides an overview of the ancient origins of interest in physical activity;

- ▶ traces the professionalisation of physical education throughout the late 19th century and early 20th century;

- ▶ chronicles the international emergence of a discipline of human movement studies from within the profession of physical education;

▶ outlines the emergence of specialist subdisciplines within human movement studies; and

▶ examines some of the future directions, challenges, and opportunities available in the field.

This chapter focuses only on the historical development of the broad field of human movement studies. The more specific developments in each of the five major biophysical subdisciplines of human movement studies are discussed separately in each of the section introductions in part II. Our coverage and categorisation of key periods in the development of a discipline of human movement studies draw heavily on the excellent 1981 review paper (detailed in the Further Reading section) by the American sport historian Roberta Park.

SCHOLARLY WRITINGS ON HUMAN MOVEMENT FROM ANCIENT CIVILISATIONS (CA. 1000 BC-350 AD)

Although there is evidence that organised physical activity existed in both China and Egypt before 1000 BC (in the form of wrestling, tumbling, swimming, and ball games in Egypt and ritualised physical training for the enhancement of physical prowess and moral strength in the Chou dynasty in China), a scholarly focus on physical activity began in ancient Greece in the period beginning around 450 BC. It can be claimed that the framework for both a profession of physical education and a discipline of human movement studies were founded during this classical antiquity period. Organised physical training in the form of wrestling, boxing, gymnastics, swimming, running, and discus and javelin throwing was used in the principal city-states of Sparta and Athens from as early as 900 BC as a means of conditioning boys for military service. Regular supervised physical training was provided to Spartan girls as part of their preparation for motherhood. This long tradition of physical activity as an integral part of ancient Greek society provided the platform for a number of the famous scholarly developments of relevance to the understanding of human movement.

Hippocrates (ca. 460-360 BC), the founder of modern medicine, argued for the study of the body (*physis*) as a natural object rather than as a spiritual or mystical object and made a number of observations on the relationship between diet, exercise, and fatigue that remain relevant today. The philosopher Plato (427-347 BC) criticised the **mind–body dualism** implicit in Hippocrates' work, arguing rather for a holistic view of the individual. Plato's writings contained numerous opinions and observations on the beneficial effects of physical education, including reference to the important roles played by exercise in growth and development and by play and physical activity in the genesis of social skills and a socially responsible community. Plato's pupil Aristotle (384-322 BC) promulgated a positive view of regular physical activity in his writings and considered gymnastics (interpreted much more broadly than in its current narrow meaning) to be a complete science that had as its goal the discovery of the exercises and types of training that were of most benefit to health. Two of Aristotle's books, *Movement of Animals* and *Progression of Animals*, although lacking experimental evidence, described many of the phenomena (such as flexion–extension, action–reaction, centre of gravity, and base of support) that are still central to modern functional anatomy and biomechanics. Plato, Aristotle, and Socrates were all critical of athletics, which placed adverse emphasis on winning as opposed to all-around harmonious development of the individual, thus anticipating by more than 2,000 yr the concerns of modern sport psychologists regarding the potentially detrimental effects on personal development and cooperation that may arise from an excessive emphasis on competition. Of particular relevance to our discussions in chapter 1 about disciplines and professions is the distinction recognised by Aristotle in his writings between the systematic, scholarly understanding of exercise (*gymnastics*) and the practical applications of training techniques (*paedotribes*), which were often based on premises of questionable validity. This distinction (a precursor to the modern discipline–profession division) was also apparent in the writings of the physician Claudius Galen (ca. 129-200 AD), whose contribution to the development of the subdiscipline of functional anatomy is examined in more detail later in this text.

THE MIDDLE AGES AS A PERIOD OF SUPPRESSION OF THE STUDY OF HUMAN MOVEMENT (CA. 350-1350 AD)

In the Middle Ages—the 1,000 yr period from approximately 350 to 1350 AD—scholarly attention to physical matters was essentially nonexistent. The Middle Ages was dominated by two philosophical beliefs, **asceticism** and **scholasticism**, both of which focused exclusive attention on matters of the spirit and the mind to the complete neglect, and indeed active suppression, of scholarly attempts to understand anything related to the body or human movement. Asceticism was a religious preoccupation with extreme self-denial of the physical in order to exclusively focus on spiritual matters as presented in the evolving Christian faiths. Scholasticism placed a premium on the development of the mind, largely through disciplined study of early Christian writings and the writings of some of the early Greek philosophers, especially Aristotle. In both instances the religious underpinnings of all scholarly work in the Middle Ages actively excluded any systematic study and investigation with respect to the biophysical properties of the human body and its capacity for movement.

SCHOLARLY WORKS ON HUMAN MOVEMENT FROM THE RENAISSANCE AND REFORMATION PERIODS (CA. 1350-1650 AD)

The Renaissance (or "rebirth") was the period from the late 14th century through the mid-16th century in which there was a great revival in learning, literature, and the arts in Europe. The Renaissance period, which marks the transition from the medieval to the modern world, was one of renewed scholarly interest in the natural world, including the human body and its capability for movement. The Reformation was the great religious movement of the 16th century that had as its objective the reform of the Roman Catholic Church and that ultimately led to the establishment of the various Protestant churches. The major advances in knowledge about human anatomy and physiology made during the Renaissance and Reformation periods were important in the upheaval of much existing dogma about the human body. In the context of a discipline of human movement studies, these advances were especially significant for the subdisciplines of functional anatomy and exercise physiology.

During the Renaissance and Reformation periods, Leonardo da Vinci (1452-1519) and Michelangelo (1475-1564) created anatomically precise drawings and paintings using knowledge acquired from cadaver dissections. Similarly, the definitive anatomical work, Vesalius' *De Humani Corporis Fabrica* (English translation: *The Structure of the Human Body*), arose from extensive, meticulous observations from cadaver dissections. Galileo's (1564-1642) discoveries in physics at this time were central to the establishment of a foundation for the study of the mechanics of human motion (biomechanics), and William Harvey's treatise *De Motu Cordis* (English translation: *On the Motion of the Heart*), which introduced the proposition of blood circulating through the body, was a cornerstone for modern physiology.

SCHOLARLY WORKS ON HUMAN MOVEMENT DURING THE PERIOD 1650-1885

The late 17th century and early 18th century saw some of the ideals for movement, physical activity, and exercise espoused earlier by the Greek philosophers re-emerging in the writings of prominent philosophers such as John Milton and John Locke in England and Jacques Rousseau in France. Locke's (1693) *Some Thoughts Concerning Education* contained the famous dictum *mens sana in corpore sano* (a sound mind in a sound body), which clearly represented a complete reversal of the prevailing philosophy of the Middle Ages. It followed implicitly from Locke's dictum that a dualism of mind and body was inappropriate and that, logically, the study and nurture of one necessitated the understanding and development of the other. Rousseau's *Emile* in 1762 advanced the view that movement, in the form of free play, was critical to cognitive, perceptual, and motor development, thus anticipating one of the predominant themes of modern research in motor control and pedagogy.

The 19th century was a period of great scientific discoveries, many of which laid the foundations for a modern discipline of human movement studies. Among the influential physiological discoveries during this period were those of the electrical properties of muscle, the heat and force properties of muscle fibres, the role of oxygenated blood in respiration, the transfer of oxygen across cell membranes, the function of the liver in carbohydrate metabolism, the role of the cell in energy exchange, and the chemical basis of physiological processes. Major contributors to original knowledge were Duchenne with *Physiologie des Mouvements* in 1865, du Bois-Reymond with *Physiology of Exercise* in 1885, Marey, Mayer, and Pflüger. Understanding of the neural bases of movement was significantly advanced by Bell's discovery of the respective sensory and motor functions of dorsal and ventral root ganglia in the spinal cord (articulated in his 1830 text *The Nervous Systems of the Human Body*) and the studies by the German psychologist Hermann von Helmholtz and others on nerve conduction velocity. Similarly the subdiscipline of biomechanics was ushered into a new period of measurement precision by the release in 1887 of Eadweard Muybridge's monumental, 11-volume *Animal Locomotion*, containing for the first time techniques for the high-speed, sequential photographic analysis of human and animal gait.

While these major advances in knowledge relevant to the understanding of human movement were taking place, the field of physical education was beginning to take shape in Western Europe. In Germany, Sweden, and England, formalised, school-based programs in physical education had either emerged or were in the process of emerging, although the programs were of somewhat different form and had different purposes in each of the countries. These programs had a profound effect on the emerging shape of the physical education profession in North America in particular and, from there, many other parts of the world. To this end, tracing the development of the profession of physical education and the discipline of human movement studies in North America from the late 19th century to the current day is enlightening in terms of understanding the current international form of both the discipline and the profession.

PROFESSIONALISATION OF PHYSICAL EDUCATION DURING THE PERIOD 1885-1929

Although physical education and sport were well established in the curriculum of many European schools (such as the famed Rugby and Eton schools of England) by the early 19th century and in a number of North American public schools and colleges by the 1860s, it was not until 1885 that the first professional organisation for physical education, the American Association for the Advancement of Physical Education (AAAPE), was founded. This body was the forerunner of the current American Alliance for Health, Physical Education, Recreation, and Dance (AAHPERD; see p. 9). Prime movers in the formation of AAAPE were Edward Hitchcock (a medical doctor from the Harvard Medical School who was director of the department of hygiene and physical education at Amherst College), Dudley Sargent (a medical doctor from Yale who was director of the Hemenway Gymnasium at Harvard; see figure 2.1), and Edward Hartwell (a medical graduate from Ohio and a PhD graduate from Johns Hopkins University who was director of the gymnasium at Johns Hopkins University).

Hitchcock collected extensive systematic anthropometric data, culminating in the publication of *Anthropometric Manual*, which first appeared in 1887. Sargent also collected extensive anthropometric data, although his principal legacy is in the area of strength development and the vertical power test that bears his name. Hartwell contributed survey research on physical education. The work of people such as Hitchcock, Sargent, and Hartwell, collected under the rubric of the annual AAAPE conferences, plus the work of others such as Norman Triplett, who made early (1898) observations on **social facilitation** effects in competitive track cyclists, provided the framework for the early development of a research basis for physical education practice. Some of this research work was disseminated to practitioners through AAAPE's own professional journal, *American Physical Education Review*, first published in 1896.

The first quarter of the 20th century saw a marked increase in the publication of books related

Figure 2.1 Dr. Dudley Sargent and exercise stations at Hemenway Gymnasium at Harvard University in 1885.

to the profession and especially the discipline of physical education. Noteworthy among these were *Principles of Physiology and Hygiene* by George Fitz in 1908, *Exercise in Education and Medicine* by the famed Canadian doctor and physical educator R. Tait McKenzie [see "R. Tait McKenzie (1867-1938)"], first published in 1909, *Gymnastic Kinesiology* by William Skarstrom in the same year, *Physiology of Muscular Exercise* by F.A. Bainbridge in 1923, and *Massage and Therapeutic Exercise* by Mary McMillan in 1921, the latter being one of the first texts for the emerging profession of physiotherapy. Perhaps even more important for the ultimate establishment of a credible discipline, original research on aspects of human movement began to appear in prestigious medical journals of the day; Schneider's 1920 paper on cardiovascular ratings of physical fatigue and efficiency in *Journal of the American Medical Association* and A.V. Hill's study on oxygen debt in *Quarterly Journal of Medicine* are classical examples. There was a clear concern in the professional body in the 1920s about the need to increase research activity to make physical education more scientific.

R. TAIT MCKENZIE (1867-1938)

Robert Tait McKenzie (figure 2.2) was born in eastern Ontario, Canada, on May 26, 1867, and throughout his 71 yr of life made unparalleled contributions not only to the profession of physical education but to the broad understanding and appreciation of human movement in general. McKenzie, in a manner reminiscent of the Renaissance man Leonardo da Vinci, made outstanding international contributions in not one but three fields, namely, medicine, physical education, and the arts. He gained a medical degree from McGill University in 1892 and from there held the position of medical director of physical training at McGill before moving to the United States in 1904 to begin a new position as full professor in physical education at the University of Pennsylvania. In his joint roles as physician and physical educator, McKenzie made outstanding original contributions to both the discipline and the profession. He published some 24 professional articles on various aspects of physical activity and health at a time when scholarly publishing was an unusual rather than expected activity. His first text, *Exercise in Education and Medicine*, was extremely influential in the medical profession and ultimately also in physical education, a field he guided to professionalisation through his proactive role in the leadership of the American Physical Education Association and in the formation of the American Academy of Physical Educators. A later text, *Reclaiming the Maimed—A Handbook of Physical Therapy*, based on his experiences as an army physician during World War I, laid

Figure 2.2 Dr. R. Tait McKenzie.

From *The R. Tait McKenzie memorial addresses*, edited by S.A. Davidson and P. Blackstock, 1980. Copyright 1980 by the Canadian Association for Health, Physical Education and Recreation. Reprinted with permission.

the foundations for the practice of physical therapy. For all these pivotal contributions to the field of human movement studies, R. Tait McKenzie's most enduring contributions are his many world-renowned sculptures of athletes and of the beauty of the human body in motion (figure 2.3). His sculptures reflected not only a rare skill and appreciation of art but also the extent of the precision of his observations and anthropometric measurements of athletes, undertaken as part of his scholastic work. Among other tributes following his death on April 28, 1938, the American Alliance for Health, Physical Education, and Recreation (AAHPER) established in 1955 the R. Tait McKenzie Memorial Address lecture series.

Figure 2.3 "The Sprinter" by R. Tait McKenzie.

From *The R. Tait McKenzie memorial addresses*, edited by S.A. Davidson and P. Blackstock, 1980. Copyright 1980 by the Canadian Association for Health, Physical Education and Recreation. Reprinted with permission.

Sources

Day, J. (1967). Robert Tait McKenzie: Physical education's man of the century. *Journal of the Canadian Association for Health, Physical Education, and Recreation, 33*(4), 4-17.

Davidson, S.A., & Blackstock, P. (Eds.). (1980). The R. Tait McKenzie memorial addresses. Ottawa, Ontario, Canada: CAHPER.

Kozar, A.J. (1975). *R. Tait McKenzie: The sculptor of athletes*. Knoxville, TN: The University of Tennessee Press.

ORGANISATION OF RESEARCH EFFORTS IN PHYSICAL EDUCATION DURING THE PERIOD 1930-1959

The year 1930 was noteworthy in the development of both the discipline and the profession as the professional body by then known as the American Physical Education Association launched two new journals. These were *Journal of Health and Physical Education* (the predecessor to the current *Journal of Health, Physical Education, Recreation, and Dance*), which was a forum for the discussion of professional issues, and *Research Quarterly* (the predecessor to the current *Research Quarterly for Exercise and Sport*), which heralded a growing commitment to research. Research work in the 1930s was primarily conducted in the area of tests and measurement and included the development (particularly by Brace, McCloy, and others) of physical fitness and motor ability tests for use in the public school, college, and university education systems. These tests were initially general in nature but, with refinement, became more multidimensional in design. The main biophysical subdiscipline of human movement stud-

ies in which original research was conducted in the 1930s was exercise physiology. The excellent work of Dill, Margaria, and associates at the Harvard Fatigue Laboratory added considerable new understanding to the mechanisms limiting exercise performance and underlying recovery from exercise.

The outbreak of World War II in 1939 brought to a temporary halt much of the basic research work of relevance to human movement but provided an immediate incentive for applied, problem-driven research, and this benefited understanding in some spheres of human movement studies. In particular, problems created by a large number of draftees failing to meet minimum physical fitness standards (in the United States about one third of the draftees examined were found to be unfit for service) led to renewed interest in population-based physical fitness and a demand for professionals with a knowledge of, and training in, the prescription of exercise. The instrumental requirements of war also resulted in a great interest in the use of abilities tests to aid in the selection of people for specific tasks (what became known as "man–machine matching") and in optimal procedures for the rapid acquisition of new movement skills. Both these concerns provided much-needed research impetus to the motor control field. The dominant research figure in the 1940s and especially the 1950s was Franklin

FRANKLIN HENRY (1904-1993)

Franklin M. Henry (figure 2.4) came to the discipline of human movement studies in an unusual fashion. He first completed formal training (a BA in 1935 and a PhD in 1938) in psychology at The University of California, Berkeley, before joining the staff of the department of physical education at the same university in 1938. He remained at that same institution until his retirement in 1971, contributing, in the interim, landmark studies in both the motor control and exercise physiology fields and helping to define the discipline (see chapter 1). His principal scholarly contributions were to understanding of metabolism and cardiovascular functioning during exercise, the specificity of training, and the preplanning of the control of rapid movements. During his career he published more than 120 original research papers in prestigious journals such as *Science, Journal of Applied Physiology, Journal of Experimental Psychology*, and *Research Quarterly* and supervised more than 80 research postgraduate students who, in turn, went on to contribute substantially to original knowledge in a number of the subdisciplines of human movement studies. Like many scientists of his era, Franklin Henry had to construct his own equipment in order to undertake experimental work. In this regard, his early training in electronics was advantageous. In addition to arguing strongly for an academic discipline of physical education, Franklin Henry contributed professionally by taking a lead role in the 1966 formation of the North American Society for the Psychology of Sport and Physical Activity. Professor Henry passed away on September 13, 1993.

Figure 2.4 Professor Franklin M. Henry.

Reprinted from C.W. Snyder and B. Abernethy, 1992. *The creative side of experimentation* (Champaign, IL: Human Kinetics), 37.

Sources

Henry, F.M. (1992). Autobiography. In C.W. Snyder Jr. & B. Abernethy (Eds.), *The creative side of experimentation* (pp. 37-49). Champaign, IL: Human Kinetics.

Park, R.J. (1994). A long and productive career: Franklin M. Henry—Scientist, mentor, pioneer. *Research Quarterly for Exercise and Sport, 65*, 295-307.

Henry, working at the University of California, who contributed original research on both the physiology of exercise and the acquisition and control of motor skills [see "Franklin Henry (1904-1993)"]. His work demonstrated, among other things, the highly specific nature of movement skills and, hence, the inappropriateness of general motor ability testing.

The 1940s also saw the initiation of a number of ambitious longitudinal research programs aimed at understanding normative motor development. Of these, the University of California adolescent growth study launched by Harold Jones, Nancy Bayley, and Anna Espenschade was probably the first, but it was followed soon after by comparable programs at the University of Oregon (the Medford boys' growth study), the University of Wisconsin–Madison, Michigan State University, and the University of Saskatchewan in Canada. While these research activities continued through the 1950s into the 1960s and beyond, the profession of physical education turned its focus somewhat away from its emerging research arm and in the 1950s became preoccupied with professional practice issues related to teaching and coaching.

BEGINNINGS OF A DISCIPLINE OF HUMAN MOVEMENT STUDIES DURING THE PERIOD 1960-1970

The foundations for a discipline of human movement studies were well laid by the early 1960s. Research activity, which had been sporadic and had taken second place in importance to practical issues related to the profession of physical education throughout the first half of the century, had developed in both quantity and quality by the 1960s and demanded greater recognition and identity. The influential papers by Henry in 1964 and Rarick in 1967 (described in chapter 1), which attempted to define a unique and unifying theme for the growing but diverse research work conducted under the banner of physical education, provided an important catalyst for the beginnings of a discipline of human movement studies and for the encouragement of research in physical education not directly anchored to the solving of practical problems. Henry's writings in the United States were paralleled by similar efforts by John Whiting and others in England who went to great lengths to describe and detail human movement studies as a legitimate field of study [see "H.T.A. (John) Whiting (1929-2001)"]. In 1975, Whiting was responsible for the formation of a new journal bearing the disciplinary title *Journal of Human Movement Studies*. Fuelling the emergence of a discipline worldwide was a search for academic credibility and worthy recognition by those undertaking fundamental research on human movement. With the initial attempts of Henry, Rarick, Whiting, and others to define a discipline came a transition in the course offerings and department names of tertiary institutions offering studies away from physical education to alternatives in the field, as discussed in chapter 1, such as human movement studies or kinesiology.

H.T.A. (JOHN) WHITING (1929-2001)

John Whiting (figure 2.5) was born in London in 1929 and was employed as a school physical education teacher for a number of years following his initial teacher training at Loughborough College in 1953. Whiting joined the department of physical education at the University of Leeds in 1960 and from this position pursued higher qualifications, completing a PhD in experimental psychology in 1967. His doctoral research on the acquisition of ball skills was a complete break from the laboratory-based research that dominated the subdiscipline of motor control at the time. The 1969 textbook that stemmed from his PhD, *Acquiring Ball Skill; A Psychological Interpretation*, became standard reading for generations of students of motor control and skill acquisition. In the 1970s, Whiting initiated a blueprint for the foundation of an academic discipline of human movement studies plus a series of state-of-the-art-reviews on aspects of the field, including the emerging field of sport psychology. In 1975 he launched *Journal of Human Movement Studies* and, in 1982, after moving to the faculty of human movement sciences at Free University, Amsterdam, he became the foundation editor of *Human Movement Science*. Both of these journals provided much-needed outlets for the publication of original research in the discipline. Whiting retired back to England in 1989 to an honorary position in the department of psychology at the University of York and continued active involvement in the field as ongoing editor of *Human Movement Science*. He passed away on October 7, 2001.

Figure 2.5 Professor H.T.A. Whiting.

Photo courtesy of Professor H.T.A. Whiting.

Sources

Savelesbergh, G., & Davids, K. (2002). "Keeping the eye on the ball": The legacy of John Whiting (1929-2001) in sport science. *Journal of Sports Sciences, 20,* 79-82.

van Wieringen, P.C.W., & Bootsma, R.J. (1989). *Catching up: Selected essays of H.T.A. Whiting.* Amsterdam: Free University Press.

Whiting, H.T.A. (1992). Autobiography. In C.W. Snyder Jr. & B. Abernethy (Eds.), *The creative side of experimentation* (pp. 79-97). Champaign, IL: Human Kinetics.

EMERGENCE OF SUBDISCIPLINES AND SPECIALISATIONS, 1970-PRESENT

Following the initial efforts to define the scope and unifying nature of the discipline in the mid-1960s and beyond, the 1970s and the decades since have been characterised by the emergence of specialised subdisciplines (as depicted in figure 1.1), each with its own professional bodies, meetings, and research journals. Examples of the proliferation of specialist journals emerging during this time include *Sport Biomechanics*, *Journal of Motor Behavior*, *Medicine and Science in Sport and Exercise*, *Journal of Sport and Exercise Psychology*, *Journal of Physical Activity and Health*, *Journal of Sport History*, *International Review of Sport Sociology*, and *Journal of Teaching in Physical Education*. Although the emergence of specialist subdisciplines is obviously a positive sign for the discipline of human movement studies in terms of adding extensive depth to the knowledge base about human movement, it does create the potential for fragmentation (as noted in chapter 1).

Coupled with the maturation of the discipline has been a proliferation of university programs, many of which offer courses in the discipline in addition to (or sometimes as an alternative to) traditional vocational courses in the profession. The relaxation of the rigid coupling of research on human movement studies with the profession of physical education has also created the potential for the knowledge base of human movement studies to be more actively incorporated into the practice of professional groups other than physical educators.

FUTURE DIRECTIONS, CHALLENGES, AND OPPORTUNITIES

The discipline of human movement studies is presently at an exciting but critical stage in its development. An awareness of the historical antecedents to the modern field of human movement studies provides an invaluable background from which to speculate on the challenges and opportunities that face the field both now and in the future.

Like other scientific disciplines, human movement studies has had the opportunity, at various times in its development, to strengthen existing links with other fields of science and its professions and to forge new partnerships. In the early stages of development of basic research about human movement, issues related to physical activity and health were very much in the foreground, and close interactions existed between early physical educators, physiotherapists, and medical doctors. However, in the 1970s, 1980s, and early 1990s, the major focus in human movement studies research was on issues related to sport science and performance enhancement rather than issues of health (especially the health of the general population), and interactions between human movement scientists and practitioners and those from other health professions were less common. Over the past two decades there have been clear signs of renewed interest in the health implications of human movement, and awareness of both the importance of health and preventive medicine and the role of regular physical activity in promoting health (a topic we cover in some detail in chapters 23 and 24) has grown worldwide. This awareness and recognition provide an opportunity for human movement studies as a discipline to make new links and reinforce existing links to a variety of health professions. The coming decade offers researchers and practitioners from human movement studies the opportunity to cement links with traditional areas of physical education and sport, to revitalise old links with medicine and the therapies, and to form new links with emerging fields such as human–computer interaction, robotics, biomedical engineering, genomics, pharmacology, dietetics, ergonomics, and music and the performing arts.

An emerging focus that must be of more central concern to human movement studies in the future is that of human ageing. With longer life expectancies and slowing birth rates worldwide, we are entering a century in which the number and proportion of older persons will grow at a rate never before seen. Currently, 1 out of every 10 persons is 60 yr or older and about 1 in every 100 persons is 80 yr or older. The Population Division of the United Nations predicts that by 2050 the proportion of persons older than 60 yr will be 1 in 5 and by 2150 the proportion will be 1 in 3. Likewise, the number of persons 80 yr or older is predicted to increase to around 3% of the population by 2050 and to just under 10% of the population by 2150. These figures will be more dramatic in some countries and regions of the world. Given the importance of movement and activity to healthy ageing, to the prevention of debilitating trips, falls, and chronic disease, and to the maintenance of independent living, it will be particularly important for professionals grounded in the knowledge base of human movement studies or kinesiology to build strong bridges to disciplines such as gerontology and professions such as geriatric medicine that have specialist interest in the ageing process and the ageing population.

In addition to challenges and opportunities created by linking the discipline of human movement studies to other disciplines and professions, the future poses challenges and opportunities in the field itself through the opposing forces of knowledge specialisation and fragmentation. At one level the expansion of specialist knowledge in each of the field's subdisciplines adds immeasurably to the depth of our understanding about human movement. However, such specialisation also has the potential to work against the essential integrative understanding of human movement that is ultimately sought by the field. Advances in molecular biology have so dominated biology as a whole over the past decade that studies of whole systems and integration between different subsystems (such as the anatomical, mechanical, physiological, neural, and psychological systems that collectively compose the biophysical bases of human movement) have been forced somewhat to the background. The growth of specialisation needs to be tempered by continued attempts to integrate knowledge across the different specialist subdisciplines of human movement studies in order to continue to advance our knowledge in a consolidated manner. (This is an issue that we return to in part VII.)

Undoubtedly one of the greatest challenges to attempts to integrate understanding of human movement comes from the incredible rate of information expansion experienced over the past decade. This explosion of information has occurred across all sciences but is especially pronounced in human biology. With more people now working in science than ever before, the rate of publication of new research has well exceeded the capacity of scientists to keep abreast in their own field of specialisation, let alone remain up to date with developments in other subdisciplines. Fortunately, the past decade has also seen quantum changes in the technologies used to store, retrieve, and search information, and a number of these technologies are becoming increasingly central to managing the information explosion in science.

SUMMARY

Both understanding the current status of and issues in human movement studies and predicting future trends and directions in the field requires an appreciation of the historical origins of the discipline. Scholarly interest in movement and physical activity dates back at least to the ancient Greeks (ca. 450 BC), and the writings of Plato and Aristotle, in particular, remain relevant to contemporary interest in the relationship between physical activity, health, and well-being. In the Middle Ages (ca. 350-1350 AD), study of the body was suppressed; however, the dawning of the Renaissance and Reformation periods (ca. 1350-1650 AD) laid the foundations for the scientific study of the body. The modern discipline of human movement studies grew out of the need for evidence-based solutions to issues in the profession of physical education but quickly developed to seek answers to questions about movement that were of fundamental interest in their own right and had relevance for professions other than physical education. With the growth of the discipline of human movement studies has come increased subdisciplinary specialisation. This, plus the knowledge explosion in the biological sciences, produces exciting new opportunities but also major challenges for knowledge integration.

FURTHER READING

Adams, W.C. (1991). *Foundations of physical education, exercise, and sport sciences.* Philadelphia: Lea & Febiger.

Dill, D.B. (1974). Historical review of exercise physiology science. In W.R. Johnson & E.R. Buskirk (Eds.), *Science and medicine of exercise and sport* (2nd ed.) (pp. 37-41). New York: Harper & Row.

Massengale, J.D., & Swanson, R.A. (Eds.). (1997). *The history of exercise and sport science.* Champaign, IL: Human Kinetics.

Park, R.J. (1981). The emergence of the academic discipline of physical education in the United States. In G.A. Brooks (Ed.), *Perspectives on the academic discipline of physical education* (pp. 20-45). Champaign, IL: Human Kinetics.

Rasch, P.J., & Burke, R.K. (1974). *Kinesiology and applied anatomy: The science of human movement* (5th ed.). Philadelphia: Lea & Febiger.

Ryan, A.J. (1974). History of sports medicine. In A.J. Ryan & F.L. Allman Jr. (Eds.), *Sports medicine* (pp. 13-29). New York: Academic Press.

PART II

ANATOMICAL BASES OF HUMAN MOVEMENT

FUNCTIONAL ANATOMY

Human anatomy is the study of the structure of the human body at a number of levels of organisation, from the subcellular (structures that can be seen with an electron microscope) through tissues (structures that can be seen with a light microscope) to organs (which can be seen with the unaided eye). At each level of organisation, there is an assumed relationship between structure and function. The relationships between structure and function of bones, joints, and muscles are emphasised in chapter 3. When applied to the musculoskeletal system, functional anatomy is the study of movement and the effects of physical activity on the organs and tissues of the system. Functional anatomy, therefore, is dynamic anatomy, which considers both the short- and long-term effects of activity on the musculoskeletal system. Functional anatomy overlaps with physiology, because of its functional approach, and with biomechanics, because functional anatomists consider the musculoskeletal system as mainly a mechanical system.

TYPICAL QUESTIONS POSED AND PROBLEMS ADDRESSED

The subdiscipline of functional anatomy is concerned with answering a range of questions related to physical activity and the musculoskeletal system. Some typical questions include the following.

▸ What functions do bones perform?

▸ How strong are bones?

▸ How do muscles produce movement?

▸ What prevents dislocation of joints during movement?

▸ How can the size and shape of a person be described?

▸ Are children merely scaled-down adults?

▸ Why do older women in particular experience more fractures than other groups?

▶ What adaptations occur when a person begins a regular exercise program?

▶ Is there an optimal type and level of exercise for maintaining the integrity of the musculoskeletal system?

LEVELS OF ANALYSIS

Anatomy is a very visual science. In gross or **macroscopic anatomy** classes, the unaided eyes are used to study the structure of the human body in the form of a **cadaver**, a model, a chart, or an anatomical atlas, which includes many illustrations or photos of different parts of the body. In the study of human movement, the observation of surface features is also important. Therefore, familiarity with surface anatomy, which requires the skills of observation (using the sense of vision) and **palpation** (using the senses of touch and pressure, particularly in the fingertips), is an important precursor to the analysis of human movement.

Light microscopes are used to aid visualisation of structures, tissues, and the cells that form tissues, whereas **electron microscopes** are used to aid visualisation of cells and structures within cells. Microscopes are used in functional anatomy to define the responses of tissues and cells to physical activity.

The discipline of anatomy includes many areas of study, only a few of which are covered in this text. The chapters that follow take a systemic and functional approach to anatomy by addressing the function and properties of structures that form the musculoskeletal system. The three main areas of study that are discussed are the study of the skeletal system (**osteology**), the joint system (**arthrology**), and the muscular system (**myology**).

HISTORICAL PERSPECTIVES

Anatomy is generally recognised as the oldest of the biomedical sciences. Although an understanding of history is not a prerequisite for the study of anatomy, an overview of the development of anatomy is valuable as a means of gaining a better appreciation of the discipline. Most people think that everything there is to know about anatomy must already be known; this is not so. Anatomy research is still being published.

KEY DEVELOPMENTS

Manuscripts on the anatomy of the pig produced in Salerno during the period 1000-1050 AD argued that the Greek term *anatomy*, which derives from *ana* (meaning "apart") and *tome* (meaning "cutting"), was actually meant to convey more than the straight derivation implies. According to these documents, the Greeks meant that, in anatomy, the "cuts" must be performed according to set rules.

Anatomy is still learned by many students who, following instructions written in dissection manuals, dissect human cadavers as part of their courses. For many other students, the material has already been dissected by others and the important structures are ready for inspection. Even if you never see a cadaver, you will see a model or a chart; these are derived from the investigation of human bodies. Possibly the first anatomical illustrations were drawn in 1522 AD by Berengario da Carpi.

The well-known medical historian Charles Singer has dubbed Herophilus from Alexandria, in Egypt, "the father of anatomy" because he was probably the first to dissect both human and animal bodies for the purposes of anatomical instruction. At that time, however, anatomy was only an investigative technique used by physiologists and surgeons. After a brief period of prominence in the Alexandrian school (300-250 BC) during the Hellenic period, there followed a period of Roman rule during which anatomy was almost exclusively based on animal dissection. The most famous anatomist of this period is Galen (129-200 AD), a Greek who was one of the greatest physicians of all time.

Galen wrote prolifically, based on both dissection and experiment. It was Galen who demonstrated that the **arteries** (meaning "air carriers") actually carried blood. Many terms first used by Galen in relation to bones, joints, and muscles have survived as part of modern nomenclature. In fact, most anatomical terms are derived from Latin or Greek but are now often anglicized for ease of use. After this glorious period from 50 to 200 AD, anatomy, and medical science in general, waned with the dominance of the "practical outlook" required by the Roman Empire during its famous decline. This demise is a classic illustration of how science depends on hypothesis- and curiosity-driven research, both of which may disappear when excessive demands for studies with immediately applicable results are made.

There was a great intellectual awakening in medicine during the 14th century AD, led by scientists at the University of Bologna. A professor from Bologna, Mondino (1270-1326 AD), restored the techniques of anatomy, including anatomical dissection, and gave anatomy the theoretical foundation to be considered as a scientific discipline distinct from surgery and physiology. From 1300 to 1325, Mondino personally dissected cadavers as part of the medical curriculum.

It was not only scientific anatomists who were interested in the structure of the human body. Artists also used scalpels to gain a better appreciation of the human form. Results of this understanding include the anatomical studies of Leonardo da Vinci (1452-1519 AD) and the anatomically correct paintings and sculptures of Renaissance artists such as Michelangelo (1475-1564 AD). The medical historian Charles Singer has stated that Leonardo da Vinci was not only "one of the greatest biological investigators of all time" but also possibly the greatest genius in human history. He was probably the first to question Galen's authoritative teachings. Today, his drawing illustrating the proportions of the human body (see figure) is perhaps the most popular and widely recognised symbol of the human body, representing the human as a physical and spiritual entity.

Galen's prolific writings survived to the Middle Ages, and his teachings influenced anatomical thinking to such an extent that dissections were performed merely to demonstrate his ideas. If the cadaver did not confirm Galen's teachings, then the cadaver was thought to be wrong. Of course, it was not long before people questioned this approach.

It was left to Andreas Vesalius of Brussels (1514-1564 AD) to reform the science of anatomy. He did this as a professor of anatomy at the University of Padua, when he published *On the Fabric of the Human Body*, a work of seven books, in 1543 when he was only 28 yr old. To Vesalius, *fabric* meant "workings," and the first two books on bones and muscles (reprints of which can still be purchased today) indicate that he regarded the human body as a machine designed to perform work. Vesalius did not completely reject the teachings of Galen, but he thought that dissection provided the opportunity to observe the human body and make comparisons with textbook descriptions. In this way previous knowledge could be confirmed or questioned. This major advance in the biological sciences was paralleled by a major publication in the physical sciences. In the same year, Nicholas Copernicus, also from the University of Padua, published *On the Revolutions of the Celestial Spheres*, in which he argued that the earth was not the centre of the universe.

In the history of anatomy, teaching has had equal prominence with discovering new structures or techniques. Possibly two of the greatest teachers of human anatomy in the English language were brothers William Hunter (1718-1783), an obstetrician, and John Hunter (1728-1793), a surgeon. The Hunterian Museum and Art Gallery at Glasgow University houses books, pictures, and specimens donated by William, and the Hunterian Museum and Archives at the Royal College of Surgeons in London still includes some of the 13,000 specimens of human and animal material donated by John. Possibly the best-known modern anatomy textbook in the English language is *Gray's Anatomy*, which is now in its 40th British edition. Henry Gray (1827-1861), who was a lecturer in anatomy and curator of the anatomy museum at St. George's Hospital, London, wrote the first two editions starting in 1858.

KEY PEOPLE

Most histories of anatomy concentrate on gross anatomy and, later, microscopic anatomy and tend to overlook many of the key innovators in the field of functional anatomy. In 1577 Girolamo Mercuriale published *The Practice of Gymnastics* in which he outlined the medical, athletic, and military aspects of exercise. The book provided instructions on correct exercise technique and claimed that exercise could improve health. Mercuriale discussed the effects of physical activity on healthy and diseased individuals and stressed that exaggerated exercise can cause damage, especially in competitive athletics. All these issues are still topical today and are addressed in the chapters that follow. Vesalius, too, had a dynamic approach to the musculoskeletal system, and Galileo noted in 1638 that there was

A drawing of the Vitruvian Man by Leonardo da Vinci titled, in English, "The Proportions of the Human Figure" (circa 1487 AD).

a direct relationship between body mass, physical activity, and bone size.

This idea of the relationship between form and function was extended by a number of 19th-century German scientists, culminating in the publication in 1892 of *The Law of Bone Transformation*, written by German engineer and anatomist Julius Wolff. This text included what is now known as Wolff's law, which is based on engineering analyses of the small bony rods (trabeculae) that form spongy bone. Wolff summarised the work of a number of researchers by concluding that there is a strong correlation between the direction of the trabeculae and the lines of forces acting through the bone. The law relates to the architecture of different bones, but others realised that there was also an adaptive relationship that is still of major interest. Research continues today on the mechanisms responsible for the biological adaptation of bone to various levels of exercise (see chapter 5).

Most actions of muscles had been determined from knowledge of muscle attachment sites, and lines of pull of the muscle related to the joint axes of rotation. Guillaume Duchenne (1806-1875), a French physician, attempted to confirm these hypotheses. His experiments and observations over many years culminated in the publication in 1865 of his influential text *Physiology of Motion: Demonstrated by Means of Electrical Stimulation and Clinical Observation and Applied to the Study of Paralysis and Deformities*. Of major interest to modern functional anatomists is the recording of the actions of almost all of the skeletal muscles in the human body.

Particularly since World War II, the converse of Duchenne's technique has been used. When a muscle is stimulated by its nerve it responds in two ways: by producing an electrical signal and by producing a mechanical force (see chapter 3). No one has yet directly measured the force in the muscle tissue of a living human, although tendon forces have been measured. It is possible to detect the electrical signal produced by muscle tissue using a variety of electrodes. This technique, known as electromyography, is used to describe muscle activity during human movement. The best-known name in this field is John Basmajian (1921-2008) from North America, who held professorships in both Canada and the United States. He was president of the American Association of Anatomists. Not only did Basmajian contribute greatly to the elucidation of dynamic human muscle function in a variety of situations, he was very involved in the teaching of gross and functional anatomy through his authorship of many texts, including the well-known *Muscles Alive: Their Functions Revealed by Electromyography*.

Although the anatomy of adult humans is of major interest to anatomists, it is not their sole interest. The structural changes that occur during childhood and later have also been studied extensively by auxologists. The field of auxology is defined as the science of biological growth, and key concepts from this field are examined in chapter 5. Many of the recent researchers in this field have been trained at the Institute of Child Health at the University of London, in a program headed by the noted auxologist Jim Tanner. Among many texts published by Tanner is one on the history of human growth studies, which is listed as a resource to supplementary material in chapter 5.

PROFESSIONAL TRAINING AND ORGANISATIONS

Anatomy is a foundation course for many professions, including medicine and all the allied health professions. Many qualified professionals often find it useful to refresh their knowledge of anatomy. For example, people interested in sports medicine may attend a session in an anatomy laboratory as part of their continuing education. Likewise, surgeons, nurses, sport coaches, ergonomists, and others find it valuable to continually improve their knowledge of anatomy.

Most countries or regions have an association, comprising practising anatomists, that holds regular conferences to present members' latest research and to discuss new directions in the teaching of anatomy. The International Federations of Associations of Anatomists (FIAA) holds a conference every 5 yr. Members of the International Society for the Advancement of Kinanthropometry (ISAK) have a particular interest in the relationships between body dimensions and human performance (see chapter 4). Functional anatomists sometimes find more interest in the meetings of societies in biomechanics than in the meetings of general anatomy societies.

FURTHER READING

Agur, A.M.R., & Lee, M.J. (2009). *Grant's atlas of anatomy* (12th ed.). Philadelphia: Lippincott Williams & Wilkins.

Backhouse, K.M., & Hutchings, R.T. (1998). *Clinical surface anatomy* (2nd ed.). London: Mosby-Wolfe.

Rohen, J.W., Yokochi, C., & Lutjen-Drecoll, E. (2011). *Color atlas of anatomy: A photographic study of the human body* (7th ed.). Philadelphia: Wolters Kluwer Health/Lippincott Williams & Wilkins.

Singer, C. (1957). *A short history of anatomy and physiology from the Greeks to Harvey (The evolution of anatomy)* (2nd ed.). New York: Dover.

Standring, S. (Ed.). (2008). *Gray's anatomy* (40th ed.). Edinburgh: Churchill Livingstone.

SOME RELEVANT WEBSITES

Anatomy on the Internet: www.meddean.luc.edu/lumen/MedEd/GrossAnatomy/anatomy.htm

Arthritis Foundation: www.arthritis.org

Australasian Menopause Society: New Directions in Women's Health: www.menopause.org.au

The Bone and Joint Decade: Joint Motion 2000-2010: www.boneandjointdecade.org; www.bjd.org.au

Centre for Physical Activity in Ageing: www.cpaa.sa.gov.au/home.html

Galen: A Biographical Sketch: www.ucl.ac.uk/~ucgajpd/medicina%20antiqua/bio_gal.html

Health and Age: www.healthandage.com

Hippocrates: The "Greek Miracle" in Medicine: www.ucl.ac.uk/~ucgajpd/medicina%20antiqua/sa_hippint.html

The Hunterian Museum and Archives of the Royal College of Surgeons, London: www.rcseng.ac.uk/museums

The Hunterian Museum and Art Gallery, University of Glasgow, Scotland: www.huntsearch.gla.ac.uk/hunter/index.html

Information about Flexibility Exercises from the American Academy of Orthopaedic Surgeons: http://orthoinfo.aaos.org/fact/thr_report.cfm?Thread_ID=4&topcategory=Wellness

The International Society for the Advancement of Kinanthropometry: www.isakonline.com

National Osteoporosis Foundation: www.nof.org

Orthopedic Patient Education from the Southern California Orthopedic Institute: www.scoi.com/anat.htm

Osteoarthritis: U.S. National Library of Medicine: www.ncbi.nlm.nih.gov/pubmedhealth/PMH0001460/

Strength training information from Georgia State University: www.gsu.edu/~wwwfit/strength.html

Vesalius the Humanist: www.hsl.virginia.edu/historical/artifacts/antiqua/vesalius.cfm

The Visible Embryo: www.visembryo.com

The Visible Human Project: U.S. National Library of Medicine: www.nlm.nih.gov/research/visible/visible_human.html

World Health Organization Child Growth Standards: www.who.int/childgrowth/standards/en/

CHAPTER 3

BASIC CONCEPTS OF THE MUSCULOSKELETAL SYSTEM

The major learning concepts in this chapter relate to

- ▸ equipment used in anatomy learning and research;
- ▸ mechanical and physiological functions of the skeletal system;
- ▸ components of bone and the different types of bone cells;
- ▸ structure of long bones and the mechanical properties of bone tissue;
- ▸ the functions, shapes, and organisation of whole bones;
- ▸ classifying joints according to their structure or function;
- ▸ characteristic features of synovial joints;
- ▸ protection, lubrication, and wear of synovial joints;
- ▸ the joint as the functional unit of the musculoskeletal system;
- ▸ structural features of muscles and their main actions;
- ▸ distinguishing properties of skeletal muscle;
- ▸ the biological and mechanical bases of muscular contraction;
- ▸ the electrical and mechanical responses of muscle to stimulation; and
- ▸ the major factors that determine strength and range of joint motion.

The purpose of this chapter is to introduce key concepts related to the structure and function of the skeletal system, the system of joints (the articular system), and the muscular system and to describe the tools for measuring these systems.

TOOLS FOR MEASUREMENT

Language is very important for basic **gross anatomy** because of the descriptions required to explain the position, relative size, and relationships of each anatomical feature. Learning anatomy from a textbook can be very difficult but becomes easier with the aid of atlases that have artistic illustrations or photographs of the various structures.

The concepts introduced in this chapter are presented as general descriptions, but the reader should realise that many of these concepts have been verified experimentally using quantitative methods. Bone density in living humans can be determined using **radiological** and other more direct techniques. Bone structure can be visualised under a microscope; however, because bone is a hard tissue, special preparation techniques are required. Chemical analyses can be performed to determine the composition of bone. Movement relies on the integrity of joints, and **goniometers** are examples of instruments used to measure ranges of joint motion. Muscles produce forces that have an external effect, and these effects can be captured by different types of **dynamometers** that measure muscle strength. The electrical signal generated just before a muscle contraction can also be detected, recorded, and analysed using electronic equipment and computers. Muscles change shape when they contract; when a muscle is artificially stimulated, the amount of deformation can be measured.

SKELETAL SYSTEM

The skeletal system is exquisitely structured to fulfill both its mechanical and physiological functions. We first examine those functions and then show how both the composition and architectural structures of the skeletal system support them.

FUNCTIONS OF THE SKELETAL SYSTEM

Humans have an endoskeleton and are **vertebrate** animals. Attached to our backbone, or vertebral column, are other bones that form the framework of the human body, much like the beams and studs of a house or the chassis of a car. These analogies are imperfect, however, because they imply that the skeletal system has only mechanical functions when, as we discuss later, the physiological functions of bone are equally important.

MECHANICAL FUNCTIONS OF THE SKELETAL SYSTEM

The most obvious mechanical function of the skeletal system is providing support for weight bearing. The skeletal system also provides protection of internal organs (e.g., protection of the brain by some bones of the skull and of the heart and lungs by the ribs). The major bones of the limbs provide rigid links between joints; these bones also provide sites for muscle attachment. These functions of the skeleton, in providing linkages and sites for muscle attachment, facilitate human movement and form the basis of the mechanical models of the human body.

PHYSIOLOGICAL FUNCTIONS OF THE SKELETAL SYSTEM

Bone is a living, dynamic tissue. The framework for a house is often made of timber that was once a living tree. Similarly, the bones in an anatomy laboratory are changed from their condition in the living body. Thus, in addition to its mechanical functions, the living skeletal system has important physiological functions. When subjected to large forces over a long period, the framework of the house may start to fail, just as the metal in a car may start to fatigue or rust. But living bone has the advantage that it may heal when broken and even carry out maintenance to prevent failure. Some bone researchers believe that the stimuli to bone adaptation include microcracks that form during increased levels of physical activity. (Adaptation in response to physical activity is explored further in chapter 6.)

Bone tissue is also involved in the storage of essential minerals such as **calcium** and **phosphorous**. Additionally, bone **marrow** produces blood cells and is part of the body's immune system.

COMPOSITION OF BONE

In a car, the metal in the engine tends to be thick and rigid because it must resist deformation, but the body panels are more cosmetic and help absorb the energy of impact during a collision. As with the metal in a car, the composition of bone in the skeleton is not uniform. This makes sense because each of the constituents of bone must accommodate different physiological and mechanical functions.

MECHANICAL PROPERTIES PROVIDED BY COMPONENTS OF BONE

It has been argued that both **stiffness** and flexibility are important properties of bone. In a car the axles must be stiff to resist deformation as they transfer the forces from the engine to the wheels, but springs in the suspension deform while the shock absorbers dampen the motion. Engineers can use different materials to suit particular purposes; they choose steel, aluminum, plastic, or other alternatives depending on the purpose of the part. To produce the optimal mechanical properties, materials engineers develop alloys that consist of different types of metal bonded together.

Is bone like a homogeneous metal, such as copper or iron, or does it consist of different components? The answer is the latter; consequently, bone is described as a composite material. One useful analogy for bone is fibreglass, which consists of glass fibres cured in an epoxy resin. The final product, fibreglass, has mechanical properties that are superior to those of its individual components.

With regard to the mechanical properties of bone, comparison with other materials is useful. Bone is similar to wood from the oak tree in its properties of strength and stiffness. Bone is about as flexible as fibreglass, although weaker, but is stronger and more flexible than ordinary glass. Ceramics in general are about one-third to one-half as strong in **compression** as bone. The blocks of older car engines were manufactured from cast iron, which is similar to bone in its **tensile** strength, but bone is three times lighter (that is, less dense) and much more flexible. Bone is clearly a remarkable physical material.

About one quarter of the mass of bone in the living body is water. Adult bone, after removal from the body and drying, is about two-thirds inorganic crystals consisting mainly of calcium and phosphorous. The main mechanical components of bone are **collagen** (most of the organic component) and calcium salts (most of the inorganic component). The collagen provides toughness and flexibility, so it contributes to the tensile strength; bone's hardness and rigidity are attributable mainly to its calcium salts, which also contribute principally to the **compressive strength**. According to one calculation, the optimal mineralisation of bone, in terms of strength and flexibility properties, is two-thirds mineral to one-third organic material. This is the proportion of calcium salts to collagen in healthy adult bone. Almost all the body's calcium is stored in bone and is released from there as required by other tissues, such as skeletal muscle, which requires calcium to contract.

As already noted, collagen fibres compose most of the organic component of dry bone. Bone cells are another organic component of bone. In the matrix of bone, the basic cells are called **osteocytes**. Also associated with bone are bone-forming cells, called **osteoblasts**. Depending on local environmental conditions, osteocytes can become osteoblasts and vice versa so that adaptive responses can occur when there are mechanical stimuli. There are also bone-eroding cells, called **osteoclasts**. Remodelling of bone, which is a continual process in adults, involves organised erosion and deposition of bone tissue by the various cells. A full cycle of remodelling takes about 3 mo.

TYPES OF BONE

Metallurgists and engineers have at their disposal different building materials, whereas the two types of bone in the human skeleton are made of the same material. **Compact** and **spongy** bone differ mainly in their porosity, although compact bone is more organised than spongy bone.

Spongy bone (also termed **cancellous** or **trabecular bone**) is a lattice meshwork of bony rods (**trabeculae**; figure 3.1*a*). This meshwork arrangement makes spongy bone much more springy than compact bone. In spongy bone, each osteocyte is close to a nutrient supply because the bony tissue is surrounded by blood vessels and associated material.

Compact bone (also called **cortical** or **dense bone**) is much more solid than spongy bone. When bone is remodelled to become compact, bone cells may be too far from their nutrient supply because the blood vessels are surrounded by bony tissue. If compact bone were not organised in a specific way, this lack of nutritive supply would result in the death of each cell. To overcome this problem, compact bone contains a basic structural unit (Haversian system or osteon) that is repeated many times. Haversian canals, longitudinally arranged in the shafts of long bones, carry blood vessels and are surrounded by layers of bone (**lamellae**). In each lamella, osteocytes are contained in **lacunae**. There is a limit to the number of lamellae so that bone cells are not too far from their supply of nutrition. Figure 3.1*b* illustrates the microscopic organisation of this

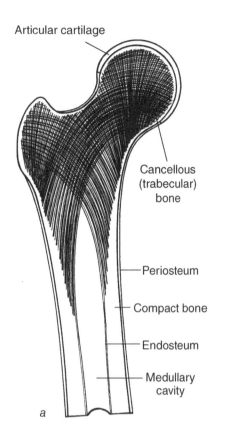

Articular cartilage

Cancellous
(trabecular)
bone

Periosteum

Compact bone

Endosteum

Medullary
cavity

a

compact bone in contrast with the organisation of spongy bone (figure 3.1*a*). Most of the calcium in bone is in the compact bone, but the calcium in the spongy bone is more easily released into the blood when required.

ARCHITECTURE OF BONE

In engineering, an important concept is the relationship between strength and mass. Elite athletes have a good **power**-to-weight relationship, which is the same idea. The analysis of bone architecture allows us to conclude that the bones in our body are very **efficient** in that they are able to withstand large forces and at the same time are relatively light.

BONE SHAPE AND ORGANISATION

When different functional requirements exist, there are variations in the arrangements of the two types of bone as well as differently shaped bones. Bone shape and predominant function have a specific structure–function relationship. The long bones of the limbs have many attached muscles and act as rigid links between major joints (figure 3.2*a*). The flat bones of the skull help protect the brain. Their structure, consisting of two layers of compact bone with spongy bone in between (figure 3.2*b*), is similar to that of some protective helmets. The ribs protect vital organs in the chest such as the heart. The short bones of the hindfoot have a thin layer of cortical bone around the outside and are filled with spongy bone (figure 3.2*c*). These weight-bearing bones resist the large compressive forces generated during running and help cushion **ground reaction forces**. All the bones illustrated in figure 3.2 consist of both compact and spongy bone in differing proportions and structural arrangements.

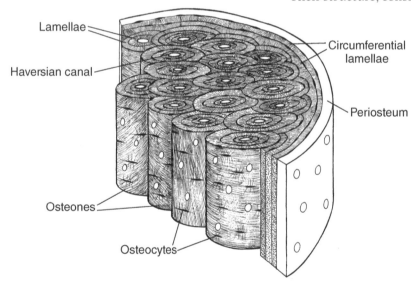

Lamellae

Haversian canal

Circumferential
lamellae

Periosteum

Osteones

Osteocytes

b

Figure 3.1 *(a)* The trabeculae of the spongy bone and *(b)* the architecture of compact bone in the shaft of a long bone, shown in longitudinal and transverse sections.

Adapted, by permission, from J. Watkins, 1999, *Structure and function of the musculoskeletal system* (Champaign, IL: Human Kinetics), 112-113.

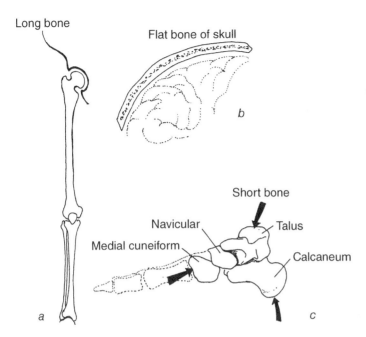

Figure 3.2 Examples of bones with different shapes to serve different functions. Muscles are attached to most bones to facilitate movement. *(a)* Long bones of the limbs, *(b)* flat bones of the skull, and *(c)* short bones of the hindfoot.

Note the arrangements of compact and spongy bone illustrated in the different parts of a **vertebra** (figure 3.3). A vertebra is classified as an irregular bone, but is more like a composite bone consisting of three main components. The vertebral body is the main weight-bearing part of the vertebra and is constructed like a short bone; in section one can see the thin layer of compact bone surrounding a mass of spongy bone. The **lamina** of the vertebral, or neural, arch protects the spinal cord and is constructed like a flat bone; in section it looks like one of the ribs that help protect the heart and lungs. The spinous and transverse processes have areas of attachment for ligaments and muscles, providing leverage for these structures, and they are constructed like long bones.

Amount, **density**, and distribution of bone material have major effects on the mechanical properties of a whole bone. Compare the resistance of a long plank of wood to bending in the vertical direction when it is supported at both ends and placed on its narrow side, where it does not deform, with the resistance when it is on its wide side, where it bends easily like a springboard. Also compare the deformation of high- and low-density foam mattresses under the weight of a person.

ARCHITECTURE OF LONG BONES

The bones of the lower limb are often under compression during standing, whereas the bones of the upper limb may be subjected to **tension** when one is holding a load in the hand or performing a giant swing on the high bar (see figure 3.4). During forward bending of the trunk, there are **shear** forces between adjacent vertebrae (see figure 3.4). During walking, the long bones of the lower limb may be acted on by **bending** and torsional (longitudinal twisting) forces. The compressive forces on the ends of the femur in the thigh are not aligned in the same straight line, causing bending, which also occurs because of the bend of the shaft of the femur. Twisting about the long axis of the foot immediately after heel strike in jogging, commonly called hindfoot pronation, produces **torsion** of the main leg bone, the tibia. These patterns of loading require optimal

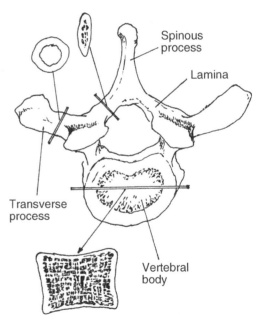

Figure 3.3 A vertebra, classified as an irregular bone, has three main components: the weight-bearing vertebral body, the flat lamina, and the long processes.

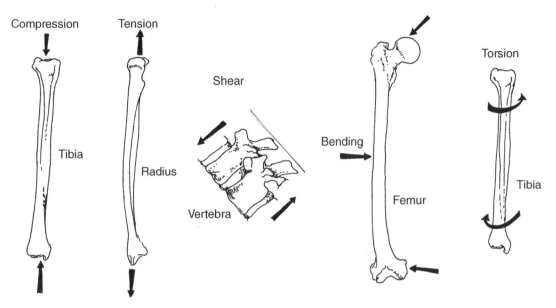

Figure 3.4 Examples of the different types of forces that may act within and between bones. Major forces that the long bones of the limbs, particularly the lower limbs, must withstand are compression along their long axes, bending, and twisting. The bones of the upper limb may be subjected to tension.

arrangement of the bone material, but what is the most efficient architecture? The shafts of long bones are hollow, giving them mechanical advantages over a solid rod of the same mass. If a pipe and a solid rod contain the same amount of material, the pipe is stronger in bending than the rod, so a hollow shaft is more efficient in terms of the relationship between strength of the whole bone and its weight. A hollow shaft also resists twisting better than a solid rod. It appears from mechanical analyses that the shapes of our bones are close to optimal for the activities that most humans perform.

During movement, especially when landing from a jump, forces are transferred from one bone to the next via the joints. A large contact area between the bones results in less pressure on the ends of the bones, so expanded ends are advantageous. Much of the material forming the expanded ends is spongy bone that absorbs energy during impact. Everybody knows the difference between jumping on a trampoline and jumping off the trampoline onto a concrete floor. Compact bone is like concrete because it does not deform much but is very strong, whereas spongy bone is more like a trampoline that cushions the force of landing as it deforms. In fact, spongy bone is 10 to15 times more flexible than compact bone. At each end of a typical long bone, a thin outer layer of compact bone protects the overlying cartilage during impact and transfers the forces to the underlying spongy bone (figure 3.5). The deformable **articular**

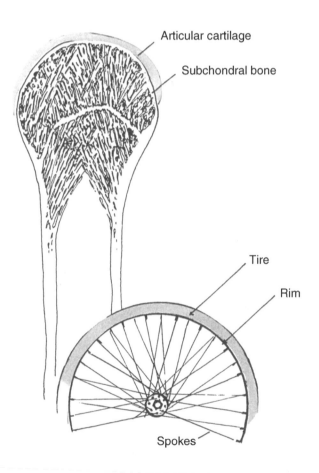

Figure 3.5 The expanded end of a long bone is compared with a bicycle wheel consisting of spokes, rim, and tire.

cartilage acts like an air-filled bicycle tire, which is supported by a light but solid rim, like the thin layer of compact **subchondral bone**. The trabeculae of the expanded ends line up along the major lines of force to perform a function similar to that of the spokes of a wheel, which help support and maintain the shape of the rim and cushion the forces produced by bumps in the road. Recall that the composition of compact bone and spongy bone is the same; just the organisation is different.

ARTICULAR SYSTEM

The system of the **joints** between the bones is called the articular system because the union of bones is an **articulation**. Joints are important because they allow movement, but they must remain stable during movement. The study of joints is called **arthrology**.

CLASSIFICATION OF JOINTS

We all know that joints allow cars and bicycles to move. These are relatively simple joints compared with the synovial joints found in the human body; in fact, engineers are still attempting to explain the functions of biological joints. The design of artificial joints for replacement of human joints such as hips and knees is progressing rapidly, but replacement joints still do not work as well as the natural joints.

To give some idea of the complexity of natural joints, we can contrast them with the joints in a car. The bearing surfaces in the artificial joints of a car are usually smooth, hard metal, whereas the articular cartilage is less smooth but quite deformable. Artificial joint surfaces are very regular, usually part of a circle or sphere, whereas human joint surfaces are ovoid (egg shaped), resulting in much more complex movement patterns. To remove debris in the joints of a car, the car must be taken to the local garage for an oil change. Any effects of wear result in rapid deterioration of artificial joints; however, in a synovial joint, cell debris can be removed continuously.

The materials between the bones forming the joint may differ, providing the basis for the structural classification of joints. All anatomical joints may be described as **fibrous**, **cartilaginous**, or **synovial**. Most of the major joints we are familiar with, such as the shoulder, elbow, knee, and ankle, are examples of synovial joints, so this section focuses exclusively on these joints.

The major function of joints is to allow movement but, at the same time, remain stable. Synovial joints allow a relatively large amount of movement; consider, for example, the ranges of motion that are possible at both the shoulder and elbow joints.

CHARACTERISTIC FEATURES OF SYNOVIAL JOINTS

A number of characteristic features are associated with a typical synovial joint (figure 3.6). On the ends of the bones forming the joint there is articular cartilage, which consists of collagen fibres in a liquid matrix. Articular cartilage has a water component of about 80% and has been likened to a sponge from which water can be squeezed. When not being deformed, it absorbs water because of the dissolved chemicals that it contains. Articular cartilage forms a relatively smooth bearing surface and acts, through its capacity to deform, to cushion forces, like a tire on a bicycle wheel (figure 3.5)

A **joint capsule** forms part of the boundary of the joint. The joint capsule contains a high proportion of collagen fibres and provides some intrinsic

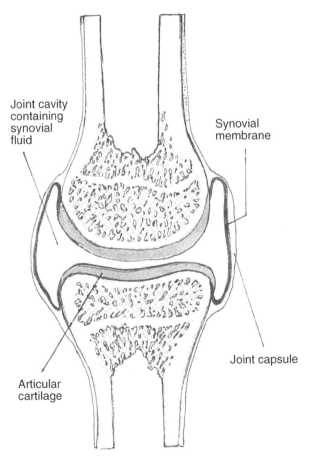

Figure 3.6 Representation of a typical synovial joint indicating its characteristic features.

stability to the joint as well as some resistance to motion. Forming the inner layer of the joint capsule is the **synovial membrane**, which has a number of important functions. It produces the fluid in the joint and removes the cell debris that results from wear and tear in the joint.

A major characteristic of a synovial joint is the cavity bounded by the articular cartilage and the synovial membrane of the joint capsule. In this joint cavity is a small amount of **synovial fluid** that contains constituents of blood, substances secreted by the synovial membrane, and some products resulting from abrasive wear within the joint.

The synovial fluid has three important functions: lubrication, protection, and nutrition. The **viscosity** of the fluid can change according to local environmental conditions, so it may be thick and protective, like grease, or thin and lubricating, like oil. Because articular cartilage does not receive an adequate blood supply, nutrients are supplied via the synovial fluid in contact with the cartilage. Pressure during movement helps squeeze fluid out from the cartilage. When the pressure is removed, the liquid can seep back into the cartilage. Physical activity thus promotes the nutritional function of synovial fluid.

To help maintain the integrity of the synovial joint, associated **ligaments** attach from bone to bone and cross the joint. Ligaments consist of collagen fibres, also a constituent of bone. In ligaments, the collagen forms about 90% of the structure, and the fibres tend to run parallel to each other. The ligaments help the joint capsule provide stability and they function to guide the joint's movements. In doing so, they provide some resistance to joint motion. Ligaments are basically passive structures that resist tensile forces, which tend to separate the bones forming the joint.

CLASSIFICATION OF SYNOVIAL JOINTS

One can use the classical anatomical view to classify synovial joints on a descriptive basis or use the view of engineers, who are interested in movement between joint surfaces in contact, to produce a system to explain joint lubrication and wear. It is not our aim in this section to provide the details of the alternative classification systems; instead, we summarise the structural and functional criteria on which each system is based.

Anatomically, each synovial joint can be classified on the basis of the approximate geometric form of its articulating surfaces. For example, the hip looks like a ball in a socket (figure 3.7a), much like the joint between a car and a trailer, so it is called a ball-and-socket joint. Synovial joints can also be classified according to the gross movements permitted. For example, because the ankle joint allows movement basically in only one plane, it is called a hinge joint (figure 3.7b). The ankle could also be classified as a uniaxial joint because it allows movement about only one principal axis. The joints forming the knuckles of the fingers are classified as biaxial because they allow motion about two principal axes (figure 3.7c). Note that the fingers can be moved in two directions at right angles—backward and forward and from side to side.

The complexity of organisation of the joint structures is another criterion for classification. If a joint consists of two bones and one pair of articulating surfaces, it can be classified as a **simple joint**; an example is the knuckles of the fingers (figure 3.7c). If there is more than one pair of articulating surfaces, such as in the elbow joint capsule (where there are three pairs), the joint is classified as **compound** (figure 3.7d). Sometimes synovial joints contain intra-articular structures, such as a cartilaginous disk or **meniscus**; these joints can be classified as **complex** (figure 3.7e).

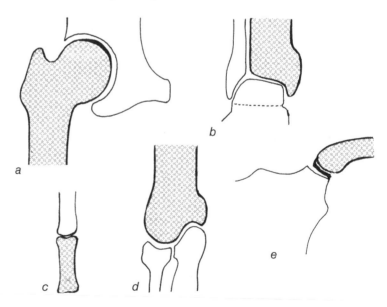

Figure 3.7 Representations of some types of synovial joints: (a) spheroidal (ball-and-socket) joint, (b) hinge-like joint, (c) simple joint, (d) compound joint, and (e) complex joint containing an intra-articular disk.

RANGES OF MOVEMENT ALLOWED BY SYNOVIAL JOINTS

Movements that occur at synovial joints have traditionally been described in terms of the major planes of the body (figure 3.8). The sagittal plane is a vertical plane dividing the body into left and right parts. The coronal (frontal) plane is a vertical plane dividing the body into front (anterior) and back (posterior) parts. The transverse plane is a horizontal plane dividing the body into top (superior) and bottom (inferior) parts. Anatomically, movements that occur in the sagittal plane have been termed **flexion** (when the angle between the limb segments decreases) and **extension** (when the angle between the limb segments increases; figure 3.8). These gross descriptions of directions of movements of body segments can be contrasted with the more mechanical approach in which the terminology relates to the movement that occurs between the articular surfaces in contact.

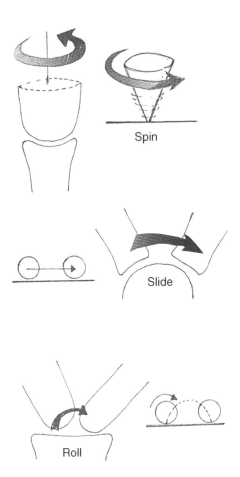

Figure 3.9 The motions of spin, slide, and roll between articular surfaces in a synovial joint.

Using the engineering approach, we can describe the relative motion between articular surfaces as a spin, a slide, or a roll (figure 3.9). The movements can be likened to the spinning of a top, the skidding of a tire on the road, and the normal rolling of a tire on the road, respectively. These movements have an effect on the frictional resistance. For example, resistance is much less when the wheels of a car are rolling freely than when the brakes are applied and the tires slide along the road. Often combinations of all movements occur during joint motion, such as during flexion and extension at the knee.

JOINT PROTECTION, LUBRICATION, AND WEAR

The structures in contact in a synovial joint are the articular cartilages on the ends of each bone. The **cartilage** is smooth and deformable, so it can cushion forces applied to its surface. The subchondral bone, which is the thin layer of compact bone under the cartilage, provides a solid base and helps

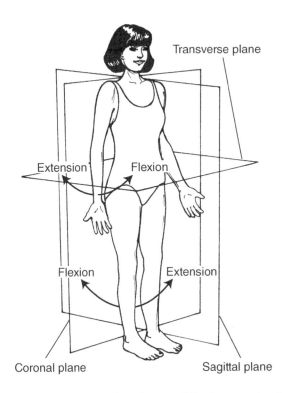

Figure 3.8 The major planes of the human body with respect to the anatomical position. The movements of flexion and extension of the elbow and knee are illustrated.

Reprinted, by permission, from J.H. Wilmore and D.L. Costill, 2004, *Physiology of sport and exercise*, 3rd ed. (Champaign, IL: Human Kinetics), 679.

protect the cartilage from damage. This thin layer of compact bone sits on a more deformable network of bony rods forming the spongy bone (figure 3.5). This organisational structure is similar to that of a bicycle wheel with spokes, where the spokes maintain the shape of the rim but also help cushion forces from the ground. (A solid bicycle wheel would be too stiff for general purposes.)

The synovial fluid acts as a lubricant so that friction between articular cartilages in a synovial joint is less than between, for example, two blocks of smooth ice. Synovial joints have a number of important characteristics that differentiate them from artificial joints. Movement at a synovial joint can best be described as oscillatory (backward and forward), but in most load-bearing joints in machines the angular motion occurs in only one direction (around and around). The load-bearing surface in a synovial joint is deformable cartilage, which is both elastic and porous, and the synovial fluid has particular chemical properties that allow it to act as a lubricant. This arrangement is much more complicated than that of most joints in cars, in which the bearing surfaces are very hard.

The Joint as the Functional Unit of the Musculoskeletal System

Human movement studies and **functional anatomy** emphasise movement; therefore, the joint is the focus of functional musculoskeletal anatomy. The characteristic features of a synovial joint have been listed (see figure 3.6), but each joint has associated with it a number of other structures. These include the following.

▶ On either side of the joint are bones that act as levers and aid force cushioning, as explained earlier (see figure 3.5).

▶ **Skeletal muscles** have a role in movement because they cross joints and thus initiate and control movement. The forces they produce across the joint also stabilise it. Thus, muscles are secondary stabilisers, in addition to the associated ligaments. If you contract all the muscle groups about a joint simultaneously so that no movement occurs, the joint becomes much more stable. The joint is also stiffer, as you would realise if somebody else attempted to move it.

▶ Muscle forces are transmitted to bony attachments via tendons; thus, it is the muscle–tendon unit that is of major significance.

▶ Nerves are also associated with joints. As discussed in chapter 15, motor nerves in the **central nervous system** provide some control over the muscles producing actions at a joint. Sensory nerves provide feedback about joint position and movement from a variety of sensors located in the joint capsule and ligaments as well as in the muscles and tendons.

Injury to any intrinsic joint structure or to the associated structures will result in functional impairment of the entire joint and adjacent body segments. Functional impairment of one joint often results in a chain reaction that affects adjacent joints. For example, an ankle or knee injury will affect the whole lower limb and therefore adversely affect performance during walking or running. Because of altered function of the injured joint, and possibly because of pain associated with the injury, the other joints must attempt to compensate, usually resulting in a limping **gait**.

MUSCULAR SYSTEM

The muscular system comprises the main effector structures for human movement. It is an important part of the musculoskeletal system because it produces joint motion (see part II on biomechanics), but it is also an important part of the neuromuscular system in that it is controlled by the central nervous system (see part IV on motor control).

STRUCTURE OF THE MUSCULAR SYSTEM

We can identify major muscles that lie just beneath the skin, such as the pecs, lats, and deltoid and biceps muscles. Understanding the mechanisms of action of these muscles relies on knowledge of muscular tissue at a number of levels of organisation, including its microscopic structure.

ASSOCIATION OF MUSCLES WITH OTHER STRUCTURES

As with bones and joints, the structure of muscles is related to their function. Muscles cross joints, and the major skeletal muscles of the trunk and limbs are attached to bones at both ends. The muscle–tendon unit therefore consists of a chain of structures, typically bone–tendon–muscle–tendon–bone.

The attachment sites of muscle and **tendon** are important because they determine the action of each muscle. Whenever a muscle shortens, it tends to pull the two attachment sites closer together, so the relationship between direction of pull of the muscle and the axis of rotation of the joint determines the resulting joint action. This type of analysis is the basis used to define the actions listed in tables in all basic texts on gross anatomy. Functional anatomists, starting with Duchenne in the late 19th century, have used other techniques to attempt to unravel the complexity of muscle contraction when different muscles act simultaneously or when joint position is changed from the normal anatomical position illustrated in figure 3.8.

STRUCTURAL FEATURES OF MUSCLES

So far in this discussion, we have used the term *muscle* in relation to only one type of muscle—skeletal muscle. There are, however, other types of muscle tissue. **Smooth muscle** is found in the walls of the digestive system and certain blood vessels. **Cardiac muscle** forms the major part of the walls of the heart. In this and subsequent functional anatomy chapters, we restrict our discussion to skeletal muscle.

Not all muscles look the same. The typical muscle has a belly and tapers toward the tendon that attaches it to bone at each end, but there are other shapes. Muscles attach to bone either directly or via a tendon. Examples of muscles with different architectural arrangements are shown in figure 3.10.

Muscle fibres are oriented in the direction of pull of the whole muscle (e.g., spindle) or at an angle to this direction (e.g., unipennate). The hip and thigh region contains many examples of these shapes. Semitendinosus (one of the **hamstring muscles** in the back of the thigh) has a spindle shape, whereas another hamstring muscle (semimembranosus) is a unipennate muscle. Rectus femoris (one of the four muscles forming the quadriceps muscle group at the front of the thigh) is a bipennate muscle. Underneath the large gluteus maximus in the buttock is the fan-shaped gluteus medius.

Skeletal muscle cells are elongated and contain many nuclei; hence, muscle cells are called *fibres*. The alternating thick and thin filaments produce characteristic striations that one can clearly see when viewing muscle tissue through an electron microscope. This appearance is produced by tens of thousands of repeating units in series that form each fibre. These repeating units, called **sarcomeres**, are visible in a photograph taken by an electron microscope. The sarcomere is the structural and functional unit of muscle. You can gain an idea of its structure by looking at figure 3.11. The function of the components of the sarcomere is discussed in "Muscle Contractions" later in this chapter. The connective tissue element of the muscle consists of thin sheets surrounding each fibre, each bundle of fibres, and the whole muscle itself.

The structural and functional unit of the neuromuscular system is a **motor unit** consisting of a

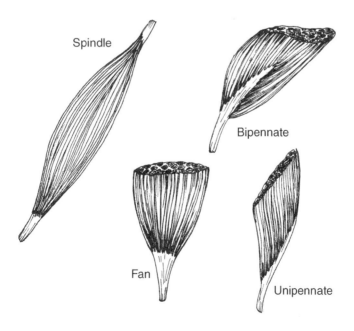

Spindle

Bipennate

Fan

Unipennate

Figure 3.10 Examples of muscles with different shapes and different arrangements of fibres.

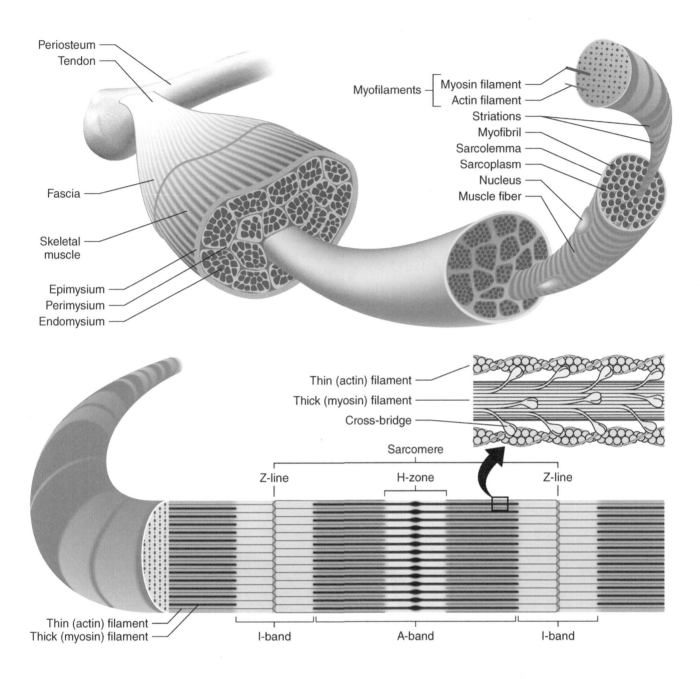

Figure 3.11 A typical muscle comprises bundles (fascicles) of muscle fibres. Each muscle fibre, in turn, contains many myofibrils that are made of repeating series of sarcomeres. In the A band at the middle of each sarcomere are myosin filaments. Overlapping slightly with each end of every myosin filament are actin filaments. The actin filaments stretch from their attachment to the Z disk at either end of the sarcomere toward the H zone in the middle of the sarcomere.

Adapted, by permission, from R.S. Behnke, 1999, *Kinetic anatomy* (Champaign, IL: Human Kinetics), 13.

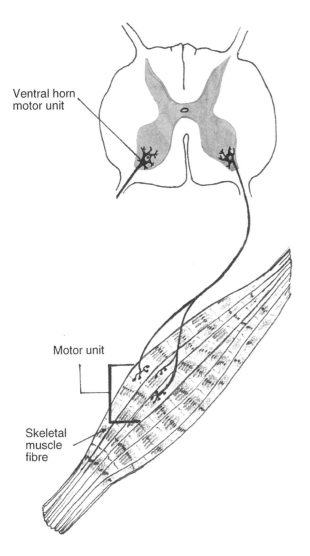

Ventral horn
motor unit

Motor unit

Skeletal
muscle
fibre

Figure 3.12 Illustration of a motor unit in a limb muscle where the nerve fibre originates in the spinal cord.

nerve and the muscle fibres that it controls (figure 3.12). Muscles over which we have fine control, such as the small muscles in the hand, have relatively few muscle fibres per motor unit, whereas the large muscles of the lower limb, over which we have only relatively coarse control, have many more muscle fibres per motor unit.

DISTINGUISHING PROPERTIES OF MUSCLES

The macro- and microstructure of muscle give it five properties that are crucial to its function.

Muscle as a tissue has three main properties:

1. its excitability in response to nerve stimulation,

2. its contractility in response to the stimulation, and

3. its conductivity, which allows the electrical signal produced by the muscle fibres in response to neural stimulation to travel along those fibres.

A whole muscle has two additional properties that arise primarily from its structure and the mechanical characteristics of the connective tissue within it:

1. extensibility, and

2. elasticity.

Because muscle fibre tissue responds to neural activation by producing a force, it is regarded as an active tissue, in contrast to the passive connective tissue component, which can only resist applied forces.

MUSCLE CONTRACTIONS

Muscle fibres operate by producing a force that tends to shorten the muscle. Knowledge of the microstructure of muscle tissue is necessary for understanding this process.

In the lengthened position there is little overlap between the longitudinal thick and thin protein filaments contained in each sarcomere. The muscle shortens (contracts) by increasing that overlap. Thus, researchers proposed the **sliding filament hypothesis** of muscle contraction, which states that the shortening of the sarcomere results from the thick and thin filaments sliding toward one another. Further research led to the **cross-bridge hypothesis**, which states that muscles shorten when cross-bridges are formed as the thick **myosin** filaments attach themselves to the thin **actin** filaments connected to the **Z disk** on either end of the sarcomere. These bridges are the means by which the myosin filaments pull the actin filaments toward themselves, shortening the muscles. Both hypotheses, which are generally accepted, indicate that there are both upper and lower limits to sarcomere (and hence muscle) length.

Activation by a nerve supplying a muscle prompts both an electrical response and a mechanical response. The electrical response spreads over the surface of each muscle fibre, causing a series of steps to occur. Here we highlight only some of the major steps in a process known as **excitation–contraction coupling**.

1. In the neuromuscular (nerve–muscle) junction, a chemical is released from the end of the nerve fibre, which causes a rapid change in voltage in the muscle.

2. The electrical signal travels over the surface of, and along, the muscle fibres.

3. The electrical signal causes the release of calcium ions into the cytoplasm (intracellular fluid) of the muscle fibre. This is an important part of the process because it is the link between excitation and contraction.

4. The calcium ions expose active sites on the actin (thin) myofilaments, to which the myosin filaments immediately attach.

5. By means of these attachments (cross-bridges), each myosin filament pulls the actin filaments that overlap with it at either end toward its centre, producing cross-bridge cycling, which is the mechanical response to the electrical signal (figure 3.13).

The whole process takes a couple milliseconds. Although almost all the calcium in the body is stored in bones, the 1% that is unbound in skeletal muscle is essential for muscle contraction.

Cross-bridge cycling involves four steps (figure 3.13): (1) attachment of the myosin to the active site of the actin filament (when the active site is exposed in the presence of calcium ions), (2) pivoting of the myosin head to pull the actin toward it (power

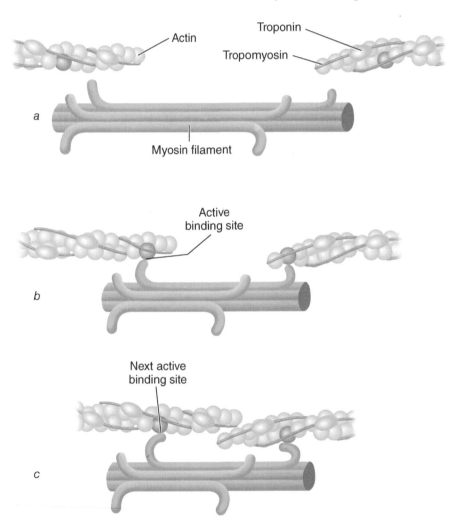

Figure 3.13 The cross-bridge cycle. The thick filament binds to the active site on the thin filament, pulls the thin filament toward it, and then detaches, ready to repeat the process.

Reprinted, by permission, from J.H. Wilmore and D.L. Costill, 1999, *Physiology of sport and exercise*, 3rd ed. (Champaign, IL: Human Kinetics), 42.

stroke), (3) release of the myosin head from the actin, and (4) reactivation of the myosin (return to the straight-head configuration of myosin).

Electrical activity of a muscle can be detected using electrodes in a technique called **electromyography** (EMG), which involves detecting the small electrical signal produced by muscle slightly before it contracts and then using appropriate hardware and software to record and analyse the signal. A basic tool in functional anatomy, EMG has been used to determine muscle roles in movement situations, to provide biological feedback to a person attempting to improve motor performance, and to investigate the effects of strength training.

As a muscle shortens it also thickens, which can be observed on the skin overlying superficial muscles. This mechanical response of muscle can be detected using a variety of techniques, collectively termed **mechanomyography** (MMG). In MMG, a muscle is artificially suprastimulated and the muscle response is detected and analysed. The size of the maximum muscle deformation and the time it takes to reach this maximal deformation indicate some of the physiological functions of the muscle. The effects of muscle fatigue and injury can be quantified using MMG.

Mechanics of Muscular Action

The force produced by muscle activation places tension on the attached tendon or bone to produce joint motion. As we shall see, muscles do not always shorten when they produce force.

Types of Contractions

When muscles contract they can perform three main actions. It is the muscular system that provides the power for the human body to perform work. Muscles cross joints so that their contractions produce movement, and whenever a muscle contracts it tends to shorten and pull the two bony attachments closer together. This action of a muscle is termed **concentric**, in contrast to an **eccentric** action that occurs when a muscle is activated but is lengthening. In this situation other forces (such as external loads) prevent the muscle from shortening. During eccentric actions, muscles are controlling the movement produced by the other forces. (For example, when lowering a weight, a muscle is controlling the movement produced by gravity working on the weight.) When a muscle relaxes it is not producing any force and is therefore not controlling the movement produced by gravity. During an **isometric** action, the muscle is activated but the overall length of the muscle–tendon unit does not change, so this type of action is important for the stabilisation of joints. Muscles may therefore act concentrically to produce movement, eccentrically to control movement, or isometrically to maintain posture and enhance joint stability. Joint stability is also a byproduct of the dynamic types of action. In figure 3.14, the elbow-flexor muscles act concentrically to move the load up against the resistance of gravity and then act eccentrically to control the downward movement. The descriptions of joint action that often appear in tables in anatomy books are based on the assumption that the joint movement is produced by a concentric muscle action.

Explaining Joint Actions

The joint actions caused by muscles would be relatively simple to predict and explain if each muscle crossed only one joint; that is, if all muscles were

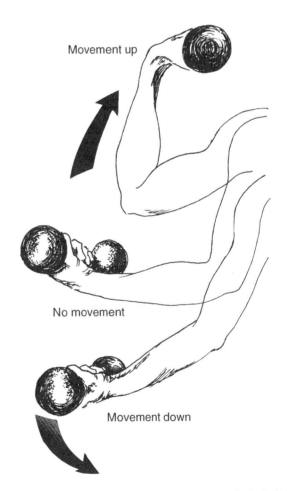

Figure 3.14 The three major types of skeletal muscle action are isometric, concentric, and eccentric. Isometric actions occur in static situations (no movement), whereas concentric (muscle shortening) and eccentric (muscle lengthening) actions occur during dynamic tasks.

monoarticular. However, in addition to monoarticular muscles, there are muscles that cross two joints (**biarticular**), such as the hamstring muscle group at the back of the thigh, and others that cross more than two joints (**polyarticular**), such as the muscles in the forearm that bend the fingers. A basic rule for predicting joint movement is that when a muscle is activated it shortens and tends to produce all the joint actions of which it is capable. If these actions do not occur, it must be that external forces are present or that other muscles are preventing some of the actions. In these situations even a purely mechanical analysis of the muscular system is quite complex.

In a whole muscle, the shortest length is determined by how much the actin and myosin filaments in each sarcomere of the muscle tissue can overlap, but the upper limit is determined mainly by the extensibility of the muscle's connective tissue component. When joints are moved through their full range, most muscles in the human body normally operate well within their available length range. In some cases, however, the limits are reached. Biarticular and polyarticular muscles may be limited because their lengths are determined by motion at a series of joints.

Try these movements to demonstrate both the upper and lower length limits of muscles crossing more than one joint. Stand on one foot and bend the unsupported limb as far back as possible at the hip (extend the hip). Now try to bend (flex) the knee as far as possible (figure 3.15a). You will notice that the range of knee flexion is less than normal when the hip is extended. (You can verify this by trying the same knee movement with the hip flexed.) Now lie on your back and bend (flex) one hip so that the thigh is vertical, and then attempt to straighten the knee (figure 3.15b). Most people are unable to do this. If you can straighten your knee in this position, flex the hip more and try again. These two examples principally indicate the lower and upper length limits of the hamstring muscle group that crosses both the hip and knee joints.

To demonstrate the lower limit of the muscles that bend the fingers, bend (flex) the wrist so the palm of the hand is brought as close as possible to the forearm and then try to make a fist. Notice this is a very weak grip compared with your strongest power grip. Notice the position of the wrist when the grip is strongest. The muscles that provide most of the strength for bending the fingers have their bellies in the forearm and tendons that cross a number of joints, including the wrist and others in the hand. When the wrist is flexed, the finger-flexing muscles are already shortened; they have almost run out of the capacity to shorten any further to produce finger flexion.

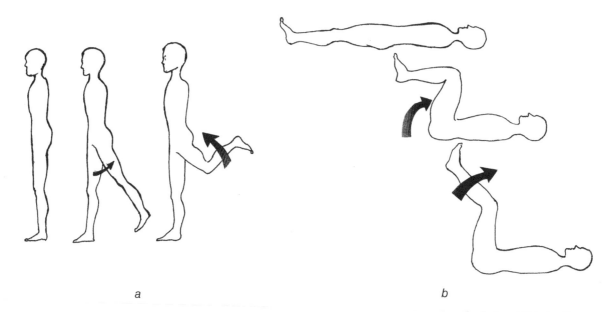

a   *b*

Figure 3.15 Exercises used to demonstrate the (*a*) shortened and (*b*) lengthened positions of the hamstring muscle group.

THE JOINT-STABILISING ROLE OF MUSCLE–TENDON UNITS

Bioengineers are fascinated with the human body because many are involved in the design of replacement parts, such as artificial hips and knees. A human synovial joint is analogous to a tent that relies on the interaction between the following structures for its stability: The bones with cartilaginous ends are like the poles of a tent that resist compressive forces, so the mechanical properties of bone and cartilage have been measured using specially designed equipment. A tent that relied solely on the poles would be very unstable; it would simply blow over in windy conditions. For extra stability, guy ropes are attached to the poles, and these ropes must be flexible (or attached to springs) to perform their functions optimally.

Ligaments are flexible structures that stabilise joints. Their tension varies through the range of joint motion because of the complex shapes of joint surfaces, so joint stability is dependent on joint position. Joints must not be completely stable because they must also allow movement. Stabilising features of joints, which tents do not share, are the muscle–tendon units that are under neural control and that produce the movements at the joints, control the movements, or both.

What is the primary mechanism that helps prevent our ligaments from tearing and our joints from dislocating? It is the stabilising influence of the contraction of surrounding musculature. In terms of injury prevention, the time required to damage a ligament is less than the time of a simple muscular reflex response to muscle stretch. This means that the stabilising muscles must be activated before the imposition of external deforming forces or they cannot prevent joint dislocation and ligament tears. When you step on the edge of a hole that you have not seen, you are likely to sustain a sprained ankle because you have relied on the ligaments for stability and, unaided by surrounding musculature, the ligaments are usually not up to the job in that situation. If you see the same hole ahead of time, however, the surrounding muscles are activated before you step; the muscle–tendon units add stability and you are much less likely to sprain your ankle joint. The activation patterns of muscles during movement are a major area of research.

LIMITATIONS TO RANGE OF JOINT MOTION

When a joint reaches the end of its range of motion, this limitation has several possible causes in addition to the intrinsic features of the joint illustrated in figure 3.6. When joint motion occurs, the body segments bend such that the tissues on one side are compressed while those on the opposite side are stretched. The most obvious limiting factor is therefore the tension in the joint capsule and its associated ligaments on the side of the joint where stretching is occurring. In addition, and most commonly, stretch of the associated muscles and tendons restricts the range of joint motion. Some of this resistance is also provided through stretching of the skin. Sometimes apposition of soft tissues also restricts movement and, rarely, apposition of bony parts forming the joint restricts movement. This bony contact would occur only at the extreme end of range and is potentially injurious in dynamic situations.

DETERMINANTS OF STRENGTH

What is **muscular strength**? Mechanically, strength is determined by both the force generated by muscular contraction and the leverage of the muscle at the joint. This concept, which is termed **moment of force**, is examined in detail in chapter 7. Muscle force is proportional to the cross-sectional area of the muscle.

SUMMARY

The musculoskeletal system consists of bones, joints, and muscles. Bone tissue is both hard and tough. Compact bone is particularly strong, whereas spongy bone is better for shock absorption. Individual bones have a variety of shapes; each shape is particularly suited for performing one of the functions of bone.

Joint mobility and stability tend to be competing requirements, so the structure of each joint is a compromise. The many types of joints are classified on the basis of their structure, the gross movements they allow, or the motions that occur between the articular surfaces in contact.

Skeletal muscle tissue looks striated in a longitudinal section under a microscope. A muscle fibre consists of up to tens of thousands of sarcomeres in series (joined end to end). Connective tissue, an important component of whole muscles, provides some of the properties of the muscle–tendon unit. The sliding filament hypothesis of muscle contraction describes what happens when muscle changes length and is based on electron micrographs of muscle at different lengths. According to the cross-bridge hypothesis, cross-bridges on the thick myofilaments pull the thin myofilaments toward them after excitation from the nerve that supplies the muscle. Length changes resulting from neural activation are dependent on the net effect of a number of forces. The muscle–tendon unit may act isometrically to stabilise joints, concentrically to produce movement, and eccentrically to control movement. The electrical response of muscle to neural stimulation can be detected and used as a tool in the study of normal muscle contraction. The mechanical response of muscle to artificial stimulation can be detected and used as a tool for measuring maximum muscle performance. Joint motion may be limited by the length of muscle–tendon units, which may be increased by flexibility exercises. Strength is determined by muscle cross-sectional area (related to the number of fibres in parallel) and leverage of the muscle that produces joint motion.

FURTHER READING

Basmajian, J.V., & de Luca, C.J. (1985). *Muscles alive: Their functions revealed by electromyography* (5th ed.). Baltimore: Williams & Wilkins.

Jenkins, D.B. (2009). *Hollinshead's functional anatomy of the limbs and back* (9th ed.). St. Louis: Saunders/Elsevier.

Levangie, P.K., & Norkin, C.C. (2011). *Joint structure and function: A comprehensive analysis* (5th ed.). Philadelphia: F.A. Davis.

Martini, F., Timmons, M.J., & Tallitsch, R.B. (2012). *Human anatomy* (7th ed.). Boston: Pearson Benjamin Cummings.

Nordin, M., & Frankel, V.H. (2001). *Basic biomechanics of the musculoskeletal system* (3rd ed.). Philadelphia: Lippincott Williams & Wilkins.

Oatis, C.A. (2009). *Kinesiology: The mechanics and pathomechanics of human movement* (2nd ed.). Baltimore: Lippincott Williams & Wilkins.

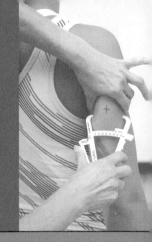

CHAPTER 4

BASIC CONCEPTS OF ANTHROPOMETRY

The major learning concepts in this chapter relate to

- ▶ the definition of anthropometry,
- ▶ the basic equipment used for physical measurement,
- ▶ quantifying body size and shape by determining the relationships between different body segments,
- ▶ the composition of the human body,
- ▶ describing physique through use of a shorthand method, and
- ▶ human variation in the musculoskeletal system and in physical dimensions.

In this chapter we examine the field of anthropometry and discuss the measurement of, and variation in, human body size, shape, and composition. We also discuss the implications of these variations for movement capability.

DEFINITION OF ANTHROPOMETRY

Anthropometry is the science that deals with the measurement of the size, proportions, and composition of the human body. In most cases it is the sizes that are directly measured, and these direct measurements can be combined to indicate the shape of the

whole body or body segments (figure 4.1). Body composition usually involves using direct anthropometric results to predict the relative amount of a particular component in the whole body. The best-known example is the use of skinfold thicknesses to predict the percentage of fat in the body (see "In Focus: Predicting Body Density and Body Composition").

One specialised branch of anthropometry of particular relevance to the study of human movement is kinanthropometry. **Kinanthropometry**, a term proposed by Canadian academic Bill Ross, has been defined by the International Society for the Advancement of Kinanthropometry as the scientific specialisation dealing with the measurement of

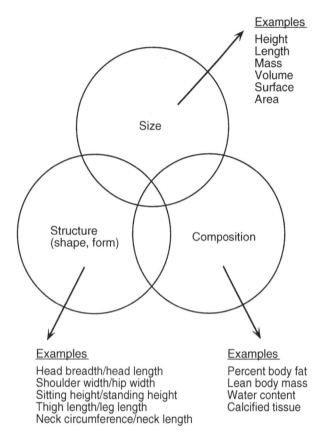

Examples
Height
Length
Mass
Volume
Surface
Area

Size

Structure
(shape, form)

Composition

Examples
Head breadth/head length
Shoulder width/hip width
Sitting height/standing height
Thigh length/leg length
Neck circumference/neck length

Examples
Percent body fat
Lean body mass
Water content
Calcified tissue

Figure 4.1 Venn diagram indicating that the three components of anthropometry overlap one another. Relative sizes indicate shapes of body segments, and simple measurements may be used to predict body composition.

people in a variety of morphological perspectives, its application to movement, and the factors that influence movement, including

- components of body build,
- body measurements,
- proportions,
- composition,
- shape,
- maturation,
- motor abilities,
- cardiorespiratory capacities, and
- physical activity, including recreational activity as well as highly specialised sport performance.

Defined as such, kinanthropometry is a scientific specialisation closely allied to physical education, sport science, sports medicine, human biology, auxology, physical anthropology, gerontology, ergometry, and several medical disciplines.

TOOLS FOR MEASUREMENT

In anthropometry, **stadiometers** are used to measure height (**stature**), **anthropometers** to measure lengths of body segments, tapes to measure body segment circumferences, **bicondylar calipers** to measure bone diameters, **skinfold calipers** to measure thicknesses of skin plus **subcutaneous** fat, and scales to measure masses.

Simple or complex mathematical manipulation can be used to derive indices to describe the shape of a body segment. The cross-sectional shape of the thorax (**thoracic index**) can be described by dividing the anteroposterior (from front to back) diameter by the transverse (from side to side) diameter. Statistical procedures are generally used to formulate prediction equations when the direct measurement of a parameter presents difficulties such as cost, time, or technical complexity.

In kinanthropometry, the basic measurement tools may be anthropometric equipment, but the researchers require important skills for interpretation of the results. Required knowledge includes the anatomy and biomechanics of the human musculoskeletal system. Because of the breadth of knowledge required by kinanthropometrists, this aspect is not emphasised here; however, examples are provided in chapter 6.

BODY SIZE

The number of measurements of the dimensions of the human body that one could take is almost infinite. Anthropometry manuals recommend a core number of measurements and include a longer list of optional measurements. In most cases, the manuals describe techniques for measuring the size of the whole body and its various segments. Body size includes height and mass, and segment sizes include lengths, circumferences, diameters, and so on. Individual segment masses cannot be directly measured, so these are predicted from previous research. Because the techniques of anthropometry are critical to measurement **validity**, the measurer or experimenter must rigorously follow predetermined measurement protocols. Different manuals may describe different protocols for apparently the same body segment, so it is crucial that the measurer choose one particular protocol and adhere to it closely.

DETERMINATION OF BODY SHAPE

Body shape can best be described by the proportions between the sizes of different body segments. These proportions indicate the structure of the body. The relationship between different measurements may be termed *dimensionality*. The relationship between height and weight is a common example. In modern nutritional surveys, the **body mass index** (BMI) is often quoted. This is the mass (in kilograms) divided by the height (stature in meters) squared, or BMI = mass/height^2.

Many indices have been used to describe shape. Have you ever noticed that sometimes when people stand up there is a large difference in their heights but when they sit down the difference is very small? This observation is related to their body proportions. Some people have relatively long lower limbs and others have relatively short limbs. To describe the relationship between lower limb length and height, the sitting height is divided by the standing height and multiplied by 100, so it is expressed as a percentage. If the lower limbs are investigated, one can calculate the ratio between the lengths of the thigh (limb segment between hip and knee) and the leg (limb segment between knee and ankle; in biomechanics, known as the shank). Sprinters tend to have relatively short, muscular thighs and long,

thin legs. The performance relationships of body proportions rely on mechanical knowledge for interpretation of the results and are consequently a topic of considerable interest to biomechanists.

TISSUES COMPOSING THE BODY

One can contrast anthropometric models and anatomical models of the composition of the human body. Anatomical **dissection**, in which one separates the body into the five masses of skin, muscle, bone, tissues from other organs (undifferentiated), and fat (see table 4.1), has been used to describe the various components. Anthropometric models seek to estimate body composition using noninvasive means. A two-component model, in common use, seeks to separate lean tissue (such as bone and muscle) from fat tissue, given the realisation that excess fat rather than excess lean tissue is a prime risk factor for premature death. Measurements of skinfold thickness have been introduced in an attempt to derive an indirect, noninvasive measure of body fatness (see "In Focus: Predicting Body Density and Body Composition"). More recently, it has been determined that the fat around the organs in the abdomen (visceral fat) is more related to ill health than is subcutaneous (under the skin) fat (see table 4.1).

IN FOCUS: PREDICTING BODY DENSITY AND BODY COMPOSITION

In the late 1980s, two articles published by researchers from Europe and North America indicated the relationships between body measurements and density and percentage body fat. Researchers from the department of human sciences, University of Technology, Loughborough in England and the human nutrition unit, Istituto Nazionale della Nutrizione, Rome in Italy took 21 body measurements from 138 male Italian shipyard workers. Examples of the mean values obtained were 1.73 m (5 ft 8 in.) for height, 76.1 kg (167 lb) for mass, 10.4 mm (0.4 in.) for triceps skinfold thickness, 37.7 cm (14.8 in.) for calf girth, and 7.0 cm (2.8 in.) for bony width across the elbow. A group of 7 researchers from 5 universities in the United States used underwater weighing to determine the body density of 310 subjects aged between

8 and 25 yr. The subjects, from Illinois and Arizona, represented both sexes and both major racial groups. As an example, the mean total body mass of the pubescent girls was 43.2 kg (95 lb).

The main aim of these projects was to compare the results of measuring percentage body fat by underwater weighing with the results obtained using existing prediction equations based on anthropometric measurements. Hardly any of the previous equations from the literature showed good correlations with the measured body densities. The results from both studies tended to confirm the general conclusion that equations that predict body density are population specific.

The authors of these studies therefore developed their own predictive equations using a statistical

(continued)

In Focus *(continued)*

technique known as **stepwise multiple regression.** Variables included in the European equation for men were age and two skinfold thicknesses. The American authors concluded that because chemical maturity has not been reached during puberty, adult prediction equations lead to an underestimate of body density and, hence, an overestimate of percentage body fat. For the 59 American pubescent subjects, the sum of two skinfolds (triceps + calf or triceps + subscapular) plus an indication of physical maturation was able to predict percentage body fat reasonably well. In a more recent review article by an experienced researcher from Loughborough University in England, the author concluded that "Users want simple, inexpensive, rapid, safe, accurate methods to measure body composition but speed and simplicity come at the expense of accuracy." (Norgan, 2005, p. 1108).

Sources

Norgan, N.G. (2005). Laboratory and field measurements of body composition. *Public Health Nutrition, 8*(7A), 1108-1122.

Norgan, N.G., & Ferro-Luzzi, A. (1985). The estimation of body density in men: Are general equations general? *Annals of Human Biology, 12,* 1-15.

Slaughter, M.H., Lohman, T.G., Boileau, R.A., Horswill, C.A., Stillman, R.J., van Loan, M.D., & Bemben, D.A. (1988). Skinfold equations for estimation of body fatness in children and youth. *Human Biology, 60,* 709-723.

Table 4.1 Proportions of Human Tissue Weights Determined From 3 Dissection Studies Involving 34 Cadavers Ranging in Age From 16 to 94 Years

Tissue	Total body weight (%)	Lean body weight (%)
Adipose	34.5	
Muscle	33.0	50.0
Bone	13.5	21.0
Skin	5.5	8.0
Undifferentiated organs	13.5	21.0

Visceral fat expressed as a percentage of total body fat was 17% in men and 13% in women.

Data from Clary et al., 1984; Martin et al., 2003.

In the two-component anthropometric model used in the calculation of body density, lean tissue and fat tissue each have an assumed **specific gravity.** Basically, lean tissue sinks in water because of its higher density, and fat floats on water. In the underwater weighing technique, which has been used as the criterion measure for the estimation of body composition, body weight in water is compared with body weight in air. Allowances are made for the volumes of air in the lungs and digestive tract. With this technique, calculations have shown that some elite power athletes had a negative percentage body fat, which is biologically impossible and incompatible with life. This measurement error arose mainly because of the assumed specific gravity of nonfat tissue (about 1.1), which includes bone and muscle.

The tissues composing the lean tissue component of the body, however, are variable in their densities and relative quantities. Also, the specific gravity of the body varies between one region and another and may vary within the one body segment. If underwater weighing can lead to incorrect estimates of body density and, therefore, of percentage body fat, it is not a valid criterion for comparison with skinfold measurements. Therefore, many researchers now recommend that skinfold measurements be reported directly and not be used in an equation that predicts percentage body fat.

Radiographic techniques can be used in a three-component model of the human body. In **dual energy X-ray absorptiometry** (DEXA), a radiation source allows estimation of the masses of calcified tissue, fat, and nonfat tissue in the body. The modern imaging techniques of **computed tomography** (CT) and **magnetic resonance imaging** (MRI) have the potential to define proportions of many types of tissues in regions of the body.

SOMATOTYPING AS A DESCRIPTION OF BODY BUILD

In the past it was thought that body types were genetically determined and therefore did not change throughout life. It is now agreed that body type does have a genetic component but that it is also a product of environment. **Somatotypes** were originally described as **genotypes**, but they are now considered **phenotypes** because of the environmental influences on body shape and because somatotype can change throughout life (see chapter 5). Anthropometric somatotypes use measurements to describe shape and composition of the body at a particular point in time. Size must be reported separately.

The **Heath-Carter anthropometric somatotype** is a useful example of most commonly used somatotyping methods. It is calculated using 10 recorded anthropometric measurements (see "Measurements for Calculating the Heath-Carter Anthropometric Somatotype"), but the protocol requires each measurement to be repeated at least once. Checking the **reliability** of measurements is a fundamental principle in any scientific experimentation.

In the Heath-Carter anthropometric somatotype, people are described by three numbers, each representing one somatotype component. The sum of three skinfolds from the arm and trunk is used to calculate the first component, which may be termed *endomorphy*. This component defines the relative fatness of the person, especially when related to the person's height. The second component, termed *mesomorphy*, represents musculoskeletal development. The skeletal development is estimated by the diameters of the bones at the elbow and knee. Muscular development is estimated from the circumferences of the arm and calf after the estimated fat component of these segments has been accounted

MEASUREMENTS FOR CALCULATING THE HEATH-CARTER ANTHROPOMETRIC SOMATOTYPE

1-4

Four skinfolds measured on the right side of the body with skinfold calipers and expressed in millimeters

1. Triceps (on the back of the arm)
2. Subscapular (just below the lowest part of the shoulder blade)
3. Suprailiac (just above the bony projection of the pelvis at the approximate level of the umbilicus)
4. Medial calf (inside of the widest part of the calf)

5

5. Height measured with a stadiometer and expressed in millimeters (for calculation of mesomorphy) or centimeters (for calculation of ectomorphy)

6-7

Two bone breadths measured on both sides of the body with bicondylar calipers and expressed in centimeters.

6. Humeral epicondylar (elbow width)
7. Femoral epicondylar (knee width)

8-9

Two limb segment circumferences measured on both sides of the body with an anthropometric tape and expressed in centimeters

8. Arm (elbow bent with biceps muscle maximally contracted)
9. Calf (largest part of calf when standing normally on both feet)

10

10. Weight mass measured on scales and expressed in kilograms

for. Bone diameters are determined more by genetics than are the segment circumferences, which are very much influenced by the type and quantity of physical activity (see figure 6.3) in which a person engages. The third component, termed *ectomorphy*, is calculated from height and mass using a cubic relationship. A person who rates highly on this component is light for his or her height.

Every person is rated on each of the three components and can be positioned on a special diagram called a **somatochart** (figure 4.2). At the centre of this triangle-shaped diagram is the unisex "phantom," which is the statistical average of a normal man and woman; males and females are not plotted on different somatocharts. An average man is more mesomorphic than the phantom, whereas the average woman is more endomorphic than the phantom (see figure 4.2). A typical endomorph is a sedentary middle-aged man or a Sumo wrestler (figure 4.3*a*). A typical mesomorph is any elite power athlete, especially weightlifters, boxers, and wrestlers (figure 4.3*b*). Typical ectomorphs are ballet dancers and female gymnasts. High jumpers and marathon runners also tend to fit into this category (figure 4.3*c*). These examples indicate one aspect of kinanthropometry: Defining relationships between body size,

a

b

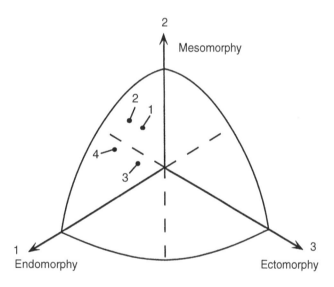

Figure 4.2 Somatochart indicating the somatotypes of four groups in the Canadian adult population. The four groups are males younger than 30 yr of age (3.5-5-2; endomorphy-mesomorphy-ectomorphy), males older than 30 yr (4-5.5-2), females younger than 40 yr (4.5-4-2.5), and females older than 40 yr (5-4.5-2).

c

Figure 4.3 Typical extreme somatotypes in different sports: *(a)* endomorph, *(b)* mesomorph, and *(c)* ectomorph.

shape, and composition and human performance and then explaining the relationships.

HUMAN VARIATION

There is a wide range of normal in almost all body measurements. Some individual differences may be explained by more general differences, such as differences between males and females, ethnic differences, age differences, and differences in types and amounts of activity. This short list implies two main sources of differences: genetics and environment. Environmental factors include (a) the mechanical environment related to the magnitude and regularity of muscular forces and (b) metabolic factors that may include deficiencies of certain nutrients in the diet. In other chapters we address differences related to age (chapter 5) and activity (chapter 6).

HUMAN VARIATION IN THE MUSCULOSKELETAL SYSTEM

Bone sizes differ among individuals, but differences in bone density can also be correlated with race. In North America, for example, it has been shown that African Americans have a higher bone density than Caucasians.

Peak bone mass in women is about 30% less than in men. This difference between the sexes is an example of sexual dimorphism, which in this case is attributable mainly to differences in physical size. Any differences in bone density between the sexes are small and are site specific rather than general.

Differences in muscle volume between males and females are also mainly related to differences in body size. However, it appears that some differences in muscle size are related to the presence of the different sex hormones and their interaction with physical activity. Generally, the muscle volume of adult males is 40% more than that of adult females; however, this is not a large difference when muscle volume is considered as a proportion of total body volume. Relative muscle masses in the lower limbs of males and females are similar; however, there are larger differences in the upper limbs. Researchers who have analysed the tissues of the arm using CT (which provides an image of the cross-section of the arm) have found the average cross-sectional area of muscle plus bone to be 62 cm^2 (10 in.^{2}) in a 30-yr-old male compared with 35 cm^2 (5 in.^{2}) in a female of the same age (see "In Focus: Relationships Among Thigh Circumference, Quadriceps Muscle Cross-Sectional Area, and Knee Extensor Strength").

HUMAN VARIATION IN PHYSICAL DIMENSIONS

There are many examples of **sexual dimorphism** in body dimensions, but often the overlap between the two sexes is so large that it would be difficult to predict the sex of an individual on the basis of body measurements alone. Males tend to be taller and heavier than females, but these are not distinguishing features. At about 20 yr of age, the ratios between males and females are 1.08:1 for height, 1.25:1 for weight, and 1.45:1 for lean body mass. These differences have important implications for both nutritional requirements and athletic performance.

The difference in shape between adult males and females is well known and is based mainly on differences in shoulder size. Measuring the width of the shoulders in centimeters and multiplying this by 3, then subtracting the width of the pelvis, enables one to correctly identify almost 90% of the adult population as male or female. This relationship between shoulder and hip width, known as the **androgyny index**, is a good example of the use of proportions to describe shape and, in this case, emphasise differences.

Some sports (such as basketball) favor tall players, whereas jockeys are usually short because they must be very light. Figure 4.4 illustrates the positions of a typical 200 cm (6 ft 7 in.) basketball player and a 155 cm (5 ft 1 in.) jockey relative to the young adult male population of Queensland in Australia. National Basketball Association centers are often more than 6 standard deviations above the average height. With regard to height comparisons, the average North American man is taller than 95% of Asian men, which probably explains the restricted interior space in Japanese cars when they were first exported during the 1960s. It is predicted that the height difference will be smaller in future generations. We explain the basis of this prediction in chapter 5.

A number of proportions, including the sitting height:standing height ratio, also show ethnic differences. The group with the smallest sitting height:standing height ratio is the Australian Aborigine; that is, people in this group have the longest lower limbs relative to their height. On average this figure is less than 50%, compared with more than 53% for people from Southeast Asia. In a photographic atlas of Olympic athletes, Jim Tanner emphasised proportional differences between African American and Caucasian male athletes. The relatively longer lower limbs of the

IN FOCUS: RELATIONSHIPS AMONG THIGH CIRCUMFERENCE, QUADRICEPS MUSCLE CROSS-SECTIONAL AREA, AND KNEE EXTENSOR STRENGTH

Researchers from the department of anatomy and human biology at University of Western Australia and Royal Perth Hospital examined thigh cross-sectional images produced by CT in 15 males who had previously incurred knee injuries. The investigators compared uninjured and previously injured limbs (on average, the knee injury had occurred 7 yr before testing). They also measured knee extension strength in a static situation (i.e., during an isometric contraction; see figure 3.14).

Repeated measurements indicated that the measurements of muscle cross-sectional area were reliable. Cross-sectional areas of the knee-extensor muscles on the previously injured side were 11% less than those on the uninjured side. These sides were termed *atrophic* and *normal*, respectively. The atrophic side was 21% weaker than the normal side. The relationship between knee extensor cross-sectional area and force varied between subjects and changed for each person during a strengthening program. Analysis of muscle density indicated that there was greater fat deposition in the atrophic muscles.

The authors discussed the clinical use of anthropometric measurements and advised that thigh circumference measured with a tape is not a valid indicator of muscle cross-sectional area. The circumference represents total amount of bone, muscle, and fat and thus does not allow distinction among these three components. Even an **anterior** thigh skinfold measurement may not provide a good indication of

fat content because of the pattern of fat deposition within atrophied muscles.

Recently, a group of researchers in Taiwan, from the National Taiwan University and Hospital and the National Health Research Institutes, performed a study on 69 subjects (33 men and 36 women) over 65 yr of age to see whether they could predict musculoskeletal function from anthropometric measurements. The researchers found that they could predict quadriceps muscle power by measuring thigh muscle volume using MRI. They also produced a prediction equation using anthropometric measurements that could estimate thigh muscle volume. This indicated that thigh muscle circumference measured with a tape can predict muscle volume when combined with other factors (age, gender, and body mass). The authors concluded that use of this linear regression model allowed clinical use of simple, noninvasive measurements to predict muscle function.

Sources

Chen, B.B., Shih, T.T., Hsu, C.Y., Yu, C.W., Wei, S.Y., Chen, C.Y., Wu, C.H., & Chen, C.Y. (2011). Thigh muscle volume predicted by anthropometric measurements and correlated with physical function in the older adults. *Journal of Nutrition Health and Aging, 15*(6), 433-438.

Singer, K.P., & Breidahl, B. (1987). The use of computed tomography in assessing muscle cross-sectional area, and the relationship between cross-sectional area and strength. *Australian Journal of Physiotherapy, 33*(2), 75-84.

black athletes and other differences related to the lower limb are thought to partly explain differences seen in general performance in the 100 m (109 yd) sprint. Although this example is appealing, one must realise that body dimensions are only one factor in performance and that there are many examples of individuals who do not conform to common perceptions of physical requirements. An example is the basketball player "Spud" Webb, who was able to win a National Basketball Association

slam dunk competition even though he was only 170 cm (5 ft 7 in.) tall.

Statistical trends relate body size and shape to athletic performance, and certain generalisations follow from these trends. One statistician has argued that taller athletes tend to perform better in almost all Olympic athletic events. Graphs of the relationships between height and performance certainly provide evidence in support of this concept.

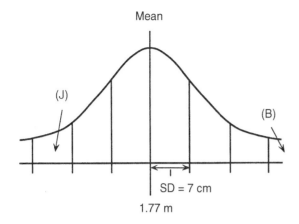

Mean

SD = 7 cm

1.77 m

Figure 4.4 A curve indicating the normal distribution of self-reported heights among almost 100,000 young adult males (aged 17-30 yr) in Queensland, Australia. Heights of typical jockeys (J) and professional basketball players (B) are indicated. The normal distribution is drawn from data supplied by the Queensland Department of Transport (1985).

Data from the Queensland Department of Transport.

SUMMARY

Anthropometry involves using relatively simple equipment and techniques and following strict protocols to take various measurements of the sizes of body segments. These measurements can be combined to indicate the shapes of different parts of the body, and simple measurements can be used to predict the more complex indicators of body composition. Kinanthropometry is an important aspect of functional anatomy because it applies the basic body measurements, and their mathematical combinations, to the movement situation. Knowledge of biomechanics and physiology aids the analysis and prediction of human movement.

Proportions between body measurements can be used to indicate the structure and composition of the body. Indices are used to describe physique and are used in public health surveys to determine risk factors for problems such as cardiovascular disease. Somatotyping is a shorthand way of defining shape and composition of the body. Endomorphy, based on skinfold thicknesses, indicates the relative amount of fat in the body; mesomorphy, based on bone diameters and muscle circumferences, indicates the relative development of the musculoskeletal system; and ectomorphy, based on the basic measurements of height and mass, provides an idea of gross body shape. A person's somatotype is a phenotype, meaning that although it has a genetic base, it is highly dependent on lifestyle factors such as nutrition and activity levels. The concept of "normal" is important in that we must define what is normal so that we can describe what is abnormal. The normal range may be different for males and females, different ethnic groups, and people of different ages.

FURTHER READING AND REFERENCES

Hall, J.G., Froster-Iskenius, U., & Allanson, J. (2007). *Handbook of normal physical measurements* (2nd ed.). Oxford: Oxford University Press.

Lohman, T.G., Roche, A.F., & Martorell, R. (Eds.). (1988). *Anthropometric standardization reference manual*. Champaign, IL: Human Kinetics.

Marfell-Jones, M., Stewart, A., & Olds, T. (Eds.). (2006). *Kinathropometry IX: Proceedings of the 9th International Conference of the International Society for the Advancement of Kinanthropometry*. New York: Routledge.

Norton, K.I., & Olds, T.S. (Eds.). (1996). *Anthropometrica: A textbook of body measurement for sports and health courses*. Sydney: University of New South Wales.

Norton, K.I., & Olds, T.S. (1999). *Lifesize*. [CD-ROM]. Champaign, IL: Human Kinetics.

Ross, W.D., Carr, R.V., & Carter, J.E.L. (2003). *Anthropometry illustrated*. [CD-ROM]. Surrey, British Columbia, CANADATurnpike Electronic Publications.

CHAPTER 5

MUSCULOSKELETAL CHANGES ACROSS THE LIFE SPAN

The major learning concepts in this chapter relate to

- ▶ auxology and gerontology;
- ▶ modern medical imaging techniques;
- ▶ embryological development of the human musculoskeletal system;
- ▶ physical maturation during the postnatal years and physical changes related to ageing;
- ▶ the various stages involved in the development of bone;
- ▶ the processes involved in the growth in length and width of bone;
- ▶ changes in bone composition across the life span;
- ▶ osteoporosis and its effects;
- ▶ the potential effects of repetitive muscular forces on developing bone;
- ▶ the development of joints and muscles;
- ▶ changes in body dimensions, size, and shape across the life span;
- ▶ intergenerational changes in body size and shape;
- ▶ physical differences between males and females at various stages of life; and
- ▶ methods of determining biological age.

In this chapter we use the basic concepts related to the musculoskeletal system and anthropometry introduced in chapters 3 and 4 to focus on the issue of life span changes. This chapter gives attention to three major issues: description of the general process of physical growth, maturation, and ageing; examination of age-related changes in the skeletal, articular, and muscular systems; and examination of changes in body dimensions across the life span.

DEFINITIONS OF AUXOLOGY AND GERONTOLOGY

Auxology and **gerontology** can be defined as the sciences of growth and ageing, respectively. Of particular interest in auxology are the ages of onset of changes, the magnitudes of these changes, and the duration of the change period. All these factors are highly variable between individuals, but the order in which major changes occur is relatively constant. The knowledge available from auxology provides a basis for determining whether people are physically older or younger than their chronological age.

Gerontology becomes an increasingly important field of study as the percentage of aged individuals increases. One focus in gerontology, particularly important in the subdiscipline of functional anatomy, is the relative effects of ageing, as opposed to the effects of inactivity, on the musculoskeletal system. It is necessary to determine the relative impacts of physical activity programs.

TOOLS FOR MEASUREMENT

Growth involves the measurable change in either the quantity of tissue or the size of body segments. Therefore, many large-scale studies on human growth and ageing use anthropometric techniques (see chapter 4). Most pregnant women have at least one **ultrasound scan** during their pregnancy. The measurements of the foetus, taken by the obstetrician, include the width of the skull and the length of the femur. These measurements are taken to confirm the age of the **foetus**. During postnatal growth, physiological measurements may also be taken (see chapter 13).

Researchers are often interested in the skeletal development of children and the skeletal changes associated with ageing. Studying these phenomena requires the use of radiological techniques. Skeletal age of children can be determined by hand and wrist radiographs, and the bone density of adults can be measured using **computed tomography** scans or **dual-energy X-ray absorptiometry** scans.

PHYSICAL GROWTH, MATURATION, AND AGEING

The major stages of the human life cycle are listed in table 5.1. About one quarter of the human life span is spent growing. Humans have a long period of infant dependency followed by a unique period of extended childhood growth between infancy

Table 5.1 Stages in Human Development

Landmark events	Major stages	Periods in life cycle	Approximate ages
Conception	Prenatal	Pre-embryonic Embryonic Fetal	Wk 1-3 after conception Wk 4-8 after conception Mo 3-9 after conception
Birth	Postnatal	Infancy Childhood	Yr 1-2 Yr 3-10
Peak height velocity		Puberty Young adult Middle-aged Elderly adult	Yr 11-17 Yr 18-40 Yr 41-60 Yr 61-
Death			

and puberty. **Puberty** is the delayed period of rapid growth leading to sexual and physical maturity. Some claim that ageing begins as soon as the peak in any given parameter is reached. This would imply a 40 to 50 yr period of ageing. Ideally, the decline in structural and functional parameters should be as gradual as possible.

Embryological Development

Development of a human begins when the **ovum** is fertilised by a **spermatozoan**, forming a **zygote**, which is a single fertilised cell. The zygote quickly begins to divide, and during the first 3 wk of development, 3 primary germ layers differentiate: the **ectoderm**, which eventually forms nervous tissue and outer part of the skin; **endoderm**, which forms the visceral organs of the body; and **mesoderm**, which forms the organs and tissues and structures of the musculoskeletal system and the inner layers of skin. Here we consider in detail only the mesoderm.

The embryonic period is a time of rapid development of the mesoderm, characterised by a series of orderly and irreversible stages through which each organism passes (figure 5.1). From relatively undifferentiated cells and tissues, there is a progression of changes leading to highly specialised cells and tissues. By the end of the embryonic period, the **embryo** is about 35 mm (1.4 in.) long, measured from the top of the head to the buttocks, and all the fingers and toes can be seen. Most tissues differentiate during this period and a number of landmark events occur.

The fetal period is a time of rapid growth of the tissues and structures that have formed during the preceding embryonic period. Growth of tissues can occur by an increase in the number of cells forming the tissue, called **hyperplasia**, or by an increase in the size of each cell, called **hypertrophy**. Hyperplastic growth is a feature of the fetal period.

Postnatal Years

Infancy, during which the baby learns to sit up, crawl, stand, and walk, is a period of rapid motor development. During childhood, growth continues at a steady rate. Another period of rapid growth occurs during puberty, which is associated with the development of secondary sexual characteristics.

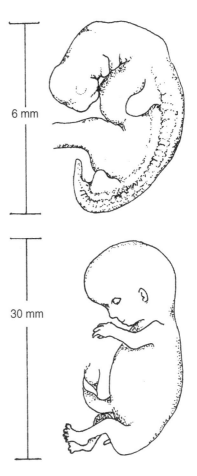

Figure 5.1 Illustrations of an embryo at 30 d and 55 d after conception, indicating sizes by the crown–rump length (i.e., the length between the top of the head and the buttocks). At 30 d the upper-limb buds are visible and the lower-limb buds are beginning to develop. By the end of the embryonic period, all essential external and internal structures are present.

The term *maturation* can be used to describe the physical changes that occur between birth and maturity. These internal processes appear to be genetically determined in contrast to external, environmental factors that affect maturation in the process of adaptation.

During early adulthood many physical characteristics exhibit their peak. The ageing process begins at this time and continues through the middle years to old age. One aim of physical activity is to delay some of the physical changes associated with ageing; some of the effects of age are related more to decreased physical activity than to some inbuilt ageing mechanism.

AGE-RELATED CHANGES IN THE SKELETAL AND ARTICULAR SYSTEMS

For the musculoskeletal system, the major periods of growth are the fetal and pubertal stages. During the early fetal growth period there is rapid multiplication of cells (**hyperplasia**), whereas during the pubertal period growth is caused mainly by development and enlargement of existing cells (**hypertrophy**). Development can also occur via the replacement of one type of tissue by another.

STAGES IN THE DEVELOPMENT OF BONE

Adult human bones appear in a variety of shapes and sizes. Their basic shape is genetically determined, but their final shape is influenced greatly by the environment in which they develop. Environmental influences include mechanical factors, such as muscle forces acting on the developing bone, and metabolic factors, which include the supply of nutrients.

Typical long bones, such as the humerus in the arm and the femur in the thigh, develop via a process of **endochondral ossification** in which a cartilaginous model precedes the bone formation. The cartilage is eventually replaced by bone. The cartilage model initially is small, and its replacement by bone begins at an early stage of development. The primary ossification centres appear near the middle of the shaft of the future long bone at about 8 wk after fertilisation, when the embryo is about 35 mm (1.4 in.) long (see figure 5.1).

The cartilage model of the future bone grows before it is replaced by bone. This replacement occurs in a series of stages. Initially, the cartilage model grows by two processes: an increase in the number of cartilage cells and then an increase in the size of each cell. Next, the gel-like matrix surrounding the cartilage cells is calcified. This hardening of the tissue surrounding the cells effectively cuts off their supply of nutrients and prevents the removal of metabolic waste products. These imprisoned cartilage cells eventually die so that the calcified cartilage has a honeycomb appearance. Invading blood vessels bring nutrients and bone-forming cells (osteoblasts) to the calcified cartilage to lay down new bone. This new bone has a disorganised appearance. If ossification were the final stage, a long bone would not develop its hollow structure. What happens instead is that some of the newly laid bone is removed from internal sites by special bone-eroding cells. These osteoclasts work with osteoblasts during remodelling of the bone, causing the cortex to drift away from the central axis of the shaft during growth. Changes in shaft diameter and compact bone thickness can occur at any age by a remodelling process. Bone can respond to changes in mechanical stimuli by changing its structure.

One type of remodelling that occurs is the replacement of the originally laid bone by compact bone that has an organised structure (figure 3.1). Most primary centres of ossification (in the developing shaft) appear before birth, whereas most secondary centres of ossification (in the developing ends of the bone) appear after birth. The times of appearance of these secondary centres of ossification vary widely. The centre in the femur near the knee is normally present at birth, but the centre in the **clavicle** (collarbone) closest to the **sternum** (breastbone) does not appear until about 18 yr of age.

GROWTH IN LENGTH AND WIDTH OF BONE

When both the shaft and ends of a long bone are developing they are separated by a growth plate, or **epiphyseal plate**. In this region, all the stages of bone development can be seen under a microscope (see figure 5.2). Each zone in figure 5.2 represents a stage in the growth of bone and then replacement of cartilage by bone; that is, the zones illustrated do not migrate downward, but rather retain their position and change their nature. This effectively increases the distance between the growth plates at each end of the bone. The process illustrated continues for almost 20 yr. Growth in length of the long bones ceases when the process stops and epiphyseal plates fuse and disappear. Actually, the growth stops because the cartilage cells no longer respond to hormonal influence.

Although growth in height ceases at a certain age, changes in thickness of the compact bone of the shaft and the density of spongy bone can occur at any time of life (see chapter 6). Growth in the thickness and diameter of long bones occurs by **appositional growth**. In this process, bone is normally added on the outside of the shaft and removed from the inside of the shaft.

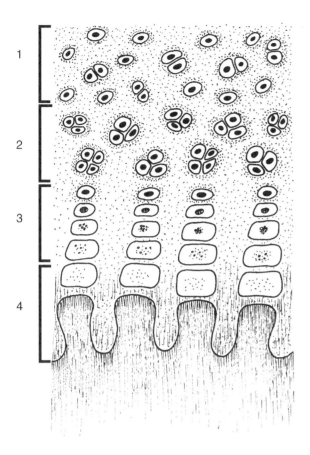

Figure 5.2 Zones in a section of the growth plate of a developing bone as seen under a light microscope. The zones, numbered from the end of the bone toward the shaft: (1) resting cartilage, (2) cartilage proliferation (hyperplasia), (3) cartilage maturation (hypertrophy), (4) cartilage calcification.

SKELETAL COMPOSITION CHANGES ACROSS THE LIFE SPAN

The relative proportions of the inorganic salt crystals and the organic collagen change throughout the life span. In a child, the flexibility of the bone is related to the large proportion of collagen. It is thought that the rapid growth of children's bones and the consequent amount of remodelling results in less-than-optimal mineralisation. In a newborn baby, the skeleton accounts for about 13% of the total body weight, but about two-thirds of the skeleton is cartilaginous.

In bone of a young adult the inorganic material is close to the optimal two-thirds mineralisation, so the bone is strong and tough. Though the proportion of skeletal material in the mature adult is also about 13% of total body weight, just over 10% of the skeleton is cartilaginous. Of the bone, four-fifths is compact and one-fifth is spongy.

Eventually, aged bone becomes brittle. This change relates partly to the increased proportion of the inorganic salts, but of more significance is the reduced total mass of bone material in the aged musculoskeletal system. Bone mass decreases mainly because of a decrease in the number of trabeculae in spongy bone (see "In Focus: Why Trunk Height Decreases With Ageing") combined with resorption and consequent thinning of the trabeculae. In compact bone, the size of the **Haversian canals** (see figure 3.1) increases and the cross-sectional area of the bone surrounding each Haversian canal

IN FOCUS: WHY TRUNK HEIGHT DECREASES WITH AGEING

Researchers from the School of Physiotherapy at the Western Australia Institute of Technology (now Curtin University) and the department of anatomy at the University of Western Australia measured the dimensions of thin sagittal sections of the third lumbar vertebrae removed from 93 adult cadavers. The gender and age of death for each cadaver had been recorded. To calculate vertebral body bone density, the measured dry mass of each section was divided by the calculated volume of each specimen.

As another measure of bone density, the amount of light transmission through each section was analysed. Size and orientation of trabeculae were also analysed (see figure 5.3). To define the shape of each vertebral body section, the midvertebral height of the section was divided by the anterior vertebral height. Average height of each lumbar intervertebral disk was calculated as the mean of three measurements—anterior, middle, and posterior—of disk height. To describe disk shape, the middle disk

(continued)

height was divided by the sum of the anterior and posterior disk heights.

All three methods of measuring bone density confirmed a loss of bone density (or increase in porosity) related to ageing. The decline in the number of vertical trabeculae was not statistically significant, but there was a marked decrease in the number of horizontal trabeculae, especially in the females. The thicknesses of trabeculae were not measured. Shape measurements of the vertebral bodies indicated that there was increasing concavity of the vertebral end plates with age for both sexes. Overall there was no decrease in disk height with ageing; in fact, some individuals exhibited an increase in disk height. Evidence of disk degeneration was observed more often as age increased, but the group with degenerative disks was still a minority in the aged individuals. The authors concluded that the age-related structural changes of the lumbar vertebral column are mainly related to the hard tissues of the vertebral body rather than to the soft tissues of the intervertebral disk (see figure 5.3). They considered that the results indicated the effects of normal ageing and did not propose a definition of osteoporosis based on the measurements used.

A Danish group from the Institute of Anatomy at the University of Aarhus developed a computerised method to distinguish between the physical characteristics of vertical and horizontal trabeculae in lumbar vertebral bodies of 48 cadavers (24 men and 24 women aged between 19 and 97 yr). From their analyses, they concluded that the thickness of vertical trabeculae was independent of age whereas horizontal trabecular thickness decreased with age. The number of vertical and horizontal trabeculae decreased with age, but the relative decrease was larger for the horizontal trabeculae.

The Australian authors used an engineering concept called Euler's theory to explain that the horizontal trabeculae are very important in that they act as cross-braces to support the vertical trabeculae. The loss of horizontal trabeculae with ageing leads to the collapse of vertical trabeculae; this in turn affects the shape of the end plates (at the top and bottom of the vertebral body) that become more concave. The authors cited this example of bone geometry as being more important than mass per se. Further, they suggested that the increasing concavity of the vertebral bodies explains the loss of trunk height and increased curvature of the vertebral column observed in old age. Radiological images of the lumbar spine may give the impression of disk space narrowing because the anterior and posterior disk heights may decrease.

Sources

Thomsen, J.S., Ebbesen, E.N., & Mosekilde, L.I. (2002). Age-related differences between thinning of horizontal and vertical trabeculae in human bone as assessed by a new computerised method. *Bone, 31*(1), 136-142.

Twomey, L.T., & Taylor, J.R. (1987). Age changes in lumbar vertebrae and intervertebral discs. *Clinical Orthopaedics and Related Research, 224*, 97-104.

Twomey, L., Taylor, J., & Furniss, B. (1983). Age changes in the bone density and structure of the lumbar vertebral column. *Journal of Anatomy, 136*, 15-25.

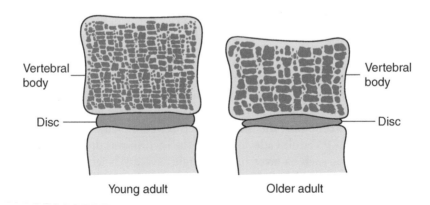

Figure 5.3 Age-related changes in dimensions of the vertebral body and intervertebral disks.

decreases. This increased porosity of bone, in turn, causes decreased bone density (mass per unit volume of tissue). In an 80-yr-old male the density of bone in the spine is often 55% of the value found in a 20-yr-old; in females who are elderly, it is often only 40% of its peak value. With ageing, bone also becomes stiffer, partly because of the collagen cross-linking. A consequence of these mechanical changes is that aged bone can absorb less energy before it fractures, so it is more liable to fail when subjected to large forces.

OSTEOPOROSIS

For most tissues in the body, their mass at any time is determined by the balance between the simultaneous formation and destruction of the tissue. **Osteopenia** ("bone poverty") is reduced bone density, particularly in women during the postmenopausal years, when the dramatic decrease in the female hormone estrogen allows the rate of synthesis to be lower than the rate of resorption. The initial effects are mainly to the spongy bone, but later the compact bone in the shafts of long bones also decreases in thickness.

According to the World Health Organization, **osteoporosis** ("bone porosity") is present when the bone density is more than 2.5 standard deviations below the average bone mass of a young adult woman. In general, the porosity of bone in the human femur doubles from 40 to 80 yr of age and people with osteoporosis show an even greater change. Thus, osteoporotic bone is even less able to absorb energy before it fails than is normally aged bone. Moreover, even though a chemical analysis of osteoporotic bone material may indicate slightly increased levels of calcium per unit mass of bone, the overall mass of bone decreases so markedly that there is less calcium stored in the skeleton overall.

During maturation, the density and calcium content of bones is increased due to genetic predisposition aided by the interaction between dietary calcium and exercise. It is now thought that peak bone mass occurs at about 16 to 20 yr of age in women, so some health authorities claim osteoporosis is a pediatric disease and have promoted the idea of "bone banks" to help prevent osteoporosis. The idea is that if a person's lifestyle during maturation and early adulthood promoted maximum deposition of bone material, then later in life, when withdrawals of bone material are inevitable, the structural integrity of the bones will be maintained longer. Financial advisers tell everybody to start saving as young as possible, and this concept is even more important for prevention of osteoporosis.

Osteoporosis is commonly associated with women, but the prevalence among men is now the same as it was in women about 25 yr ago. This trend has been attributed to lifestyle changes, which include inadequate intake of calcium and less-than-optimal levels of physical activity. The personal lifestyle changes due to osteoporosis and the public health care costs of osteoporosis are so great that they are considered further in chapter 23.

BONE FAILURE IN RELATION TO BONE DEVELOPMENT, AGE, OR ACTIVITY

The modern automotive industry attempts to protect passengers by enclosing them in a cage that is lined by material that cushions the forces involved in a collision. In similar fashion, the soft tissues of the human body, including fat and muscle, perform the role of a crumple zone. However, despite this protection, sometimes the energy involved in a mishap is enough to cause a fracture anyway. Bone failure may be related to bone development, age, or activity.

Certain types of fractures are associated with particular stages of development. The flexible bones of a child tend to splinter in a manner similar to the broken branch of a growing tree; this type of fracture is called a greenstick fracture. Late in life the fracture is more likely to be of the brittle type because of the increased stiffness of old bones. The bones of adults who are elderly are also more likely to fail because of the decreased bone mass and density in osteoporosis.

Specific fractures are related to certain ages. For example, fractures of the neck of the femur (commonly called hip fractures) are common in women in the elderly population. Incidence of forearm fracture increases near the peak of the pubescent growth spurt. This is associated with increased porosity of compact bone as the remodelling space increases to provide calcium to the rapid-growth regions. A second increase in wrist fractures is seen in women who are elderly; this is related to the combination of osteoporosis and increased incidence of falls. In developing children, the cartilaginous growth plate may also fail; this most often occurs at the zone of maturation, or hypertrophy (see figure 5.2). The development of bone has implications for potential injury to the musculoskeletal system. For example, injury to the cartilaginous growth plates can be produced by single excessive forces causing trauma or by repetitive compressive forces, or tensile forces produced by muscle–tendon pull on an area of developing bone.

In sports medicine, specific injuries have been attributed to certain activities. One of the best examples is Little League elbow, described in more detail in chapter 24. Other examples of the relationships between activity and types of growth-plate problems are wrist injuries in gymnasts and tibial tuberosity avulsions in sports involving sprinting and jumping. Pain over the tibial tuberosity (just below the knee cap) in a prepubescent child is indicative of **Osgood-Schlatter's** condition.

EFFECTS OF VARIOUS FACTORS ON RANGE OF MOTION

Range of motion at joints is affected by joint structure and the mechanical properties of the tissues associated with the joints. A general perception is that joint range of motion decreases during life. Although this is the general trend, the rate of loss is not constant. Ranges of joint motion are very large in a newborn baby. As an example, the range of ankle **dorsiflexion** in a newborn is limited only by the contact of the top of the foot against the shin. Try this movement yourself to indicate how much flexibility you have lost since birth.

Between the ages of 6 and 15 yr, there is a general trend for joint range of motion to decrease in boys, whereas for girls the effects are variable and joint dependent. Girls are generally more flexible than boys during childhood and adolescence. Changes in joint ranges of motion between adolescence and young adulthood are variable and, as discussed in chapter 6, appear to be related to physical activity. Similarly, the general trend for joint flexibility to decrease during ageing may not be completely explained by biological ageing processes.

Ageing people may suffer **arthritis** ("inflammation of joints"), which markedly restricts range of motion. **Rheumatoid arthritis** involves inflammation of the synovial membrane, while **osteoarthritis** involves degeneration of the articular cartilage (see figure 3.6). In the total adult population, about three times as many women have **osteoarthritis** as do men. Thus there is some genetic determination of osteoarthritis, to which are added the effects of environmental factors. It would appear that the risk factors for rheumatoid arthritis are mainly related to a family history of the condition.

AGE-RELATED CHANGES IN THE MUSCULAR SYSTEM

Mesoderm, which forms toward the end of the third week of development, eventually forms the muscle and connective tissues of the body. Some of the mesoderm is segmented such that it is in lumps on either side of the midline, and these lumps form the muscles of the trunk. Other mesoderm forms the muscles of the limbs; this tissue migrates into the limbs when they first appear. Limb muscles migrate toward their final position during further development.

Certain precursor cells differentiate into **myoblasts** (muscle-forming cells) that fuse longitudinally to form the long muscles and muscle fibres we can observe. The number of muscle fibres appears to be genetically determined so that large people often have more muscle fibres than persons who are smaller in stature. As muscles grow in length, the number of sarcomeres increases. Although direct evidence is sparse, many believe that the stimulus for growth of muscle is the bone growth that pulls on the attached ends of the muscle. Developmental growth occurs in both muscle length and cross-sectional area, and these factors may also be affected by activity. The effects of physical training on muscle is summarised in chapters 6, 10, and 13.

In an infant, muscle tissue accounts for about 25% of total body weight, but in a young adult the proportion of muscle is more than 40%. In terms of growth, muscle increases from about 850 g (1.9 lb) at birth to about 30 kg (66 lb) in a young adult male weighing 70 kg (154 lb).

Strength, which may be defined as the capacity to produce force against an external resistance, is related to the cross-sectional areas of the muscles producing the force. Muscle volume peaks at about 30 yr of age and then gradually decreases. The atrophy of skeletal muscle associated with disuse is much faster than any ageing-related atrophy. Muscle elasticity also decreases during ageing so that muscles are stiffer and less extensible. This is one contributing factor to the loss of joint range of motion described earlier.

CHANGES IN BODY DIMENSIONS ACROSS THE LIFE SPAN

Because size measurements can be reported directly, information on the average heights (figure 5.4) and weights of children is available. Height can be plotted against age, but plotting height gain against age more obviously illustrates the growth spurt (figure 5.4). Within this period is the **peak height velocity** (PHV), which is a major landmark in pubertal growth and a reference for many other changes. In boys, the relative amount of fat in the body may decrease during this period of rapid growth. The peak weight velocity is delayed relative to PHV by a few months. Peak strength velocity may be delayed for more than a year. Unfortunately, PHV and other peaks can be determined only retrospectively; they cannot be easily or accurately predicted. Age at PHV can be used to measure maturational stage, and this in turn can be used to determine the effects of early versus late maturation on adult body dimensions (see "In Focus: The Pattern of Growth in Swedish Children").

Overall dimensions of the body such as stature and volume (related to body mass) do not increase at a constant rate during growth, nor do they change at a constant rate during ageing. This idea is illustrated in figure 5.4 and is recognised in the common expression *growth spurts*. If one has a specific interest in growth spurts and PHV, mentioned earlier, one can plot the height versus age as height gain per year against age, as illustrated in figure 5.4*b*.

COMBINING SIZE MEASUREMENTS TO PROVIDE INFORMATION ABOUT SHAPE

Galileo first described a square–cubic relationship that is seen in the human body. Assume that two objects have the same shape. If one object is twice as tall as the other, the cross-sectional area will be 4 times as large (squared relationship) and its volume will be 8 times as large (cubic relationship). Remember that the area of a circle is πr^2 and the volume of a sphere is $4/3\pi r^3$. Let's see if these relationships are likely to describe a normal pattern of growth. Suppose that a baby is 50 cm (20 in.) and 3 kg (6.6

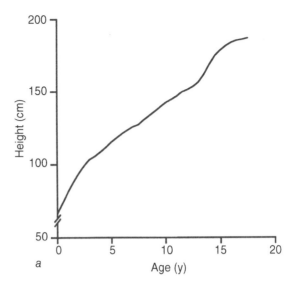

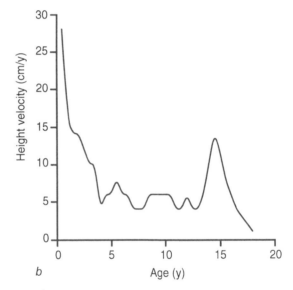

Figure 5.4 Two growth charts for the same individual derived from the same data. Part *(a)* illustrates the normal relationship between height and age. When *(b)* growth in height is expressed as height gain per year, the PHV (at 15 yr) can be determined. These curves are derived from measurements taken by de Montbeillard of his son from the year of his birth in 1759 to 1777.

lb) when born, and assume she grows 3 times longer, to 150 cm (59 in.). Assuming that volume is directly related to mass, we would expect her adult mass to be 3 multiplied by 3 kg³. The calculated result of 81 kg (179 lb) is very heavy for a person who is 150 cm tall. What does this tell us about growth? Basically,

IN FOCUS: THE PATTERN OF GROWTH IN SWEDISH CHILDREN

Commencing in 1955, 103 boys and 80 girls were followed from birth to adulthood. Patterns of maturation were analysed by researchers from the department of orthodontics, University of Lund, Malmo, and the department of pediatrics, University of Goteborg. The subjects were divided into 3 maturity groups, according to their age at PHV. On average, the early-maturing girls were 10.7 yr at PHV and the boys were 12.6 yr. The average maturers were 12.1 yr (girls) and 14.0 yr (boys) at PHV, and the late maturers were 13.6 yr (girls) and 15.6 yr (boys). Just over half of the subjects were classified as average maturers, almost one quarter were early maturers, and about one fifth were late maturers.

Between the ages of 5 and 14 yr, the early-maturing girls were taller than the late maturers [maximum of 13.1 cm (5.2 in.) difference at 13 yr of age]; however, the 3 groups were very similar in height at 25 yr of age. The early-maturing boys were taller than the late maturers between the ages of 12 and 15 yr, but were actually shorter as adults. The early maturers were 11.8 cm (4.6 in.) taller than the late maturers at 14 yr but were 6.5 cm (2.6 in.) shorter at 25 yr of age. The late maturers were also taller than the average maturers at 25 yr, by 4.2 cm (1.7 in.).

A recent analysis of 527 subjects in the Gothenburg osteoporosis and obesity determinants study, by researchers from the Center for Bone and Arthritis Research Institute of Medicine and the Gothenburg Paediatric Growth Research Center at Gothenburg University, showed similar general results but provides extra detail. They found that the differences in adult height between groups that matured at different ages were mainly attributable to length of the lower limbs. In addition, their analysis showed that the subjects with a low body mass index in childhood had a shorter adult trunk length, measured by sitting height.

Sources

Hagg, U., & Taranger, J. (1991). Height and height velocity in early, average and late maturers followed to the age of 25: A prospective longitudinal study of Swedish urban children from birth to adulthood. *Annals of Human Biology, 18,* 47-56.

Lorentzon, M., Norjavaara, E., & Kindblom, J.M. (2011). Pubertal timing predicts leg length and childhood body mass index predicts sitting height in young adult men. *Journal of Pediatrics, 158*(3), 452-457.

it demonstrates that the relationship between height and weight during growth is not cubic and so all dimensions are not changing at the same rate. In other words, the shape of the human body changes during growth. We can see this change in proportionality most clearly by considering the relative size of the head at different ages (figure 5.5). This illustrates that the brain and surrounding cranial bones are closer to adult size at an earlier age than other body parts are, demonstrating the concept of differential growth (that different parts of the body grow at different rates during different periods of development). Obviously, human growth is not explained by a squared relationship either because, in this situation, the baby girl would have grown to 150 cm (59 in.) and 27 kg (60 lb)!

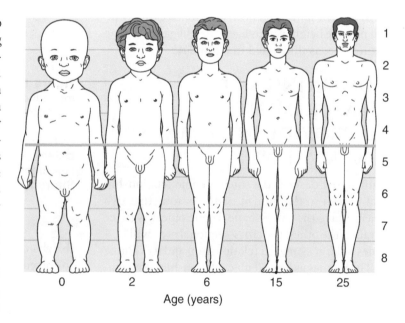

Figure 5.5 Relative sizes of the head at five ages; the total body is scaled to the same height.

Reprinted from P.S. Timiras, 1972, *Developmental physiology and aging* (New York: Macmillan Publishers), 284.

SECULAR TREND IN BODY DIMENSIONS

You may be familiar with the idea that there are differences in size between people of different generations. It appears to be common for teenagers to be taller than their parents. This generational change, known as a **secular trend**, can be illustrated by differences in the heights of Australian school children from 1911 to 1976 (figure 5.6). This figure illustrates a number of interesting changes that have occurred in just over 2 generations. In 1911, 7- and 8-yr-old boys and girls were similar in height, but in 1976 the boys were considerably taller. Note that 10-yr-old boys in 1911 were taller than the girls, but this was reversed in 1976. Earlier maturation of girls is indicated by the height differences of the 12-yr-old children. Twelve-year-old girls were still taller than boys of the same age; therefore, there must be an age at which the girls overtake the boys. This was 11 yr of age in 1911, but it was 10 yr of age by 1950. The age of crossover between the male and female heights provides some indirect evidence for another good example of secular trend.

Studies from a number of countries show that the age at menarche—the age at which menstruation first occurs—has decreased by up to 4 yr during the past century. Girls appear to be maturing earlier relative to the boys than they were in previous generations.

If the heights of 80-yr-old women were compared with those of 20-yr-old women, would this provide evidence of the secular trend in height of adult females? Unfortunately, the answer is no because there are confounding variables. One effect of ageing is decreased stature due to two main factors: the collapse of the vertebrae as a result of decreased bone density (see "In Focus: Why Trunk Height Decreases With Ageing"), and postural changes, including exaggerated curvature of the thoracic region of the vertebral column and flexion at the joints of the lower. Some researchers have calculated that structural and postural changes account for a 4.3 cm (1.7 in.) decrease in stature; the extra 3.0 cm (1.2 in.) difference observed between 20- and 80-yr-old adult females is attributable to the secular trend in height.

GROWTH RATES OF BODY SEGMENTS

An important concept in growth is the differential growth of tissues and body segments. Total body height, usually termed stature, is the total of lower-limb length, trunk length, and head height. These three components do not grow at the same rate. Head height approximately doubles from birth to maturity, whereas trunk length approximately triples. The limbs grow faster; during maturation, the length of the upper limb increases fourfold and the length of the lower limb increases fivefold.

Differential changes in the sizes of body segments mean changes in physique because the shape of the body is defined by proportions. The sitting height:standing height ratio changes from birth to adulthood, but the rate of change is not constant, and even the sign (direction) of the ratio changes. At 2 yr of age, the ratio is close to 60%. It then decreases to a minimum near puberty, indicating the relatively rapid growth of the lower limbs, after which it increases slightly because of some catch-up growth by the trunk. At the age of 9 to 10 yr, the sitting height:standing height ratio is similar to the adult ratio of about 52%.

GROWTH RATES OF BODY TISSUES

The various components of the human body are listed in table 4.1. Earlier in this chapter we noted that the proportion of the body made up of muscle tissue increases during growth and maturation.

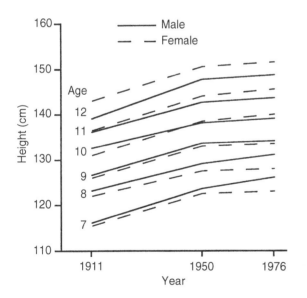

Figure 5.6 Secular trend in height of Queensland schoolchildren from 1911 to 1976 showing an increase in all age groups.

Data from A.E. Dugdale, V. O'Hara, and G. May, 1983 "Changes in body size and fatness of Australian children 1911-1976," *Australian Paediatric Journal* 19: 14-17.

Consequently, the relative amounts of other tissues must decrease, thus indicating that the various tissues must grow at different rates. The differential growth in some body tissues is illustrated in figure 5.7. Note again that the nervous tissue, such as the brain, is closer to adult size at an earlier age than other tissues. The rapid increase in the reproductive tissues is associated with the development of secondary sexual characteristics and the growth spurt around PHV. As well as changes in muscle and bone, there are changes in major organs such as the brain and liver. In the newborn baby, 13% of the body weight is brain and 5% is liver. In the adult, these relative weights have decreased to 2% and 2.5%, respectively. Obviously, there is a good correlation between the size of the skull and the size of the brain. The protruding abdomen observed during infancy and early childhood is indicative of the relatively large digestive organs, including the liver.

SEXUAL DIMORPHISM IN GROWTH

Figure 5.6 shows sexual dimorphism in growth. At any given age, the average of girls' heights is closer to their average adult height than is the case for boys. This phenomenon is particularly notice-

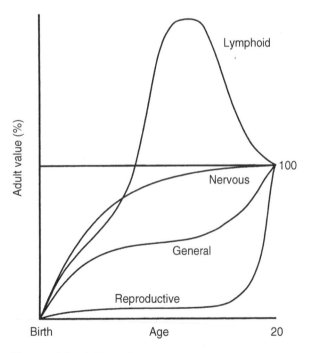

Figure 5.7 Differential growth in nervous, reproductive, and lymphoid tissues compared with the general growth of tissues.

able during the upper elementary and lower high school years when girls are often taller than their male classmates; the growth spurt usually occurs 2 yr earlier in the female than in the male. A rule of thumb is that girls reach half their adult height at about 2 yr of age and boys reach this height at 2.5 yr of age.

There are three main reasons that adult males tend to be taller than adult females. First, because the pubescent growth spurt starts later in males (see "In Focus: The Pattern of Growth in Swedish Children"), they are already taller when they enter the period of rapid growth. Second, the male growth spurt is more intense, meaning that the PHV is greater. Third, the growth spurt lasts longer in males. Of these factors, the predominant one is the delay in entering the growth spurt.

Sexual dimorphism is also seen in divergences in the average sitting height:standing height ratio between boys and girls of the same age. During childhood the ratios are very similar, but from about 11 to 12 yr onward the ratio is higher for girls. One explanation is that during the period of rapid growth the lower limbs of boys grow relatively faster than those of girls. During ageing, the ratio between sitting and standing heights decreases again because of the shortening of the vertebral column, related to both structural and postural changes mentioned earlier.

Relative amounts of fat tend to be greater in females at all ages, but the major divergence between males and females occurs when girls enter puberty (figure 5.8). The size of the triceps skinfold between the ages of 1 and 20 yr exhibits different time courses for males and females. At 20 yr of age, the skinfold thickness is greater than during infancy for females but is less than at 1 yr of age for males. The minimum skinfold thickness occurs at 7 yr of age for females; it then gradually increases, and there is almost a plateau for about 2 yr during puberty. For males, there are 2 periods of minimum skinfold thicknesses, at 7 and at 15 to 16 yr of age. The second minimum occurs because there is a decrease in skinfold thickness for about 4 yr during puberty. Especially for females, the terms *puppy fat* or *baby fat* are probably inappropriate because there is no period during puberty when the skinfold thicknesses are decreasing.

Muscle and bone masses tend to be similar in males and females until about 13 yr of age. At this age the hormonal influence on males is such that they have a greater increase in bone and, particularly, muscle mass.

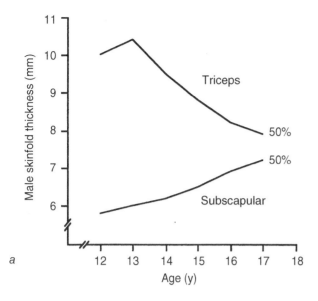

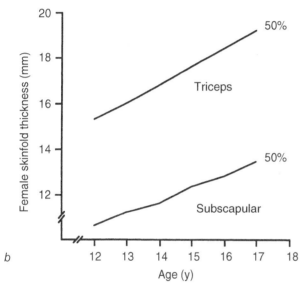

Figure 5.8 Changes in skinfold thicknesses during growth and maturation for 50th percentile Australian *(a)* males and *(b)* females. The triceps skinfold is measured at the back of the arm, and the subscapular skinfold is measured just below the shoulder blade.

Data from J.M. Court et al., 1976, "Growth and development of fat in adolescent school children in Victoria, Part I. Normal growth values and prevalence of obesity," *Australian Paediatric Journal* 12: 296-304.

SOMATOTYPE CHANGES DURING GROWTH, MATURATION, AND AGEING

A longitudinal study of somatotypes of European males between the ages of 11 and 24 yr indicated that there are 2 stages of puberty. In the first stage

between about 11 and 15 yr, endomorphy decreases because a decrease in skinfold thicknesses occurs at the same time as an increase in height. This increase in height is rapid, so the growth tends to be linear, resulting in an increase in ectomorphy. The second stage of puberty is marked by an increase in mesomorphy, which results from relative increases in transverse diameters of the skeleton and increases in muscle volume. Ectomorphy decreases between 15 and 24 yr of age because of the increased mass of the musculoskeletal system.

A study of almost 14,000 Canadian adults showed that males over 30 yr of age are slightly more endomesomorphic than those under 30 and that there are similar differences between women over and under the ages of 40 yr (see figure 4.2).

METHODS OF DETERMINING AGE

Of some interest is the concept of chronological age versus biological age. We all know children who are early maturers and teenagers who are late maturers. Masters athletes are very pleased when told by their doctor that they have the body of a person 10 yr younger. The time since birth is not the only method of representing age.

Examples of maturation include deciduous dentition ("baby teeth") followed by the permanent dentition, the gradual replacement of cartilage by bone, the onset of menarche (beginning of menstruation in girls), and the appearance of secondary sexual characteristics. These changes tend to occur in a similar order across individuals; the age of onset of the changes and the period of time required to complete the changes are more variable. Therefore, stages in development can be used as a basis for determining biological age during maturation.

In young people, skeletal or bone age can be calculated by inspection of hand and wrist X-ray images. This site is the most suitable because the appearance of ossification centres and fusion of bony parts takes place over many years in a fairly regular pattern. Bone age can be determined to the nearest month and then compared with chronological age to ascertain whether the person is ahead of or behind the normal developmental pattern. Clinically, bone age can be used, in conjunction with other measurements, to predict adult height. **Radiographic** techniques are not the most suitable to use in large studies because of the ethical issues involved with the exposure to X-radiation. It is preferable to use techniques that are less potentially hazardous.

Much of the musculoskeletal development that occurs around puberty is related to hormonal changes associated with sexual maturation. Therefore, sexual maturity itself can be determined using stages proposed by Tanner (1978, pp 197-201). During the development of secondary sexual characteristics, all individuals pass through a common series of body changes that differ for males and females. The order of development is relatively constant, but the age of onset of the first stage of maturation and the duration of progress through all stages show wide variation. Genital and associated development is classified to determine which of the five Tanner stages best describes a particular person.

Sexual maturity rating has been used in male junior sports in North America because it relates to the musculoskeletal development induced by the increased levels of testosterone. In a collision sport such as American football, musculoskeletal development is important. School administrators in New York State during the early 1980s decided that no high school boy could play football until he had entered stage 3 on the sexual maturity rating. This recommendation was seen as an important sports medicine intervention that protected boys who were perceived as most likely to suffer injury during participation in the sport. Because sexual maturity rating involves estimation of genital development, it is psychologically invasive. Other less-embarrassing methods are being developed that one hopes will be equally valid.

During ageing, such parameters as bone density, range of joint motion, and muscle strength decrease. If these values are greater than expected for a person of a given age, the person could be described as biologically young for his or her chronological age. There are no formalised equations that group parameters for people who are elderly as there are for children.

SUMMARY

Most tissues and structures develop during the 8 wk after conception. The fetal period therefore involves growth of tissues and organs, mainly by increase in the number (hyperplasia) and size (hypertrophy) of cells.

Maturation involves the changes that occur between birth and maturity. This period may vary depending on the specific physical parameter being considered. Most of the long bones of the limbs develop by a process of replacement of a cartilage model by bone, which takes about 2 decades. Around puberty, the cartilaginous growth plates may be susceptible to injury related to large applied muscle forces. During maturation, calcium is laid down in the bone; the peak amount may determine an individual's probability of becoming osteoporotic in later years. After birth, muscle mass increases more than 30 times, and during maturation the relative muscle mass almost doubles. Differential growth is an important concept. All tissues do not grow at the same rate at the same times, and organs do not necessarily maintain their shape as they grow.

Ageing involves decrements in a number of physical parameters, but some of the decrease in physical ability may be related to lifestyle changes rather than to natural biological ageing processes. Quality of life, associated with general mobility and physical independence, will become more important as people live longer. Relative muscle mass and strength decrease in old age, and the negative effects on lifestyle can be severe. The prevention and treatment of osteoporosis and osteoarthritis are becoming increasingly important.

Each generation exhibits some physical differences from the previous generation. Children have been getting progressively taller for the last few generations. Unfortunately, the incidence of osteoporosis has also been increasing rapidly so that now the incidence in men is the same as it was in women in the previous generation.

FURTHER READING

Eveleth, P.B., & Tanner, J.M. (1990). *Worldwide variation in human growth* (2nd ed.). Cambridge: Cambridge University Press.

Roche, A.F., & Sun, S.S. (2003). *Human growth: Assessment and interpretation.* Cambridge: Cambridge University Press.

Shephard, R.J. (1997). *Aging, physical activity, and health.* Champaign, IL: Human Kinetics.

Shephard, R.J., & Parizkova, J. (Eds.). (1991). *Human growth, physical fitness, and nutrition.* Basel, Switzerland: Karger.

Sinclair, D., & Dangerfield, P. (1998). *Human growth after birth* (6th ed.). Oxford: Oxford University Press.

Tanner, J.M. (1978). *Foetus into man: Physical growth from conception to maturity.* Ware, U.K.: Castlemead.

Tanner, J.M. (1989). *Foetus into man: Physical growth from conception to maturity* (2nd ed.). Ware, U.K.: Castlemead.

CHAPTER 6

MUSCULOSKELETAL ADAPTATIONS TO TRAINING

The major learning concepts in this chapter relate to

▶ effects of activity and injury on bone;

▶ effects of activity on joints and their ranges of motion;

▶ effects of activity on muscle structure and function;

▶ effects of training and other lifestyle factors on body size, shape, and composition; and

▶ relationships between body sizes and types and different types of sport.

The purpose of this chapter is to examine the changes in the skeletal, articular, and muscular systems and in overall body shape, size, and composition that occur as an adaptation in response to physical activity.

EFFECTS OF PHYSICAL ACTIVITY ON BONE

Charles Darwin (1809-1882) recognised the relationship between physical activity and bone mass in his groundbreaking text *The Origin of Species*, first published in 1859. A modern version, abridged by the physical anthropologist Richard

Leakey, includes the statement that "with animals the increased use or disuse of parts has a marked influence; thus in the domestic duck the bones of the wing weigh less and the bones of the leg more, in proportion to the whole skeleton, than do the same bones in the wild-duck; and this may be safely attributed to the domestic duck flying much less, and walking more, than its wild parents" (Darwin, 1979, p. 50).

Human clinical cases and animal experimentation have shown similar adaptability of bone. The major weight-bearing bone in the human leg is the tibia, which is in contact with the femur at the knee. In cases where the tibia is congenitally missing (not present at birth), the fibula has been surgically moved over to establish the necessary contact between it and the femur. The increases in size and strength of the fibula in this situation are quite spectacular.

According to one version of Wolff's law (p. 13) proposed by Sir Arthur Keith in a lecture to the Royal Society in London in 1921, "Every change in the form and function of a bone or of their function alone, is followed by certain definite changes in their internal architecture, and equally definite secondary alterations in their external conformation, in accordance with mathematical laws" (Tobin, 1955, p. 57). This translation of the law implied that bones sustain a maximum force with a minimum of bone tissue and that bone reorganises to resist forces most economically.

EFFECTS OF ACTIVITY LEVEL ON BONE

Not all adaptation of bone is positive; certain circumstances may lead to maladaptation (see "In Focus: How Does Exercise Affect the Bones of Growing Animals?"). For example, clinical evidence from elite junior athletes (see chapter 23) suggests that intense physical activity can produce maladaptation in growing bones. Stress fractures may occur in young adults, especially when the onset of intense

IN FOCUS: HOW DOES EXERCISE AFFECT THE BONES OF GROWING ANIMALS?

Researchers from The University of Queensland in Australia studied the effects of a 1 mo intensive exercise program on the structural and functional properties of 17 pubescent male rats. An equal number of rats that were not specifically trained were used as controls for comparison with the experimental group.

For 5 d of each of the 4 wk of the experiment, the experimental rats ran on a treadmill for 1 h/d and swam for 1 h. It was estimated that the rats were working at about 80% of their maximal oxygen consumption. The researchers studied the structure and function of the major lower-limb bones, measuring bone length and width and examining bone slices under a light microscope. Whole bones were also mounted in a specially manufactured torsion-testing machine to measure the strength of the bones when twisted at a physiological rate.

During the month, the experimental rats ate more than the controls, but at the end of the training period the experimental rats were lighter. The long bones of the rat hindlimb were shorter and lighter in the experimental animals, and the epiphyseal plates were thinner. As a result of the intense exercise program the femur was not significantly affected, but the tibia exhibited a significant decrease in the amount of energy it could absorb before it fractured.

The authors postulated that the repetitive cyclical loading caused an accumulation of microcracks in the bone that resulted in maladaptation. The authors therefore questioned the proposal that exercise is always beneficial for young animals, indicating that the literature related to stress fractures in humans reflected the potential problem.

Another research group from the University of Melbourne and Royal Melbourne Hospital in Australia studied the effects of resistance training on 20 young female rats. The rats were encouraged to use a climbing rack 3 times per week for 10 weeks. A gradually increasing load (commencing at 20% of body weight and finishing at 150% of body weight) was attached to the rats. The authors' hypothesis that resistance training would increase the rate of gain in bone mineral content and bone strength was not supported by the results, leading to speculation that more intensive exercise may be required to exhibit the hypothesized changes.

After the previous two papers were published, a group of German and New Zealand researchers representing the disciplines of biochemistry, biomechanics, and sport science studied the effects of different levels of exercise on bone composition and function. Thirty female growing rats were randomly assigned to an unlimited exercise group (involving a running wheel), a limited exercise group, and a sedentary control group. Running tended to increase the height of the growth plate of the femur (thigh bone), and there was greater mineralisation of the hypertrophic zone (zone 3 in figure 5.2) of rats that performed unlimited exercise, but there were no significant differences in mechanical properties among the growth plates from the three experimental groups.

These three studies indicate how difficult it is to show unequivocal exercise-induced changes in the bones of growing animal models, such as rats.

Sources

Bennell, K., Page, C., Khan, K., Warmington, S., Plant, D., Thomas, D., Palamara, J., Williams, D., & Wark, J.D. (2000). Effects of resistance training on bone parameters in young and mature rats. *Clinical and Experimental Pharmacology and Physiology, 27*(1-2), 88-94.

Forwood, M.R., & Parker, A.W. (1987). Effects of exercise on bone growth: Mechanical and physical properties studied in the rat. *Clinical Biomechanics, 2*, 185-190.

Niehoff, A., Kersting, U.G., Zaucke, F., Morlock, M.M., and Bruggemann, G.-P. (2004). Adaptation of mechanical, morphological, and biochemical properties of the rat growth plate to dose-dependent voluntary exercise. *Bone, 35*(4), 899-908.

physical activity is rapid, as it is among recruits in the armed forces.

One can find a number of good examples of bone adaptation in response to external forces. It was realised soon after animals and humans went into space that they incurred a rapid reduction in bone mass, probably related to the decreased muscle activity in a zero-gravity environment. Scientists are still trying to develop the optimal exercise regimen to prevent loss of bone in space. Likewise, when a limb is immobilised in a plaster cast, evidence of disuse osteopenia can be observed on X-ray images. Osteopenia, the mechanism by which bone mass decreases, is attributable to the imbalance between bone resorption and deposition. The result is that the porosity increases and therefore the bone is less dense and appears less radio opaque on an X-ray image (bone absorbs more X-radiation than any other tissue, except teeth). The bone loss is related to the absence of external forces on the bone, which are produced mainly by muscle activity. It would appear that bone has a genetically determined baseline mass that is less than that required for normal functioning. Therefore, a certain level of physical activity is necessary for good bone health. Complete rest results in rapid loss of both bone and muscle mass.

Generally, lack of physical activity results in a loss of bone mass and increased activity results in a gain. Some of the best examples of the relationship between physical activity and bone mass involve comparisons between the humerus bone of the playing and nonplaying arms of professional tennis players. For example, radiological studies demonstrated not only that the bone was denser in the playing arm but that the diameter of the shaft was larger and the compact bone forming the shaft was thicker. Reanalyses of the original data have shown that the greatest observed increases occurred in individuals who began training at a young age (i.e., before puberty).

These results have implications for young adults, who are advised to optimise their bone density. Cross-sectional studies indicate that adults with active lifestyles have greater bone mass than sedentary adults, and better-controlled longitudinal studies indicate that exercise can cause an increase in bone mass. Even in postmenopausal women, exercise can add bone; exercise is certainly important in maintaining bone mass that would otherwise be lost through ageing. Modest additions of bone can occur in ageing individuals with adequate levels and types of exercise, but the biggest contribution of exercise is reducing the bone loss that would occur with insufficient levels of physical activity.

EFFECTS OF ACTIVITY TYPE ON BONE

Weight-bearing activities such as walking and running are associated with the addition of bone, but this effect may not be so great in other activities such as swimming and cycling. Elite swimmers have lower bone densities than other elite athletes; weightlifters have the highest bone density. As indicated in figure 6.1 the initial positive changes occur slowly whereas the initial response to decreased activity is rapid; therefore, an exercise program must be regular and must be maintained over a long period to be beneficial. As noted in chapter 3, a remodelling cycle is about 3 mo. About 3 or 4 remodelling cycles are required to reach a new steady state in response to increased physical activity.

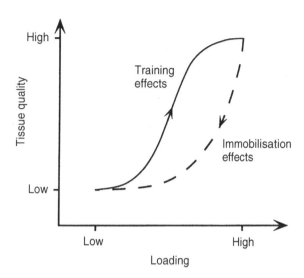

Figure 6.1 Possible causes of musculoskeletal tissue adaptation related to training (solid line) and immobilisation (broken line). There is a delay in improvements related to training, but the negative effects of immobilisation are almost immediate.

Adapted from Kannus et al., 1992, "The effects of training, immobilization and remobilization on musculoskeletal tissues. 1. Training and immobilization," *Scandinavian Journal of Medicine and Science in Sports* 2: 100-118.

BONE REPAIR AND PHYSICAL ACTIVITY

Our discussion thus far indicates that if exercise level gradually increases from an initial baseline level of activity, positive adaptation will result. If activity is too intense, however, there may be negative changes that could lead to injuries such as stress fractures. The following period of immobilisation will result in loss of bone and decreased density so that the bones are weaker. During rehabilitation, after repair of the injured site, bone will regain its former properties. However, because the loss of strength is rapid in comparison with its return, it is important for coaches and sports medicine personnel to ensure that this scenario does not become a recurring cycle for an elite athlete. This concept applies to ligament sprains and muscle strains as well. Figure 6.1 indicates that there is an optimal level of activity for each person, and the challenge for prescribers of exercise is to determine this level on an individual basis.

Remember that bone is more like a tree than like a piece of wood because it is a living, dynamic, metabolically active tissue with a good blood supply. It undergoes continual remodelling and can therefore recover from damage. The major stages during fracture healing are illustrated in figure 6.2. During a fracture, the bone and the surrounding tissues bleed when the blood vessels are ruptured. Subsequently, a blood clot forms, and capillaries later invade the clot, bringing cells and nutrients. A fibrous **callus** fills the space between the ends of broken bone; then a bony callus replaces the fibrous callus. Appropriate physical rehabilitation is important because the final strength of the bone will depend on remodelling of the bony callus, and this remodelling is affected by physical activity.

EFFECTS OF PHYSICAL ACTIVITY ON JOINT STRUCTURE AND RANGES OF MOTION

A warm-up is usually recommended before any physical activity. Many people believe that a warm-up is performed only for cardiovascular reasons, including increasing the heart rate, but warming up also has positive effects on the musculoskeletal system.

SYNOVIAL FLUID, ARTICULAR CARTILAGE, AND LIGAMENTS

During cyclical exercise, such as jogging, the lower-limb joints are continually moving within their ranges of motion. The short-term effect on articular cartilage is that it thickens and thus can better perform its function of force dissipation during dynamic activities. The cartilage thickens because it absorbs synovial fluid, and the flow of synovial fluid into and out of the cartilage improves the supply of nutrients and the removal of waste products of metabolism. Chronic exercise results in long-term thickening except when compressive forces on the cartilage are excessive, as may occur during repeated downhill running or during heavy workouts involving a great deal of jumping.

Normally the quantity of synovial fluid in a joint is very small. The knee has the largest joint cavity in the body, but it contains only about 0.2 to 0.5 mL of synovial fluid at rest. After a 1 or 2 km (0.6 or 1.2 mile) run, the volume may have increased by up to 2 or 3 times. One can argue that the synovial fluid

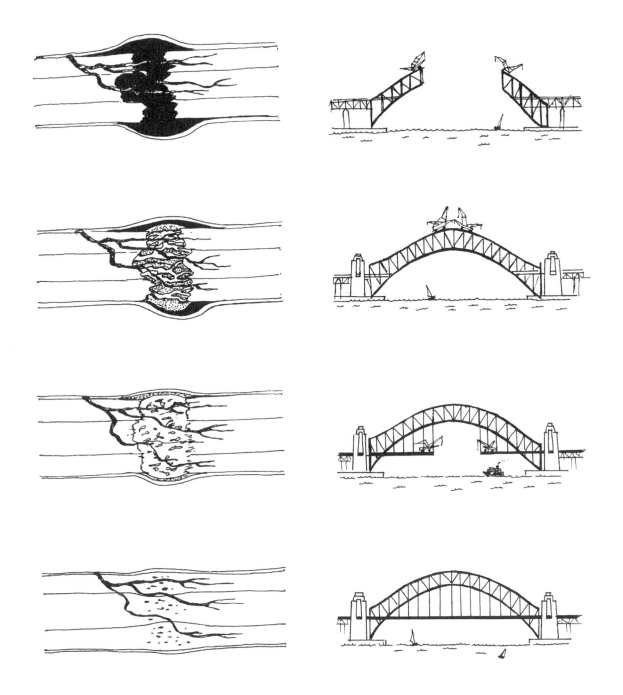

Figure 6.2 The major stages in bone healing compared with the construction of the Sydney Harbor Bridge in Australia, which was completed in 1932.

can then better perform its functions of lubrication and nutrition. Certainly, short-term exercise makes synovial fluid less viscous and thus more like oil than grease. It may therefore be a better lubricant, but does it sacrifice this property for the role of protection? Because the articular cartilage soaks up the thin synovial fluid, the joint is probably better protected by the thickened cartilage after a warm-up period.

Ligaments are passive stabilising structures of joints, and the loads applied to them vary when a joint is moved through its full range of motion. The oscillating increases and decreases in tensile forces on ligaments during physical activity result in adaptations. The size of the ligament increases because of an increase in collagen synthesis, so the ligament becomes stronger. Moreover, an increase in cross-linking in the ligament makes it stiffer. There is evidence that endurance exercise has more positive effects on ligament strength than sprint training.

DEGENERATIVE JOINT DISEASE AND EXERCISE

Indications of degenerative joint disease, or osteoarthritis, are thinning of the articular cartilage and thickening of the thin layer of compact bone under the articular cartilage. It has been hypothesized that changes in this subchondral bone precede the cartilage thinning. Animal studies have shown that repetitive impulsive loading causes stiffening of the thin subchondral bone so that the cartilage may be more susceptible to damage. The human knee is among the joints commonly affected by osteoarthritis, which increases in incidence with age. To these possible genetic and ageing factors may be added environmental factors such as obesity, which is associated with a greatly increased risk of osteoarthritis.

Although clinicians report that osteoarthritis is among the long-term effects of jogging, epidemiological studies of the relationship between exercise and degeneration of the articular cartilage have shown that regular runners do not experience an increased incidence of osteoarthritis. Recent studies indicate that the connection between exercise and later degenerative changes in the articular cartilage relates to previous synovial joint injury such as ligament sprains and tears. Ligament damage may allow abnormal motion within the synovial joint, resulting in excessive localised loading.

EFFECTS OF PHYSICAL ACTIVITY ON MUSCLE–TENDON UNITS

If you have experienced a major joint injury, you know how quickly muscle size decreases with disuse. A major goal of rehabilitation after such an injury is to increase the size and contraction force of the muscles involved. If you go to the gym for regular strength training, you expect to see increased muscle size; you would also hope that the attached tendons would become stronger in

order to withstand the greater muscle forces being produced.

FLEXIBILITY

Flexibility is a term commonly used to indicate ranges of joint motion. The term *joint range of motion* seems to indicate that some of the specific structures of synovial joints, illustrated in figure 3.6, restrict motion. In most cases, however, flexibility or lack of it depends on the muscle–tendon units that cross joints. In fact, most flexibility exercises stretch muscle–tendon units more than joint capsules or ligaments. Flexibility that relates to muscle–tendon length is positive. When some ligaments are stretched, however, the joint may be classified as loose, or lax. This characteristic is associated with increased risk of injury. Joint laxity may in fact require physical therapy or orthopaedic surgery.

Flexibility is often listed as one of the factors composing physical fitness, which gives the impression that flexibility is a general characteristic of an individual. In reality, flexibility is highly joint specific because it is related to normal activity patterns. The flexibility of ballet dancers provides an excellent example. A range of motion study in ballerinas showed that the range of external rotation of the hip was much greater in ballet dancers than in the normal population but that the range of internal hip rotation was less than normal. This variability suggests that activity patterns in ballet, involving externally rotated postures of the hip, produce structural changes that adapt to the forces but that then restrict motion in the opposite direction.

Studies on teenage gymnasts and aged persons indicate that regular performance of flexibility exercises can prevent or even reverse the normal decreases in ranges of joint motion. Certainly, stretching exercises performed regularly for a period of weeks, months, or years can increase joint flexibility. When this increase occurs, how does the whole muscle respond to regular static stretching? Remember that whole muscles consist of both muscle and connective tissue. Does the number of sarcomeres in series forming each muscle fibre increase, or do the extensibility and elasticity of the connective tissue component increase? In other words, does a muscle grow longer or become more extensible as a result of flexibility exercises? The answer is probably that increased extensibility of the connective tissue is the predominant factor, although sarcomere number will adapt rapidly to any habitual length change.

A perception is that weight trainers are muscle bound; that is, that their muscle development leads to a decrease in the range of joint motion. This general idea appears to be ill founded. With correct weight-training techniques, including flexibility exercises, range of motion can be maintained and even enhanced, as evidenced by the celebratory gymnastics of some Olympic weightlifters after winning a gold medal.

Understanding what is behind flexibility helps clarify why a general warm-up, even when combined with massage, seems to have little effect on joint flexibility. Clearly, specific joint flexibility exercises that stretch specific muscle–tendon units must be a separate part of the warm-up routine.

STRENGTH

People who undertake a weight-training program often experience a rapid increase in measured strength followed by a plateau effect and then further, gradual increases. Electromyographic (EMG) studies have indicated that during the initial 6 to 8 wk, the muscular activation pattern becomes more efficient and thus the strength gains are related to improved control of the neuromuscular system. In the next phase, the size of the muscle fibres increases due to an increase in the number of myofibrils within each fibre. That is, an initial phase of neural adaptation (**neurotrophic phase**) is followed by a longer period of muscle hypertrophy (**hypertrophic phase**) in which the cross-sectional area of each muscle fibre increases. The major effect of weight training is hypertrophy of the muscle fibres, as discussed further in chapter 14. It appears that increases in the connective tissue component of whole muscle parallel those of the muscle tissue during strength-training programs. Long-term effects of endurance exercise are related mainly to changes within the muscle fibre to make the muscle less fatigable, but there is also an increase in the relative amount of connective tissue within the whole muscle.

TENDON ADAPTATION

Exercise has a positive effect on tendon; however, as with other connective tissue structures, the rate of adaptation is much slower than that of muscle. In tendon, collagen synthesis increases and the collagen fibres line up more regularly in a longitudinal direction. As with other tissues, training that is too strenuous may be injurious (figure 6.1). When muscle strains occur, they often happen at the junction between the muscle and its tendon.

EFFECTS OF PHYSICAL ACTIVITY ON BODY SIZE, SHAPE, AND COMPOSITION

It was emphasised in chapters 4 and 5 that a person's somatotype is as much a phenotype as it is a genotype, implying that the somatotype can be modified by factors such as training. Although amount of body fat and its distribution are under a certain amount of genetic control, the combination of increased activity and reduced energy intake via the diet will decrease the fat content of the body. A person following such a regimen will become less endomorphic. All studies that have compared athletes with the sedentary population have found that athletes have less body fat and a larger lean body mass, partly because of an increased skeletal mass but mainly because of a larger muscle mass. Thus, exercise may not alter a person's ectomorphy rating because the decrease in endomorphy (fat) may be balanced by an increase in mesomorphy (muscle and bone). In other words, the skinfolds decrease but the muscular circumferences of the arms and legs increase. If body mass does not decrease as a result of an exercise program, this does not necessarily mean that the program has been ineffective.

ROLE OF LIFESTYLE FACTORS IN DETERMINING PHYSIQUE

Athletes such as bodybuilders, weightlifters, and ballet dancers appear to have unusual body shapes that relate more to their adaptation to training than to differences determined before birth. When measurements of weightlifters and ballet dancers are compared with values for the normal population, interesting similarities and differences can be observed (figure 6.3). The measurements of bone sizes, such as circumferences of the wrist, knee, and ankle, are relatively normal. The bone in the shaft of a weightlifter is likely to be denser and heavier, but these parameters are not measured in normal anthropometric protocols. Width of the expanded ends of long bones is less susceptible to training.

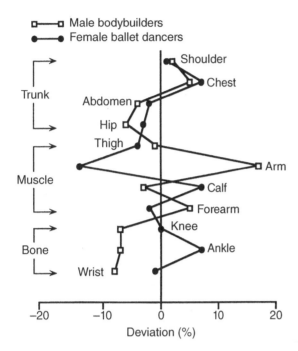

Figure 6.3 Differences in relative body-segment circumferences between male bodybuilders and their reference population and between female ballet dancers and their reference population. Both reference populations are represented by the solid vertical line.

Reprinted, by permission, from V.L. Katch, F.L. Katch, R. Moffatt, and M. Gittleson, 1980, "Muscular development and lean body weight in body builders and weight lifters," *Medicine and Science in Sports and Exercise* 12: 340-344; and F.A. Dolgener, T.C. Spasoff, and W.E. St. John, 1980, "Body build and body composition of high ability female dancers," *Research Quarterly for Exercise and Sport* 51: 599-607.

Notice particularly that the muscular circumferences (e.g., biceps) are relatively large in weightlifters as a result of muscle hypertrophy. In contrast, muscular circumferences are relatively small in ballet dancers; this is related to the demands of the dancers' training, including type of physical activity (e.g., demands on the muscles of the arm are far less than on the muscles of the calf) and dietary intake. Training can affect physical and physiological factors at any time of life. This topic is a major focus of chapter 14.

RELATIONSHIP OF BODY SIZES AND TYPES TO SPORT

One good example of **kinanthropometry** is provided by the following quote: "Javelin throwers and gymnasts have practically identical somato-

types, although the javelin throwers, at 179.5 cm and 76.7 kg, are much bigger than the gymnasts, at 167.4 cm and 67.1 kg" (Hebbelinck & Ross, 1974, p. 545). After reading the next section of this book on **biomechanics** (chapters 7-10) you should be able to explain these similarities and differences. One important aspect is the large difference in the size measurements, such as height and weight. Height is predominantly genetically determined and under normal conditions is not greatly affected by environmental factors. Therefore, sport performance is affected by genetically determined factors as well as training and the physical adaptations to training. The physical dimensions that cannot be changed are often used in talent identification, a topic we explore more fully in part III.

In a given sport, particular events may require mainly strength or endurance. In running, for example, the 100 m (109 yard) sprint and the marathon are at two ends of a strength–endurance continuum. In running, the athlete's own body weight must be transported, and runners tend to be lighter than the average person; they get progressively lighter as the distance of the event increases. Sprint and middle-distance runners tend to be taller than the average person, but long-distance runners tend to be shorter than the average person.

Events such as the shot put require even greater strength than sprinting. Height is also an advantage because it means that the shot can be released from a greater height (and hence travel further for any particular velocity and angle of release). Shot putters and discus throwers tend to be taller and heavier than other track and field athletes. Body weight is not such a limiting factor for these athletes because they do not have to transport their own body over a long distance or lift their centres of mass very far, so they tend to be very heavy. Also, a large lean body mass is advantageous for the development of power required in explosive movements.

Somatotypes also change as the requirements for strength and endurance change. When there is a high strength component, such as in the shot put and discus, the athletes are very mesomorphic. Athletes tend to move from this position on the somatochart downward and to the right as the endurance component increases (see "In Focus: Are There Physical Differences Between Players in Different Sports?").

A research group from the School of Education at the Flinders University of South Australia used the Heath-Carter anthropometric somatotype method (see chapter 4) to characterise 206 national standard sportsmen in 17 sports and 127 female representative squad members in 10 sports. Endomorphy was corrected for height. Age, height, and mass were reported separately.

The male athletes were described generally as being balanced mesomorphs. Their mesomorphy ratings were above average, and the ratings for endomorphy and ectomorphy were both below average. When compared with an age- and sex-matched group of untrained Canadian males, the South Australian athletes were taller, lighter, less endomorphic, more mesomorphic, and more ectomorphic. The overall mean for relative body fat of the athletes was estimated to be 10%.

The most mesomorphic sportsmen were the powerlifters, and the least mesomorphic were the long-distance runners. The gymnasts were more mesomorphic than the track and field athletes when considered as a single group. The tallest sportsmen were the basketball players and rowers, and the shortest groups were the powerlifters and the gymnasts. The heaviest group was the Rugby Union football players, and the lightest group was the gymnasts.

The authors compared their results with those of previous studies and provided a good kinanthropometric explanation of the differences between groups. The differences confirm the relationship between somatotype and relative requirements for strength and endurance (see figure 6.4).

Compared with a reference sample of 135 women, the female athletes were less endomorphic, slightly more mesomorphic, and more ectomorphic. Other studies have found a tendency for endomorphy to decrease as the level of physical activity increases. The basketball and netball players were taller than those playing soccer, lacrosse, cricket, hockey, and softball, and the softball players were lighter than the basketball players. There was a tendency for the netball, squash, and volleyball players to be less

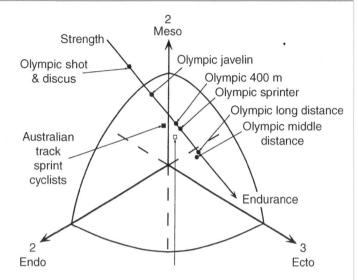

Figure 6.4 Somatochart illustrating the somatotypes of some Olympic track and field athletes. Notice the strength–endurance continuum from a predominance of mesomorphy to a predominance of ectomorphy. Note also that none of these elite athletes is endomorphic.

Data from J.E.L. Carter, S.P. Aubry, and D.A. Sleet, 1982, Somatotypes of Montreal Olympic athletes. In *Physical structure of Olympic athletes. Part I. The Montreal Olympic games Anthropological Project. Medicine and Sport, vol. 16,* edited by J.E.L. Carter (Basel: S Karger), 53-80; and B.D. McLean and A.W. Parker, 1989, "An anthropometric analysis of elite Australian track cyclists," *Journal of Sports Sciences* 7: 247-255.

endomorphic and more ectomorphic than those playing badminton, basketball, cricket, hockey, lacrosse, soccer, and softball.

The authors emphasise that there are many factors other than body size and shape that determine successful sport performance. This caveat is particularly important in the case of field games that require a range of skills, especially perception–action coupling, as well as various components of fitness.

Sources

Withers, R.T., Craig, N.P., & Norton, K.I. (1986). Somatotypes of South Australian male athletes. *Human Biology, 58,* 337-356.

Withers, R.T., Whittingham, N.O., Norton, K.I., & Dutton, M. (1987). Somatotypes of South Australian female games players. *Human Biology, 59,* 575-584.

SUMMARY

Biological tissues, in general, respond to their environment. The tissues of the musculoskeletal system, in particular, respond to mechanical forces. The genetic baseline bone mass is much less than normal, so physical activity is required just to maintain normal levels. Appropriate exercise prescription can lead to increased bone mass, joint flexibility, and strength, all of which can be beneficial for general health and exercise performance.

The processes of adaptation of the different tissues of the musculoskeletal system occur at different rates and may involve stages, such as the improved neural control followed by muscle hypertrophy during a weight-training program. Training results in a change of somatotype as the relative amounts of the major tissues of the body alter in response to training. Generally, athletes have reduced amounts of fat and increased amounts of bone and muscle when compared with the general population. Excessive levels of activity can lead to tissue breakdown and a decrease in bone mass in women.

FURTHER READING AND REFERENCES

Ackland, T.R., Elliott, B.C., & Bloomfield, J. (Eds.). (2009). *Applied anatomy and biomechanics in sport* (2nd ed.). Champaign, IL: Human Kinetics.

Burr, D.B., & Milgrom, C. (2001). *Musculoskeletal fatigue and stress fractures.* Boca Raton, FL: CRC Press.

Darwin, C. (1859/1979). *The illustrated origin of species.* London: Book Club Associates.

Delavier, F. (2010). *Strength training anatomy* (3rd ed.). Champaign, IL: Human Kinetics.

Elliott, B. (Ed.). (1998). *Training in sport: Applying sport science.* Chichester, NY: Wiley.

Stone, M.H., Stone, M., & Sands, W.A. (2007). *Principles and practice of resistance training.* Champaign, IL: Human Kinetics.

Tanner, J.M., Whitehouse, R.H., & Jarman, S. (1964). *The physique of the Olympic athlete: A study of 137 track and field athletes at the XVIIth Olympic Games, Rome, 1960: And a comparison with weight-lifters and wrestlers.* London: Allen and Unwin.

Tobin, W.J. (1955). The internal architecture of the femur and its clinical significance: The upper end. *Journal of Bone and Joint Surgery, 37A,* 57-72.

PART III

MECHANICAL BASES OF HUMAN MOVEMENT

BIOMECHANICS

Biomechanics is the application of mechanics to the study of living systems. Having evolved from the days of Aristotle in 384 to 322 BC, today biomechanics sits at the intersection of biology, medicine, and quantitative mechanics. Basic scientists use the principles of mechanics to understand the mechanisms underlying function in the healthy state. Clinicians, on the other hand, are more interested in formulating and solving problems related to injury and disease. The four chapters that follow show how the principles of biomechanics may be used to perform quantitative analyses of movement in the healthy state and to examine how musculoskeletal changes introduced by injury, training, growth and maturation, and ageing affect motor performance. The focus, at all times, is at the systems level, although many biomechanists also work at the cellular and molecular levels, attempting to understand the mechanics of molecules and how they apply to the morphology and motility of cells.

TYPICAL QUESTIONS POSED AND LEVELS OF ANALYSIS

Why can some people jump higher than others? Why do some run faster and throw farther? These are questions typically addressed by sports biomechanists. Why do children and healthy older adults walk differently from healthy young adults? How does rupture of the anterior cruciate ligament alter function of the remaining knee ligaments during activity? What are the best approaches for reconstructing a torn anterior cruciate ligament? How does muscle function of the leg change to compensate for the presence of spasticity resulting from cerebral palsy, stroke, or traumatic brain injury? How does a total hip replacement alter the mechanics of gait? These are all questions that concern the clinical biomechanist, orthopaedic surgeon, and physical therapist.

The approach taken to answer a question depends on the question being asked. For example, if one is interested in learning how high Michael Jordan can jump, the answer can be obtained by performing a relatively simple, noninvasive biomechanical experiment. A high-speed video system may be used to track the positions of surface markers mounted on various parts of the body, and these data can then be used to calculate the trajectory of the centre of mass during a jump. Alternatively, a device called a force plate may be used to record the time history of the vertical component of force exerted on the ground during the jump. A fundamental principle of **mechanics** called the impulse-momentum principle may then be applied to calculate the vertical velocity of the centre of mass at the instant of liftoff and, hence, jump height.

On the other hand, if one is interested in understanding *why* Michael Jordan can jump higher than most people his size, then a much deeper knowledge of the interaction between physiology and mechanics is needed. Specifically, information about the architectural, physiological, and mechanical properties of Michael Jordan's leg muscles must be correlated with records of the forces developed by his leg muscles during the ground-contact phase of a jump. This information cannot be obtained simply by performing a biomechanical experiment on Michael Jordan. Records of muscle forces, in particular, cannot be obtained by direct measurement, so another approach involving computer modelling is commonly used. In this approach, the geometry and properties of the musculoskeletal system (i.e., muscles, tendons, ligaments, bones, and joints) are represented in a **model** of the body, and a problem is formulated to estimate those quantities that cannot be obtained directly from experiment. If the problem entails the calculation of muscle forces, the method most commonly used is inverse **dynamics**. Here, measurements of body motions and ground forces are used to determine the net muscle moments exerted about the joints. A mathematical theory called **optimisation** is then applied to distribute the net moments about the various muscles crossing each joint.

HISTORICAL PERSPECTIVES

The history of science begins with the ancient Greeks. Aristotle was probably the first biomechanist. He wrote the first book related to movement, *De Motu Animalium* (English translation: *On the Movement of Animals*), and actually viewed the bodies of animals as mechanical systems. Nearly 2,000 yr later, Leonardo da Vinci took up where Aristotle left off. Da Vinci, who was born in 1452, was a self-educated man. He became famous as an artist, but worked mostly as an engineer. His contributions to science and technology were prodigious by any measure, and he is credited with numerous inventions ranging from water skis to hang gliders. Da Vinci understood concepts such as the components of force vectors, friction, and how a falling object accelerates through the air. By combining his knowledge of human anatomy and mechanics, da Vinci was able to analyse relatively detailed systems of muscle forces acting on body parts.

The man regarded by many as the father of biomechanics is Giovanni Alfonso Borelli, born in 1608 in Naples, Italy. He was the first to understand that the levers of the musculoskeletal system (i.e., moment arms) cause muscles to develop very large forces during physical activity. Using his intuition about the principle of statics, Borelli figured out the forces needed to hold a joint in equilibrium. He also determined the position of the centre of gravity of the whole body in a human and calculated inspired and expired air volumes.

There would be little to show for the next 250 yr, until the appearance of the French physiologist Étienne Marey. Marey was the first to actually film people moving. He constructed a camera to expose multiple images of a subject on the same photograph at set intervals. Marey and his colleague Eadweard Muybridge were the first to use high-speed film recordings to analyse movement.

PROFESSIONAL ORGANISATIONS

The major international organisations representing biomechanics are the International Society of Biomechanics (ISB), the American Society of Biomechanics (ASB), the European Society of Biomechanics (ESB), the Canadian Society of Biomechanics (CSB), the Australian and New Zealand Society of Biomechanics (ANZSB), and the Gait and Clinical Movement Analysis Society (GCMAS). These societies regularly organise and hold major international scientific meetings. The ISB meets every 2 yr, and the meeting is generally very well attended by biomechanists from all over the world. The ASB and CSB hold their annual meetings in the United States and Canada, respectively, and they also hold a combined meeting every 2 yr. The ANZSB and GCMAS meet once a year in Australia or New Zealand and North America, respectively. The focus of the GCMAS is clinical gait analysis, so its members tend to be a mix of movement scientists, engineers, clinicians, and physical therapists. The ANZSB meetings are heavily focused on applications of biomechanics in orthopaedics and are therefore well attended by scientists and clinicians working in this area. Other professional organisations with major initiatives in biomechanics include the Society for Experimental Biology and the American Society of Mechanical Engineers, which holds its Summer Bioengineering Conference annually in the United States.

FURTHER READING

Burkett, B. (2010). *Sport mechanics for coaches* (3rd ed.). Champaign, IL: Human Kinetics.

Enoka, R.M. (2008). *Neuromechanics of human movement* (4th ed.). Champaign, IL: Human Kinetics.

McGinnis, P.M. (2013). *Biomechanics of sport and exercise* (3rd ed.). Champaign, IL: Human Kinetics.

Panjabi, M.M., & White, A.A., III. (2001). *Biomechanics in the neuromuscular system*. New York: Churchill Livingstone.

Zatsiorsky, V.M. (2002). *Kinetics of human motion*. Champaign, IL: Human Kinetics.

SOME RELEVANT WEBSITES

ANZSB: www.anzsb.asn.au

ASB: www.asbweb.org

ESB: www.esbiomech.org

GCMAS: www.gcmas.org

ISB: http://isbweb.org

CHAPTER 7

BASIC CONCEPTS OF KINEMATICS AND KINETICS

The major learning concepts in this chapter relate to

- ▶ adding two vectors together to produce a resultant vector,
- ▶ calculating linear velocity (the time rate of change in displacement) and acceleration (the time rate of change in velocity) of a particle,
- ▶ calculating angular velocity (the time rate of change in angular displacement) and acceleration (the time rate of change in angular velocity) of a body segment,
- ▶ calculating velocity and acceleration of any point on a body segment,
- ▶ identifying the degrees of freedom of a rigid body,
- ▶ distinguishing between external and internal forces and their relationship to free-body diagrams,
- ▶ using static equilibrium analyses to determine unknown forces acting on body segments, and
- ▶ modelling movement with computers.

This chapter introduces some fundamental concepts of kinematics and kinetics as they relate to human movement. The emphasis is on application of the basic laws of mechanics. Vectors underlie much of what is known about mechanics, so we begin with a brief review of these quantities and then illustrate how they may be used to describe forces and motion in human activity. We focus on displacement, velocity, and acceleration of a point on a rigid body; force and its relationship to moment of force about a point in space; application of free-body diagrams to analyses of force systems; and the Newton-Euler method for generating dynamical equations of motion of biomechanical systems. The chapter concludes with a brief introduction to computer modelling and simulation of human movement.

VECTORS

Any physical quantity that is fully described by a number is called a **scalar**. Examples of scalars commonly encountered in biomechanical studies of movement are length, mass, and moment of inertia. A physical quantity that is described by both its magnitude and direction is called a **vector**. The position, velocity, and acceleration of a point on a body (e.g., the **centre of mass**), a muscle force, the moment of force of a muscle about a joint, forces arising from contact of body parts with the environment (e.g., forces exerted by the ground on the feet during walking and running), and forces acting between bones that meet at a joint (i.e., joint-contact forces) are all examples of vectors. In this and subsequent chapters, any quantity that is underlined is a vector; otherwise, it is a scalar.

Vectors cannot simply be added to, subtracted from, and multiplied or divided by each other in the same way that scalars can. There are special rules for operating on vectors. To understand some of the most basic concepts of kinematics and kinetics, it is important to understand first how two vectors can be added together to produce a resultant vector. Consider moving a book from one position to another as shown in figure 7.1. The direction of $\underline{r}$ indicates the direction of the displacement of the book in moving from location 1 to location 2, and the actual distance moved is given by the magnitude of the displacement represented by $|\underline{r}|$. (The symbol | | denotes the magnitude of the vector.) Let the displacement of the book in moving from location 2 to location 3 be

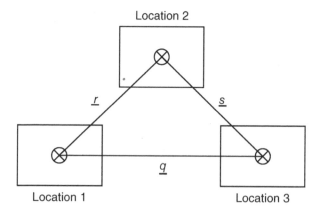

Figure 7.1 Diagram showing how vector addition may be used to calculate the displacement of a rigid body in space.

represented by the vector $\underline{s}$ and the corresponding distance moved by $|\underline{s}|$. The resultant displacement of the book in moving from location 1 to location 3 is (see figure 7.1)

$$q = \underline{r} + \underline{s}. \qquad 7.1$$

In the special case when the vectors $\underline{r}$ and $\underline{s}$ are perpendicular, the magnitude of the resultant displacement vector is given by $|\underline{q}| = \sqrt{|\underline{r}|^2 + |\underline{s}|^2}$, which follows from Pythagoras' theorem for a right-angle triangle. The direction of the resultant displacement is found from $\theta = \tan^{-1}\left(\dfrac{|\underline{s}|}{|\underline{r}|}\right)$, where θ is the angle formed between the resultant and the first displacement vector $\underline{r}$ (figure 7.1). Vector addition is also often used to find the resultant of force systems (e.g., muscle forces) acting on a body part (see "In Focus: How Much Force Do the Ankle Muscles Develop in Normal Walking?").

IN FOCUS: HOW MUCH FORCE DO THE ANKLE MUSCLES DEVELOP IN NORMAL WALKING?

A critical phase of the walking cycle is the push-off phase. This is the time when the ankle-plantarflexor muscles of the stance leg are heavily activated in order to push the centre of mass of the body forward and upward just before the heel strike of the **contralateral** (swing) leg. The gastrocnemius (GAS) and soleus (SOL) both exert forces to plantarflex the ankle during push-off. Both muscles insert on the foot via the Achilles tendon. The forces developed by these muscles have been estimated using sophisticated computer models of normal walking (Anderson & Pandy, 2001). A recent analysis showed that GAS and SOL develop their peak forces at the same time during the push-off phase. Specifically, GAS generated a peak force of $F_{GAS} = 900$ N, whereas the peak force in SOL was estimated to be $F_{SOL} = 2000$ N. The line of action of each muscle was calculated from anatomical measurements of its origin and insertion sites. Thus, at the instant under consideration, the line of action of GAS was $\theta_{GAS} = 20°$ relative to the vertical (Y axis), whereas the line of action of SOL was $\theta_{SOL} = 30°$ relative to the vertical (see figure 7.2).

On the basis of this information, the resultant peak force developed by the plantarflexor muscles may be found as described here. First, resolve each muscle force vector along the X and Y axes as follows:

$$F_{GASX} = F_{GAS} \sin\theta_{GAS}; \; F_{GASY} = F_{GAS} \cos\theta_{GAS}$$

and

$$F_{SOLX} = F_{SOL} \sin\theta_{SOL}; \; F_{SOLY} = F_{SOL} \cos\theta_{SOL},$$

7.1.1

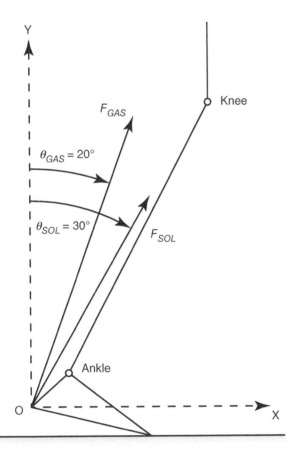

Figure 7.2 Schematic diagram of a human leg with vectors representing the forces acting in two of the ankle-plantarflexor muscles: gastrocnemius (GAS) and soleus (SOL).

where F_{GASX}, F_{GASY} are the X and Y components of GAS's force vector and F_{SOLX}, F_{SOLY} are the X and Y components of SOL's force vector. Adding the X and Y components of each force gives

$$F_{RESX} = F_{GAS} \sin\theta_{GAS} + F_{SOL} \sin\theta_{SOL}$$

and

$$F_{RESY} = F_{GAS} \cos\theta_{GAS} + F_{SOL} \cos\theta_{GOL},$$

7.1.2

where F_{RESX} and F_{RESY} are the X and Y components of the resultant force developed by the ankle plantarflexors. Substituting the magnitudes and directions of the force vectors given here, equation 7.1.2 becomes

$$F_{RESX} = 308 + 1000 = 1308 \text{ N}$$

and

$$F_{RESY} = 846 + 1732 = 2578 \text{ N}.$$

The magnitude of the resultant plantarflexor muscle force is then found using Pythagoras' theorem,

$$F_{RES} = \sqrt{F_{RESX}^2 + F_{RESY}^2} = 2891 \text{ N},$$

and the direction of the resultant muscle force is found from

$$\theta_{RES} = tan^{-1}\left(\frac{F_{RESX}}{F_{RESY}}\right) = 27°$$

(from the vertical; Y axis).

Note that the peak forces developed by GAS and SOL during a maximum isometric contraction are approximately 1700 N and 3000 N, respectively (Anderson & Pandy, 2001). Thus, the peak force developed by the ankle plantarflexors during normal walking is roughly 80% of the peak force developed by this muscle group during a maximum isometric contraction.

Source

Anderson, F.C., & Pandy, M.G. (2001). Static and dynamic optimization solutions for gait are practically equivalent. *Journal of Biomechanics, 34,* 153-161.

MOTION

Motion is the term used to describe the displacement, velocity, and acceleration of a body in space. Displacement is the change in the body's position as it moves from one location to another; **velocity** is the rate of change in the body's displacement over time; and **acceleration** is the rate of change in the body's velocity over time. The following sections provide simple examples to illustrate each of these quantities as they relate to human movement.

MOTION OF A PARTICLE

Consider a man walking at his natural speed over level ground. Let the man's body be imagined as a point P located at the centre of mass of the whole body (figure 7.3). Let O be any point fixed on the ground and $\underline{r}(t)$ be the position vector from O to the centre of mass P at time t. The displacement of P in a finite time interval Δt is the change in position of P from time t to time $t + \Delta t$. That is, the displacement of P is given by $\underline{r}(t + \Delta t) - \underline{r}(t)$, where $\underline{r}(t + \Delta t)$ is the position of P at time $t + \Delta t$. Velocity is defined as the rate of change in position with respect to time. Thus, the average velocity of P in the interval Δt is given by

$$\underline{v} = \frac{\underline{r}(t + \Delta t) - \underline{r}(t)}{\Delta t} \qquad 7.2$$

where $\underline{r}(t + \Delta t) - \underline{r}(t)$ represents the change in position, or displacement, of point P in the interval Δt. Similarly, if acceleration is the rate of change in velocity over time, the average acceleration of P relative to O in the time interval Δt is

$$\underline{a} = \frac{\underline{v}(t + \Delta t) - \underline{v}(t)}{\Delta t} \qquad 7.3$$

where $\underline{v}(t)$ is the velocity of P at time t and $\underline{v}(t + \Delta t)$ is the velocity of P at time $t + \Delta t$. The SI (Systems Internationale) unit of position and displacement is metres, so velocity and acceleration are given in metres per second and metres per second squared,

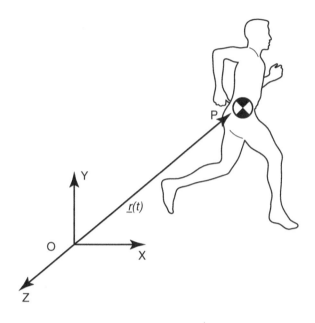

Figure 7.3 Centre of mass of the body imagined as a point in space. O is the origin of the X,Y reference frame fixed on the ground.

respectively. Equations 7.1 and 7.2 may be used to estimate the average velocity and acceleration of the centre of mass of the body in human locomotion (see "In Focus: Velocities and Accelerations of Points on Bodies Can Be Estimated From Measurements of Position").

IN FOCUS: VELOCITIES AND ACCELERATIONS OF POINTS ON BODIES CAN BE ESTIMATED FROM MEASUREMENTS OF POSITION

Equations 7.1 and 7.2 can be used to estimate the velocity and acceleration of a point on a body part if the position of that point has been measured during the activity. Consider motion of the hip joint during walking. The position of the hip can be measured by placing a reflective marker on the surface of the skin at the approximate centre of the joint. Video cameras can then be used to track the changing position of the marker over time as the person walks. This type of measurement is routinely performed in movement biomechanics laboratories throughout the world. These laboratories are equipped with video cameras to measure the changing positions and orientations of body parts and with devices (called force plates)

for measuring the forces exerted on the ground and instrumentation for monitoring and recording the pattern of muscle activity.

Table 7.1 gives the position of the hip joint in the fore–aft (X) direction at three successive instants during normal walking. The fore–aft velocity of the hip can be estimated using equation 7.3.1,

$$v = \frac{r(t + \Delta t) - r(t)}{\Delta t}, \qquad 7.3.1$$

where $r(t)$ is the position of the hip at time t, $r(t + \Delta t)$ is its position at time $t + \Delta t$, and Δt is the time step between any two successive instants during the gait cycle. For the first two time points in table 7.1,

Table 7.1 Fore-Aft Position of the Hip in Normal Walking on Level Ground

Time (s)	X position (cm)
0.272	83.12
0.286	85.69
0.399	87.77

Data from Winter 1990.

$r(t)$ = 83.12 cm, $r(t + \Delta t)$ = 85.69 cm, and Δt = 0.014 s. Therefore, from equation 7.2.1, the velocity of the hip in the horizontal (fore–aft) direction is found to be v = 183.6 cm/s or v = 1.84 m/s. Similarly, using the data for the second two time points, the velocity of the hip is v = 160 cm/s or v = 1.60 m/s.

Most adults walk at a speed of 1.35 m/s when they are free to choose **speed**, **cadence**, and **step length**. This is the speed at which the centre of mass of the whole body moves during normal walking over level ground. Thus, the previous calculations show that not all points on the body move with the same velocity during gait. Furthermore, the calculations show that the velocity of any given point on the body, such as the hip, changes over time. That is, all points on the body accelerate and decelerate during the gait cycle.

Indeed, equation 7.3 may be used to find exactly how much the hip accelerates (or decelerates) based on the data of table 7.1. From the results just obtained for the velocity of the hip, one can write $v(t)$ = 1.84 m/s, $v(t + \Delta t)$ = 1.60 m/s, and Δt = 0.013 s. Using equation 7.3, the fore–aft acceleration of the hip is found to be 18.5 m/s².

It is important to realise that the values of velocity and acceleration calculated here are average values; that is, it is assumed by using equations 7.2 and 7.3 that the values of velocity and acceleration remain constant over the interval of time defined by Δt. It is also possible to obtain instantaneous values of the velocity and acceleration of a point on a body (i.e., the values of velocity and acceleration at a given point in time), but this requires knowledge of calculus and, more specifically, knowledge of the rules of differentiation.

Source

Winter, D.A. (1990). *Biomechanics and motor control of human movement*. New York: Wiley.

ANGULAR MOTION OF A RIGID BODY

A **rigid body** is an idealised model of an object that does not deform or change its shape; that is, the distance between every pair of points on the body remains constant. Consider again the case of a man walking on level ground, but now assume that one leg is represented by a rigid link as shown in figure 7.4. The centre of mass of the body is lumped at the tip of the pendulum, which coincides with the man's hip. Let $\underline{\theta}(t)$ be the angular position vector of the leg at time t relative to a horizontal line fixed on the ground. The average **angular velocity** of the leg over a finite interval Δt is the rate of change of angular position of the leg,

$$\underline{\omega} = \frac{\underline{\theta}(t + \Delta t) - \underline{\theta}(t)}{\Delta t}, \qquad 7.4$$

where $\underline{\theta}(t + \Delta t)$ is the angular position vector at time $t + \Delta t$. The direction of $\underline{\omega}$ is perpendicular to the plane of motion of the body; in Figure 7.4 $\underline{\omega}$ points out of the plane of the paper because the body is assumed to rotate counterclockwise relative to a horizontal line fixed on the ground. The average angular acceleration of the leg is then

$$\underline{\alpha} = \frac{\underline{\omega}(t + \Delta t) - \underline{\omega}(t)}{\Delta t}, \qquad 7.5$$

where $\underline{\omega}(t)$ is the angular velocity of the leg at time t and $\underline{\omega}(t + \Delta t)$ is the angular velocity at time $t + \Delta t$. $\underline{\alpha}$ is also directed perpendicular to the plane of motion of the leg (i.e., $\underline{\omega}$ and α are assumed to have the same direction, out of the plane of the paper in figure 7.4). Note, however, that $\underline{\omega}$ and α may not always have the same direction; for example, it may happen that the body is rotating in one direction (e.g., counterclockwise) but slowing down (i.e., decelerating) as it rotates. In this case, $\underline{\omega}$ would be directed counterclockwise, but $\underline{\alpha}$ would be directed clockwise because the body is decelerating. The SI unit of angular position is radians, so angular velocity and angular acceleration are given in radians per second and radians per second squared, respectively.

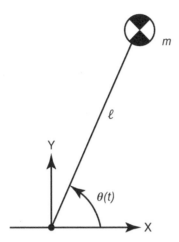

Figure 7.4 Diagram representing the human leg as an inverted single pendulum in normal walking on level ground. $\theta(t)$ is the angle that the leg makes relative to a horizontal line parallel with the ground at some time, t; l = length; m = mass of the body.

MOTION OF A POINT ON A RIGID BODY

To describe the motion of a rigid body in three-dimensional space, it is sufficient to consider the motion of a single point, such as the centre of mass, and the rotational motion of the body about that point. Consider a rigid body A rotating about a fixed point C in the plane of the paper (figure 7.5). The velocity (v) of any point P fixed on the body is given by

$$v = \omega R, \qquad 7.6$$

where ω is the magnitude of the angular velocity of the body and R is the distance from C to P. The velocity vector of P relative to the ground is in the plane of motion of the body and is perpendicular to the position vector from C to P as indicated in figure 7.5. Thus, v and ω are also perpendicular to each other because ω is directed out of the page.

It happens more often that all points of a body are moving at each instant in time, rather than one point being fixed to the ground as shown in figure 7.5. Thus, point C usually has some velocity relative to the ground. In this case, the velocity of P relative to C is still given by equation 7.6; however, the velocity of C relative to some point fixed on the ground, called O, is no longer zero. The velocity of P relative to O (the ground) can be found by adding two vectors as follows:

$$^{O}\underline{v}^{P} = {}^{O}\underline{v}^{C} + {}^{C}\underline{v}^{P}, \qquad 7.7$$

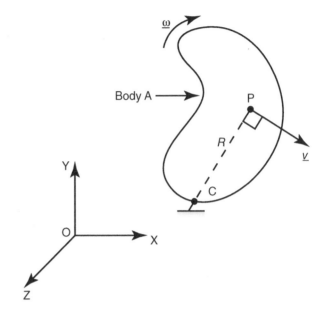

Figure 7.5 Diagram illustrating motion of a rigid body A about a point C fixed in space. X, Y, Z is a ground-fixed reference frame with origin O. Because C is fixed relative to the ground, its absolute velocity is zero.

where $^{O}\underline{v}^{P}$ is the velocity of P relative to O, $^{O}\underline{v}^{C}$ is the velocity of C relative to O, and $^{C}\underline{v}^{P}$ is the velocity of P relative to C. As described in the opening section of this chapter, special rules must be followed when operating on vector quantities because a vector is defined by both its magnitude and direction. The magnitude of the resultant velocity, $|^{O}v^{P}|$, can be found by using the cosine rule in trigonometry. Thus,

$$\left|{}^{O}v^{P}\right| = \sqrt{\left|{}^{O}v^{C}\right|^{2} + ({}^{C}v^{P})^{2} - 2\left|{}^{O}v^{C}\right|\left|{}^{C}v^{P}\right|\cos\beta}, \qquad 7.8$$

where $|^{C}v^{P}| = \omega R$ from equation 7.6 and β is the angle formed between the vectors $^{O}v^{C}$ and $^{C}v^{P}$. Although the velocity vector of P relative to O remains in the plane of motion of the body (i.e., perpendicular to the angular velocity vector of the body), its direction is not in general perpendicular to the position vector from C to P.

The acceleration of point P relative to any point O fixed on the ground follows from the equation defining the velocity of P relative to O. In the special case when one point C of the body is fixed relative to the ground, the magnitude of the acceleration of P relative to O is given by

$$\left|{}^{O}a^{P}\right| = \sqrt{(\alpha R)^{2} + (\omega^{2}R)^{2}}, \qquad 7.9$$

where α is the angular acceleration of the body at any instant. The acceleration vector of P relative to O remains in the plane of motion of the body (i.e., perpendicular to the angular velocity vector), but it has two components, indicated by the two terms on the right side of equation 7.9. The first term, αR, is called the **tangential acceleration** and is directed perpendicular to the position vector from C to P. The second term, $\alpha^2 R$, is called the **centripetal acceleration** and is directed radially inward from P to C. Note that the tangential and centripetal accelerations are always perpendicular to each other, and therefore the magnitude of the resultant acceleration is found simply by squaring each component, adding them together, and taking the square root of the result, as indicated in equation 7.9.

GENERALISED COORDINATES AND DEGREES OF FREEDOM

The minimum number of independent coordinates needed to specify the position and orientation of a rigid body in space is referred to as the generalised coordinates of that body. The number of gener-

alised coordinates is usually equal to the number of **degrees of freedom** of the body. Six generalised coordinates are needed to completely specify the position and orientation of a rigid body in space, so a rigid body has six degrees of freedom for unconstrained motion of the body in space. Three generalised coordinates are needed to describe the translation of any point on the body, and another three generalised coordinates are needed to describe three independent rotations of the body about three mutually perpendicular axes.

A rigid body has fewer than six degrees of freedom whenever it is acted on by kinematic constraints. Kinematic constraints exist when forces act on the body to restrict the motion of the body at each instant in time. Consider, for example, a woman landing from a vertical jump as illustrated in figure 7.6. When the body is in the air, six generalised coordinates are needed to completely specify its position and orientation with respect to a reference frame fixed on the ground. Once the toes make contact with the ground, the number of degrees of freedom is reduced to four because the ground then applies a force at the toes that prevents the foot from slipping. Next, the heel comes into contact with the ground and the number of degrees of freedom is then reduced to three; only

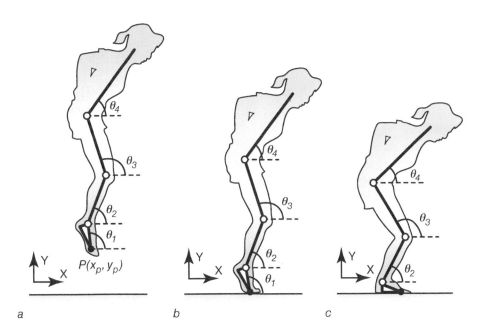

Figure 7.6 Diagram illustrating the number of degrees of freedom in a model for landing from a jump. All body segments are assumed to move only in the sagittal plane. X,Y is a reference frame fixed on the ground. Point P marks the tip of the toes segment in the model. *(a)* The jumper is airborne. *(b)* Only the toes are in contact with the ground. *(c)* The foot is flat on the ground.

Adapted, by permission, from F. Zajac and M. Gordon, 1989, "Determining muscle's force and action in multi-articular movement," *Exercise and Sport Sciences Review* 17(1): 187-230.

three coordinates are needed to fully specify the position and orientation of the shank, thigh, and trunk because the foot is now firmly planted on the ground and is effectively fixed to it.

The concepts of generalised coordinates and degrees of freedom are particularly important when characterising joint motion. In the case of the human knee, for example, the femur and tibia are held together mainly by the cruciate and collateral ligaments. Because the ligaments are elastic, they stretch as the muscles apply forces to move the bones. Thus, the femur and tibia displace relative to each other so that the knee, in theory, has six degrees of freedom of joint movement. These six movements are often characterised in anatomical terms as three translations of the tibia relative to the femur (anterior–posterior translation, medial–lateral translation, and proximal–distal distraction) plus three rotations (flexion–extension, abduction–adduction, and internal–external rotation). Most of the other joints in the body have fewer than six degrees of freedom. The hip, for example, has only three degrees of freedom, all of which are rotations (i.e., flexion–extension, abduction–adduction, and internal–external rotation of the femur relative to the pelvis). There are no translational degrees of freedom at the hip because the femoral head sits tightly inside the acetabulum, preventing any relative translation of the bones. Most joints of the fingers and toes have just one degree of freedom, allowing only flexion and extension of the metacarpal and metatarsal bones.

FORCE

Force is a measure of the amount of effort applied. It is a vector quantity, requiring both the magnitude and direction of the force to be specified. The direction of a force is often referred to as the **line of action of the force**. The SI unit of force is **Newton** (N).

INTERNAL AND EXTERNAL FORCES

There are many kinds of forces in nature, but most of the forces encountered in biomechanical studies of movement fall into one of two categories: external forces and internal forces. A given body is subjected to an **external force** if the force is applied by a different object. An **internal force** is any force applied by one part of a body on another part of the same body. Some types of forces are always external forces; for example, gravitational forces act on all parts of a body and are proportional to the body's mass. Other types of forces usually appear as internal forces (e.g., the forces applied by the muscles and the forces transmitted by the bones that meet at a joint). Caution is required in distinguishing between external and internal forces because the distinction is intimately related to the definition of the system being analysed. Specifically, whether a force is treated as external or internal depends on how the boundary of the system is defined, which is determined by a concept known as the free-body diagram.

FREE-BODY DIAGRAM

The external forces acting on a system are identified by drawing a free-body diagram. There are three steps involved in constructing such a diagram. First, the system to be isolated must be identified. Second, a sketch must be drawn of the system isolated from its surroundings. Third, all external forces acting on the system must be drawn and labeled as vectors.

An exercise commonly prescribed for maintaining thigh muscle strength subsequent to knee ligament injury or reconstruction is isokinetic knee extension. Isokinetic exercise means that the joint is flexed and extended at constant speed. The subject sits comfortably in a dynamometer and the subject's torso and thigh are strapped firmly to the seat (see figure 7.7a). The lower leg (shank plus foot) is strapped to the arm of the machine, which moves at a predetermined (constant) speed. A free-body diagram of the lower leg is shown in figure 7.7b. In this example, because the lower leg is isolated from the thigh and the machine arm, any force that breaks the system boundary is treated as an external force. Thus, the forces applied by the quadriceps (via the patellar tendon), the hamstrings, and the gastrocnemius muscles are treated as external forces, whereas those applied by all the other muscles (e.g., soleus) are treated as internal forces and are therefore not shown in figure 7.7b because they do not break the imaginary system boundary.

MOMENT OF FORCE

The moment of a force about a point is a measure of the **turning effect** of the force about that point. One may also think of the moment of a force about a point as a measure of the force to cause rotation

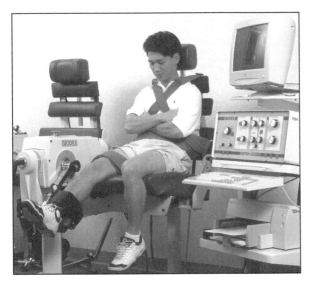

a

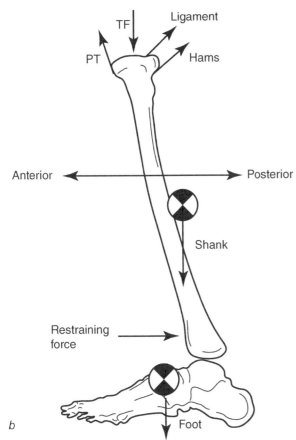

b

Figure 7.7 *(a)* Man seated in a Biodex dynamometer. The strap on the machine arm is attached distally on the subject's leg (i.e., just above the ankle). *(b)* Free-body diagram showing the external forces acting on the lower leg (shank plus foot) during isokinetic knee-extension exercise.

Free-body diagram courtesy of Dr. Kevin B. Shelburne. Copyright Kevin B. Shelburne.

of a body about that point. Moments are vectors, requiring both magnitude and direction to be specified. Consider a force vector $\underline{F}$ and any point O as shown in figure 7.8. The moment vector, $\underline{M}$, of force $\underline{F}$ acting about O is

$$\underline{M} = \underline{r} \times \underline{F},$$

where $\underline{r}$ is a vector directed from O to any point on the line of action of the force $\underline{F}$, and the symbol x indicates the cross-product of two vector quantities. The magnitude of the moment about O is

$$M = Fd, \hspace{3cm} 7.10$$

where F is the magnitude of the force vector and d is the perpendicular distance from point O to the line of action of the force. It should be clear from equation 7.10 that the SI unit of moment is Newton metre.

Muscles develop force and cause rotation of body segments about the joints. The moment of a muscle force follows directly from equation 7.10, where d, the perpendicular distance between the line of action of the muscle and the axis of rotation of the joint, is called the *muscle moment arm*. Muscle moment arms are important because they describe how closely the muscles pass relative to the axes of rotation of the joints. The closer a muscle is to the axis of rotation, the less effective it is in causing rotation of the body segments because the smaller the moment arm the smaller the moment of force (from equation 7.10) (see "In Focus: How Much Force Can the Quadriceps Develop in a Maximal Contraction?").

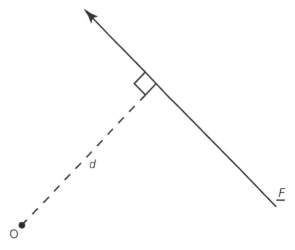

Figure 7.8 Force $\underline{F}$ acting about any point O in space. The force has a magnitude of F and a line of action as indicated in the diagram; d is the perpendicular distance from the line of action of $\underline{F}$ and point O.

IN FOCUS: HOW MUCH FORCE CAN THE QUADRICEPS DEVELOP IN A MAXIMAL CONTRACTION?

The quadriceps is the strongest muscle in the body. This can be demonstrated by performing an isometric contraction in a Cybex or Biodex dynamometer as shown in figure 7.7. For peak force to be developed in the quadriceps, the hip must be flexed to approximately 60° and the ankle must be held roughly in the neutral (standing) position. The leg is strapped to the arm of the machine, which is then locked in place, so that the muscles crossing the knee contract isometrically (see figure 7.7a). Under these conditions, healthy, young male subjects develop maximum moments in the range of 250 to 300 Nm (figure 7.9a). As one might expect, the maximum moment developed by the quadriceps varies with the knee-flexion angle. In fact, peak knee-extensor moment occurs in most people when the knee is flexed to 60°, and it decreases significantly as the knee is moved toward extension. This means that the knee-extensor muscles are strongest (i.e., they exert the most leverage) when the knee is bent to 60° and that their leverage is severely compromised when the knee reaches full extension. Why?

To answer this question, consider the two factors that contribute to knee-extensor moment: the moment arm of the knee-extensor mechanism and the amount of force developed by the quadriceps (equation 7.10). The moment arm of the knee-extensor

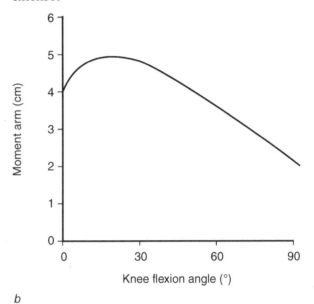

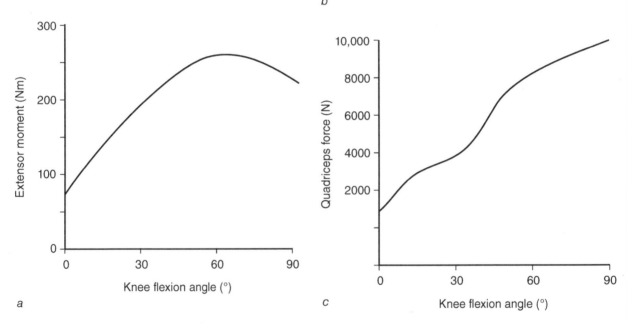

Figure 7.9 *(a)* Maximum isometric moment of force plotted against knee-flexion angle for maximum isometric knee-extension exercise, *(b)* the corresponding knee-extensor moment arm, and *(c)* quadriceps force plotted against the flexion angle.

Reprinted from *Journal of Biomechanics* Vol. 30, K.B. Shelburne and M.G. Pandy, "A musculoskeletal model of the knee for evaluating ligament forces during isometric contractions," pgs. 163-176, Copyright 1997, with permission from Elsevier.

mechanism is the perpendicular distance from the line of action of the patellar tendon to the axis of rotation of the knee (which is located roughly at the centre of the joint). This distance is determined by the shapes of the distal femur, proximal tibia, and patella and by the way the relative positions of these bones change during flexion and extension of the knee. The moment arm of the knee-extensor mechanism has been measured in a number of studies using intact human cadaver specimens, and the result is given in figure 7.9b. Perhaps the most striking feature of this curve is that it does not vary much as the knee flexes and extends between 0° and 90°. The peak moment arm of the knee-extensor mechanism is about 5 cm and occurs when the knee is flexed to about 20°. The minimum value is about 4 cm and occurs when the knee is either fully extended or flexed to 50° (figure 7.9b).

From equation 7.10, the moment of a muscle force is equal to the magnitude of the muscle force multiplied by the muscle's moment arm. Thus, dividing the muscle moment curve by the moment arm curve in figure 7.9 gives quadriceps force as a function of the knee-flexion angle (figure 7.9c). Peak quadriceps force is about 9500 N for maximum isometric knee extension. Peak force occurs with the knee bent to 90° and it decreases as the knee is moved toward extension. Quadriceps force decreases as the knee extends because the muscle then moves down the ascending limb of its force–length curve (i.e., muscle length decreases). Indeed, quadriceps force is minimum when the knee is fully extended because the muscle fibres are too short to develop much force in

this position. Conversely, quadriceps force peaks when the knee is bent to 90° because this position corresponds with the length at which the maximum number of actomyosin cross-bridges are formed.

The quadriceps exert the least leverage (i.e., smallest moment) when the knee is fully extended because these muscles cannot generate much force in this position. At full extension, the effectiveness of the quadriceps to cause rotation is determined by the force–length properties of these muscles and not by the geometry of the knee (i.e., the moment arm of the knee-extensor mechanism). At 60° of knee flexion, the effectiveness of the quadriceps is maximum because the force developed by the muscle and the moment arm of the extensor mechanism are both relatively large, and their product is then optimal. At 90° of knee flexion, quadriceps moment once again decreases, even though the force developed by the muscle continues to increase (figure 7.9c, compare force at 60° and 90°). In this case, quadriceps moment decreases because the moment arm of the extensor mechanism decreases. Thus, the effectiveness of the quadriceps at large knee-flexion angles is governed by the geometry of the knee (i.e., the moment arm of the extensor mechanism).

Sources

Pandy, M.G., & Shelburne, K.B. (1998). Theoretical analysis of ligament and extensor-mechanism function in the ACL-deficient knee. *Clinical Biomechanics, 13,* 98-111.

Shelburne, K.B., & Pandy, M.G. (1997). A musculoskeletal model of the knee for evaluating ligament forces during isometric contractions. *Journal of Biomechanics, 30,* 163-176.

FORCE ANALYSES

A body is in **equilibrium** if every point of the body has the same velocity. If the velocity and acceleration of every point on the body are zero, the body is said to be in **static equilibrium**. In mathematical terms, a body is in static equilibrium if the sum of all the external forces in the X, Y, and Z directions and the sum of all the moments of force about the X, Y, and Z axes are all simultaneously zero. That is,

$$\sum_{i=1}^{N} F_{xi} = 0; \quad \sum_{i=1}^{N} F_{yi} = 0; \quad \sum_{i=1}^{N} F_{zi} = 0 \qquad 7.11.1$$

and

$$\sum_{i=1}^{N} M_{xi} = 0; \quad \sum_{i=1}^{N} M_{yi} = 0; \quad \sum_{i=1}^{N} M_{zi} = 0, \qquad 7.11.2$$

where F_{xi}, F_{yi}, F_{zi} are the ith forces acting in the X, Y, and Z directions, respectively, and M_{xi}, M_{yi}, M_{zi} are the ith moments exerted about the X, Y, and Z axes, respectively. All six conditions given in equations 7.11 must hold if static equilibrium is to apply. For example, a body may be at rest (zero velocity) but not in static equilibrium. Consider the simple case of a boy throwing a ball straight up into the air. When the ball reaches its maximum height, its vertical velocity is zero yet the ball continues to accelerate. Indeed, the vertical acceleration of the ball from the moment it leaves the boy's hand is 9.81 m/s² (32.2 ft/s²) if gravity is the only force acting while the ball remains in the air. In this case, the ball is not in static equilibrium at the top of its flight, even though it is momentarily at rest, because the net acceleration of the ball is not zero.

Consider once again the knee-extension exercise illustrated in figure 7.7, but assume now that the exercise is performed with the muscles held isometrically (i.e., at constant length). Assume also that movement of the lower leg is permitted only in the sagittal plane and that only the quadriceps muscles are being activated (i.e., the hamstrings and gastrocnemius muscles are assumed to be fully deactivated). The free-body diagram for the isometric knee-extension exercise is shown in figure 7.10 (compare with figure 7.7b). It is assumed that the knee is bent to 90°. The external restraining force applied by the machine arm, F_{ext}, acts perpendicular to the leg. The thigh is held horizontal as shown. F_Q is the magnitude of the resultant force developed by the quadriceps, which is applied to the leg via the patellar tendon. For the position shown, the quadriceps is assumed to be directed 15° anterior to the long axis of the leg. F_{Jx} and F_{Jy} are the X and Y components of the unknown articular contact force acting between the tibia and femur.

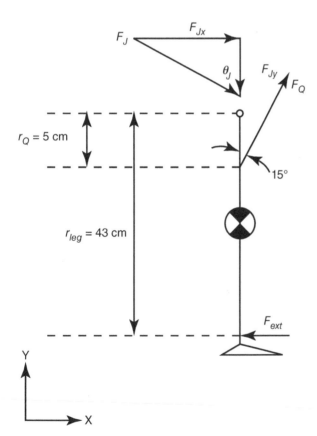

Figure 7.10 Free-body diagram showing the external forces acting on the lower leg during isometric knee-extension exercise.

Equations 7.11 may be used to find the magnitude of the quadriceps force needed to hold the lower leg in static equilibrium during the exercise. Summing forces in the X and Y directions gives

$$\sum F_x = 0; \quad F_{Qx} + F_{Jx} - F_{ext} = 0 \qquad 7.12.1$$

and

$$\sum F_Y = 0; \quad F_{Qy} - F_{Jy} = 0, \qquad 7.12.2$$

where $F_{Qx} = F_Q \sin 15°$ is the component of quadriceps force in the X direction and $F_{Qy} = F_Q \cos 15°$ is the component of quadriceps force in the Y direction (see figure 7.10). Summing moments about the centre of the knee joint gives

$$\sum M_Z = 0; \quad F_{Qx}\, r_Q - F_{ext}\, r_{leg} = 0, \qquad 7.12.3$$

where r_Q is the distance from the insertion of the patellar tendon to the centre of the knee joint and r_{leg} is the distance from the point of application of the external restraining force from the centre of the knee. Notice that the resultant joint-reaction force, F_J, does not appear in equation 7.12.3. The reason is that this force passes through the centre of knee and therefore contributes nothing to the moment about the joint. If the magnitude of the external restraining force is known (F_{ext} is assumed to be 250 N here, which corresponds to the peak value typically exerted by young healthy men), the value of F_{Qx} can be found from equation 7.12.3 as follows: $F_{Qx} = F_{ext} \dfrac{r_{leg}}{r_Q} = 250 \times \dfrac{43}{5} = 2150$ N. Thus, the resultant force developed by the quadriceps is given by $F_Q = \dfrac{F_{Qx}}{\sin 15°} = 8307$ N.

The X and Y components of the joint-reaction force are then obtained from equations 7.12.1 and 7.12.2 as follows: $F_{Jx} = F_{ext} - F_{Qx} = -1900$ N and $F_{Jy} = F_{Qy} = F_Q \cos 15° = 8024$ N. Finally, the magnitude and direction of the resultant joint reaction force are found from $F_J = \sqrt{F_{Jx}^2 + F_{Jy}^2} = 8246$ N, and $\theta_J = \tan^{-1}\left(\dfrac{F_{Jx}}{F_{Jy}}\right) = 13°$, respectively. The magnitude of the quadriceps force is roughly 12 times greater than the weight of an average person (an average adult man weighs about 700 N). Indeed, muscles routinely develop very large forces in activity; maximum isometric knee-extension exercise is an extreme case. In normal walking, for example, the quadriceps muscle group develops a peak force of approximately 1000 N, which is still larger than the average person's body weight.

EQUATIONS OF MOTION

The relationships between the forces applied to a rigid body and the resulting motion of that body are described by the equations of motion of the biomechanical system. Different methods may be used to derive equations of motion; for example, the Newton-Euler method explicitly considers the forces acting on the system, and the Lagrangian method considers the total mechanical energy of the system at any instant in time. In this section, we describe how the Newton-Euler method can be applied to derive the dynamical equations of motion of a biomechanical system.

Consider a representative limb segment, segment i, with forces and moments acting as shown in figure 7.11. The first step in deriving the equations of motion of any system is to write equations for the position, velocity, and acceleration of the centre of mass of each segment. Assuming X and Y define the horizontal and vertical directions, respectively, we can write the following equations for the positions, velocities, and accelerations of the mass centre of segment i:

$$x_{ci} = x_i + r_{ci}\cos\theta_i;\ y_{ci} = y_i + r_{ci}\sin\theta_i,\quad 7.13.1$$

$$\dot{x}_{ci} = \dot{x}_i - r_{ci}\sin\theta_i\dot{\theta}_i;\ \dot{y}_{ci} = \dot{y}_i + r_{ci}\cos\theta_i\dot{\theta}_i\quad 7.13.2$$

and

$$\ddot{x}_{ci} = \ddot{x}_i - r_{ci}(\sin\theta_i\ddot{\theta}_i + \cos\theta_i\dot{\theta}_i^2);$$
$$\ddot{y}_{ci} = \ddot{y}_i - r_{ci}(\cos\theta_i\ddot{\theta}_i - \sin\theta_i\dot{\theta}_i^2),\quad 7.13.3$$

where x_i, y_i are the horizontal and vertical positions of joint i (see figure 7.11); x_{ci}, y_{ci} are the horizontal and vertical positions of the centre of mass of segment i; θ_i is the angle that segment i makes with the horizontal axis, X; r_{ci} is the distance from the distal joint i to the centre of mass of segment i; $\dot{x}_i$, $\dot{y}_i$ are the horizontal and vertical velocities of joint i; $\dot{x}_{ci}$, $\dot{y}_{ci}$ are the horizontal and vertical velocities of the centre of mass of segment i; $\ddot{x}_i$, $\ddot{y}_i$ are the horizontal and vertical accelerations of joint i; $\ddot{x}_{ci}$, $\ddot{y}_{ci}$ are the horizontal and vertical accelerations of the centre of mass of segment i; and $\dot{\theta}_i$, $\ddot{\theta}_i$ are the angular velocity and angular acceleration of segment i. Note that the horizontal velocity of joint i, $\dot{x}_i$, is the derivative of the horizontal position of joint i with respect to time, whereas the horizontal acceleration of joint i, $\ddot{x}_i$, is the derivative of the horizontal velocity of joint i with respect to time.

The second step in deriving the equations of motion is to apply Newton's second law of motion, which states that the sum of all forces acting on a body is equal to the mass of the body multiplied by its acceleration; thus, $\sum \underline{F} = m\,\underline{a}$. Referring to figure

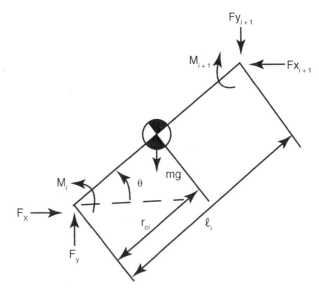

Figure 7.11 Free-body diagram of a generic (ith) body segment showing all external forces and moments acting.

7.11, we can write force balances in the X and Y directions as follows:

$$F_{xi} - F_{xi+1} = m_i\ddot{x}_{ci}\quad 7.14.1$$

and

$$F_{yi} - F_{yi+1} - mg = m_i\ddot{y}_{ci},\quad 7.14.2$$

where F_{xi}, F_{yi} are the horizontal and vertical components of the joint reaction force acting at the distal joint i, F_{xi+1}, F_{yi+1} are the horizontal and vertical components of the joint reaction force acting at the proximal joint $i + 1$, m_i is the mass of segment i, and g is the value of gravitational acceleration near the surface of the earth (9.81 m/s² or 32.2 ft/s²).

The final step in deriving the equations of motion is to again apply Newton's second law of motion, but this time for rotation of the body. That is, $\sum \underline{M} = I\,\underline{\alpha}$, where I is the moment of inertia of the body and α is its angular acceleration. Summing moments about the centre of mass of segment i gives

$$M_i - M_{i+1} + F_{xi}r_{ci}\sin\theta_i - F_{yi}r_{ci}\cos\theta_i +$$
$$F_{xi+1}(l_i - r_{ci})\sin\theta_i - F_{yi+1}(l_i - r_{ci})\cos\theta_i = I_i\ddot{\theta}_i,\quad 7.15$$

where M_i, M_{i+1} are the net moments acting at joints i and $i + 1$, respectively; r_{ci} is the distance from joint i to the center of mass of the segment, and l_i is the length of segment i. The equation of motion of the system is found by first substituting equations 7.13 into equations 7.14 and then eliminating the unknown joint reaction forces by substituting equations 7.14 into equation 7.15. A simple example is provided in "In Focus: Equation of Motion for the Single Pendulum Model of Walking."

The simplest model of human walking is the single inverted pendulum with the mass of the whole body lumped at the tip of the pendulum as shown in figure 7.4. In this model, flexion of the hip and knee of the stance leg are neglected, as are the inertial effects of the contralateral leg. As indicated in figure 7.4, one generalised coordinate, θ, is sufficient to describe the position and orientation of the mass, m, at any instant during the gait cycle. The pendulum model therefore has one degree of freedom, so one equation of motion can be written to describe the relationship between the forces applied to the pendulum and its resulting motion in the sagittal plane. Figure 7.12 is a free-body diagram of the pendulum showing the external forces acting on the system. The Newton-Euler method can be used to derive the equation of motion for the pendulum model.

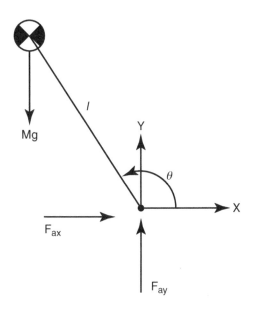

Figure 7.12 Free-body diagram of the pendulum model of walking showing the forces acting on the body during the stance phase.

Step 1: Write equations for the position, velocity, and acceleration of the centre of mass.

The horizontal and vertical positions, velocities, and accelerations of the centre of mass can be expressed as follows:

$$x_c = l\cos\theta; \; y_c = l\sin\theta,$$

$$\dot{x}_c = -l\sin\theta\dot{\theta}; \; \dot{y}_c = l\cos\theta\dot{\theta},$$

and
$$\ddot{x}_c = -l(\sin\theta\ddot{\theta} + \cos\theta\dot{\theta}^2);$$
$$\ddot{y}_c = l(\cos\theta\ddot{\theta} - \sin\theta\dot{\theta}^2)$$
7.16.1

Step 2: Write force balance equations in the horizontal and vertical directions.

Applying Newton's second law of motion (i.e., $\sum \underline{F} = m\,\underline{a}$) in the X and Y directions gives

$$F_{Gx} = m\ddot{x}_c \text{ and } F_{Gy} - mg = m\ddot{y}_c.$$
7.16.2

where F_{Gx} and F_{Gy} are the horizontal and vertical components of the ground reaction force, respectively.

Step 3: Sum moments about the centre of mass to obtain a relationship between the external moments acting on the pendulum and its angular motion. Applying Newton's second law of motion for rotation (i.e., $\sum \underline{M} = I\,\underline{\alpha}$) gives

$$M_a + F_{Gx}l\sin\theta - F_{Gy}l\cos\theta = I\ddot{\theta},$$
7.16.3

where M_a is the net muscle moment acting about the ankle joint, l is the length of the leg (i.e., the distance from the ankle to the centre of mass of the body), and I is the moment of inertia. Substituting equations 7.16.1 into equations 7.16.2 gives

$$F_{Gx} = -ml(\sin\ddot{\theta} + \cos\ddot{\theta}^2)$$
$$\text{and } F_{Gy} - mg = ml(\cos\ddot{\theta} - \sin\ddot{\theta}^2).$$
7.16.4

Finally, substituting equations 7.16.4 into equation 7.16.3 and then rearranging terms yields the following equation of motion for the pendulum model:

$$(ml^2 + I)\ddot{\theta} + mgl\cos\theta = M_a.$$
7.16.5

Given the time history of the net ankle moment, M_a, together with initial conditions of the system [i.e., $\theta(0) = \theta_0$, $\dot{\theta}(0) = \dot{\theta}_0$, where θ_0, $\dot{\theta}_0$ are initial values of the angular displacement and angular velocity of the pendulum], equation 7.16.5 can be solved numerically using a computer to determine the time histories of the angular displacement, velocity, and acceleration of the pendulum during walking.

COMPUTER MODELLING OF MOVEMENT

Mathematical (computer) models are widely used today, mainly because they give information that is not easily obtained from experiments. In biomechanical studies of movement, for example, muscle, ligament, and joint-contact forces cannot be measured noninvasively, so these quantities are often estimated using computer models of the body. Indeed, detailed models of the lower limb have been used to determine musculoskeletal forces in various activities, including walking, running, jumping, cycling, and throwing.

Figure 7.13 shows a model that was used to estimate muscle, ligament, and joint-contact forces in a knee-extension exercise. The lower limb was represented as a kinematic chain with three segments and four degrees of freedom. The hip and ankle were each modelled as a hinge with one degree of freedom that allowed only flexion and extension in the sagittal plane. The knee was modelled as a joint with two degrees of freedom that allowed rolling and sliding of the tibia on the femur in the sagittal plane.

The geometry of the distal femur, proximal tibia, and patella was based on digitised slices averaged across 23 cadaveric knees. Two-dimensional profiles for the lateral femoral condyle and the trochlear groove were obtained by fitting splines to the mid-parasagittal slice of each surface. The tibial plateau was assumed to be flat and sloped 8° posteriorly in the sagittal plane. The patella was assumed to be

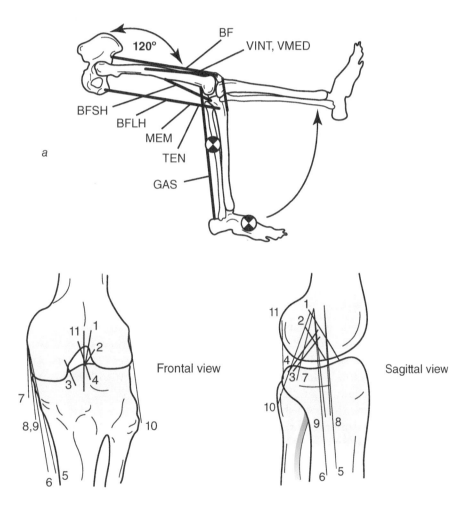

Figure 7.13 Schematic of the model leg used to calculate the maximum isometric moment, moment arm, and quadriceps force developed during isometric knee-extension exercise. *(a)* Muscles actuating the model leg. *(b)* Shapes of the bones and locations of the knee ligaments assumed in the model.

Reprinted from F. Serpas, T. Yanagawa, and M.G. Pandy, 2002, "Forward-dynamics simulation of anterior cruciate ligament forces developed during isokinetic dynamometry," *Computer Methods in Biomechanics and Biomechanical Engineering* 5(1): 33-43, reprinted by permission of Taylor & Francis Ltd, http://www.tandf.co.uk/journals.

massless and rectangular in shape. All bones were assumed to be rigid; the behaviour of the menisci and cartilage were neglected in the model.

Eleven bundles were used to model the geometry and mechanical properties of the cruciate ligaments, the collateral ligaments, and the posterior capsule. The anterior cruciate ligament (ACL) and the posterior cruciate ligament were each represented by two bundles: an anterior bundle and a posterior bundle. The medial collateral ligament was separated into two portions: the superficial fibres, represented by an anterior bundle, an intermediate bundle, and a posterior bundle; and the deep-lying fibres, represented by an anterior bundle and a posterior bundle (see figure 7.13). The lateral collateral ligament and posterior capsule were each represented by one bundle. The path of each ligament was approximated as a straight line in the model. Each ligament was assumed to be elastic, and its mechanical behaviour was represented by a nonlinear force–length curve.

The model knee was actuated by 11 musculotendinous units (figure 7.13). Each unit was modelled as a three-element entity in series with tendon. The mechanical behaviour of muscle was described by a three-element Hill-type model (see "In Focus: Modelling the Mechanics of Muscle Contraction" in chapter 8). Tendon was assumed to be elastic, and its properties were described by a linear force–length curve.

The quadriceps muscles can exert up to 9500 N of force when fully activated and contracting isometrically (figure 7.14). Peak isometric force is developed when the knee is bent to 90°, and it decreases as the knee moves toward extension (figure 7.14, quads). Although the quadriceps force increases from full extension to 90° of flexion, the force transmitted to the ACL increases and then decreases as the knee is flexed from full extension (figure 7.14, aAC and pAC). The computer model predicts that the ACL is loaded only from full extension to 80° of flexion and that the resultant force in the ACL reaches 500 N at 20° of flexion, which is lower than the maximum breaking strength of the human ACL (2000 N). The calculations also show that load sharing in the knee ligament may not be uniform. For example, the force transmitted to the anteromedial bundle of the ACL (aAC) increases from full extension to 20° of flexion, where peak force occurs, and aAC force then decreases as knee flexion increases (figure 7.14, aAC). On the other hand, the force transmitted to the posterolateral bundle of the ACL (pAC) is quite different (compare aAC and pAC in figure 7.14).

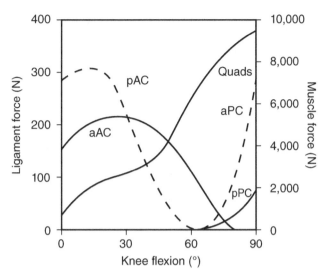

Figure 7.14 Resultant force developed by the quadriceps (thick solid line) and the corresponding cruciate ligament forces calculated for maximum isometric knee extension.

Reprinted from *Journal of Biomechanics* Vol. 30, K.B. Shelburne and M.G. Pandy, "A musculoskeletal model of the knee for evaluating ligament forces during isometric contractions," pgs. 163-176, Copyright (1997), with permission from Elsevier.

The forces exerted between the femur and patella and between the femur and tibia depend mainly on the geometry of the muscles that cross the knee. For maximum isometric extension, peak forces transmitted to the patellofemoral and tibiofemoral joints are around 11000 N and 6500 N, respectively (i.e., approximately 16 and 9 times body weight, respectively; figure 7.15). As the knee moves faster during isokinetic extension, joint-contact forces decrease in direct proportion to the drop in quadriceps force. Quadriceps force decreases as the knee extends faster because the muscle then shortens more quickly and, from the force–velocity property of muscle, an increase in shortening velocity leads to less force (see "In Focus: At What Speed Must a Muscle Shorten to Develop Maximum Power?" in chapter 10).

The computer model predictions of figures 7.14 and 7.15 have significant implications for the design of exercise regimens aimed at protecting injured or newly reconstructed knee ligaments. For maximum isolated contractions of the quadriceps, figure 7.14 shows that the ACL is loaded at all flexion angles less than 80°. Quadriceps-strengthening exercises should therefore be limited to flexion angles greater than 80° if the injured or newly reconstructed ACL

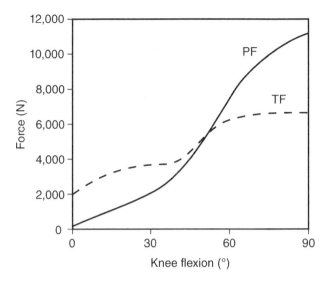

Figure 7.15 Resultant forces acting between the femur and tibia (TF) and between the femur and patella (PF) during maximum isometric knee-extension exercise.

Reprinted, by permission, from M.G. Pandy and R.E. Barr, 2002, "Biomechanics of the human musculoskeletal system." In *Biomedical engineering and design handbook*, Vol. 1, 2nd ed., edited by M. Kutz (New York, NY: McGraw-Hill), 629. © The McGraw-Hill Companies.

is to be protected from strain. In this region, however, very high contact forces can be applied to the patella (figure 7.15), so much so that limiting this exercise to large flexion angles may result in patellofemoral pain. This scenario, known as the paradox of exercise, is the reason why surgeons and physical therapists now prescribe closed-chain exercises, such as squatting, for strengthening the quadriceps muscles after ACL injury or repair.

SUMMARY

Any physical quantity that is described by its magnitude and direction is called a vector. Vector algebra is fundamental to analyses of human movement; for example, displacement, velocity, and acceleration vectors are used to study how body segments move under the action of force systems created by the actions of muscles and the environment. The number of generalised coordinates of a rigid body is equal to the minimum number of independent coordinates needed to specify the position and orientation of that body in space. Any body segment has at most six degrees of freedom: three for the translation of any point on the segment, such as the centre of mass, and three for the rotation of the segment about a line passing through that point. Body segments have fewer than six degrees of freedom when their motion is constrained by the actions of external forces (e.g., contact of the foot with the ground during walking). Free-body diagrams and the concept of equilibrium are used to analyse the resultant effect of force systems acting on body parts. Analyses based on the principle of static equilibrium are often used to estimate the magnitudes and directions of muscle, ligament, and joint-reaction forces present during activity. Dynamical equations of motion represent the relationships between the forces applied to a rigid body and the resulting motion of the body. Equations of motion are solved numerically using a computer to simulate human movement.

FURTHER READING

Bedford, A., & Fowler, W. (1995). *Engineering mechanics: Dynamics.* Reading, MA: Addison-Wesley.

Bedford, A., & Fowler, W. (1995). *Engineering mechanics: Statics.* Reading, MA: Addison-Wesley.

Lieber, R.L. (1992). *Skeletal muscle structure and function: Implications for rehabilitation and sports medicine.* Baltimore: Williams & Wilkins.

Nordin, M., & Frankel, V.H. (1989). *Basic biomechanics of the musculoskeletal system* (2nd ed.). Philadelphia: Lea & Febiger.

Pandy, M.G. (2001). Computer modeling and simulation of human movement. *Annual Review of Biomedical Engineering, 3,* 245-273.

Pandy, M.G., & Andriacchi, T.P. (2010). Muscle and joint function in human locomotion. *Annual Review of Biomedical Engineering, 12,* 401-433.

Winter, D.A. (1990). *Biomechanics and motor control of human movement.* New York: Wiley.

CHAPTER 8

BASIC CONCEPTS OF ENERGETICS

The major learning concepts in this chapter relate to

- ▶ calculating the kinetic energy (the amount of mechanical energy a body segment possesses due to its motion) of a body segment,
- ▶ calculating the potential energy (the amount of mechanical energy a body segment possesses by virtue of its height above the ground) of a body segment,
- ▶ calculating the total mechanical energy (the sum of kinetic energy and gravitational potential energy) of a body segment,
- ▶ calculating the instantaneous power (the rate at which work is done on the body segment) of a body segment,
- ▶ calculating elastic strain energy (the amount of mechanical energy stored in the elastic tissues of muscle and tendon during movement),
- ▶ describing qualitatively and quantitatively how metabolic energy is consumed during movement, and
- ▶ defining the efficiency of movement (the ratio of mechanical energy output to metabolic energy input).

One of the fundamental laws of mechanics, as elementary as Newton's laws of motion, is the **principle of work and energy**. This principle, which states that the work done on an object is equal to the change in its kinetic energy, can be derived from Newton's famous second law $\underline{F} = m\underline{a}$. The work done on an object is equal to the net force acting on the object multiplied by its displacement (change in position) as a result of the force. The mechanical energy of the object—its capacity to do work—may be visible as kinetic energy, gravitational potential energy, and, unless the object is rigid, elastic strain energy, which is stored as a result of the deformation created by the applied force.

This chapter describes the roles that kinetic energy, potential energy, and elastic energy play in the performance of human movement. Whereas kinetic and potential energy are related to the mass of the body, elastic **strain energy** is associated with the properties of muscle and tendon tissue. Voluntary movement is made possible by the development of muscle force, which is fueled by chemical energy made available through the oxidation of foodstuffs. The last two sections of this chapter describe the energetics of muscle contraction and show how the metabolic cost of movement is governed by the interplay of kinetic energy, gravitational potential energy, and strain energy stored in the elastic tissues of muscle and tendon.

KINETIC ENERGY

The kinetic energy of a body is the amount of mechanical energy the body possesses due to its motion. Consider a rigid body translating and rotating in space (figure 8.1). Let the mass of the body be represented by m and the moment of inertia of the body about its centre of mass be given by I_c. The moment of inertia of a body is a measure of the ability of the body to resist changes in its angular velocity. More simply, it may be regarded as a measure of the body's resistance to rotation. The moment of inertia depends on the point of the body about which it is calculated. It is minimal about the centre of mass of the body. The Systeme International (SI) unit of moment of inertia is kg $\times$ m² (lb $\times$ ft²). At some time t during the body's motion, $\underline{v}$ represents the velocity of the centre of mass of the body and $\underline{\omega}$ is its angular velocity. The kinetic energy of the body is made up of two parts: the kinetic energy due to translation of the centre of mass and that due to rotation about its centre of mass. The translational kinetic energy is given by

$$T_t = \frac{1}{2}m\underline{v}^2,\qquad 8.1$$

whereas the rotational kinetic energy is given by

$$T_r = \frac{1}{2}I_c\underline{\omega}^2.\qquad 8.2$$

Note that T_t and T_r are both scalars because any vector multiplied by itself is a scalar. That is, $\underline{v}^2 = \underline{v} \times \underline{v} = v^2$, and v^2 is a scalar. Thus, the total kinetic energy of the body is

$$T = \frac{1}{2}m\underline{v}^2 + \frac{1}{2}I_c\underline{\omega}^2,\qquad 8.3$$

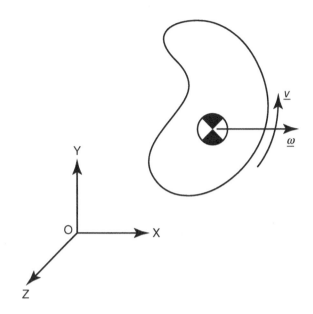

Figure 8.1 Rigid body rotating with an angular velocity $\underline{\omega}$ and translating with a linear velocity of its centre of mass of $\underline{v}$.

and because $\underline{v}^2 = \underline{v} \times \underline{v} = v^2$ and $\underline{\omega}^2 = \underline{\omega} \times \underline{\omega} = \omega^2$, equation 8.3 can be rewritten more simply as

$$T = \frac{1}{2}mv^2 + \frac{1}{2}I_c\omega^2.\qquad 8.4$$

The SI unit of kinetic energy is the **joule**, which is equivalent to Newton metre (ft $\times$ lb).

POTENTIAL ENERGY

Potential energy is the amount of mechanical energy a body possesses due to a change in its position. It results from the gravitational acceleration experienced by the body as it moves from one position to another near the surface of the earth. Consider the rigid body of figure 8.1 moving from position 1 to position 2 (see figure 8.2). The change in gravitational potential energy of the body is given by

$$U_g = mg(y_2 - y_1),\qquad 8.5$$

where y_1 is the vertical position (height) of the centre of mass of the body in position 1, y_2 is its height in position 2, and g is the gravitational acceleration constant near the surface of the earth (equal to 9.81 m/s² or 32.2 ft/s²). Potential energy is a scalar and its SI unit is the joule [Newton metre (ft $\times$ lb)].

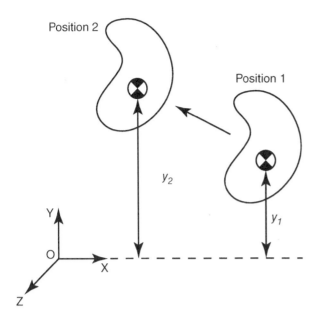

Figure 8.2 Displacement of a rigid body between two positions. The change in potential energy of the body is proportional to the displacement of the centre of mass from position 1 to position 2.

TOTAL MECHANICAL ENERGY

The total mechanical energy of a rigid body is equal to the sum of its kinetic energy and gravitational potential energy:

$$E_T = \frac{1}{2}mv^2 + \frac{1}{2}I_c\omega^2 + mg(y_2 - y_1). \qquad 8.6$$

When the total mechanical energy of the system remains constant (i.e., the value of E_T does not change over time), energy is conserved and the system is said to be *conservative*. In the real world, no system is completely conservative; some energy is always lost by virtue of the system's interaction with the environment. Take the pendulum that is commonly found in a grandfather clock. In order for the clock to keep perfect time, the pendulum bob must swing with exactly the same amplitude from cycle to cycle. The small amount of friction in the pendulum's bearings decrease (by a very small amount) the amplitude of the pendulum's swing over time. To ensure that the pendulum swings with precisely the same amplitude from cycle to cycle, a small moment is exerted on the pendulum to compensate for the slowing effect of friction.

It is worthwhile to look at the fluctuations in kinetic and potential energy of the ideal pendulum (in the absence of friction) during each cycle of its swing because this model illustrates many of the same biomechanical features evident in human walking. Figure 8.3 shows a single pendulum of length l and mass m. At any instant during its motion, the kinetic energy of the pendulum can be written as

$$T = \frac{1}{2}mv^2, \qquad 8.7$$

where v, the velocity of the tip of the pendulum, is given by $v = \omega l$, and ω is the angular velocity of the pendulum. Note that the pendulum has no rotational kinetic energy here because its moment of inertia is assumed to be zero. The gravitational potential energy is given by

$$U = mg(l - l\cos\theta), \qquad 8.8$$

where the zero position is (arbitrarily) taken to be the lowest point reached by the pendulum during its motion. Thus, the total mechanical energy of the pendulum at any instant is

$$E_T = \frac{1}{2}mv^2 + mg(l - l\cos\theta). \qquad 8.9$$

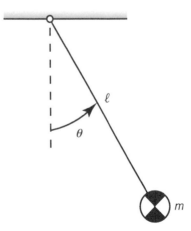

Figure 8.3 Diagram of a single pendulum. All the mass of the pendulum is concentrated at its tip, and the moment of inertia of the pendulum is neglected. The hinge about which the pendulum pivots is assumed to be frictionless.

As shown in figure 8.4, kinetic energy is maximum when the pendulum reaches its lowest point because its velocity is then maximum. Conversely, potential energy is minimum at the lowest point. Thus, kinetic energy is maximum when potential energy is minimum and vice versa. The fluctuations of kinetic and gravitational potential energy are exactly equal and opposite in phase during each cycle because energy losses due to friction are neglected here. Thus, the total mechanical energy, E_T, of this ideal pendulum remains constant for all time (see figure 8.4).

Unfortunately, the human body is not a conservative system. If it was, the total mechanical energy of the body would remain constant and no additional energy would be needed from the muscles to sustain movement. This is not the case because muscles consume metabolic energy during contraction and do mechanical work to move the joints. Fluctuations in kinetic and potential energy of the whole body usually do not cancel each other out during each cycle of movement, so additional energy—energy that must be supplied by the muscles—is needed to keep the body moving (see "In Focus: Is Mechanical Energy Conserved in Walking?").

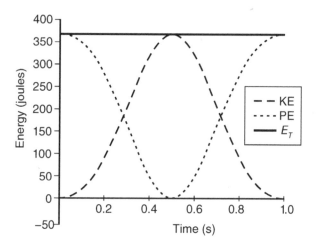

Figure 8.4 Kinetic energy (KE), gravitational potential energy (PE), and total mechanical energy (E_T) calculated for the single-pendulum model of figure 8.3.

IN FOCUS: IS MECHANICAL ENERGY CONSERVED IN WALKING?

The changes in kinetic and potential energy of the centre of mass that occur during locomotion can be calculated from the measured force exerted on the ground. Force-measuring devices called force plates are used for this purpose. A common type of force plate consists of a flat metal plate supported by four triaxial strain-gauge transducers. Within each of these transducers is a calibrated metal plate or beam that undergoes a very small deflection (called a strain). The deflection causes a change in the resistances of an electrical circuit, resulting in a change in voltage that is proportional to the force applied to the plate. These devices can measure force very accurately to within 1 or 2 N, depending on whether the force is applied statically (constant in time) or dynamically (changing in time).

The position and velocity of the centre of mass of the body can be found by integrating the ground force recorded from the force plate. For example, an equation that relates the vertical component of the ground force to the vertical acceleration of the centre of mass can be written as

$$F_{gy} = m\ddot{y}_{cm} + mg \qquad 8.10$$

where F_{gy} is the measured vertical ground force, $\ddot{y}_{cm}$ is the vertical acceleration of the centre of mass, m is the mass of the body, and g is the value of gravitational acceleration at the surface of the earth. Solving equation 8.10.1 for the vertical acceleration of the mass centre and then integrating gives

$$\dot{y}_{cm} = \int_0^{t_f} \ddot{y}_{cm}\, dt \qquad 8.10.1$$

and

$$y_{cm} = \int_0^{t_f} \dot{y}_{cm}\, dt \qquad 8.10.2$$

where $\ddot{y}_{cm} = (F_{gy} - mg)/m$ from equation 8.10.1; $\dot{y}_{cm}$ and y_{cm} are the vertical velocity and position of the centre of mass, respectively; and t_f defines the interval of time over which the integration is carried out. Equations 8.10.1 and 8.10.2 give the change in vertical velocity and vertical position of the centre of mass, respectively. Substituting these quantities into equation 8.6 then gives the change in total mechanical energy of the centre of mass over one cycle, where ω in equation 8.6 is taken to be zero because the centre of mass is a point (i.e., only rigid bodies can rotate and have angular velocities; points on bodies can only translate).

If the preceding analysis is carried out for walking at a normal speed, the fluctuations in kinetic and gravitational potential energy are found to be nearly equal in magnitude and opposite in phase (see figure 8.5 and compare with the results shown in figure 8.4). In figure 8.5, all energies were calculated from force plate data recorded from humans walking at self-selected speeds (i.e., at speeds of approximately 1.35 m/s). In walking, the centre of mass reaches its highest point at midstance; gravitational potential energy is therefore maximum at this point. Almost all of the kinetic energy in walking is due to the forward velocity of the body. The changes in vertical velocity are much smaller, so the fluctuations in vertical kinetic energy are negligible. The forward velocity of the body is maximum when the body is at its lowest point (when both legs are in contact with the ground) and it is least when the body is at its highest point (at midstance). Thus, kinetic energy is maximum during double support and minimum in midstance, precisely opposite to the pattern of changing gravitational potential energy (figure 8.5).

When the curves representing kinetic and potential energies in figure 8.5 are added numerically (which is possible because energy is a scalar quantity), the fluctuations in total mechanical energy of the centre of mass are seen to be relatively small (figure 8.5, E_T). This result indicates that the mechanical energy of the body is nearly conserved when people are free to choose the speed at which they walk. That is, at the speed at which metabolic energy consumption is minimised, there is almost a complete exchange of kinetic and gravitational potential energy, similar to what the ideal single-pendulum model of walking predicts.

Sources

Cavagna, G.A., Thys, H., & Zamboni, A. (1976). The sources of external work in level walking and running. *Journal of Physiology*, 262, 639-657.

Farley, C.T., & Ferris, D.P. (1998). Biomechanics of walking and running: Centre of mass movements to muscle action. *Exercise and Sport Sciences Reviews*, 26, 253-285.

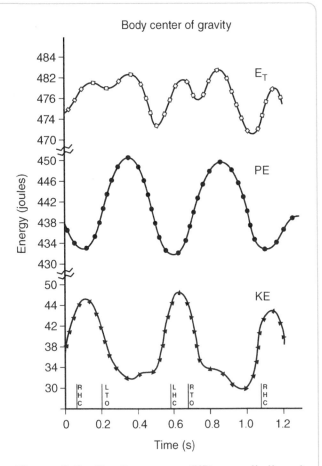

Figure 8.5 Kinetic energy (KE), gravitational potential energy (PE), and total mechanical energy (E_T) calculated for normal walking on level ground. RHC = right heel contact; RTO = right toe-off; LHC = left heel contact; LTO = left toe-off.

Reprinted, by permission, from D.A. Winter, 1979, "A new definition of mechanical work done in human movement," *Journal of Applied Physiology: Respiratory, Environmental, and Exercise Physiology* 46: 79-83.

POWER

Power is the rate of doing work. The work done by any external force $\underline{F}$ acting on a mass during a very small (infinitesimal) displacement $d\underline{r}$ is $\underline{F} \cdot d\underline{r}$. The power P, which is a scalar quantity, is found by dividing the work done by the interval of time, dt, during which the displacement takes place. Thus,

$$P = \frac{\underline{F} \cdot d\underline{r}}{dt} = \underline{F} \cdot \underline{v}, \qquad 8.11$$

where $\underline{v}$ is the velocity of the point on the rigid body at which the force is applied. The analogous relation for rotational motion is

$$P = \frac{\underline{M} \cdot d\theta}{dt} = \underline{M} \cdot \underline{\omega}, \qquad 8.12$$

where $\underline{M}$ is the net moment applied to the body and $\underline{\omega}$ is the angular velocity of the body. In SI units, power is expressed in **watts** (which is equivalent to Newton metres per second).

Power can be positive or negative depending on whether it is being transferred to or taken from the mass. A muscle produces energy when it contracts concentrically (i.e., shortens) against an external load; in this case the value of muscle power is taken to be positive. Conversely, muscles that contract eccentrically (i.e., lengthen) absorb energy, and the value of muscle power is then taken to be negative.

If muscle force and the rate of muscle shortening (or lengthening) are known, equation 8.11 can be used to estimate the power produced (or absorbed) by the muscle during contraction (see "In Focus: Modelling the Mechanics of Muscle Contraction"). Indeed, equations 8.11 and 8.12 are the basis on which detailed analyses have been performed to determine how muscles contribute power to the body segments in various tasks, including jumping, pedaling, and walking (see "In Focus: Which Muscles Are Most Important to Vertical Jumping Performance?").

IN FOCUS: MODELLING THE MECHANICS OF MUSCLE CONTRACTION

In the late 1930s, the famed muscle physiologist A.V. Hill postulated a conceptual model of muscle contraction. This theoretical model integrates three of the most important force-producing properties of muscle: the force–length and force–velocity properties and muscle's active state. In the late 1950s, another well-known scientist, Sir Andrew Huxley, proposed the first complete mechanistic model to explain how the actin and myosin filaments interact with each other in the development of muscle force. This theory has come to be known as the sliding filament theory of muscle contraction. Although Hill's model cannot explain the mechanisms by which force and energy are produced during a contraction, it yields much insight into the mechanophysiological relationships between length, velocity, activation, and force.

Figure 8.6 is a model of a musculotendon actuator that is often used in theoretical studies of movement. The musculotendon actuator is represented as a three-element muscle in series with tendon. The mechanical behaviour of muscle is described by a Hill-type contractile element (CE), which models the muscle's force–length and force–velocity properties;

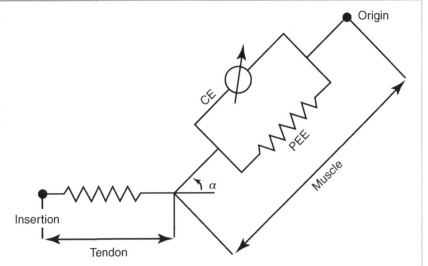

Figure 8.6 Schematic representation of a model of musculotendon actuation. Three components are used to model the mechanical behaviour of muscle: a CE, an SEE, and a PEE. Tendon is represented as a nonlinear spring.

Reprinted, by permission, from F.E. Zajac, 1989, Muscle and tendon: Properties, models, scaling, and application to biomechanics and motor control. In *Critical reviews in biomedical engineering*, edited by J.R. Bourne (Boca Raton, FL: CRC Press LLC), 367.

a series elastic element (SEE), which models muscle's active stiffness; and a parallel elastic element (PEE), which models muscle's passive stiffness. The active stiffness is thought to arise from the cross-bridges formed by actomyosin binding, so this property is present only when the muscle is activated. The passive stiffness resides in the material properties of the collagen molecules that are the building blocks of each muscle fibre. Because tendon (and, for that matter, ligament) comprises collagen as well, its force–length property is similar to that of passive muscle.

Passive muscle develops a force that is proportional to its stretch because the behaviour of the muscle is fully described by its force–length curve in the passive state. However, when muscle is activated, the force it develops depends on the instantaneous values of its length, velocity, and activation level. A first-order differential equation can be derived to describe the dynamics of the musculotendon actuator shown in figure 8.6 (Zajac, 1989):

$$\frac{d F^{MT}}{dt} = f\left[F^{MT}, l^{MT}, v^{MT}, a(t)\right]; 0 \le a(t) \le 1. \quad 8.13$$

Equation 8.13 indicates that the time rate of change of musculotendon force, $\frac{d F^{MT}}{dt}$, depends on musculotendon length, l^{MT}, musculotendon velocity, v^{MT}, muscle activation level, a, and musculotendon force, F^{MT}. If the values of musculotendon length, musculotendon velocity, muscle activation level, and musculotendon force are known at one time instant, equation 8.13 can be integrated to find the value of musculotendon force at the next time instant.

Indeed, if the trajectories of musculotendon length, musculotendon velocity, and muscle activation are known for all time during a movement and the value of musculotendon force is given at the initial state (i.e., at time $t = 0$), the time history of musculotendon force can be found by integrating equation 8.13 for the duration of the motor task. The problem of integrating a differential equation given a forcing function and a set of initial conditions is referred to as the forward-dynamics problem in biomechanics.

Sources

Pandy, M.G., Zajac, F.E., Sim, E., & Levine, W.S. (1990). An optimal control model for maximum-height human jumping. *Journal of Biomechanics, 23*, 1185-1198.

Pandy, M.G. (2001). Computer modeling and simulation of human movement. *Annual Review of Biomedical Engineering, 3*, 245-273.

Zajac, F.E., & Gordon, M.E. (1989). Determining muscle's force and action in multi-articular movement. *Exercise and Sport Sciences Reviews, 17*, 187-230.

Zajac, F.E. (1989). Muscle and tendon: Properties, models, scaling, and application to biomechanics and motor control. In J.R. Bourne (Ed.), *CRC critical reviews in biomedical engineering 19* (pp. 359-411). Boca Raton, FL: CRC Press.

IN FOCUS: WHICH MUSCLES ARE MOST IMPORTANT TO VERTICAL JUMPING PERFORMANCE?

Because muscle force cannot be measured noninvasively, computer models are often used to estimate muscle force in tasks such as walking, running, jumping, and cycling. The model of figure 8.7 was used to simulate maximum-height jumping in humans. The skeleton was represented as a planar linkage with four segments and four degrees of freedom that was joined to the ground at the toes and articulated at the ankle, knee, and hip by frictionless hinge joints. The model skeleton was actuated by eight lower-extremity musculotendinous units; each unit was modelled as a three-element muscle in series with tendon (see "In Focus: Modelling the Mechanics of Muscle Contraction").

A mathematical theory called optimisation was used to simulate the biomechanics of maximum-height squat jumping. In this task, the body begins from a static squatting position with the hip, knee, and ankle angles all flexed to 90°. Comparison of the model results with kinematic, force plate, and muscle electromyographic data obtained from experiment showed that the model reproduced the major features of the ground-contact phase of the jump. The simulation results were then analysed to determine how muscles accelerate and contribute power to the body segments during jumping. Specifically, equations similar to equations 8.11 and 8.12 were used to calculate the power developed by the muscles and the amount of energy subsequently transferred to the skeleton during the ground-contact phase of a maximum-height squat jump.

The vasti and gluteus maximus were found to be the major energy producers—the prime movers—of the lower limb in vertical jumping. These muscles contributed most significantly to the total energy made available for propulsion (note the area under

(continued)

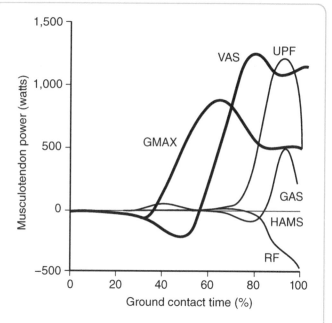

Figure 8.8 Mechanical power generated by the musculotendon actuators in the model of figure 8.7 during the ground-contact phase of a maximum-height squat jump.

Reprinted from *Journal of Biomechanics*, Vol. 24, M.G. Pandy and R.E. Zajac, "Optimal muscular coordination strategies for jumping," pgs. 1-10, copyright 1991, with permission from Elsevier.

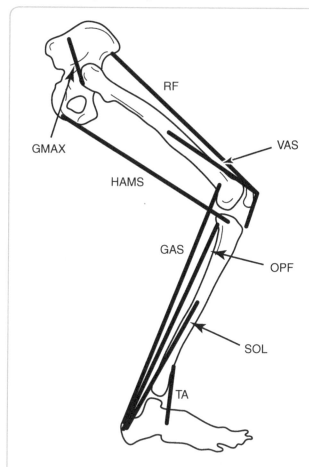

Figure 8.7 Diagram showing a musculoskeletal model of the body that was used to simulate a maximum-height squat jump.

Reprinted from *Journal of Biomechanics*, Vol. 23, M.G. Pandy et al., "An optimal control model for maximum-height human jumping," pgs.1185-1198, copyright 1990, with permission from Elsevier.

each graph in figure 8.8). However, in the final 20% of ground-contact time, just before liftoff, the ankle plantarflexors, soleus and gastrocnemius, also contributed significantly to the total energy delivered to the skeleton.

Figure 8.9 is a plot of the total instantaneous power delivered to each body segment during the jump. The total area under each curve is equal to the total mechanical energy (kinetic plus gravitational potential energy) of each body segment at the instant the body leaves the ground. A large proportion of the total energy developed by the muscles was delivered to the trunk segment. In fact, the combined energy of the thigh, shank, and foot amounted to only 30% of the total energy available at liftoff. The remainder (approximately 70%) of the input muscle energy

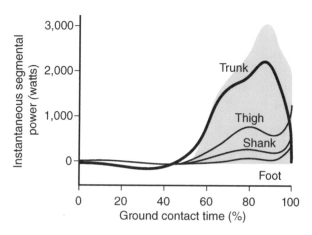

Figure 8.9 Instantaneous power delivered to each body segment in the model during the ground-contact phase of the simulated squat jump.

Reprinted from *Journal of Biomechanics*, Vol. 24, M.G. Pandy and R.E. Zajac, "Optimal muscular coordination strategies for jumping," pgs. 1-10, copyright 1991, with permission from Elsevier.

was transferred to the trunk. This is not a surprising result given that the trunk segment represents approximately 70% of the total body mass.

The model simulation results also showed that muscles dominate the instantaneous power of the trunk segment for most of the ground-contact phase of the jump (figure 8.10, compare muscle with total, which is represented by the shaded region). Only near liftoff do centrifugal forces (i.e., forces arising from motion of the joints) become so important that they dominate the power delivered to the trunk (figure 8.10, centrifugal). Centrifugal forces are large only near liftoff because the velocities of the joints increase greatly just before the body leaves the ground. According to the model calculations, gravitational forces contribute little to trunk energy in maximum-height jumping.

Figure 8.11 shows the relative contributions of individual leg muscles to trunk power during the jump. The shaded region is the total power delivered to the trunk by all the muscles in the model. The dashed line is the contribution of all the biarticular muscles (muscles that span two joints) in the model (hamstrings and rectus femoris). The area under each curve represents the total energy delivered by that muscle (or group of muscles) to the trunk segment at the instant the body leaves the ground.

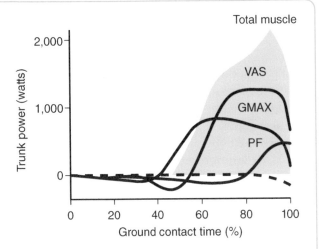

Figure 8.11 Contribution of each muscle to the total power delivered to the trunk segment during the ground-contact phase of the simulated squat jump. VAS = the contribution of the vasti; GMAX = the contribution of the gluteus maximus; and PF = the contribution of all the ankle plantarflexors (soleus [SOL], gastrocnemius [GAS], and other plantarflexors [OPF]) in the model (see figure 8.7).

Reprinted from *Journal of Biomechanics*, Vol. 24, M.G. Pandy and R.E. Zajac, "Optimal muscular coordination strategies for jumping," pgs. 1-10, copyright 1991, with permission from Elsevier.

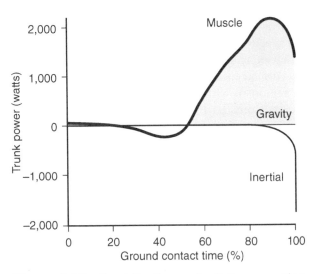

Figure 8.10 Contributions of all the muscles (muscle), gravitational forces (gravity), and inertial forces (inertial) to the instantaneous power of the trunk for the ground-contact phase of the jump.

Reprinted from *Journal of Biomechanics*, Vol. 24, M.G. Pandy and R.E. Zajac, "Optimal muscular coordination strategies for jumping," pgs. 1-10, copyright 1991, with permission from Elsevier.

Of all the muscles, the vasti and gluteus maximus contributed most significantly to the power delivered to the trunk (figure 8.11). The energy delivered by these muscles amounted to nearly 90% of the total energy delivered to the trunk during the jump. Thus, the vasti and gluteus maximus appear to be the most important muscles for maximum-height jumping. The ankle plantarflexors, soleus and gastrocnemius, are also important, but only in the last 20% of ground contact. Near liftoff, the ankle plantarflexors deliver as much power to the trunk as either the vasti or the gluteus maximus. Finally, the model simulation results also suggest that the biarticular muscles are relatively unimportant to overall jumping performance.

Sources

Pandy, M.G., & Zajac, F.E. (1991). Optimal muscular coordination strategies for jumping. *Journal of Biomechanics, 24*, 1-10.

Zajac, F.E. (2002). Understanding muscle coordination of the human leg with dynamical simulations. *Journal of Biomechanics, 35*, 1011-1018.

ELASTIC STRAIN ENERGY

Potential energy may be stored in an elastic form rather than a gravitational form. In explosive movements such as running and jumping, considerable amounts of strain energy may be stored in the elastic tissues of muscle and tendon. If some of this stored elastic energy can be returned to the skeleton, less metabolic energy will be needed to keep the body moving. Storage and utilisation of elastic strain energy may therefore reduce the amount of metabolic energy consumed by the muscles.

As explained in "In Focus: Is Mechanical Energy Conserved in Walking?," calculations based on force-plate measurements of human walking show that changes in forward kinetic energy are nearly perfectly out of phase with changes in gravitational potential energy, so that the total mechanical energy of the centre of mass is kept nearly constant during a step. This is because the centre of mass is highest in midstance, when the forward kinetic energy is least, and is lowest in double support, when the forward kinetic energy is greatest. The opposite is true for running, where changes in forward kinetic and gravitational potential energy are substantially in phase, leading to large changes in the total mechanical energy of the mass centre during each step. Thus, in running, the centre of mass is lowest in midstance, when the forward kinetic energy is least, and is highest during the flight phase, when the forward kinetic energy is greatest.

Two very different mechanisms explain the difference in mechanical energy between walking and running. The almost complete transfer of potential and kinetic energies in walking means that not all the energy required to lift and accelerate the centre of mass has to come from the muscles; storage and utilisation of gravitational potential energy allow the muscles to do less work, as is the case in the pendulum. In running, however, an exchange of gravitational and forward kinetic energy is not possible because the centre of mass goes through an increase in both height and speed at the same time during a step. The large fluctuations in total mechanical energy during each cycle suggest that the pendulum model of walking does not apply for running. Nonetheless, metabolic energy is still saved—not by the exchange of gravitational potential and kinetic energy, but by storage and utilisation of elastic strain energy.

Although the amount of elastic energy stored in muscle and tendon cannot be calculated precisely from force-plate measurements, it can be estimated approximately by knowing the efficiency with which muscles convert chemical energy into mechanical work. The kangaroo is perhaps the best example of an animal that is able to exploit the benefits of elastic energy storage. A 40 kg kangaroo has an Achilles tendon measuring approximately 1.5 cm in diameter and 35 cm in length. It is likely, then, that substantial quantities of elastic energy are stored in the tendons of these animals after impact with the ground. Indeed, when a kangaroo hops at a speed of 30 km/h, it has been estimated that the contractile machinery supplies only one third of the energy required to lift and accelerate the centre of mass while the legs are on the ground. The remaining two thirds of the energy required for each hop is thought to come from strain energy stored in the animal's tendons. Although this fraction is likely to be much smaller in a running man, some estimates put it as high as one half; that is, one half of the mechanical energy needed to lift and accelerate the centre of mass during ground contact, plus the energy needed to move the limbs, is supplied by strain energy stored in the elastic tissues of muscle and tendon.

More quantitative estimates of elastic energy storage require knowledge of the mechanical properties of muscle and tendon and of the forces they exert. Although all biological tissues exhibit nonlinear force–extension curves, there is always a region in which extension varies linearly with force. Consider the elastic behaviour of tendon as represented by a simple linear spring. Let the stiffness of the tendon (spring) be given by k^T and the amount of stretch by Δl^T. The elastic strain energy stored in the tendon spring is then

$$U_S = \frac{1}{2} k^T (\Delta l^T)^2. \qquad 8.14$$

This equation has been used to estimate the amount of strain energy stored in the elastic tissues of muscle and tendon when humans jump to their maximum height (see "In Focus: Does Storage and Utilisation of Elastic Strain Energy Increase Vertical Jumping Performance?").

Several experiments have shown that humans jump higher when propulsion is preceded by a preparatory countermovement (i.e., downward movement of the centre of mass). One hypothesis given to explain this result is that lengthening of the extensor muscles during a countermovement leads to an increase in the amount of energy stored in the elastic tissues, which in turn increases the energy delivered to the skeleton during the propulsion phase. To examine whether storage and utilisation of elastic strain energy increases jumping performance and to understand how the elastic tissues enhance muscle performance, it is necessary to determine the amount of energy contributed by the elastic tissues during the ground-contact phase of jumping. This requires knowledge of the forces developed by the muscles and the length changes associated with the various elastic tissues.

The computer model of figure 8.7 was used to simulate two types of maximum vertical jumps: a squat jump (SJ), which is a maximum-height jump that begins from a static squatting position, and a counter-movement jump (CMJ), which begins from a standing position and involves significant downward motion of the centre of mass before upward propulsion. The biomechanics of each jump were simulated using opti-misation theory (see "In Focus: Which Muscles are Most Important to Vertical Jumping Performance?"). For both the SJ and the CMJ, the optimisation solution predicted the pattern of muscle activations, the motion of the body segments, and ground-reaction forces needed to produce a maximum-height jump. A comparison of the model's response with kinematic, ground-reaction force, and muscle electromyographic data collected in the laboratory showed that the model predictions were closely similar to the experimental results obtained for the SJ and the CMJ.

The model simulations gave the time histories of the leg muscle forces and the associated length changes of the muscles and tendons during the ground-contact phase of each jump. The amount of strain energy stored in the elastic tissues of muscle and tendon and the amount of work done by the contractile machinery were calculated based on the time histories of the leg muscle forces. Equation 8.14 was then used to calculate the elastic energy stored

in muscle and tendon. Finally, the work done by the contractile elements was found by integrating the power developed (or absorbed) by each leg muscle during the ground-contact phase of the jump.

Surprisingly, the model results showed that the elastic tissues delivered nearly the same amount of energy to the skeleton during an SJ and a CMJ (figure 8.12). The reason is that nearly the same amount of

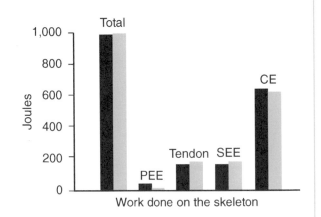

Figure 8.12 Work done on the skeleton by the leg muscles of the model shown in figure 8.6. The black bars are the results obtained for a maximum-height countermovement jump, and the gray bars are corresponding results for a squat jump. Total = the work done by all the musculotendon actuators in the model; CE = the work done by all the contractile elements that model muscle's force–length and force–velocity properties; PEE = the work done by all the parallel elastic elements that model the passive properties of muscle; SEE = the work done by all the series elastic elements that model the active stiffness properties of muscle; and Tendon = the work done by all the tendons in the model. The total work done during the countermovement jump is practically the same as that done during the squat jump. In both jumps, the CE contributed 65% of the total energy delivered to the skeleton, whereas the elastic tissues (tendon plus SEE) accounted for 35%. This shows that elastic energy storage is significant in both jumps.

Reprinted from *Journal of Biomechanics*, Vol. 26, F.C. Anderson and M.G. Pandy, "Storage and utilization of elas-tic strain energy during jumping," pgs. 1413-1427, copyright 1993, with permission from Elsevier.

(continued)

115

elastic strain energy was stored in the SJ as in the CMJ. Only actuators with relatively long tendons are able to store significant amounts of elastic energy when stretched. If there is no difference in the forces exerted by the lower-limb muscles during the SJ and the CMJ, then the amount of elastic energy stored in the two jumps should also be the same. This is precisely what the model predicted.

Nearly the same amount of elastic strain energy was stored during the SJ and the CMJ because the forces developed by those muscles that had relatively long tendons in the model were practically the same in both jumps. Given that humans jump slightly higher during a CMJ than during an SJ, the model results suggest that the difference in performance between countermovement and squat jumps cannot be explained by storage and utilisation of elastic strain energy. Instead, it appears that countermovements are performed to increase the ground-contact time of the jump. Increasing ground-contact time increases the amount of time that the muscles are permitted to shorten and do positive work on the skeleton, and increasing the work done on the skeleton means an increase in jump height.

Sources

Anderson, F.C., & Pandy, M.G. (1993). Storage and utilization of elastic strain energy during jumping. *Journal of Biomechanics, 26,* 1413-1427.

Pandy, M.G., Zajac, F.E., Sim, E., & Levine, W.S. (1990). An optimal control model for maximum-height human jumping. *Journal of Biomechanics, 23,* 1185-1198.

METABOLIC ENERGY CONSUMPTION

When a muscle is stimulated, heat is liberated, and the amount of heat produced can be measured by the temperature change in the muscle. If there is also a change in length during a contraction, then mechanical work is done by the muscle. According to the **first law of thermodynamics**, the total rate of energy production during muscle contraction, $\dot{E}$, is equal to the rate at which heat is liberated, $\dot{H}$, plus the rate at which work is done to move the external load, $\dot{W}$. Thus,

$$\dot{E} = \dot{H} + \dot{W}. \qquad 8.15$$

The rate at which mechanical work is done is equal to the power developed by the muscle. Power, in turn, is given by the force exerted by the muscle multiplied by its shortening (or lengthening) velocity (see equation 8.11). The rate of heat production is more difficult to estimate. This quantity depends on a number of factors related to the mechanics and physiology of muscle contraction, including muscle length, mass, contraction velocity, and activation level; muscle fibre recruitment rate; and muscle fibre type. At least qualitatively, though, the rate at which heat is produced during a contraction can be estimated from four quantities: activation heat, maintenance heat, shortening heat, and resting heat.

When muscle is excited, a relatively large amount of heat is produced early in the contraction. This activation heat reflects the energetics of calcium release and reaccumulation, and it can account for as much as 25% to 30% of the total energy consumed during a contraction.

The continued heat production observed once steady state has been reached is called **maintenance heat**. This portion of the heat rate is thought to represent the chemomechanical events associated with steady-state actomyosin interaction, steady turnover of calcium, and reuptake by the sarcoplasmic reticulum. The maintenance heat accounts for the largest fraction of the total heat liberated during contraction, and its magnitude is a strong function of muscle's force–length property.

Shortening heat was first discovered by A.V. Hill in the middle to late 1930s. Hill defined this quantity as the difference between the heat liberated by a muscle when it shortens and the heat liberated by the same muscle when it contracts isometrically. The shortening muscle liberates more energy because it does external work to move a load and because it liberates more heat. From his early experiments on frog muscle, Hill concluded that the amount of shortening heat was directly proportional to the distance the muscle shortened. Shortening heat represents only a small (often insignificant) fraction of the total energy consumed during contraction.

All muscles produce a little heat as a consequence of being alive. **Resting heat** is the heat released as chemical energy, derived from oxidation of foodstuffs, consumed in the resting state. The rate of heat production in the resting state (i.e., sitting quietly or lying down) has been found from experiment to be about 1.04 J/(kg × s). Interestingly, the cost associated with quiet standing is roughly 30% higher: approximately 1.5 J/(kg × s).

As soon as an individual begins to walk, a great increase in energy expenditure occurs, reflecting the metabolic cost to the muscles of moving the body against gravity and of accelerating and decelerating the limbs. A large number of experimental studies have shown that metabolic energy expenditure increases with walking speed in the manner indicated in figure 8.13. The relationship is expressed fairly well by a parabolic equation of the form

$$\dot{E} = 32.0 + 0.005 v^2, \qquad 8.16$$

where $\dot{E}$ is the rate of metabolic energy consumption, expressed in J/(kg × s), and v is walking speed. The constants in equation 8.16 were determined on the basis of data obtained from nearly 100 adults (both males and females) who walked on level ground at a frequency and step length of their choosing.

Dividing equation 8.16 by walking speed, v, gives the rate at which metabolic energy is consumed per unit distance travelled:

$$\dot{E}_d = \frac{32.0}{v} + 0.005\, v, \qquad 8.17$$

where $\dot{E}_d$ is expressed in J/(kg × m). Whereas $\dot{E}$ increases monotonically with walking speed, $\dot{E}$ has a well-defined minimum at about 1.3 m/s (figure 8.13). As shown in figure 8.13, the minimum in metabolic energy expenditure predicted by equation 8.17 coincides with that obtained from oxygen consumption measurements made on people.

Energy consumption rate increases as a curvilinear function of speed for walking, but it becomes a straight-line function of speed for running (figure 8.14). At a given speed, the rate of energy consumption increases as the slope of the ground increases and decreases as the slope decreases, reaching a minimum at a gradient of –10% (figure 8.14). For gradients steeper than –10%, energy consumption increases again because of the postural changes necessary for continued locomotion and the significant braking action needed from the muscles.

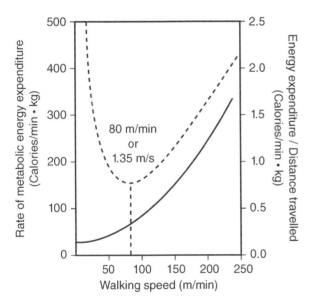

Figure 8.13 Metabolic energy expenditure plotted as a function of walking speed. The rate of metabolic energy consumption increases parabolically with walking speed (solid line). When the rate of metabolic energy consumption is normalised by the distance travelled, an optimal walking speed is predicted at roughly 80 m/min or 1.35 m/s (dashed line).

Reprinted, by permission, from H.J. Ralston, 1976, Energetics of human walking. In *Neural control of locomotion*, edited by R.M. Herman et al. (Heidelberg, Germany: Springer).

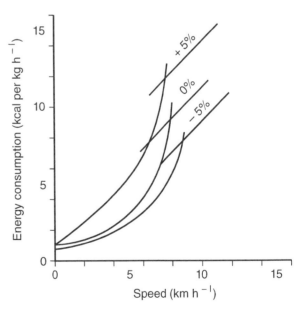

Figure 8.14 Rate of metabolic energy consumption plotted against speed for walking and running on level ground (0%), on a 5% incline (5%), and on a 5% decline (–5%).

Reprinted, by permission, from R. Margaria et al., 1988, "Energy cost of running," *Journal of Applied Physiology* 18(2): 367-370.

EFFICIENCY OF MOVEMENT

Muscles create and absorb mechanical energy by shortening and lengthening during flexion and extension of the joints. When mechanical energy is absorbed, muscles are said to do negative work; this is the work done by a muscle when it is developing an active force at the same time as it is being lengthened by some externally applied force. If the muscle is shortening as it develops a force, it is said to do positive work. The efficiency with which a muscle operates under these conditions can be defined as

$$\text{efficiency} = \frac{\text{mechanical work done}}{\text{metabolic energy consumed}}, \quad 8.18$$

where the mechanical work done on the muscle is regarded as negative work and that done by the muscle is positive work.

The efficiencies of walking and running on negatively and positively sloped surfaces have been calculated and are plotted in figure 8.15. The lines emanating from the zero point on the graph represent constant efficiency of transport. Walking up steeper and steeper inclines requires more and more energy for the same distance travelled. Notice that the curve for walking up inclines approaches the limit of 25% efficiency, which represents the maximum efficiency with which a muscle can convert chemical energy into mechanical work that is needed to move the centre of mass. For walking down inclines, however, the limit of efficiency is given to be –120%. This negative value of efficiency is explained by the fact that the leg muscles do mostly negative work (i.e., absorb mechanical energy) when walking down a steep incline.

Notice that the curve for running lies above the curve for walking at the same speed. In walking, metabolic energy is saved by converting gravitational potential energy into forward kinetic energy, as explained by the single pendulum model. Although some amount of elastic strain energy is stored and reutilised in each cycle of running, the results of figure 8.15 suggest that this mechanism is less efficient than the pendulum-like mechanism present in walking. That is, in relation to the cost of transport, storage and utilisation of elastic energy in running is not as efficient as the conversion of gravitational potential energy into forward kinetic energy in walking.

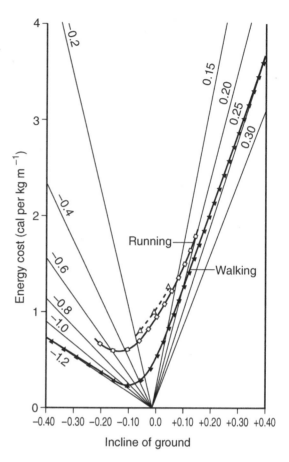

Figure 8.15 Metabolic energy consumed per unit distance moved plotted against inclination of the ground. The data were obtained from humans walking and running on a treadmill with varied orientation to the ground.

Reprinted, by permission, from R. Margaria et al., 1988, "Energy cost of running," *Journal of Applied Physiology* 18(2): 367-370.

SUMMARY

Muscles hydrolyse high-energy phosphate molecules (adenosine triphosphate) to produce chemical energy and to do mechanical work. Roughly 75% of the total metabolic energy consumed during activity is dissipated as heat; the remaining 25% is used to perform useful mechanical work (i.e., to support the body and to swing the limbs back and forth during locomotion). Mechanical energy is visible in two forms: kinetic energy and potential energy. Kinetic energy is the energy the body possesses due to its motion; it is a function of the rate of change of position (i.e., velocity) of the body at each instant. Potential energy may be stored in its gravitational form or as strain energy in the elastic tissues of muscle and tendon. Storage and utilisation of elastic strain energy can increase the efficiency with which muscles convert chemical energy to mechanical work.

FURTHER READING

Bedford, A., & Fowler, W. (1995). *Engineering mechanics: Dynamics*. Reading, MA: Addison-Wesley.

Margaria, R. (1976). *Biomechanics and energetics of muscular exercise*. Oxford: Clarendon.

McMahon, T.A. (1984). *Muscles, reflexes, and locomotion*. Princeton, NJ: Princeton University Press.

McNeill Alexander, R., & Goldspink, G. (1977). *Mechanics and energetics of animal locomotion*. London: Chapman & Hall.

Rose, J., & Gamble, J.G. (1994). *Human walking* (2nd ed.). Baltimore: Williams & Wilkins.

Winter, D.A. (1990). *Biomechanics and motor control of human movement*. New York: Wiley.

CHAPTER 9

BIOMECHANICS ACROSS THE LIFE SPAN

The major learning concepts in this chapter relate to

▸ the kinematics and kinetics of normal walking,

▸ major differences in the gait patterns of children and healthy young adults, and

▸ major differences in the gait patterns of elderly persons and healthy young adults.

As life expectancy continues to increase, so does the population of elderly adults. Maintaining independent function is an important goal for older adults because these persons are prone to fall-related injuries. Independence in activities such as shopping, using public transportation, and visiting friends requires adequate muscle strength, adequate flexibility (i.e., joint ranges of motion), and adequate neuromotor performance (i.e., reaction times and proprioception). The incidence of falling is known to significantly increase when musculoskeletal performance is degraded by ageing.

This chapter describes how the natural ageing process affects biomechanical aspects of the performance of motor skills. Because the ability to walk is critical to maintaining an independent lifestyle, the discussion is centred on the effects of ageing on the biomechanics of gait. Less is known about the effects of growth and development on movement in the very young. Some studies have investigated how growth and maturation affect gait mechanics in young children, and the results of these studies are summarised here.

BIOMECHANICS OF NORMAL GAIT

In order to appreciate the changes that occur in walking across the life span, it is necessary to first understand the biomechanics of gait in healthy young adults. The gait cycle is usually defined as the time between successive heel strikes of the same leg. Because normal walking is very nearly symmetrical, one half of the gait cycle is defined from heel strike of one leg to heel strike of the other (figure 9.1). The normal gait cycle is usually divided into two distinct phases: stance and swing. The stance phase, which is the time from heel strike to toe-off of one leg, can be divided into five subphases: initial contact, loading response, midstance, terminal stance, and preswing. The swing phase is the portion of the cycle from toe-off to heel strike of the same leg, and it is usually divided into three subphases: initial swing, midswing, and terminal swing (see figure 9.1). The stance phase can also be divided into periods of single and double support. There are two periods of single and double support in each gait cycle; each double-support phase occupies 12% of the cycle time and single support occupies 38%. Each leg is in contact with the ground for 62% of the gait cycle (stance) and spends 38% of the time in the air (swing).

Much of what is currently known about the biomechanics of normal walking is based on the results of gait-analysis experiments performed over the past 100 yr. The quantities recorded in these experiments are the relative movements of the joints, the pattern of force exerted on the ground, and the sequence and timing of leg-muscle activity. The relative movements of the joints are usually obtained using high-speed video cameras that track the three-dimensional positions of reflective markers mounted on the body parts. Ground forces are recorded using force plates that measure the three components of force (anterior–posterior, vertical, and medial–lateral) exerted by the leg on the ground, and muscle activations are usually monitored noninvasively by attaching electromyography (EMG) electrodes to the surface of the skin.

KINEMATICS OF NORMAL GAIT

Kinematic measurements have shown that the centre of mass describes a smooth, sinusoidal path when projected on the sagittal plane (figure 9.2). The total amount of vertical displacement is 5 cm when healthy young adults walk at their preferred speeds. The peaks of the vertical oscillations occur at midstance of the supporting limb, and the troughs occur at roughly the middle of the double-support phase. The curve is smooth and fluctuates regularly between its peaks and troughs. Interestingly, at midstance, when the vertical displacement of the centre of mass is greatest, the top of the head is actually lower than the standing height of the person. Thus, people are slightly shorter when they walk than when they stand fully erect. This explains why one is able to walk freely through a tunnel that is exactly the same height as one's standing height without the fear of bumping one's head.

The centre of mass is also displaced laterally in the transverse plane (i.e., in a direction perpendicular to the sagittal plane; figure 9.2). The amplitude of the path traced in the transverse plane is also about 5 cm, but its frequency is only one half of the

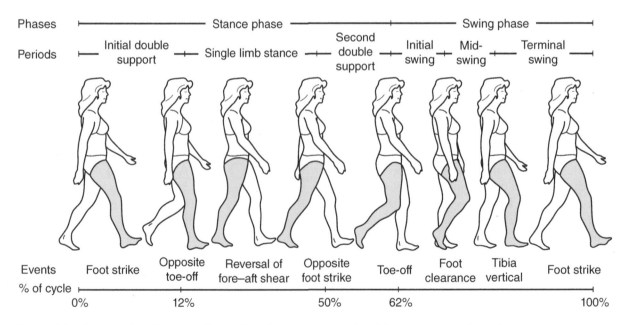

Figure 9.1 Schematic diagram illustrating the major events of the normal gait cycle.

Based on Vogler and Bove, 1985.

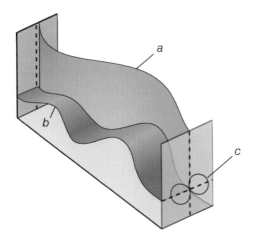

Figure 9.2 Schematic diagram showing the displacements of the centre of mass of the body in *(a)* the transverse plane, *(b)* the sagittal plane, and *(c)* the frontal plane.

Reprinted, by permission, from J. Rose and J.G. Gamble, 1994, *Human walking*, 2nd ed. (Philadelphia, PA: Lippincott, Williams, and Wilkins), 5.

frequency of the vertical oscillations. When viewed from the back, the centre of mass is seen to move up and down and to swing from side to side during each cycle. Indeed, when the curves describing the vertical and lateral displacements are projected onto the frontal plane, the resulting curve resembles the number 8 lying on its side (figure 9.2).

The upper body moves forward throughout the gait cycle, but its speed does not remain constant. It moves fastest during the double-support phase and slowest when the supporting leg is near midstance. The trunk twists about its long axis as the shoulder and pelvis rotate in opposite directions. The arms also swing in opposition to the movements of the legs; thus, the right arm and right side of the shoulder move forward in unison with the left leg and left side of the pelvis. The total displacements of the shoulder and pelvis are relatively small, measuring 7° and 12°, respectively, during the course of a full cycle. Similar to the displacement of the centre of mass, the trunk rises and falls twice and moves from side to side once during each cycle. In addition to twisting about its long axis in the transverse plane, the pelvis tips slightly backward and forward (in the sagittal plane) and from side to side (in the frontal plane).

The hip flexes and extends once during the gait cycle (figure 9.3, hip). The limit of hip flexion is reached near midswing, after which it remains flexed to this position until initial contact. Peak hip extension is reached before the end of the stance phase (initial swing).

The knee shows two flexion and extension peaks during each gait cycle (figure 9.3, knee). Flexion–extension movements of the knee remain relatively small during the stance phase compared with during the swing phase. The knee is nearly fully extended at initial contact, and it flexes slightly (to no more than 20°) during loading response. It extends again during midstance and terminal stance, so much so that it is nearly fully extended just before contralateral heel strike. Knee flexion increases rapidly during preswing and continues well into initial swing, reaching a peak of nearly 60°. The knee then extends rapidly during midswing and terminal swing and becomes nearly fully extended once again in preparation for initial contact.

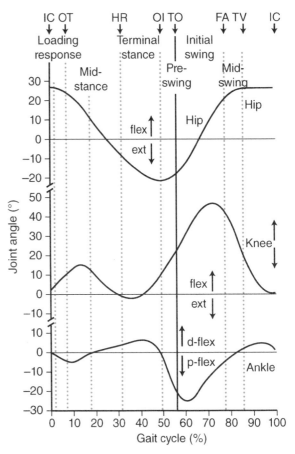

Figure 9.3 Mean sagittal-plane joint angles measured in healthy young adults walking at their preferred speeds. Data are plotted for the right hip (flexion positive), right knee (flexion positive), and right ankle (dorsiflexion positive). The gait-cycle events are shown at the top of the figure. IC = initial contact; OT = toe-off of contralateral leg; HR = heel rise; OI = initial contact of contralateral leg; TO = toe-off; FA = feet adjacent; TV = tibia vertical; d-flex = dorsiflexion; p-flex = plantarflexion.

Reproduced from *Gait Analysis: An Introduction*, 2nd ed., M.W. Whittle, editor, copyright 1996, with permission from Elsevier.

The ankle is very close to the neutral position (i.e., its position during normal standing) at the time of initial contact, after which it plantarflexes rapidly to bring the foot flat on the ground (figure 9.3, ankle). During midstance, the tibia rotates forward relative to the foot, moving the ankle into a dorsiflexed position. Just before contralateral heel strike, the ankle plantarflexes rapidly, which serves to the lift the centre of mass into the air and smooth the transition from single to double support. During the swing phase, the ankle dorsiflexes in order to provide toe clearance for the freely swinging leg. Once the toes have cleared the ground, the ankle maintains a roughly neutral position in preparation for initial contact.

Although the major movements of the lower-limb joints occur in the sagittal plane, joint displacements in the transverse and frontal planes are also important for producing the sinusoidal displacement of the centre of mass shown in figure 9.2. Indeed, three of the six major determinants of gait are associated with movements of the bones in the transverse and

IN FOCUS: THE MAJOR DETERMINANTS OF NORMAL GAIT

About 60 yr ago, a group of scientists at the University of California at Berkeley in the United States proposed six kinematic mechanisms to explain how movements of the lower-limb joints contribute to the pathway of the centre of mass when humans walk at their preferred speeds. These six mechanisms, termed the major determinants of gait, are hip flexion, stance-knee flexion, ankle plantarflexion, pelvic rotation, pelvic list, and lateral pelvic displacement (Saunders et al., 1953).

The simplest form of walking is described by a **compass gait** in which all joints, except the hips, are locked in the anatomical position. Flexing and extending the hips means that the centre of mass moves on the arcs of circles, the radii of which correspond to the lengths of the legs. To produce reasonable step lengths, the centre of mass must rise and fall much more in the compass gait than it does in normal walking. Apart from the vertical displacement of the centre of mass being too large, the other main objection to a compass gait is the discontinuity that it produces in the path of the centre of mass during the transition from single- to double-leg stance (see figure 9.4).

The second determinant permits bending of the knees in the sagittal plane. At normal speeds of walking,

the amplitude of knee flexion during stance is roughly 20°. Stance knee flexion flattens the arcs defined by compass gait, as does the third determinant, ankle plantarflexion. However, ankle plantarflexion also eases the transition from double- to single-leg stance by producing a smoother trajectory of the centre of mass in the vicinity of the two intersecting compass-gait arcs. In normal walking, the ankle plantarflexes by more than 30° during terminal stance.

Two of the remaining three determinants, pelvic rotation and pelvic list, further flatten and smooth the arcs traced by the centre of mass. Pelvic rotation

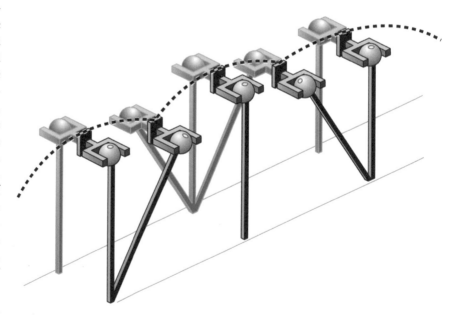

Figure 9.4 Diagram depicting the compass gait in which only hip flexion and extension is present. The path of the centre of mass is a series of arcs.

Reprinted, by permission, from J.B. Saunders, V.T. Inman, and H.D. Eberhart, 1953, "The major determinants in normal and pathological gait," *Journal of Bone and Joint Surgery* 35-A: 543. http://jbjs.org

permits movement of the pelvis in the transverse plane, which allows for an increase in step length; the amplitude of transverse pelvic rotation is about 8° on either side of the central axis of the body. The fifth determinant, pelvic list, is the downward movement of the pelvis in the frontal plane, brought about by abducting the hip. Transverse pelvic rotation raises the points of intersection between two compass-gait arcs, whereas pelvic list further flattens the trajectory by lowering the centre of mass as the pelvis tilts from the weight-bearing leg to the leg that is about to enter swing. The sixth determinant, lateral pelvic displacement, is concerned with movement of the centre of mass in the transverse plane. As support is alternated from one leg to the other, the centre of mass is displaced from side to side. Shifting the centre of mass from side to side is made possible by transverse rotation of the tibia about the subtalar joint.

Source

Saunders, J.B., Inman, V.T., & Eberhart, H.D. (1953). The major determinants in normal and pathological gait. *Journal of Bone and Joint Surgery, 35-A,* 543-558.

frontal planes (see "In Focus: The Major Determinants of Normal Gait"). The pelvis rotates as much as 20° in the transverse plane and lists up to 10° in the frontal plane. The tibia also rotates about its long axis, courtesy of the subtalar joint at the ankle. Pelvic rotation and pelvic list reduce the vertical displacement of the centre of mass and therefore lower the metabolic cost of transport in each step. Ankle inversion–eversion also keeps the metabolic cost of walking low by allowing the centre of mass to pass closer to the hip of the supporting leg in midstance.

MUSCLE ACTIONS DURING NORMAL GAIT

Kinematic and ground-force measurements may be used to calculate the net muscle moments exerted about the joints in gait. This procedure is known as the inverse dynamics method. The pattern of muscle moments for normal walking is illustrated in figure 9.5. Shortly after initial contact, the ankle dorsiflexors exert a small moment to decelerate the foot and prevent it from slapping the ground (figure 9.5, ankle; 0-5% of the gait cycle). Once the foot is placed flat on the ground, the quadriceps contract eccentrically and exert an extension moment at the knee, which peaks just after the contralateral leg leaves the ground (figure 9.5, knee; 5%-20% of the gait cycle). Thus, the main function of the quadriceps in early single-leg stance is to limit stance knee flexion (i.e., accelerate the knee into extension).

Knee extension in early stance (loading response) is also brought about by the action of the ankle plantarflexors. The soleus muscle exerts a plantarflexor moment at the ankle that restrains forward rotation of the tibia and simultaneously extends the knee.

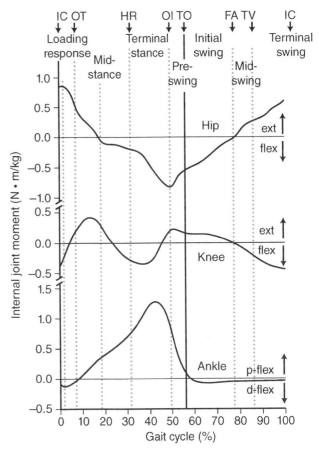

Figure 9.5 Mean sagittal plane net muscle moments applied at the lower-limb joints during one cycle of normal walking. Data are shown for the right hip (extensor moment positive), right knee (extensor moment positive), and right ankle (dorsiflexor moment positive). The gait-cycle events are shown at the top of the figure. IC = initial contact; OT = toe-off of contralateral leg; HR = heel rise; OI = initial contact of contralateral leg; TO = toe-off; FA = feet adjacent; TV = tibia vertical; d-flex = dorsiflexion; p-flex = plantarflexion.

This plantarflexor moment increases through midstance and peaks during terminal stance, just before the contralateral leg makes contact with the ground (figure 9.5, ankle; 10%-50% of the gait cycle). The peak moment created by the ankle plantarflexors in terminal stance is often referred to as push-off. Subsequent to contralateral heel strike, the plantarflexors (with assistance from the hip flexors, iliacus, and psoas) continue to lift and accelerate the leg forward (figure 9.5, hip; 50%-60% of the gait cycle).

Perhaps the most distinguishing feature of the swing phase is the lack of muscle activity in the lower limb (figure 9.5; compare joint moments at the ankle, knee, and hip during stance and swing). Very little muscle activity is needed in swing because the leg behaves very much like a double pendulum, swinging freely through the air under the action of gravity and inertia alone. Only small moments must be exerted at the hip (extensor) and knee (flexor) to decelerate the leg in preparation for heel strike.

Vast improvements in computer-processing power are also allowing researchers to use sophisticated computer models of the body to learn more about function of individual muscles during gait. As described in chapters 7 and 8, because models can be used to estimate muscle forces during activity, the model simulations may be analysed to understand muscle function at a much deeper level than is presently possible with gait-analysis experiments (see "In Focus: Using Computer Models to Study Muscle Function in Gait").

IN FOCUS: USING COMPUTER MODELS TO STUDY MUSCLE FUNCTION IN GAIT

There is great range in the complexity of computer models used to study human gait. The simplest of all models is the inverted single pendulum, which has been used to explain the changes in kinetic and potential energy that occur when humans walk at their preferred speeds. At the other end of the spectrum, complex models have been built to learn more about how muscles coordinate movement of the body segments during the gait cycle.

Figure 9.6 shows a three-dimensional model of the body that has been used to simulate the biomechanics of normal walking. This model includes all six major determinants of normal gait: hip flexion, stance knee flexion, ankle plantarflexion, pelvic rotation, pelvic list, and lateral pelvic displacement (see "In Focus: The Major Determinants of Normal Gait"). The skeleton is represented as a 10-segment, 23 degrees-of-freedom mechanical linkage. The first 6 degrees of freedom define the position and orientation of the pelvis relative to the ground. The head, arms, and torso are lumped together and represented as a single rigid body, which articulates with the pelvis by means of a 3 degrees-of-freedom ball-and-socket joint located roughly at the level of the third lumbar vertebra. Each hip is modelled as a ball-and-socket joint having 3 degrees of freedom: hip flexion, internal–external rotation (which permits pelvic rotation), and abduction–adduction (which permits pelvic list). Each knee is modelled as a hinge joint with a single degree of freedom, allowing only flexion and extension. Each ankle–subtalar complex is represented as a universal joint with 2 degrees of freedom: ankle plantarflexion and subtalar inversion–eversion. Each foot is represented by 2 segments in the model: a hindfoot and a toes segment, hinged together by a metatarsal joint with a single degree of freedom. Five springs with damping are placed under the sole of each foot to simulate the interaction of the foot with the ground.

Relative movements of the body segments in the model are controlled by 54 muscles: 24 muscles per leg plus 6 abdomen and back muscles. Each muscle is given realistic force–length and force–velocity properties, and series and parallel elastic elements with active and passive stiffness properties (see "In Focus: Modelling the Mechanics of Muscle Contraction" in chapter 8). Tendon is modelled as an elastic material. The delay between the incoming neural excitation and muscle activation (force) is also taken into account.

Muscle forces during walking were calculated by minimising the cost of transport during walking (i.e., metabolic energy per unit of distance moved). Metabolic energy was estimated by summing the heat liberated during muscle contraction and the mechanical work done by the muscles to move the joints (see chapter 8). One full cycle of walking (i.e., single- and double-leg support) was simulated by solving a dynamic optimisation problem using high-performance parallel computing (Pandy, 2001).

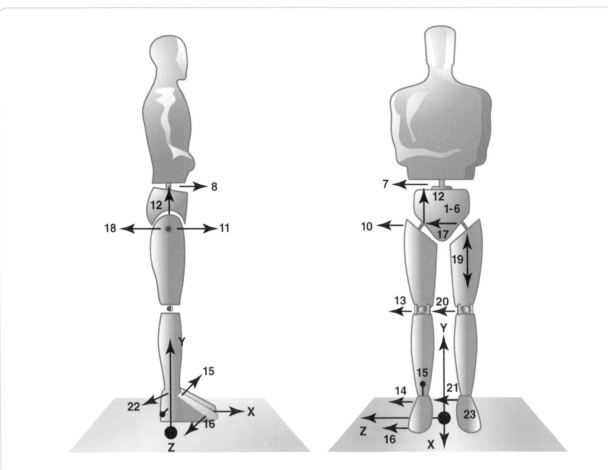

Figure 9.6 Schematic diagram of the three-dimensional model used to simulate one full cycle of normal walking.

Reprinted, by permission, from F.C. Anderson and M.G. Pandy, 2001, "Dynamic optimization of human walking," *Journal of Biomechanical Engineering* 123(5): 382. © ASME.

The joint angles, ground forces, and muscle-activation patterns predicted by the dynamic optimisation solution were similar to measurements obtained from gait-analysis experiments performed on five healthy young adult male participants.

A general principle that has emerged from analyses of **computer simulations** of movement is that muscles can accelerate joints they do not span. In other words, a muscle can contribute to the acceleration of a body segment without actually touching it. The physical explanation is as follows. When a muscle touches a body segment, it can apply a force to that segment, which is then transmitted up and down the multilink chain via the contact forces acting at the joints. Thus, a muscle like vasti, which originates on the femur and inserts on the tibia, and can accelerate not only the femur and tibia but also

the trunk, foot, and arms because the vasti induces contact forces at all of the joints (e.g., ankle, knee, hip, back, shoulder) simultaneously.

This principle of induced accelerations is nicely illustrated by the behaviour of the soleus during the midstance phase of the gait cycle. As noted previously, the soleus is activated in midstance to restrain forward rotation of the tibia about the ankle. In doing so, however, the soleus also accelerates the knee into extension with great vigor. Thus, the soleus, which crosses only the ankle joint, may accelerate the knee into extension as much as, if not more than, it accelerates the ankle into plantarflexion.

The principle of induced accelerations has been used to show how muscles provide support against the downward pull of gravity, generate forward progression to maintain a steady speed of walking,

(continued)

and control mediolateral (sideways) balance. Support and forward progression are generated mainly by the actions of five lower-limb muscles: gluteus maximus, gluteus medius, vasti, soleus, and gastrocnemius (figure 9.7, top two panels). The gluteus maximus, gluteus medius, and vasti generate the majority of support during the first half of the stance phase, so these muscles are responsible for the appearance of the first peak in the vertical ground reaction force. Similarly, the ankle plantarflexors, soleus and

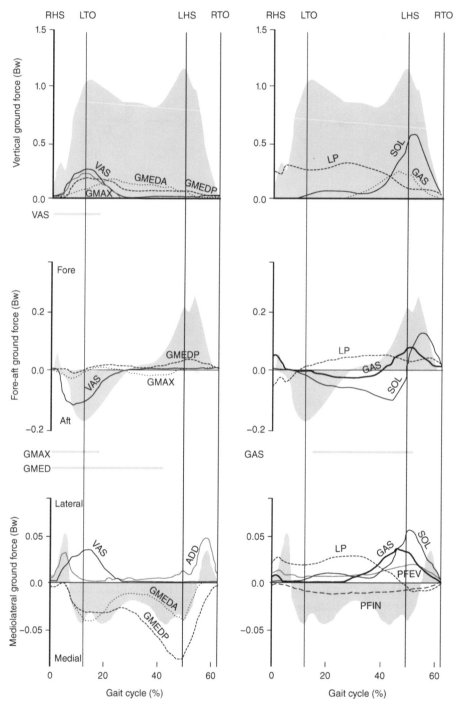

Figure 9.7 Major contributions made by individual lower-limb muscles to the vertical, fore–aft, and mediolateral ground reaction forces generated during normal walking. The results were obtained from the model illustrated in figure 9.6.

Reprinted from *Gait & Posture* Vol. 17(2), F.C. Anderson and M.G. Pandy, "Individual muscle contributions to support in normal walking," pgs. 159-169, copyright 2003, with permission from Elsevier.

gastrocnemius, generate nearly all of the support needed during the second half of stance. These muscles are therefore responsible for the appearance of the second peak in the vertical ground reaction force. The actions of these five muscle groups also explain the shape of the ground reaction force generated in the fore–aft direction. The vasti contributes most significantly to the fore–aft ground force during the first half of stance, thereby decelerating the centre of mass at this time, whereas the ankle plantarflexors, soleus and gastrocnemius, generate nearly all of the forward acceleration of the mass centre in late stance. The aforementioned muscles also control motion of the centre of mass in the mediolateral direction. The vasti, soleus, and gastrocnemius all generate laterally directed ground reaction forces, whereas the hip abductor, gluteus medius, actively controls balance by generating a ground reaction that is directed medially (figure 9.7, bottom panel). Thus, gluteus medius is the main muscle that controls balance when humans walk at their preferred speeds.

The reason these results could be obtained is that the model simulations were able to predict how much force each muscle develops at each instant during the gait cycle. Such information cannot presently be obtained from gait-analysis experiments alone.

Sources

Anderson, F.C., & Pandy, M.G. (2001). Dynamic optimization of human walking. *Journal of Biomechanical Engineering, 123,* 381-390.

Anderson, F.C., & Pandy, M.G. (2003). Individual muscle contributions to support in normal walking. *Gait and Posture 17:* 159-169.

Pandy, M.G. & Andriacchi, T.A. (2010). Muscle and joint function in human locomotion. *Annual Review of Biomedical Engineering 12:* 401–433.

Zajac, F.E. & Gordon, M.E. (1989). Determining muscle's force and action in multiarticular movement. *Exercise and Sport Sciences Reviews 17:* 187–230.

CHANGES IN MUSCLE STRENGTH WITH AGE

Virtually no data are available to show how upper- and lower-limb muscle strength changes in the developing infant and in very young children. It is not possible to elicit maximum voluntary contractions in the very young, so changes in muscle strength are usually extrapolated from measurements of the changes in fiber type and size.

Muscles increase rapidly in diameter postnatally, and this increase is a result of hypertrophy (i.e., continued growth of existing fibers) rather than hyperplasia (i.e., an increase in the number of fibers in the muscle belly). The increase in fiber size is brought about by an increase in the number of nuclei in the muscle cell. Boys show a 14-fold increase in the number of muscle nuclei from infancy to adolescence, compared with a 10-fold increase in girls. Adult-size muscle fibers are attained in adolescence.

Very few studies have used isokinetic dynamometers to determine the development of muscle strength in young children. The only data available are those for children immediately before and during puberty (i.e., 10-16 yr

of age). One longitudinal study measured isokinetic leg strength in forty-one 10- to 14-yr-old English boys and girls. No sex differences were found in the maximum voluntary strength of the quadriceps once muscle strength was normalised by body size. Furthermore, peak knee-extensor torque increased linearly with age, but the increase was not one to one (figure 9.8). In fact, the 14-yr-olds were more than 2 times stronger than the 10-yr-olds. If these

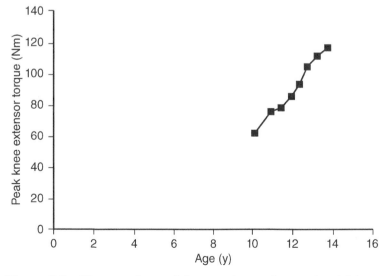

Figure 9.8 Changes in peak knee-extensor torque in children aged 10 through 14 yr of age.

Data reported by De Ste Croix et al., 2002, "Longitudinal changes in isokinetic leg strength in 10-14-year-olds," *Annals of Human Biology* 29: 50-62.

data were extrapolated back to 7-yr-old children, the curve in figure 9.8 would predict a 6-fold increase in strength for a doubling in age from 7 yr to 14 yr.

In contrast to the very young, changes in muscle strength in the elderly are well documented. Muscle strength appears to be maintained up to the fifth decade of life, but then decreases by roughly 15% in the sixth and seventh decades. The decrease in strength is most rapid beyond the seventh decade (i.e., in people 60 yr and older). For example, quadriceps strength in a group of healthy 80-yr-old men and women was found to be 30% lower than that in a population of 70-yr-olds. The reductions in static and dynamic muscle strength were quite similar, ranging from 24% to 36% in people 50 to 70 yr of age.

The decrease in muscle strength with advancing age may be explained by a loss of muscle mass or a decrease in the muscle's capacity to generate force (i.e., a reduction in motor-unit recruitment or a loss of contractile properties of the muscle). There is experimental evidence to support both possibilities. Measurements of maximum voluntary quadriceps strength have shown that strength levels in 70-yr-olds are roughly 35% lower than those in 20-yr-olds. In addition, the mean midthigh cross-sectional areas measured in the older subjects were 33% less than those measured in the younger subjects. However, when maximum quadriceps force was divided by muscle cross-sectional area, there was no significant difference between the younger and older adults, suggesting that there is no difference in the intrinsic strength of an older person's muscle.

In contrast, other studies have shown that the reduction in muscle strength with ageing is primarily the result of a decrease in muscle size (specifically cross-sectional area). In a large cross-sectional study focusing on the vastus lateralis, it was demonstrated that the reduction in muscle cross-sectional area begins before the age of 30 yr, and that by 50 yr strength is reduced by approximately 10%. The rate of decline in muscle cross-sectional area is most rapid after the age of 50 yr, so much so that between 20 and 80 yr of age people typically lose 40% of their muscle size. Most of the decline in muscle strength was attributed to the reduction in muscle cross-sectional area, which argues against the notion that skeletal muscle becomes intrinsically weaker with age.

Muscle weakness is regarded by many as a strong predictor of falling in older adults. Falls leading to injury and loss of independent living cost the health care economy billions of dollars each year.

Fortunately, there is some evidence to suggest that resistance-training programs may be effective in preserving muscle strength and size in elderly persons. In a recent study, 10 older male adults engaged in a resistance-training program consisting of bilateral isotonic (i.e., constant weight) knee-extension exercises performed at 80% of 1RM. (1RM, or one repetition maximum, is the maximum weight that can be lifted with one leg by extending the knee.) The leg-muscle exercises were performed 3 d/wk for 12 wk. On completion of the training program, the subjects were divided into two groups. Group 1 (the training group) performed further strength training 1 d/wk for the next 6 mo. Group 2 (the detraining group) returned to the sedentary lifestyle that they had led before participating in the study. Measurements of maximum knee-extensor torque and thigh cross-sectional area were made at 3 time points during the study: before the study began (T1), after the initial 12 wk of resistance strength training (T2), and after the 6 mo training–detraining period (T3).

Muscle strength increased significantly from T1 to T2 in both the training and detraining groups. Six months after the initial resistance strength-training program, muscle strength was maintained in the training group but decreased significantly (by 11% on average) in the detraining group. Changes in thigh cross-sectional area were consistent with measurements of maximum knee-extensor torque. After 12 wk of resistance strength training, the training group experienced an average increase in muscle size of 7.4%, compared with 6.5% in the detraining group. Following 6 mo of no resistance training, however, the detraining group had a 5% decrease in thigh muscle cross-sectional area, whereas the training group experienced no change. Thus, training 80% of 1RM appears to be sufficient to preserve both muscle mass and strength characteristics in older adults following 12 wk of resistance training.

GAIT DEVELOPMENT IN CHILDREN

Gait matures in a relatively short period of time. Most children begin walking within 3 mo of their first birthday. Heel strike and reciprocal swinging of the extremities are usually established by the age of 1.5 yr. The kinematic, kinetic, and muscle EMG patterns observed in adults are fully developed in children by the age of 7 yr.

Understanding the mechanics of walking in children is hampered by the fact that gait experiments are often very difficult to perform in the very young. Measurements of ground forces are difficult to obtain in children under 2 yr of age because these subjects take such small steps and usually contact the same force plate with successive strides. Muscle EMG data are also difficult to obtain for ethical reasons. Fine-wire electrodes, which are needed to access deep-lying muscles of the leg and abdomen, have yet to be used on very young children.

Nonetheless, a clear picture of at least the five major determinants of mature gait has emerged (contrast these with the six major determinants of normal gait described in "In Focus: The Major Determinants of Normal Gait").

▸ The first determinant, duration of single-leg stance, increases steadily from about 32% in 1-yr-olds to about 38% in 7-yr-olds; the most rapid increase occurs between the ages of 1 and 2.5 yr. The duration of the single-support phase during normal walking in adults is typically 39%.

▸ The second determinant, walking speed, increases steadily from 0.64 m/s at age 1 yr to 1.14 m/s at age 7 yr. The preferred speed of walking in 7-yr-olds is still significantly lower than the average walking speed of 1.34 m/s recorded in adults.

▸ The third determinant, cadence (i.e., the rate at which the legs swing back and forth), decreases steadily with age. Small children walk with a fairly high cadence (short gait cycle time), with a mean at age 1 yr of approximately 176 steps per minute. At age 7 yr, cadence decreases to around 145 steps per minute, which is still well over the average mark of 113 steps per minute measured in adults.

▸ The fourth determinant, step length, increases almost linearly with age between 1 and 4 yr, and thereafter continues to increase, but at a slower rate. Step length in 1-yr-olds is typically around 0.2 m, whereas 7-yr-olds take steps that are as much as 2.5 times bigger (approximately 0.5 m). The mean step length during normal walking in adults is around 0.7 m.

▸ The fifth determinant, **step width**, decreases with age. In figure 9.9, step width is expressed as pelvic width divided by ankle spread; ankle spread is the distance between the two ankles when both legs are in contact with the ground. Thus, as age increases, step width decreases, so the ratio of pelvic span to ankle spread increases as shown. The increase in the ratio of pelvic span to ankle spread is most pronounced between the ages of 1 and 3.5 yr. The support base is roughly 70% of pelvic width in 1-yr-olds, and decreases to 45% in 3.5-yr-olds. After the age of 3.5 yr, the ratio of pelvic span to ankle spread remains more or less constant until the age of 7 yr. The average value in adults walking at their preferred speeds is around 30%.

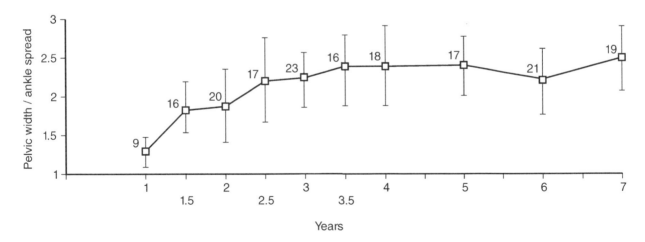

Figure 9.9 Change in the ratio of pelvic width to ankle spread with age in young children. Note that as the ratio of pelvic width to ankle spread increases, the support base (i.e., step width) decreases because ankle spread decreases. The numbers next to each point on the curve indicate the numbers for each age group represented. The vertical bars are plus and minus one standard deviation.

Reprinted, by permission, from D.H. Sutherland et al., 1980, "The development of mature gait," *The Journal of Bone and Joint Surgery* 62-A: 352. http://jbjs.org/

Thus, very young children (i.e., 1-yr-olds) walk with high cadence and take short steps to move their centres of mass at relatively slow speeds. The fact that children take shorter steps during independent walking is only partly explained by the fact that they have much shorter legs. Indeed, step length increases linearly with leg length between the ages of 1 and 7 yr. The immature child, however, takes shorter steps than is possible given his or her leg length. In connection with this, young children also spend less time supporting their centres of mass with one leg (i.e., the swing phase occupies a smaller proportion of the gait cycle in children than in adults; see figure 9.10).

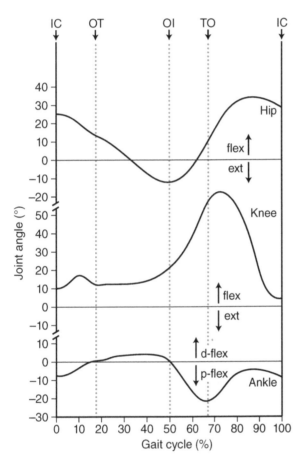

Figure 9.10 Mean sagittal-plane angles measured for the lower-limb joints in young children. Data are plotted for the right hip (flexion positive), right knee (flexion positive), and right ankle (dorsiflexion positive). The gait cycle events are shown at the top of the figure. IC = initial contact; OT = toe-off of contralateral leg; HR = heel rise; OI = initial contact of contralateral leg; TO = toe-off; d-flex = dorsiflexion; p-flex = plantarflexion.

There are three possible reasons to explain the reduction in step length and duration of single-leg stance. First, balancing skills are not fully developed, posing a threat to stability. Second, intermuscular control (i.e., muscle coordination) has yet to be learned. Third, the ankle plantarflexors are still too weak to provide the necessary thrust to lift and accelerate the body forward in late stance. With maturation and growth, step length and walking speed increase whereas cadence decreases; these three determinants reach normal adult values only around the age of 15 yr.

Other important kinematic and kinetic features that distinguish walking in the very young from normal walking in adults are as follows.

▸ Very young children (1-yr-olds) have no heel strike (i.e., initial contact with the ground is made with the entire foot).

▸ The degree of hip extension is reduced, and the hip does not remain flexed for as long as it does during the swing phase of normal walking (see figure 9.10).

▸ Young children do not extend their knees very much during single-leg stance (figure 9.10).

▸ Children keep their legs externally rotated at all times during the gait cycle.

▸ Reciprocal arm swing is absent from the gait patterns of the very young.

One-year-olds show some increase in knee flexion at initial contact but very little knee extension during midstance compared with adults. This behaviour is thought to be a result of inadequate ankle plantarflexor muscle force. In normal walking, knee extension in stance is brought about by the action of the ankle plantarflexors combined with some amount of quadriceps force (as described in the previous section). In particular, the ankle plantarflexors act to restrain forward rotation of the tibia during loading response and midstance. Because the plantarflexors are activated less during stance in 1-yr-olds, excessive forward rotation of the tibia results, leading to an increase in knee flexion. To compensate, the quadriceps are activated for a longer period of time during the stance phase. Clear patterns of difference from usual childhood gait are also apparent among children with neuromuscular conditions such as cerebral palsy (see "In Focus: Gait Abnormalities in Children With Cerebral Palsy").

Cerebral palsy (CP) is the most common cause of childhood disability throughout the developed world. It occurs with an incidence of 2 to 3 per 1,000 live births. For example, in Australia it is estimated that a child is born with CP every 18 h. CP is a neuromuscular condition that results from damage to the brain at or around the time of birth. The most significant effects are muscle spasticity, muscle weakness, loss of muscle control, balance problems, and various muscle and bone deformities, including tibial torsion and excessive femoral neck anteversion. The diagnosis and treatment of children with CP have improved greatly over the past 20 yr since the introduction of gait-analysis techniques (figure 9.11). Gait

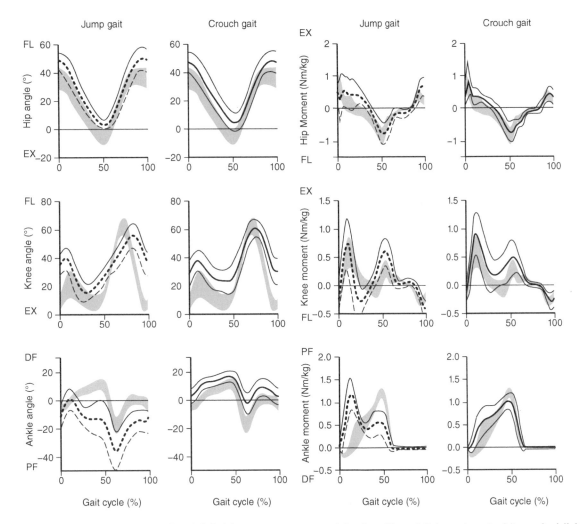

Figure 9.11 Joint angles and net joint torques measured for healthy children (controls) and children walking with crouch gait and jump gait patterns. The gray areas represent one standard deviation from the mean for the control subjects. The dashed lines represent the mean (thick) and one standard deviation from the mean (thin) for children walking with a jump gait. The solid lines represent the mean (thick) and one standard deviation from the mean (thin) for children walking with a crouch gait. Data are based on 10 control subjects, 8 crouch-gait patients, and 10 jump-gait patients.

Reprinted from *Gait and Posture*, Vol. 36(2), T. Correa, A.G. Schache, R. Baker, H,K, Graham, and M.G. Pandy, "Potential of lower-limb muscles to accelerate the body during cerebral palsy gait," pgs. 194-200, copyright 2012, with permission of Elsevier.

(continued)

analysis has the capacity to evaluate the biomechanics of gait abnormalities and provide quantitative information on limb motion, ground reaction forces, net joint torques, and the sequence and timing of muscle activity using EMG. Pharmacological and surgical interventions may be prescribed to patients with gait abnormalities. Common orthopaedic surgeries performed on children with CP include muscle-tendon lengthenings, muscle-tendon transfers, and osteotomies.

Two of the most common gait patterns observed in patients with CP are crouch gait and jump gait. These two patterns can be distinguished based on the configurations of the lower-limb joints observed during the stance phase of walking. Although excessive amounts of hip and knee flexion are present in both types of gait, crouch gait is characterised by excessive dorsiflexion at the ankle, whereas jump gait is characterised by excessive plantarflexion at the ankle (figure 9.11). Interestingly, the pattern observed in crouch gait more closely resembles that adopted by healthy control subjects walking at their preferred speeds. For example, the net torques measured about the hip, knee, and ankle joints in children with crouch gait exhibit the same patterns as those used by healthy controls, whereas distinct differences are evident in the ankle torque for patients who walk with a jump gait.

GAIT CHANGES IN OLDER ADULTS

Gait kinematics are different in healthy older adults compared with healthy young adults when subjects walk at their preferred speeds. Muscle strength and balance seems to play a major role in these changes, and the metabolic costs of walking are affected.

CHANGES IN GAIT KINEMATICS

Cross-sectional studies have shown that preferred and maximal walking speeds remain constant until the seventh decade of life (60-70 yr). Thereafter, preferred walking speed declines at a rate of 12% to 16% per decade, and maximal walking speed decreases at a rate of 20%.

Healthy older persons walk more slowly than healthy young adults because they take shorter steps and not because they move their legs at a slower rate (i.e., by decreasing cadence). Many studies have shown that healthy older adults have a reduced step length, and therefore spend less time with one foot in contact with the ground (i.e., single-support time is shorter). This also means that older persons spend more time with both legs on the ground (i.e., double-support time is longer) compared with their younger counterparts. For example, stance time has been found to increase from 59% in 20-yr-old men to 63% in 70-yr-old men. The concomitant increase in double support time is 18% to 26% from the healthy young to the healthy old. Cadence, however, remains nearly unchanged with advancing age.

Changes also occur in gait kinetics as age increases beyond the seventh decade. The most consistent finding reported in the gait literature is a decrease in ankle plantarflexor function, specifically joint torque and power developed during the push-off phase. Other changes related to muscle function have also been reported (e.g., an increase in hip-extensor and hip-flexor torque and power as well as a decrease in knee-extensor torque and power), but these changes appear to be more variable and may be a function of the design of the experimental studies reported (e.g., the pool of subjects recruited for the individual studies and the conditions under which the studies were conducted, including whether walking speed was controlled). Healthy older adults also exert smaller forces on the ground when they walk. Specifically, peak anterior–posterior forces are smaller, as is the peak vertical force exerted on the ground during single-limb stance. Smaller ground reaction forces are consistent with the knowledge that older adults prefer to walk more slowly.

CAUSES OF GAIT CHANGES IN AGEING

Why do elderly persons walk more slowly? The reasons are not fully clear, but muscle weakness and impaired balance seem to be the prime suspects underlying the decline in walking speed with age.

Qualitatively, balance is the ability to control the position of the centre of mass during movement, so a reduction in step length (and therefore walking speed) may be an appropriate strategy for increasing stability by reducing the duration of single-limb support. Weaker muscles develop less force, so less power is available for support and propulsion. Weakness of the hip-extensor muscles may reduce joint power early in the stance phase, leading to a decrease in walking speed. Walking speed may also be compromised by weak ankle plantarflexors because these muscles are responsible for providing a large thrust, both upward and forward, in terminal stance (see "In Focus: Using Computer Models to Study Muscle Function in Gait"). Thus, weak ankle plantarflexors (specifically, soleus and gastrocnemius) may reduce the power delivered to the skeleton during push-off, causing older adults to walk more slowly by taking smaller steps.

In one study, three-dimensional biomechanical analyses were performed on 26 older adults (average age of 79 yr) and on 32 young subjects (average age of 26 yr). All participants walked at their preferred cadence and step length (i.e., their natural walking speed). The older subjects chose step lengths that were 10% shorter than those of the young subjects, mainly because they had significantly reduced ankle plantarflexion in terminal stance (i.e., just before contralateral heel strike). Furthermore, the older subjects exerted less plantarflexor torque and developed less power at the ankle during terminal stance than did the younger subjects. The older subjects also developed more flexor power at the hip in terminal stance. These results suggest an adaptation in the elderly that manifests itself as a redistribution of motor patterns in the lower limb during gait.

The ankle plantarflexors are believed to have three roles in normal walking, but what these muscles precisely do during the gait cycle remains unknown. Most researchers would agree that their major function in single-limb stance is to contract eccentrically (i.e., lengthen) in order to control forward rotation of the tibia during midstance. Because the plantarflexors are almost fully activated and contract concentrically (i.e., shorten) in terminal stance, just before contralateral heel strike, some scientists believe that their role in this portion of the gait cycle is to lift and accelerate the centre of mass forward (see "In Focus: How Much Force Do the Ankle Muscles Develop in Normal Walking?" in chapter 7). However, some researchers have argued that the ankle plantarflexors work synergistically with the hip flexors in terminal stance, immediately after contralateral heel strike, to accelerate the leg forward and upward in preparation for swing. So, although the function of these muscles appears to be well defined in early stance, their role in late stance is open to question. If healthy elderly adults use their hip-flexor muscles more in order to compensate for a reduction in the power developed by their ankle plantarflexors, this suggests that the main role of the ankle plantarflexors in late stance is to lift and accelerate the leg forward in preparation for swing rather than to propel the centre of mass forward just before contralateral heel strike.

METABOLIC COST OF WALKING IN OLDER ADULTS

Recall from chapter 8 that when people are permitted to choose their cadence and step length, a walking speed is selected by which metabolic energy consumed per unit of distance traveled is minimised. This most economical speed is approximately 1.3 m/s in healthy young adults (see figure 9.12). As the speed of walking deviates from this value, the cost of transport (i.e., metabolic energy consumed per unit of distance traveled) increases. If healthy elderly persons walk more slowly than their healthy younger counterparts, does this imply that they also consume less metabolic energy to move their centres of mass a unit distance (1 m)? The answer is no. In assuming lower preferred walking speeds, healthy older adults actually show higher aerobic demands per unit of distance walked than healthy young adults.

In one study, oxygen consumption was measured as 30 healthy young adults (age 18-28 yr) and 30 healthy older adults (age 66-86 yr) walked at various speeds, including their preferred speed, on a treadmill. Starting with the treadmill at low speed, walking speed was slowly increased until each person subjectively identified the speed that was most comfortable. Both the young and older subjects showed the familiar U-shaped speed–energy curve when metabolic energy consumption was normalised by distance walked (figure 9.12). The minimum cost of transport occurred at the same speed (1.34 m/s) in both groups. This minimum, however, was significantly higher (by 7%) in the older subjects than in the younger ones. These two results show that a single speed–energy curve like that in figure 9.12 does not describe the metabolic

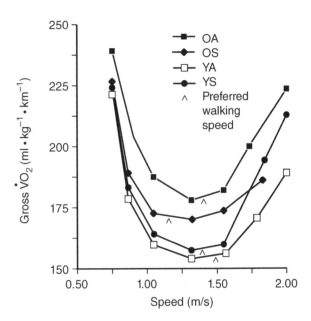

Figure 9.12 Gross oxygen consumption normalised to body mass plotted against walking speed for young and older adults. The preferred walking speed for each group is indicated. OA = old active; OS = old sedentary; YA = young active; YS = young sedentary.

Reprinted from P.E. Martin, D.E. Rothstein, and D.D. Larish, 1992, "Effects of age and physical activity status on the speed-aerobic demand relationship of walking," *Journal of Applied Physiology* 73: 200-206. Used with permission.

cost of walking in healthy young and older adults and that the effect of ageing is represented by a vertical shift upward in the speed–energy curve.

What explains the age-related difference in the relationship between metabolic cost and walking speed shown in figure 9.12? The answer appears to be due to the higher cost of generating muscle force in the elderly. It is well known that both muscle mass and muscle strength decrease beginning in the fourth decade. There is also evidence from studies in other animals (mice) that the force generated per unit of cross-sectional area of muscle is roughly 20% lower in the aged. It would appear, therefore, that healthy elderly adults need to recruit a greater number of motor units per muscle in order to generate the muscle force needed for a given motor task. It may also be that elderly people need to recruit a higher number of fast-twitch fibers to move their centre of mass at the required rate; fast-twitch fibers are known to be less economical in terms of energy consumption than are slow and intermediate muscle-fiber types. Thus, to walk at their preferred

speed, older adults must use more oxygen, implying that metabolic energy consumption is not the criterion optimised during gait.

EXERCISE AND AGEING

There is a strong correlation between walking speed and level of mobility. Elderly persons who walk very slowly and who have poor balance are prone to falling. Indeed, very slow walkers are usually homebound. Normal walking requires adequate levels of muscle strength as well as adequate neuromuscular control and joint proprioception. As noted previously, walking speed begins to decline in the seventh decade; this is usually a sign of deterioration in strength, control, or proprioceptive abilities. Interventions such as resistance strength training (which is discussed more fully in chapter 10) may reverse, or at least slow, the decline in walking speed with age and may therefore decrease the risk of falling.

Various combinations of strength and neuromuscular training have been tried, and their effects on walking speed are well documented. In one study, 34 subjects aged 75 yr and older underwent an exercise protocol that consisted of flexibility, balance, and resistance strength training 3 times/wk for 12 wk. Flexibility training began with stretching exercises performed while sitting, followed by calf and hamstrings stretches, and, finally, spinal-extension exercises. Resistance training included isometric hip-abduction and knee-extension exercises performed to fatigue; resistance was set at roughly 80% of maximum voluntary muscle contraction. Balance exercises included shifting weight from one foot to the other in the medial–lateral and anterior–posterior directions as well as performing simple tai chi movements for 5 to 10 min each day.

Walking speed and step length both correlate with knee-extensor (quadriceps) muscle strength in healthy older adults. The results show that below a knee-extensor moment of roughly 50 Nm, walking speed and step length both decrease with progressive quadriceps weakness (figure 9.13). Above this limit, the correlation is poorer (i.e., the slopes of the lines in figure 9.13 are lower beyond 50 Nm), implying that the limit of walking speed is then more closely related to the strengths of the other leg muscles.

Significant improvements in both muscle strength and walking speed were obtained from the training program. Isometric knee-extension

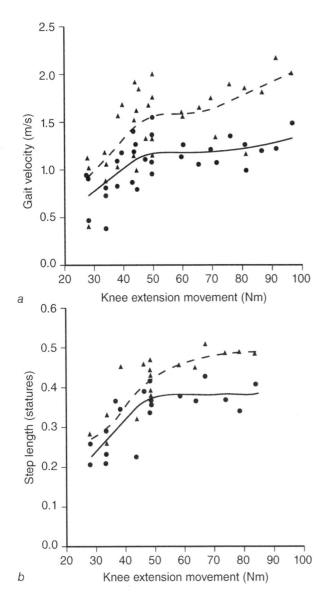

Figure 9.13 *(a)* Relationship between walking speed and maximum isometric knee-extensor moment (quadriceps strength) in older adults. The solid line and circles are data for participants walking at their preferred speeds. The dashed line and triangles are data for participants walking at much faster than their preferred speed. *(b)* Relationship between step length (normalised by height in statures) and maximum isometric knee-extensor moment (quadriceps strength) in older adults. The solid line and circles are data for participants walking at their preferred speeds. The dashed line and triangles are data for participants walking at much faster than their preferred speed.

Reprinted from *Archives of Physical Medicine and Rehabilitation*, Vol. 74, J.O. Judge, M. Underwood, T. Gennosa, "Exercise to improve gait velocity in older persons," pgs. 400-406, copyright 1993, with permission from Elsevier.

strength increased by more than 30% and walking speed increased by 8%. The improvement in walking speed in particular is clinically significant because gait velocity declines by roughly 15% per decade after the age of 60 yr. These results are not universal, however, because others have found no improvement in walking speed when moderate resistance and balance exercises were undertaken by subjects aged 60 to 71 yr. The difference in these findings may be explained by differences in subject characteristics. Subjects who did not improve their walking speed were much stronger to begin with than those who did. Thus, resistance and balance training improved muscle strength in these subjects by less than 10%, which may explain the imperceptible change in walking speed. This, in turn, implies that walking speed improvements are more likely to occur in older adults who have significant leg-muscle weakness (i.e., the frail elderly).

SUMMARY

Normal gait is characterised by six kinematical mechanisms known as the major determinants of gait: hip flexion, stance knee flexion, ankle plantarflexion, pelvic rotation, pelvic list, and lateral pelvic displacement. These mechanisms explain how movements of the lower-limb joints contribute to the sinusoidal displacement of the centre of mass when humans walk at their preferred speeds. Muscle strength increases linearly in children aged 10 through 16 yr. Muscle strength is maintained from young adulthood up to the fifth decade, but then decreases by roughly 15% in the sixth and seventh decades. The decrease in strength is most rapid in people 60 yr and older. Very young children (<7 yr) and healthy older adults walk differently compared with healthy young adults. The development of mature walking is characterised by changes in the duration of single-leg stance, walking speed, cadence, step length, and step width. The duration of single-leg stance, step length, and walking speed are all lower in very young children than in healthy young adults; cadence and step width are both higher. The trends are similar in older adults. Step length, walking speed, and the duration of single-leg stance all decrease with advancing age, whereas step width increases to widen the base of support. Regular intervals of strength and neuromuscular training may help maintain walking speed in older adults, but the effects on step length are currently unknown.

FURTHER READING

Malina, R., Bouchard, C., & Bar-Or, O. (2004). *Growth, maturation, and physical activity* (2nd ed.). Champaign, IL: Human Kinetics.

Pandy, M.G. (2001). Computer modeling and simulation of human movement. *Annual Review of Biomedical Engineering, 3,* 245-273.

Pandy, M.G., & Andriacchi, T.A. (2010). Muscle and joint function in human locomotion. *Annual Review of Biomedical Engineering, 12,* 401-433.

Rogers, M.A., & Evans, W.J. (1993). Changes in skeletal muscle with aging: Effects of exercise training. In J.O. Holloszy (Ed.), *Exercise and sports sciences reviews* (pp. 65-102). Baltimore: Williams & Wilkins.

Rose, J., & Gamble, J.G. (1994). *Human walking* (2nd ed.). Baltimore: Williams & Wilkins.

Schultz, A.B. (1992). Mobility impairment in the elderly: Challenges for biomechanics research. *Journal of Biomechanics, 25,* 519-528.

Sutherland, D.H., Olshen, R., Biden, E.N., & Wyatt, M.P. (1988). *The development of mature walking.* London: MacKeith Press.

Whittle, M.W. (1996). *Gait analysis: An introduction.* Oxford: Butterworth-Heinemann.

CHAPTER 10

BIOMECHANICAL ADAPTATIONS TO TRAINING

The major learning concepts in this chapter relate to

▸ the effects of training on muscle properties, particularly muscle strength and muscle contraction speed;

▸ how training can change neuromuscular control of movement, especially muscle coordination;

▸ how neuromuscular control and therefore muscle function can change in response to injury; and

▸ the effects of muscle speed and muscle contraction speed on motor performance.

The use of training methods to improve muscular function is widespread. Specialised training programs not only improve performance but also prevent injuries. Training can change both the contractile properties of muscle (i.e., strength and contraction speed) and the ability of the nervous system to control muscular function (i.e., coordination). Resistance training increases an individual's ability to produce maximum muscle force, whereas plyometric training, in the form of dynamic depth jumps, can enhance the ability to rapidly develop force. Plyometric training is also used as an adjunct in jump-training programs to improve neuromuscular and proprioceptive control. This chapter describes how the neuromuscular and musculoskeletal systems respond to various training methods instituted for the purpose of improving performance and decreasing the incidence of injuries in sport. Experimental and theoretical results are also presented to show how biomechanical performance may be altered through manipulation of changes in strength, speed, and control.

MUSCULAR ADAPTATIONS TO TRAINING

Training can change both muscle strength and muscle contraction speed. Strength may be increased by increasing the net neural drive to the muscle, increasing muscle size, or both. Muscle contraction speed may be altered through a change in the shape of a muscle's force–velocity curve, change in the value of its intrinsic maximum shortening velocity, or both.

Two of the most common methods used to increase muscle strength are isometric and isokinetic training. Both types of exercises are performed on a device called a dynamometer. A dynamometer is an electromechanical machine that contains a speed-controlling mechanism that accelerates to a preset speed when a body part applies an external force to the machine's arm. Once the preset speed has been reached, a loading mechanism creates an equal and opposite force that balances the applied force (see figure 7.7a).

Performance of an isometric contraction provides a measure of the muscle's maximum strength. In isometric knee extension, for example, the participant is asked to maximally activate his or her quadriceps while the machine arm is prevented from rotating. The maximum force produced by the muscle depends on the length at which the muscle is held. Thus, one can obtain a muscle's maximum isometric force–length curve by measuring the muscle's maximum isometric force at various flexion angles of the knee (see figure 7.9).

Early in a training program (2-8 wk), strength gains are found to be more attributable to neural adaptations than to increases in muscle size. This is because the ability to activate all of the available motor units is enhanced in the early phases of training, when subjects are still learning how to exert force effectively. A dramatic increase in adaptation of neural factors is observed over the first 6 to 10 wk of a resistance-training program. However, as the duration of training is extended beyond 10 wk, muscle hypertrophy begins to contribute more than neural adaptations to the strength gains observed. Eventually, muscle hypertrophy reaches a plateau where additional exercise produces no further change in maximum muscle strength.

Changes in strength are also caused by mechanisms that increase muscle size. The increase in muscle size can be attributed to an increase in the cross-sectional area of individual muscle fibres composing the whole muscle (**hypertrophy**) or to an increase in the number of muscle fibres (**hyperplasia**). Strength gains can be substantial with short-duration, high-intensity training, as has been demonstrated on the flexor and extensor muscles of the knee. Participants trained 4 times per week for a period of 7 weeks on an isokinetic dynamometer; each training bout consisted of maximal knee extensions and flexions at a constant angular velocity of 180°/s. Measurements of peak torque were obtained at speeds ranging from 0°/s (isometric) to 300°/s over a knee flexion range of 90°.

IN FOCUS: AT WHAT SPEED MUST A MUSCLE SHORTEN TO DEVELOP MAXIMUM POWER?

Maximum-height jumping, maximum-distance throwing, and maximum-speed pedaling are three examples of activities in which performance is determined by the ability to generate maximum muscle power. The speed at which a muscle develops maximum power is not immediately obvious because this value is determined by the shape of the muscle's force–velocity curve, which happens to be nonlinear. Indeed, it is not an easy feat to deduce this property by experiment.

The procedure involves a quick-release experiment that is performed using the muscle-lever system shown in figure 10.1. The muscle is taken out of the body and immersed in a salt bath to ensure that the contractile machinery is still able to function. Tendon is usually separated from the musculotendinous preparation because the **compliance** of the tendon will affect the force developed by the muscle. One end of the muscle belly is clamped to the workbench (ground), and the other end is attached to the lever as shown. A weight is also hung from one end of the lever as indicated. An electromagnetic device is used to prevent the lever from rotating as the muscle is stimulated to its maximal force initially. At some point in time, the electromagnet is released, and the lever is then free to rotate. The direction in which the lever rotates depends on the relative magnitudes of the forces applied by the hanging weight and the pull of the muscle. If the hanging weight is greater than the maximum isometric force developed by the muscle (for the length at which the muscle is contracting), the lever will rotate counterclockwise, as viewed in figure 10.1, and the muscle will lengthen (i.e., contract eccentrically). Conversely, if the isometric force developed by the muscle is greater than the force exerted by the hanging weight, the lever will

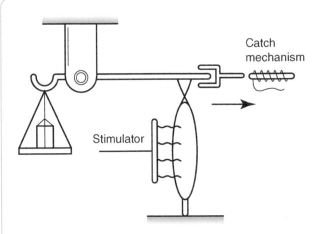

Catch
mechanism

Stimulator

Figure 10.1 Schematic diagram showing instrumentation used to estimate a muscle's force-velocity curve in a quick-release experiment. When the electromagnetic catch mechanism on the right is released, the muscle shortens or lengthens depending on whether the weight in the pan on the left is lighter or heavier than the isometric force developed by the muscle.

McMahon, Thomas, A.: MUSCLE, REFLEXES AND LOCOMOTION. Copyright © 1984 by Thomas A. McMahon. Reprinted by permission of Princeton University Press.

rotate clockwise and the muscle will shorten (i.e., contract concentrically). In either case, the muscle undergoes an **isotonic** (constant force) **contraction** because the hanging weight is the only external force applied to the muscle.

Unfortunately, the results of the quick-release experiment tend not to be dependable when muscle is forcibly stretched. As actomyosin bonds are broken, the muscle incurs some damage as it is stretched. If the hanging weight is made large enough, the muscle will eventually yield or give as it is stretched; this phenomenon is indicated by the somewhat flat line obtained in the lengthening region of the muscle's force–velocity curve (figure 10.2, solid line).

The experiment works well, however, as long as the muscle is permitted to shorten against the externally applied load. For each weight, the shortening velocity of the muscle can be measured from its length trajectory in time. Specifically, the length change of the muscle is recorded as it shortens against the force of the hanging weight. The slope of the length–time trajectory is calculated at a point when the muscle length change is no longer instantaneous. (As soon as the electromagnet is released, the muscle undergoes an instantaneous change in length, which is attributable to the stiffness of the cross-bridges; see figure 8.6.) The slope of the length–

time trajectory is taken as the shortening velocity of the muscle for the load in question. Thus, different muscle-shortening velocities can be found, depending on the amount of weight placed in the pan.

The solid line in figure 10.2 was derived by fitting a curve to the data points obtained from such a series of quick-release experiments. The relationship describing this curve is a hyperbola and is known as Hill's force–velocity equation (Hill, 1938). The equation takes the form

$$(F^M + a)(v^M + b) = (F_o^M + a)b \qquad 10.1$$

where F^M is the force developed by the muscle, F_o^M is the muscle's maximum isometric force at a given length, v^M is the shortening velocity of the

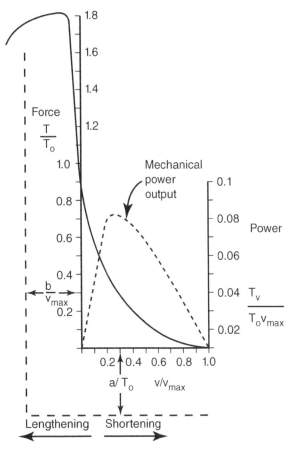

Figure 10.2 Force–velocity curve for muscle as fitted by the hyperbola given in equation 10.1 (solid line). The dashed line is the corresponding power developed by the muscle in the shortening region. Peak power occurs at about one third of muscle's maximum shortening velocity.

McMahon, Thomas, A.: MUSCLE, REFLEXES AND LOCOMOTION. Copyright © 1984 by Thomas A. McMahon. Reprinted by permission of Princeton University Press.

(continued)

muscle, and *a* and *b* are constants that vary with the temperature of the muscle. Interestingly, the shape of the force–velocity curve is the same for practically all muscles known to exist in nature. The only exception appears to be insect flight muscle, whose force–velocity data do not fit the curve predicted by equation 10.1.

The maximum shortening velocity of the muscle, v_{max}^M, can be found by extrapolating the experimental data back to the point where force is zero (i.e., the intercept of the velocity axis in figure 10.2). This defines the (maximum) velocity with which the muscle will shorten when no external load is applied. It is often regarded as a constant property of the muscle, although some evidence exists to suggest that the value of v_{max}^M depends on muscle length and muscle activation level. Nonetheless, it is a good indicator of the distribution of fibre type in a given muscle. Muscles with low maximum-shortening velocities (e.g., $v_{max}^M = 3$ muscle lengths per second for human muscle) can be expected to comprise predominantly slow-twitch, oxidative (type SO) fibres; those with intermediate values of v_{max}^M (i.e., $v_{max}^M = 10$ muscle lengths per second) will have mixed, fast-twitch, oxidative-glycolytic (type FOG) fibres; and those with high values of v_{max}^M (i.e., $v_{max}^M = 20$ muscle lengths per second) will have predominantly fast-twitch, glycolytic (type FG) fibres.

Because power is the product of force and velocity, muscle power can be found by solving equation 10.1 for muscle force, F^M, and multiplying the result by the muscle's shortening velocity. Thus,

$$power = F^M \; v^M = \frac{v^M (b \, F_o^M - a \, v^M)}{(v^M + b)}. \qquad 10.2$$

Maximum muscle power is then obtained by using the rules of differentiation from calculus. Specifically, equation 10.2 is differentiated with respect to shortening velocity and the result is then set equal to zero. Solving the resulting equation for shortening velocity gives the velocity at which the muscle develops maximum power. If the algebra is carried out, it can be shown that, in general, maximum muscle power occurs at roughly $0.3v_{max}^M$. That is, muscles must shorten at approximately one third of their maximum shortening velocity if they are to develop maximum power in each stroke. This result is used to design the gearing on bikes so that maximum power may be delivered to the crank at all times, especially while riding up and down hills.

Sources

Hill, A.V. (1938). The heat of shortening and the dynamic constants of muscle. *Proceedings of the Royal Society (London), Series B, 126,* 136-195.

McMahon, T.A. (1984). *Muscles, reflexes, and locomotion.* Princeton, NJ: Princeton University Press.

The quadriceps muscles were typically 2 times stronger than the hamstrings before training. After training, muscle strength of both the quadriceps and hamstrings increased significantly. Peak knee-extensor torque increased by as much as 16%, and the increase in peak knee-flexor torque was slightly more (i.e., an increase of 25%). Even larger strength increases have been found in other studies. One study found a 20% increase in maximum quadriceps force after 48 days of isometric strength training; another found that 6 mo of weight training increased the peak isometric force of quadriceps by nearly 30%, and muscle cross-sectional area increased by as much as one third.

Interestingly, strength training can also change the rate at which muscles shorten and lengthen over time (i.e., contraction speed). Isokinetic knee-extension training performed at relatively high speed has been shown to change the shape of the force–velocity curve for quadriceps; relatively higher quadriceps force was obtained at high velocities after training for 4 wk. Training can also increase a muscle's maximum shortening velocity. The maximum shortening velocity of adductor pollicis was found to increase by 21% after a 10 d training program consisting mainly of isokinetic exercise. Increasing a muscle's maximum shortening velocity (v_{max}) increases the power developed by the muscle because increasing v_{max} increases the area under the muscle's force–velocity curve. Changing the shape of the muscle's force–velocity curve by increasing the force that the muscle develops at high velocities also increases the power developed by the muscle because this too increases the area under the muscle's force–velocity curve (see "In Focus: At What Speed Must a Muscle Shorten to Develop Maximum Power?").

NEUROMUSCULAR ADAPTATIONS TO TRAINING

In addition to increasing the neural drive to muscle, training can change the way muscle action is coordinated during a motor task. Current interest in developing training programs to improve neuromuscular control is driven by the need to reduce muscle, ligament, and joint injuries in sport.

TRAINING TO PREVENT ANTERIOR CRUCIATE LIGAMENT INJURY

One important example of the use of training in injury prevention relates to the disproportionate number of anterior cruciate ligament (ACL) injuries sustained by female athletes in sports characterised by running, jumping, and cutting maneuvers (see "In Focus: Anterior Cruciate Ligament Force During Drop Jumping"). In fact, the incidence of serious knee injury is reported to be approximately six times higher in female athletes than in their male counterparts. Many factors, both intrinsic and extrinsic, have been identified as predictors of ACL injuries in the female athlete. Intrinsic risk factors include malalignment of the lower extremity, decreased intercondylar notch width at the knee, increased laxity of the knee joint, and hormonal influences. Extrinsic risk factors include imbalance in quadriceps and hamstrings muscle strength and inadequate neuromuscular control. Although sports medicine research has yet to isolate the cause of ACL injury in females, many studies have found that the incidence of ACL injuries is reduced through participation in training programs that focus on teaching neuromuscular control.

Neuromuscular training programs that prevent ACL injury are based on the dictum that knee stability and function are improved when postural equilibrium, intermuscular control, and leg muscle strength are all enhanced. One particular example is the jump-training program developed at the Sports Medicine and Orthopaedic Center in Cincinnati, Ohio, in the United States. This program emphasises improving both muscle strength and neuromuscular control by incorporating stretching, plyometric exercises, and weight lifting. The overall goals are to decrease landing forces by teaching neuromuscular control of the leg muscles during landing and to increase strength of the leg musculature by weight training. The program has three distinct phases:

stretching exercises, followed by jump training, and, finally, resistance strength training.

The stretching exercises focus on muscles of the leg, particularly the soleus, quadriceps, and hamstrings, but attention is also paid to the muscles of the upper limb, including the latissimus dorsi, pectoralis major, and biceps. Weight-training exercises include abdominal curls, leg presses, calf raises, bench presses, and forearm curls. The jump-training protocol emphasises proper jumping and landing techniques and gives special attention to body posture at ground contact, knee stability, and soft landings (i.e., toe-to-heel landings with the knees bent).

Research studies have been conducted to evaluate the efficacy of the jump-training program in relation to jumping performance and landing technique and its effectiveness in decreasing the incidence of ACL injuries in female athletes. High school female volleyball players were trained for 6 wk using the ACL Force During Drop Jumping protocol. The training sessions lasted approximately 2 h/d and were conducted 3 d/wk on alternating days. The control group for the study included untrained male subjects matched for height, weight, and age. The results showed that peak landing forces resulting from a volleyball block jump decreased by 22%, knee abduction and adduction muscle moments decreased by approximately 50%, and hamstrings:quadriceps muscle moment ratio increased by as much as 26%. Also, the peak hamstrings:quadriceps moment ratio was significantly higher in trained females than in untrained female athletes but was similar to that measured in the male controls.

The results of prospective studies support the notion that neuromuscular training can reduce the incidence of knee injuries in sport. When the Cincinnati training protocol was implemented on 1,263 high school volleyball, soccer, and basketball players, untrained females demonstrated a knee injury rate 3.6 times higher than that of trained females and 4.8 times higher than that of untrained male controls. The incidence of knee injury in the untrained group was 0.43 per 1,000 player exposures compared with 0.12 in the trained group. The injury rate for the male controls was 0.09 per 1,000 exposures. In all, only 14 serious knee injuries were sustained among the 1,263 athletes who participated in the training program. The decreased injury rate in the trained athletes was attributed to an increase in dynamic stability of the knee after training. Knee joint stability was increased after training by correcting the imbalance between quadriceps and hamstrings muscle strength and by increasing the hamstrings:quadriceps strength ratio in the female athletes.

Impact of the leg with the ground while landing from a jump may predispose the individual to an ACL injury. Because muscle, ligament, and joint contact forces cannot be measured noninvasively, computer models of the musculoskeletal system have been used to obtain quantitative information about musculoskeletal loading during movement.

The three-dimensional model described in chapter 9 (see figure 9.6) was used to simulate a drop-landing task in which a volunteer participant stepped off a platform that was 60 cm high and landed onto a force plate mounted on the ground. Joint motion, ground reaction forces, and muscle electromyography activity were recorded simultaneously during the task. A simulation of the drop-landing task was performed by specifying an appropriate set of muscle excitation signals that served as inputs to the model. The input muscle excitations were adjusted by trial and error until the performance of the model matched the in vivo data. In particular, the time histories of the joint angles and ground reaction forces calculated in the model were required to be within 1 standard deviation of joint angles and ground reaction forces recorded from the landing experiment. However, the model shown in figure 9.6 represents the knee as a simple hinge joint with 1 degree of freedom and takes no account of the shapes of the articulating surfaces of the bones at the knee or the presence of the knee ligaments. Therefore, a more detailed model of the knee joint was used to obtain estimates of knee ligament loading during the drop-landing task. Specifically, the lower-limb muscle forces calculated from the whole-body model shown in figure 9.6 were then applied to another, more complex, three-dimensional model of the knee joint (figure 10.3). In this model, the shapes of the distal femur, proximal tibia, and patella were reproduced from measurements obtained from cadaver knees. The

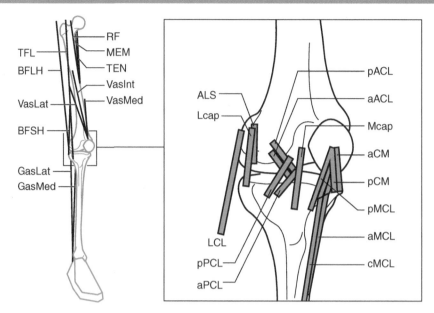

Figure 10.3 Three-dimensional model of the lower limb, including a detailed representation of the knee joint. PCL = posterior cruciate ligament; MCL = medial collateral ligament; LCL = lateral collateral ligament; Mcap = medial posterior capsule; Lcap = lateral posterior capsule; aCM = anterior deep medial collateral ligament; pCM = posterior deep medial collateral ligament; ALS = anterolateral structures; a = anterior; p = posterior; c = central.

properties of the soft tissues, including the ligaments, meniscus, and articular cartilage, were based on data obtained from tension–compression experiments performed on cadaver specimens.

The model simulation results showed that the ACL was loaded during only the first 25% of the landing task and that peak ACL force was much lower than expected—around 250 N or 0.4 times body weight. Considering that the human ACL ruptures at a load of approximately 2000 N or 3.0 times body weight, this result suggests that the risk of injury to the ACL may not necessarily be high during a drop-landing jump. ACL load remains relatively low during this task because of the large, posteriorly directed shear force applied to the leg from the ground reaction force generated on impact. Any force acting posteriorly on the leg will unload the ACL and therefore protect it. Attempting to better understand the biomechanics of ACL injury during jump-landing tasks is an active area of research, and more work is needed to identify the factors responsible for the injury rates observed in elite and recreational athletes.

This finding is important because it shows that the hamstrings:quadriceps muscle strength ratio is an important factor in preventing ACL injury. The ACL prevents excessive anterior (forward) translation of the tibia relative to the femur during activity. When the quadriceps muscles are activated, they develop a force that is transmitted to the tibia by means of the extensor mechanism comprising the quadriceps tendon, patella, and patellar tendon. The force in the patellar tendon pulls the tibia forward because it points anteriorly relative to the long axis of the tibia. Thus, a force developed by the quadriceps muscles translates the tibia anteriorly, which strains the ACL. Hamstrings muscle action reduces ACL strain because these muscles pass posterior to the knee and insert on the back of the tibia. Thus, the hamstrings can protect the ACL by creating a backward pull on the tibia, thereby decreasing the force transmitted to the ligament (see "In Focus: How Can Hamstrings Muscle Contractions Protect the ACL From Injury?").

Consider figure 10.4, which is a free-body diagram showing all the forces that typically act on the leg during activity. *PT* represents the resultant force applied by the patellar tendon, *ligament* is the resultant force of the cruciate and collateral ligaments, *hams* is the resultant force applied by all the hamstrings muscles, *ground* is the resultant ground reaction force, and *TF* is the resultant compressive force acting between the femur and tibia. Each force may be resolved into a shear component that acts

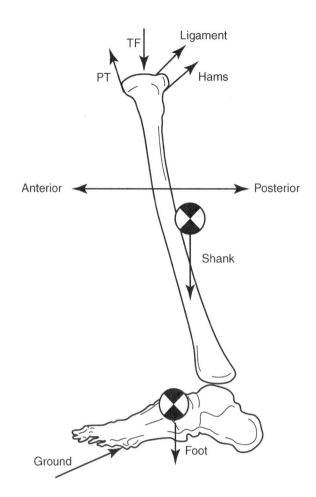

Figure 10.4 Forces that typically act on the lower leg during activity.

IN FOCUS: HOW CAN HAMSTRINGS MUSCLE CONTRACTIONS PROTECT THE ACL FROM INJURY?

Rupture of the ACL is one of the most common and debilitating injuries sustained by an athlete. Approximately 1 in every 1,000 people experiences an injury to their ACL. Deleterious effects resulting from ACL injury include degenerative changes inside the knee consistent with osteoarthritis and injury to the remaining passive structures, particularly the collateral ligaments, joint capsule, and menisci, all of which provide secondary restraint to anterior translation of the tibia. The risk of recurring injury is increased in the ACL-deficient knee because of the significant increase in anterior translation

that occurs when the ACL is absent. For example, injury to the medial collateral ligament (MCL) is common because an increase in anterior tibial translation means that the MCL will be stretched by a proportionately larger amount. An increase in anterior tibial translation may also cause tearing of the menisci, which protect the underlying cartilage from increased stress.

One effective method of protecting the ACL from injury is the use of hamstrings cocontraction. Because the hamstrings insert on the back of the tibia, these muscles are able to apply a posterior pull

(continued)

to the leg, which decreases the amount of anterior translation at the knee. The model shown in figure 7.12 has been used to explain the effect of hamstrings muscle action on ACL force in the intact knee and on anterior stability of the ACL-deficient knee (Pandy & Shelburne, 1997, 1998; Yanagawa et al., 2002). The analyses were performed on isometric and isokinetic knee extension because this exercise is commonly prescribed by physical therapists as a means of preserving thigh muscle strength subsequent to ACL injury and repair.

For maximum isolated contractions of the quadriceps in the intact knee, the peak force transmitted to the ACL is around 500 N. Furthermore, the ACL is loaded in the range 0° to 60° of knee flexion if the hamstrings muscles remain inactive throughout this task (see chapter 7). Activation of the hamstrings decreases both the peak force borne by the ACL and the flexion range over which the ligament is loaded (see figure 10.5a). As the level of hamstrings cocontraction is increased, ACL force decreases in proportion. However, the effect is not linear as indicated in figure 10.5b, which shows ACL force plotted against hamstrings cocontraction force over the full range of knee flexion angles. In fact, the relationship between hamstrings cocontraction and ACL force depends on a third variable: the flexion angle of the knee. Both the model and experimental results have shown that hamstrings cocontraction is more effective when the knee is bent than when it is straight. With the knee fully extended, hamstrings cocontraction has hardly any effect in lowering the force transmitted to the ACL (figure 10.5, 0°). However, with the knee bent to angles greater than 15°, activating the hamstrings together with the quadriceps decreases ACL load significantly (compare curves for 15°, 30°, and 45° with that for 0° in figure 10.5b).

Hamstrings cocontraction reduces ACL force by changing the balance of shear forces applied to the leg. The ACL is stretched and bears force whenever the tibia is translated anteriorly (forward) relative to the femur. Anterior translation of the tibia is ultimately a result of a net anterior shear force applied to the leg. Thus, the pattern of ACL loading during knee-extension exercise is explained by the balance of shear forces acting on the leg. Figure 10.6, *a* and *b*, shows the anterior–posterior shear forces applied to the leg during isometric knee-extension exercise

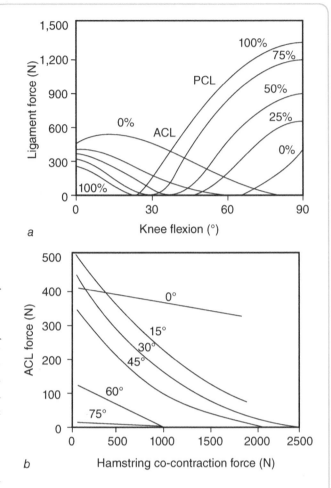

Figure 10.5 Effect of hamstrings cocontraction on ACL and PCL forces during isometric knee-extension exercise. *(a)* Ligament plotted against knee angle for various levels of hamstrings activation; *(b)* ACL force plotted against hamstrings cocontraction force at various knee angles.

Reprinted from *Journal of Biomechanics*, Vol. 30, M.G. Pandy and K.B. Shelburne, "Dependence of cruciate-ligament loading on muscle forces and external load," pgs. 1015-1024, copyright 1997, with permission from Elsevier.

performed with and without hamstrings cocontraction, respectively. The results were obtained from the two-dimensional model of the knee shown in figure 7.12. All shear forces acting on the leg were resolved perpendicular to the long axis of the leg. In the figure, *total* (thick black line) represents the net shear force applied to the leg, which was found by summing the contributions from the patellar–tendon force (PT, dashed line), the tibiofemoral compressive force (TF, dotted line), and the restraining force (restraint, gray

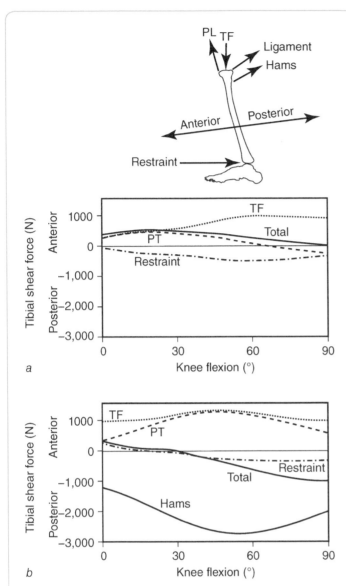

Figure 10.6 Anterior–posterior shear forces applied to the leg during isometric knee-extension exercise. *(a)* Results for isolated maximum contractions of the quadriceps muscles (i.e., with the hamstrings and gastrocnemius muscles turned off in the model); *(b)* results from simultaneous contractions of the quadriceps and hamstrings muscles.

Reprinted from *Journal of Biomechanics,* Vol. 30, M.G. Pandy and K.B. Shelburne, "Dependence of cruciate-ligament loading on muscle forces and external load," pgs. 1015-1024, copyright 1997, with permission from Elsevier.

line). *Hams* (thin black line) represents the shear force applied to the leg by the resultant force developed by the hamstrings. *Total* is equal and opposite to the net shear force applied by all the ligaments and the

capsule in the model. The tibiofemoral compressive force applies an anterior shear force to the leg at all knee-flexion angles because the tibial plateau slopes roughly 8° from front to back in the knee. The patellar tendon applies an anterior or a posterior shear force to the leg depending on its orientation relative to the long axis of the leg. The orientation of the patellar tendon relative to the long axis depends, in turn, on the flexion angle of the knee.

At all but very small knee-flexion angles, ACL force decreases with increasing hamstrings cocontraction because the net anterior shear force applied to the leg decreases (compare Total in figure 10.6, *a* and *b*). The net anterior shear force decreases because the hamstrings muscles apply a posterior force to the leg. When the knee is near full extension, however, the hamstrings all meet the tibia at relatively small angles, so they can no longer apply a large posterior shear in this position. Thus, the hamstrings cannot pull the tibia backward far enough to unload the ACL when the flexion angle of the knee is small.

The same mechanism acts to limit excessive anterior translation of the tibia and increase stability of the ACL-deficient knee. A large number of experimental studies have shown that anterior stability of the knee is compromised when the ACL is absent (this phenomenon is often referred to as "giving way"). Figure 10.7 shows the effect of ACL deficiency on anterior translation of the tibia in maximum isometric knee-extension exercise. The data points in the figure are the mean of measurements obtained from healthy subjects (top) and ACL-deficient patients (bottom) during maximum contractions of the quadriceps muscles. The shaded areas define plus and minus 1 standard deviation from the mean, and 0 mm represents the neutral position of the knee, which is the point where the tibia and femur contact each other when no external forces are applied to the leg and all the muscles are relaxed. The results show that peak anterior translation of the tibia is significantly greater in the ACL-deficient knee (mean of 15 mm) than in the intact knee (mean of 10 mm). This is because the ACL restrains forward movement of the tibia relative to the femur in the intact knee, so when it is absent, anterior translation of the tibia is then much greater.

(continued)

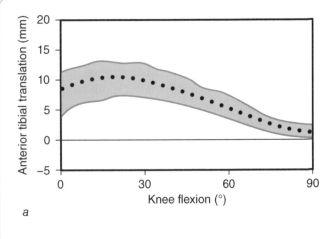

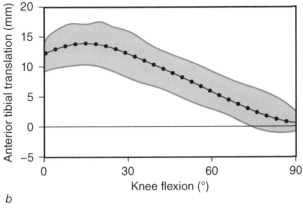

a

b

Figure 10.7 Anterior translation of the tibia relative to the femur in *(a)* the ACL-intact knee and *(b)* the ACL-deficient (ACLD) knee during knee-extension exercise.

Reprinted from *Journal of Biomechanics*, Vol. 30, M.G. Pandy and K.B. Shelburne, "Dependence of cruciate-ligament loading on muscle forces and external load," pgs. 1015-1024, copyright 1997, with permission from Elsevier.

The model of figure 7.12 has also been used to study the effect of hamstrings cocontraction on anterior tibial translation (ATT) during isokinetic extension exercise in the ACL-deficient knee. The model simulations show that ATT is inversely related to hamstrings cocontraction at all knee-extension speeds (figure 10.8). Peak ATT decreases by more than 1 cm at all speeds when the hamstrings are fully activated. Again, peak ATT decreases as hamstrings activation increases because the posterior shear force provided by these muscles overwhelms the anterior shear force supplied by the patellar tendon and that attributable to the compressive force developed as a

result of contact between the tibia and femur (see figure 10.4).

Although the mechanism by which the hamstrings limits ATT in the ACL-deficient knee is now well understood, ultimately its success depends on whether training programs can be developed to teach ACL-deficient patients to activate their hamstrings at appropriate intervals during an activity. Otherwise, excessive ATT over time may cause damage to the menisci and surrounding passive structures, resulting in further degeneration of the joint.

There is some evidence to suggest that anterior knee stability may be improved in an ACL-deficient population by training patients to alter their muscle recruitment patterns during activity. One study investigated the effects of strength training versus neuromuscular training on muscle reaction times of the quadriceps, hamstrings, and gastrocnemius muscles during isokinetic exercise. Neuromuscular training, which focused on improving coordination of the lower-extremity muscle groups that stabilise the knee during activity, was found to significantly decrease the reaction times of the knee-flexor muscles. Thus, the time to peak knee-flexor torque was

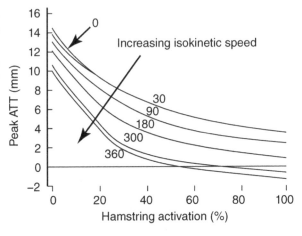

Figure 10.8 Peak ATT versus hamstrings cocontraction level for isokinetic knee extension at 0, 30, 90, 180, and 300°/s in the ACL-deficient knee.

Reprinted from *Clinical Biomechanics*, Vol. 17(9-10), T. Yanagawa, F. Serpas, M.G. Pandy and K.B. Shelburne, "Effect of hamstrings muscle action on stability of the ACL-deficient in isokinetic extension exercise," pgs. 705-712, copyright 2002, with permission from Elsevier.

decreased, allowing the flexor muscles, particularly the hamstrings, to exert more control over anterior translation of the tibia at the knee.

Sources

Pandy, M.G., & Shelburne, K.B. (1997). Dependence of cruciate-ligament loading on muscle forces and external load. *Journal of Biomechanics, 30,* 1015-1024.

Pandy, M.G., & Shelburne, K.B. (1998). Theoretical analysis of ligament and extensor-mechanism function in the

ACL-deficient knee. *Clinical Biomechanics, 13,* 98-111.

Wojtys, E.M., Huston, L.J., Taylor, P.D., & Bastian, S.D. (1996). Neuromuscular adaptations in isokinetic, isotonic, and agility training programs. *American Journal of Sports Medicine, 24,* 187-192.

Yanagawa, T., Shelburne, K.B., Serpas, F., & Pandy, M.G. (2002)). Effect of hamstrings muscle action on stability of the ACL-deficient knee in isokinetic exercise. *Clinical Biomechanics, 17,* 705-712.

in either the anterior or posterior direction and an axial component that acts along the long axis of the leg. Anterior shear forces tend to move (translate) the tibia anteriorly (forward) relative to the femur, and therefore will strain the ACL. Posterior shear forces, on the other hand, tend to move the tibia backward relative to the femur, and therefore protect the ACL from strain. However, posterior shear forces may simultaneously strain the posterior cruciate ligament (PCL). Neuromuscular training can increase knee joint stability by increasing the hamstrings:quadriceps muscle strength ratio and by fine-tuning the control exerted over the hamstrings.

BIOMECHANICAL ADAPTATIONS TO INJURY

Adaptations are common following injury and surgical treatment. One interesting example found in the orthopaedics literature is a phenomenon known as quadriceps avoidance gait. This particular adaptation occurs in ACL-deficient (ACLD) patients; that is, patients who have suffered a complete rupture of the ACL. ACLD patients usually have difficulty with movements involving either lateral thrusts (e.g., cutting from side to side) or twisting (i.e., rotation of the femur relative to the tibia in the transverse plane).

Gait analysis has been done on ACLD patients during various activities of daily living, including level walking, jogging, walking up and down stairs, and even running, cutting, and pivoting. Kinematic and force-plate data were used to estimate the net muscle moments exerted at the knee in all three planes of movement: flexion–extension, abduction–adduction, and internal–external rotation. Interestingly, the greatest functional changes between

ACLD patients and healthy subjects occurred while walking at preferred speeds on level ground. During the midstance portion of the gait cycle, the ACLD patients exhibited a net extensor moment that was significantly lower than normal (figure 10.9, solid line). One interpretation of these findings is that the ACLD patients walk by reducing the demand on their quadriceps during stance (i.e., the quadriceps muscles are activated less in the ACL-deficient patients, giving rise to a quadriceps-avoidance gait pattern).

It is important to realise that a lower net extensor moment at the knee does not necessarily mean that the moment exerted by quadriceps is lower. Because the net moment at the knee is a combination of

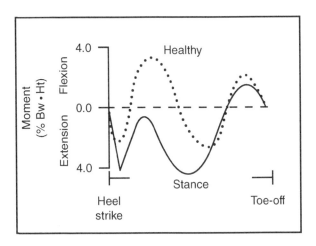

Figure 10.9 Net moment exerted about the knee by healthy subjects and by ACL-deficient patients during walking on level ground. Knee-extensor moment is normalised by body mass and height.

Adapted, by permission, from T.P. Andriacchi, R.N. Natarajan, and D.E. Hurwitz, 1997, *Basic orthopaedic biomechanics* (Philadelphia, PA: Lippincott, Williams, and Wilkins), 64.

quadriceps and hamstrings muscle action, a lower extensor moment could result from an increase in the flexor moment applied by hamstrings. However, muscle electromyography results tend to support the interpretation that quadriceps moment is lower because the activation level of the quadriceps is lower in the ACLD patients.

Gait analysis results have also shown that the ACLD patients reduce their net extensor moment by 25% during jogging compared with a 100% reduction in walking. However, the net extensor moment in stair climbing was the same in ACLD patients and normal subjects. When taken together, these results suggest that the observed quadriceps avoidance in ACLD patients is not a result of quadriceps weakness but rather is due to a change in neuromuscular control.

DEPENDENCE OF MOTOR PERFORMANCE ON CHANGES IN MUSCLE PROPERTIES

Strength and quickness are critical determinants of performance in explosive movements such as jumping, throwing, and sprinting. Together, these two factors determine the amount of power developed instantaneously during a task (i.e., muscle power is given by the product of muscle force and muscle contraction speed; see "In Focus: At What Speed Must a Muscle Shorten to Develop Maximum Power?"). As discussed earlier in this chapter, resistance strength training can alter both muscle force and fibre contraction speed. This section describes the effects that changes in the contractile properties of muscle have on motor performance, specifically, how biomechanical performance depends on muscle strength and muscle fibre contraction speed.

Vertical jumping is one of the most heavily studied motor tasks. Many scientists have studied jumping with the aim of learning more about the biomechanics and control of this task and, specifically, how performance may be influenced by training. In a recent jump-training program that incorporated stretching, plyometric exercises, and weight lifting, female high school volleyball players demonstrated a mean increase of 4 cm (1.5 in.) in jump height after training. This represented an almost 10% increase in jump height as a result of the 6 wk period of training. Even larger increases in vertical jumping performance have been documented in the literature. One extreme example is the 1984 United States Olympic gold medal volleyball team, which showed a 10 cm (4 in.) increase in jump height after 2 yr of jump training.

USING COMPUTER MODELLING TO STUDY VERTICAL JUMPING PERFORMANCE

It is gratifying to learn that jumping performance can be increased significantly by strength and neuromuscular training. However, it is difficult, if not impossible, to explain why these increases occur. Noninvasive measurements of biomechanical performance cannot pinpoint the factor or factors responsible for the increase in jump height. The reason is that several properties of the neuromuscular and musculoskeletal systems change simultaneously during the training regimen. Alternatively, computer models may be used to study the relationships between training effects and performance; that is, a model of the neuromusculoskeletal system may be used to predict how changes to specific parameters affect the performance of a motor task.

Models of the body similar to the one shown in figure 8.7 have been used to study how changes in muscle strength, muscle contraction speed, and motor unit recruitment affect jump height. The values of each of these parameters in the model were increased by amounts consistent with results obtained from strength-training programs. For example, the peak isometric strength and maximum shortening velocity of each muscle in the model were increased by 20% and the activation level of each muscle was increased by 10%. The three training effects were first applied to all muscles simultaneously and then to the ankle plantarflexors, knee extensors, and hip extensors separately. In this way, the modelling results were used to determine whether it is better to train all the leg muscles simultaneously or to isolate specific muscle groups such as the quadriceps (knee extensors) and gluteus maximus (hip extensors).

INSIGHTS INTO THE EFFECTS OF TRAINING PROVIDED BY COMPUTER MODELS

Increasing the peak isometric force of all the leg muscles in the model by 20% produced the largest increase in jump height. In this case, jump height increased by 7 cm, or 5% of that obtained before the simulated training effect. Increasing the maximum shortening velocity of each muscle by 20% or the activation level of each muscle by 10% increased jump height by only 4 cm, or 3% of that obtained for the nominal (untrained) model. When all three training effects were introduced simultaneously, jumping performance increased by nearly 17 cm, or 12% of that calculated for the untrained model. Thus, training programs that increase strength, fibre contraction speed, and motor recruitment (i.e., activation level) of all the leg muscles simultaneously are most beneficial to overall jumping performance.

The modelling results also suggest that training the knee extensors is better than training either the ankle plantarflexors or the hip extensors. When peak isometric force and maximum shortening velocity of the quadriceps were increased by 20% and quadriceps activation was simultaneously increased by 10%, jump height increased by almost 10 cm, or 7% of the value calculated for the untrained model. The same changes made to either the ankle plantarflexors or the hip extensors produced increases in jump height of only 3 cm, or 2% of the value obtained for the untrained model.

The latter result is a little puzzling because the model calculations had previously shown that the quadriceps and gluteus maximus are the major energy producers—the prime movers—of the body in vertical jumping (see "In Focus: Which Muscles Are Most Important to Vertical Jumping Performance?" in chapter 8). It should be noted here that each time a change was introduced to the model, a new optimal pattern of muscle activations was found by re-solving the optimisation problem for a maximum-height jump. Thus, one interpretation of the result is that what matters most in terms of performance is the quadriceps:gluteus maximus muscle strength ratio and not the absolute strength of these muscles. In other words, jumping performance is most sensitive to a change in the

knee-extensor:hip-extensor muscle strength ratio. Increasing quadriceps strength by 20% increases the knee-extensor:hip-extensor muscle strength ratio in the model, whereas increasing gluteus maximus muscle strength decreases this ratio. The same line of reasoning may be used to explain why increasing ankle plantarflexor muscle strength by 20% leads to an increase in jump height of just 3 cm in the model.

Training for strength is also better than training for quickness or speed. Figure 10.10 shows the effects of increasing muscle strength and muscle contraction speed on vertical jump height as predicted by the four-segment, eight-muscle, sagittal-plane model of the body described in chapter 8 (see figure 8.7). An increase in muscle strength was simulated in the model by simultaneously increasing body weight because muscle strength increases in proportion to muscle mass. Thus, changes in

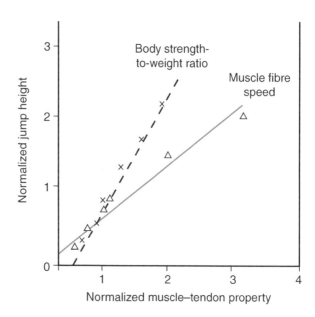

Figure 10.10 Effect of changes in muscle strength, muscle fibre contraction speed, and tendon compliance on vertical jump height. The results were obtained from the model shown in figure 8.7. Jump height is normalised by dividing by the performance predicted for the untrained (nominal) model.

Adapted from M.G. Pandy, 1990, "An analytical framework for quantifying muscular action during human movement." In *Multiple muscle systems: Biomechanics and movement organization*, edited by J. Winters and S. Woo (New York, NY: Springer-Verlag), 653-662, with kind permission from Springer Science+Business Media B.V.

body strength:weight ratio are represented in figure 10.10 rather than changes in muscle strength alone. Also, jumping performance is normalised in these results by dividing by the value of jump height calculated for the untrained model. Similarly, body strength:weight ratio and muscle contraction speed are each normalised by dividing by the value of body strength:weight ratio and muscle contraction speed in the untrained model, respectively.

The simulation results show that the slope of the line predicted for changes in body strength:weight ratio is twice as large as that obtained when changes in muscle fibre contraction speed are made (compare solid and dashed lines in figure 10.10). Thus, muscle strength has a greater effect on vertical jump height than muscle fibre contraction speed, even when the accompanying increase in body mass is taken into account.

Another important lesson learned from the modelling studies is that musculoskeletal changes must be accompanied by appropriate changes in neuromuscular control; otherwise, the expected improvement in motor performance will not be seen. In vertical jumping, if the pattern of muscle activations remains unchanged subsequent to strength training, jump height actually decreases relative to the untrained state. Figure 10.11 shows a simulated jump in which the strength of the knee-extensor muscles has been increased by 20% but the control exerted over the joints is the same as that calculated for the untrained model before the training effect was introduced. When the pattern of muscle activations was not optimised to match the changes introduced to the neuromusculoskeletal model, the body left the ground prematurely (i.e., the centre of mass was at a lower height at liftoff than is optimal for a maximal jump).

One consequence of not optimising the pattern of muscle activations (i.e., the controls) is that a larger fraction of the total work produced by the muscles goes into rotating the body segments rather than accelerating the centre of mass upward. In a maximum-height jump, approximately 90% of the total work done by the leg muscles is used to propel the centre of mass upward. This number is closer to 80% when muscle coordination (i.e., the sequence and timing of muscle activations) is not optimal. So, even though jumping performance depends heavily on muscle strength, and to a lesser extent on muscle fibre contraction speed, optimal performance is also intimately related to neuromuscular control. For this reason, and as discussed at the beginning of

Knee extensors 20% stronger, not optimised
Vertical velocity at liftoff: 2.45 m/s
Jump height: 0.31 m

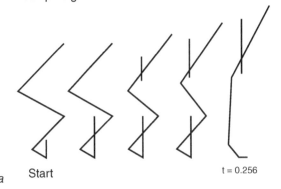

a   Start   t = 0.256

Knee extensors 20% stronger, optimised
Vertical velocity at liftoff: 2.73 m/s
Jump height: 0.43 m

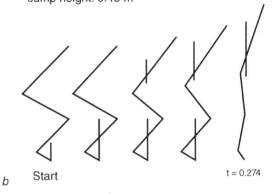

b   Start   t = 0.274

Figure 10.11 Stick figures showing the effect of changing quadriceps muscle strength in a model of jumping without reoptimisation of the control. At liftoff, the knee was hyperextended and the hip was not extended enough.

Reprinted, by permission, from M.F. Bobbert and A.J. van Soest, 1994, "Effects of muscle strengthening on vertical jump height: A simulation study," *Medicine and Science in Sport and Exercise* 26: 1012-1020.

this chapter, jump-training programs now focus on maneuvers that blend muscle strength with neural control.

SUMMARY

Training can change both muscle strength and fibre contraction speed. Strength may be increased by increasing motor unit recruitment (net neural drive to the muscle) or by increasing muscle fibre size. Contraction speed may be altered by changing the shape of a muscle's force–velocity curve or by changing the value of its intrinsic maximum shortening velocity. Early in a training program (2-8 wk), increases in muscle strength are brought by neural adaptations rather than by increases in muscle size. Training can also change the way muscle action is coordinated during activity (i.e., neuromuscular control). Neuromuscular training programs are usually designed to improve stability (balance and coordination) and proprioception (joint position sense) in addition to muscle strength. Training is vital for improving biomechanical performance and for preventing injuries during sport.

FURTHER READING

Andriacchi, T.P., Natarajan, R.N., & Hurwitz, D.E. (1997). Musculoskeletal dynamics, locomotion, and clinical applications. In V.C. Mow & W.C. Hayes (Eds.), *Basic orthopaedic biomechanics* (pp. 37-68). Philadelphia: Lippincott-Raven.

Kraemer, W.J., Fleck, S.J., & Evans, W.J. (1996). Strength and power training: Physiological mechanisms of adaptation. In J.O. Holloszy (Ed.), *Exercise and sports sciences reviews* (pp. 363-397). Baltimore: Williams & Wilkins.

Pandy, M.G. (1990). An analytical framework for quantifying muscular action during human movement. In J. Winters & S.L. Woo (Eds.), *Multiple muscle systems: Biomechanics and movement organization* (pp. 653-662). New York: Springer-Verlag.

Pflum, M., Shelburne, K.B., Torry, M., Decker, M., & Pandy, M.G. (2004). Model prediction of anterior cruciate ligament force during drop-landings. *Medicine and Science in Sports and Exercise, 36*, 1949-1958.

McMahon, T.A. (1984). *Muscles, reflexes, and locomotion.* Princeton, NJ: Princeton University Press.

PART IV

PHYSIOLOGICAL BASES OF HUMAN MOVEMENT

EXERCISE PHYSIOLOGY

Exercise physiology is a subdiscipline of physiology and a primary component of human movement studies. Physiology is the study of living organisms and how they perform various activities such as feeding, moving, and adapting to changes. **Exercise physiology** therefore is concerned with the physiological adaptations that the body makes when faced with the stress of exercise. Exercise physiology focuses on both acute (immediate) responses to exercise and chronic (long-term) adaptations to exercise training.

The basic building blocks of the body are cells, which join to form tissues such as the blood or muscular tissue. The grouping of different types of tissue forms organs, such as the brain or heart, and the organs themselves form distinct physiological systems. For example, the heart and blood vessels form the cardiovascular system.

Exercise physiology can be studied from the whole-body level to the molecular level. At the whole-body level, the exercise physiologist might ask whether a lifetime of exercise can increase life span. At the systemic level, exercise physiology aims at understanding the effects of regular exercise on, for instance, the cardiovascular system (e.g., does exercise help prevent heart disease?). Understanding the effects of exercise on the pattern of adipose tissue distribution and the implications for the prevention of obesity represents work at the tissue level. At the cellular and subcellular levels, the exercise physiologist may study the mechanisms responsible for muscular hypertrophy following strength training. Finally, work at the molecular level might focus on the extent to which elite sport performance is determined by genetics.

APPLICATIONS OF EXERCISE PHYSIOLOGY

As "Some Applications of Exercise Physiology" reveals, exercise physiology knowledge and skills are used in varied situations, including sport, physical fitness, health promotion, preventive medicine, exercise rehabilitation, the workplace, school health, physical education, and the research laboratory.

SOME APPLICATIONS OF EXERCISE PHYSIOLOGY

Health Promotion and Preventive Medicine

- Physical activity for general health
- Exercise prescription in specific conditions, such as obesity and diabetes
- Corporate and community fitness programming and health promotion

Exercise Rehabilitation

- Exercise programming in patient groups, such as those with cardiovascular diseases (e.g., ischemic heart disease, hypertension), respiratory diseases (e.g., asthma, chronic obstructive pulmonary disease), metabolic conditions (e.g., diabetes, obesity), musculoskeletal conditions (e.g., arthritis, low back pain), and neurological diseases (e.g., multiple sclerosis, Parkinson's disease)
- Injury rehabilitation (e.g., work or sport related)
- Fitness programming to prevent injury

Sport

- Fitness profiling of athletes
- Talent identification
- Designing sport-specific training programs
- Evaluating effectiveness of training programs

Worksite

- Assessment of physical fitness demands of specific occupations, such as firefighter, military personnel, and police officer
- Exercise programming to meet occupational physical fitness standards
- Prevention of occupational injuries, such as back pain

School Health, Physical Education, and Youth Sport

- Specialist teaching in health and physical education
- Coaching in community-based sport for children and adolescents
- Adapting sport for children

Research Laboratory

- Physiological, biochemical, and hormonal responses to exercise
- Mechanisms of training adaptations
- Mechanisms responsible for health benefits of physical activity

---

Exercise principles are used in designing effective assessment and training programs that enhance the fitness and performance of athletes. Talent identification involves physiological and anthropometric profiling to identify young people with potential to excel in specific sports. Exercise physiology forms the basis of physical activity prescription for healthy adults as well as for special populations such as children, the elderly, pregnant women, and individuals with conditions such as diabetes, heart disease, or arthritis.

Regular exercise also has a role in preventing and rehabilitating injury, including sport and occupational injuries. Principles of exercise physiology are used to assess the physical demands of some occupations, such as firefighting, and to develop programs to help workers achieve the fitness needed for their jobs. Coaches and physical education teachers working with children use exercise physiology to design training programs specifically for the young athlete. Finally, research in exercise physiology focuses on understanding the mechanisms underlying the physiological responses and adaptations to exercise.

In part IV, we provide an overview of the basic principles of exercise physiology. We discuss exercise metabolism in chapter 11 and the nutritional demands and needs of exercise in chapter 12. In chapter 13 we consider how these principles are applied and adapted during different stages of the lifespan. Finally, in chapter 14 we describe how the body adapts to exercise training and present recommendations for exercise prescription in adults.

HISTORICAL PERSPECTIVES

Just as the applications and levels of analysis of exercise physiology are multifaceted, so too are its historical origins. Some of the earliest writings on what we now recognise as exercise physiology are thought to come from the Indian physician Susruta around 600 BC. Susruta was an advocate of preventive medicine and promoted moderate-intensity exercise for the improvement of physical and psychological health and for the prevention of disease states. He is believed to be the first physician to prescribe exercise for health and the first to promote the concept that sedentary behaviour causes diseases such as obesity and diabetes. Ancient Greek and Roman physicians followed this application of exercise physiology, championing the importance of exercise for preserving health, advocating the necessity of balance between nutrition and exercise, and promoting various exercise training regimens. This emphasis on the use and promotion of exercise for health and longevity persists today. However, before the 19th century, the term *exercise physiology* was not used to describe the discipline. This all changed in 1855 with an article titled "On the Physiology of Exercise" by William H. Byford, published in *American Journal of Medical Sciences*. Discoveries by Fick, Tissot, and Haldane in the late 19th and early 20th centuries made possible the measurement of oxygen uptake as a marker of aerobic exercise metabolism. The seminal work of A.V. Hill on maximal oxygen uptake generated much of the foundation of modern exercise physiology and expanded the application of exercise physiology to athletes, medicine, health, and physical education.

With the reintroduction of the needle biopsy by Bergstrom in the 1960s, the focus of exercise physiology research shifted more toward biochemistry and subcellular physiology. For example, the importance of muscle glycogen stores and dietary carbohydrates to exercise performance could be studied using the biopsy procedure. Now, the emergence and availability of new imaging technologies and the adoption of cellular and molecular techniques in exercise physiology research are allowing the development of a more thorough understanding of the mechanisms underlying the exercise response.

Exercise physiology as an academic discipline and the scientific study of sport as a subject in its own right did not begin to flourish until the 1970s. Today exercise physiology is placed firmly in human movement studies (sport and exercise science, kinesiology). Where traditional technologies have proved inadequate in providing a detailed understanding of the complexity of the cellular metabolic response to exercise, exercise physiology now relies on noninvasive imaging techniques and breath-by-breath gas analysis to facilitate a more integrated understanding of the physiological responses to exercise.

PROFESSIONAL ORGANISATIONS AND TRAINING

There are a number of international and national organisations relevant to exercise physiology. As noted in chapter 1, the International Federation of Sports Medicine, the International Council of Sport Science and Physical Education, the American College of Sports Medicine, and the European College of Sports Sciences have well-established international reputations. In addition to these, there are many regional associations representing the professional interests of exercise physiology practitioners. Websites of some of these associations are listed at the end of this introduction.

Exercise physiologists are often members of other professional associations that have specialised interest in various aspects of basic physiology, physical activity, or health. These might include associations devoted to research and applications in physiology (e.g., The Physiology Society), rehabilitation (e.g., American Physiological Society), or health promotion (e.g., American Public Health Association). It is not unusual to belong to several professional associations at the same time, depending on one's training, interests, and occupation.

FURTHER READING

Tipton, C.M. (1998). Contemporary exercise physiology: Fifty years after the closure of the Harvard Fatigue Laboratory. *Exercise and Sports Science Reviews, 26,* 315-340.

Tipton, C.M. (2008). Susruta of India, an unrecognized contributor to the history of exercise physiology. *Journal of Applied Physiology, 104,* 1553-1556.

SOME RELEVANT WEBSITES

American College of Sports Medicine: www.acsm.org

American Heart Association: www.heart.org

American Society of Exercise Physiologists: www.asep.org

British Association of Sport and Exercise Sciences: www.bases.org.uk

British Heart Foundation: www.bhf.org.uk

Canadian Society for Exercise Physiology: www.csep.ca

Centers for Disease Control and Prevention (United States) Physical Activity and Health Initiative: www.cdc.gov/nccdphp/sgr/npai.htm

European College of Sport Science: www.ecss.mobi

Exercise and Sports Science Australia: www.essa.org.au

Health Canada: www.hc-sc.gc.ca

Healthy People 2020 (United States): www.healthypeople.gov

Heart and Stroke Foundation of Canada: www.heartandstroke.ca

Hong Kong Association of Sports Medicine and Sports Science: www.hkasmss.org.hk

International Federation of Sports Medicine: www.fims.org

International Society for Behavioral Nutrition and Physical Activity: www.isbnpa.org

International Society for Physical Activity and Health: www.ispah.org

Sport New Zealand: www.sparc.org.nz

World Health Organization: www.who.int

World Heart Federation: www.world-heart-federation.org

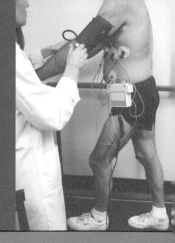

CHAPTER 11

BASIC CONCEPTS OF EXERCISE METABOLISM

The major learning concepts in this chapter relate to

▶ how energy for exercise is produced,

▶ how the body provides a continual supply of oxygen to the working muscles for sustained exercise,

▶ how exercise capacity is measured, and

▶ how human skeletal muscle cells are classified.

One important question asked by exercise physiologists is "What factors limit performance?" That is, what prevents a runner from running faster, what causes an athlete to become fatigued, and what limits how much weight a powerlifter can lift?

Exercise capacity is determined by how much energy the muscle cell can produce and how quickly this energy can be made available to the contractile elements in skeletal muscle. For example, maximum sprinting speed can be maintained for only 100 to 200 m (10-20 s at world-record pace). After the first 10 to 20 s, running pace slows because the muscle cells cannot maintain the required rate of energy supply. Fatigue occurs when the rate of energy demand exceeds the rate of production in skeletal muscle.

If we know what limits exercise capacity in a particular activity we can use this knowledge to improve performance. For example, it is well known that depletion of stored glucose (**glycogen**) from skeletal muscle causes fatigue in long-duration events such as marathon running. It is now common practise for endurance athletes to consume a high-carbohydrate diet to enhance muscle glycogen stores and delay the onset of fatigue during training and competition.

Metabolism is the collective term given to the many chemical changes that are constantly occurring within cells, such as the breakdown of glucose and fats for energy. Some of these changes, known as catabolism, involve the breakdown of large molecules to smaller ones. Other changes, known as anabolism, involve the synthesis of large molecules from smaller ones.

In this chapter we examine a range of fundamental concepts related to exercise metabolism. We discuss how energy for exercise is produced in the muscle, how energy metabolism during exercise is measured, and how different physical activities require different amounts of energy.

PRODUCTION OF ENERGY FOR EXERCISE

As the body goes from rest to maximal exercise, the metabolic rate in human skeletal muscle can increase by up to 50 times. The muscle must be able to produce energy at the needed rate. It is metabolically inefficient for cells to produce more energy than needed, and muscle cells do not store vast quantities of energy. During sustained exercise, chemical energy must be continually supplied to the working muscles.

PRODUCTION OF ADENOSINE TRIPHOSPHATE

Muscular work requires transfer of chemical energy to mechanical energy. This chemical energy is supplied in the form of **adenosine triphosphate** (ATP). Adenosine triphosphate is called a high-energy phosphate molecule because energy is released when it is split into adenosine diphosphate (ADP) and inorganic phosphate (P_i) in a reversible reaction (see figure 11.1). The energy released during this reaction is required for cross-bridge interaction between the thin and thick filaments of skeletal muscle resulting in the production of force in the muscle (see figure 3.11). The reaction is reversible because ADP can be recombined with P_i to resynthesise ATP and maintain the energy supply needed during prolonged exercise. When ATP resynthesis does not keep pace with ATP demand, exercise must slow and fatigue occurs.

a Energy for muscular work

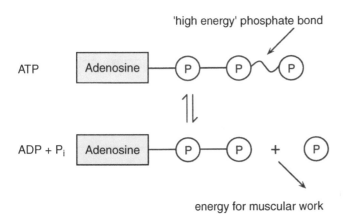

b PCr splitting to resynthesise ATP

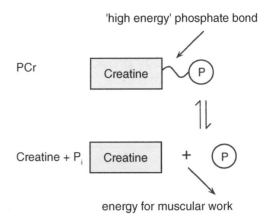

c PCr and adenosine diphosphate resynthesis of ATP

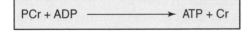

d Adenosine diphosphate resynthesis of ATP

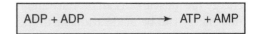

Figure 11.1 Schematic representation of high-energy phosphagens. *(a)* When ATP is split into ADP + P_i, cleavage of the terminal (high energy) phosphate bond yields energy needed for generation of tension in skeletal muscle. *(b)* Cleavage of the high-energy phosphate bond of PCr (phosphocreatine) yields energy that can be used to resynthesise ATP from ADP + P_i. *(c)* At the onset of exercise and for very brief high-intensity exercise, PCr provides the major means of regeneration of ATP from ADP + P_i.

Table 11.1 The Three Energy Systems

	Immediate phosphagen	Anaerobic glycolytic	Oxidative
Substrate	ATP, PCr, or glucose	Glycogen or glucose	Glycogen, fat, or protein
Relative rate of ATP production	Very fast	Fast	Slower
Duration at maximal pace	0-30 s	20-180 s	>3 min
Limiting factors	PCr depletion	Lactic acid accumulation	Glycogen depletion
Examples of activities	Power or weight lifting, short sprints, jumping, throwing	Longer sprints, middle-distance team sports, ball games	Endurance events, team sports, ball games

ATP = adenosine triphosphate; PCr = phosphocreatine.

Three energy systems resynthesise ATP: the immediate energy system, or **phosphagen system**; the anaerobic system, sometimes called the **anaerobic glycolytic system** or **lactic acid system**; and the aerobic or **oxidative system**, known otherwise as **oxidative respiration**. The three systems combine to make ATP available for all types of exercise, from very short bursts such as a 10 m sprint to sustained activity such as a triathlon. All three systems operate simultaneously, and their relative contributions to ATP resynthesis depend on the intensity and duration of the exercise.

PHOSPHAGEN ENERGY SYSTEM

Figure 11.1 illustrates how the phosphagen energy system enables the regeneration of ATP via the transfer of high-energy phosphate molecules. Stored muscle PCr provides an immediate source of energy for exercise. At the start of exercise, PCr is rapidly degraded to creatine plus P_i. The phosphate is then donated to ADP to resynthesise ATP. ADP molecules can also combine to synthesise ATP and adenosine monophosphate (AMP). AMP is broken down, resulting in the production of ammonia. Although this last reaction does not result in the resynthesis of ATP, it is important for understanding the bioenergetics underpinning phosphate transfer in the exercising muscle, which enables sufficient energy to be supplied despite depletions in ATP.

PCr is a major energy source for very intense exercise such as sprinting or jumping (table 11.1). The phosphagen system can resynthesise ATP at extremely high rates, but phosphocreatine is rapidly depleted during maximal-effort exercise and can be completely depleted within 10 s. Therefore,

the energetic capacity of the phosphagen system is largely dependent on the concentration of muscle PCr (figure 11.2).

Many sports require repeated maximal-effort sprints. The ability to support these energetically comes from the recovery of PCr. Replenishment of PCr can take as little as 5 min or in excess of 15 min, depending on the severity of PCr depletion and the muscle fibre type characteristics. How exercise training influences PCr recovery is discussed further in chapter 14. "In Focus: Gaining Insight Into Cellular Metabolism During Exercise—The Role of New Scanning Technologies" describes one of the ways PCr can be measured during exercise.

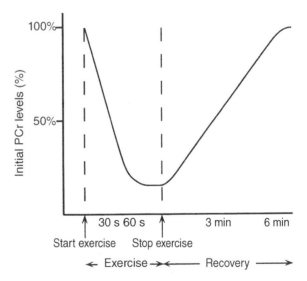

Figure 11.2 Phosphocreatine depletion and replenishment. PCr is rapidly depleted during high-intensity exercise. Up to 7 min may be required to fully replenish muscle PCr stores after depletion.

IN FOCUS: GAINING INSIGHT INTO CELLULAR METABOLISM DURING EXERCISE—THE ROLE OF NEW SCANNING TECHNOLOGIES

Our understanding of cellular metabolism during exercise has, for a long time, been hampered by the need for an invasive muscle biopsy. Magnetic resonance spectroscopy (MRS) is a noninvasive technique that enables the study of high-energy phosphates in human skeletal muscle. This technique provides a measure of metabolic activity of skeletal muscle via examination of PCr, P_i, and intracellular pH and has been validated in both adults and children.

Increases in P_i and decreases in PCr during exercise of varying intensities provide markers of mitochondrial function. A muscle with a greater oxidative capacity will show a lower change in P_i and PCr for a given increase in power output and a higher PCr recovery rate. Changes in pH provide an indication of muscle glycolytic activity, although this is not a direct measure of glycolysis.

Several studies using MRS have shown that PCr recovery rates are much quicker in long-distance runners compared with sprinters, confirming that endurance athletes have a greater oxidative capacity. This technique has also been used to explore cellular metabolism in various disease states. For example, some recent studies with patients suffering from peripheral vascular disease have shown larger changes in PCr during exercise and slower recovery rates, reflecting the impaired oxidative capacity of these patients.

PCr activity determined by MRS is providing fascinating insight into cellular metabolic processes during exercise and how these processes adapt to exercise training. However, MRS use is limited by cost and technical complexity. PCr activity determined by MRS has been shown to mirror pulmonary oxygen uptake responses; this has helped confirm that measures of gas exchange can provide valid estimates of energy utilisation at the muscular level.

Sources

McCully, K.K., Vandenborne, K., DeMeirleir, K., Posner, J.D., & Leigh, J.S. (1992). Muscle metabolism in track athletes, using ^{31}P magnetic resonance spectroscopy. *Canadian Journal of Physiology and Pharmacology, 70,* 1353-1359.

Kemp, G.J., Roberts, N., Bimson, W.E., Bakran, A., Harris, P.L., Gilling-Smith, G.L., Brennan, J., Rankin, A., & Frostick, S.P. (2011). Mitochondrial function and oxygen supply in normal and in chronically ischemic muscle: A combined ^{31}P magnetic resonance spectroscopy and near infrared spectroscopy study in vivo. *Journal of Vascular Surgery, 34,* 1103-1110.

Jones, A.M., & Poole, D.C. (2005). Oxygen uptake dynamics: From muscle to mouth—An introduction to the symposium. *Medicine and Science in Sports and Exercise, 37,* 1542-1550.

ANAEROBIC GLYCOLYTIC SYSTEM

Anaerobic glycolysis provides the major source of ATP for exercise lasting longer than a few seconds (table 11.1). As the name implies, this system produces energy anaerobically, or without oxygen, and uses glucose as the fuel (substrate). This immediate use of carbohydrate is initiated by the production of AMP during reactions in the phosphagen energy system. The glucose comes from muscle glycogen stores and from an increase in blood glucose uptake into the muscle. Glycogen is the storage form of glucose, consisting of a polymer of many glucose molecules.

One molecule of glucose, a 6-carbon sugar, is degraded via a series of 10 reactions to pyruvic acid (figure 11.3). This series of reactions is called glycolysis. Because it involves more steps, the rate of ATP synthesis is slower than the phosphagen system, but is markedly faster than oxidative respiration. When a particular activity is performed at a rate greater than maximal oxygen consumption, the activity is supported energetically largely via glycolysis, which can support activity at this intensity for about 2 to 3 min.

To maintain the high rate of glycolysis, as much pyruvate as possible must be removed. This is achieved by removal of pyruvate from the contracting muscle cells, conversion of pyruvate to lactic acid, and pyruvate being utilised by oxidative respiration. Lactic acid is a byproduct of anaerobic glycolysis; that is why this system is sometimes called the lactic acid system. Although originally viewed as a waste product that simply caused muscle fatigue, lactate is now known to be metabolically beneficial and an essential component of glycolysis. The conversion of pyruvate to lactic acid enables the high rate of glycolysis to be maintained via the recycling of protons released from the reactions of glycolysis. These protons need to be removed or metabolic acidosis results. We now know that lactic acid acts as a proton buffer and actually represses metabolic acidosis rather than causes it. Without lactic acid, high-intensity exercise could last for only about 15 s.

OXIDATIVE SYSTEM

Oxidative production of ATP occurs in the mitochondria, which are membrane-bound subcellular organelles found in most cells. Pyruvic acid, fatty acids, and amino acids are further degraded via the Krebs cycle, producing carbon dioxide, electrons, and hydrogen ions (H$^+$; figure 11.3). The carbon dioxide diffuses out of muscle cells into the blood and is transported to the lungs, where it is exhaled. The electrons and hydrogen ions enter the electron transport chain, a series of enzymes that eventually combine the electrons, hydrogen, ions, and oxygen to produce water. This transfer of electrons and hydrogen ions provides chemical energy to resynthesise ATP from ADP and P$_i$. Continued production of ATP via the electron transport chain requires a constant supply of oxygen to the muscle cell.

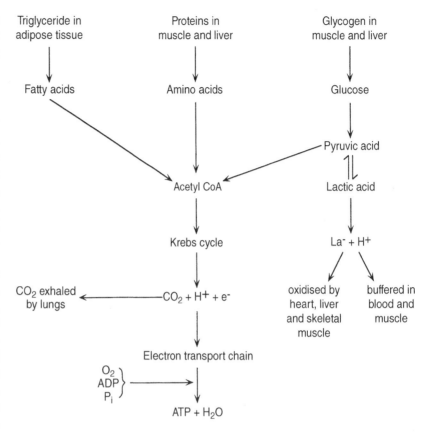

Figure 11.3 Simplified scheme of energy metabolism. Fats (fatty acids), proteins (amino acids), and carbohydrate (glycogen or glucose) can all be metabolised to produce ATP. Carbohydrate can be metabolised either aerobically or anaerobically, but fats and protein require oxygen. The end products are carbon dioxide, lactic acid, ATP, and water.

THREE ENERGY SYSTEMS AS A CONTINUUM

It is important to note that the three energy systems operate as a continuum. Each system is always functioning, even at rest. What varies is the relative contribution each system makes to total ATP production at any given time (figure 11.4).

Even in the extremes of activity, such as marathon running or brief sprinting, all three systems are used. For example, in a marathon, the phosphagen system provides ATP at the start of the race. The

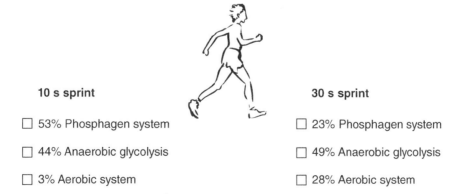

10 s sprint

☐ 53% Phosphagen system

☐ 44% Anaerobic glycolysis

☐ 3% Aerobic system

30 s sprint

☐ 23% Phosphagen system

☐ 49% Anaerobic glycolysis

☐ 28% Aerobic system

Figure 11.4 The energy system continuum. The relative contribution of each system to ATP resynthesis depends on exercise duration and intensity.

anaerobic glycolysis provides much of the needed ATP for the first few minutes until the oxidative system reaches steady state. The anaerobic system also provides a significant amount of ATP during the race (e.g., for uphill running and for sprinting at the end). Over the course of the entire marathon, the oxidative system provides most of the ATP needed.

FUELING OF ATP PRODUCTION BY CARBOHYDRATE, FATS, AND PROTEINS

The nutritional demands of exercise are discussed more thoroughly in chapter 12, but in brief, ATP can be synthesised via metabolism of carbohydrate, fats, and proteins. It is important to recognise that these nutrients are not transformed into ATP. Rather, the body breaks down these nutrients to release energy from their chemical bonds, which is then used to synthesise ATP.

Carbohydrate (i.e., glucose) can be used to synthesise ATP either anaerobically or aerobically. In contrast, fats in the form of fatty acids and proteins in the form of amino acids can be used to synthesise ATP only via oxidative respiration. At any given time the body metabolises a mixture of these nutrients to synthesise ATP. However, the relative contribution of each nutrient to ATP synthesis varies with exercise intensity and, thus, metabolic rate.

At rest and during low-intensity exercise, fatty acids predominate over glucose usage. As exercise intensity increases, ATP production progressively relies more on glucose and less on fatty acids. During maximal exercise, the muscle primarily metabolises glucose, derived from muscle glycogen. Amino acids usually contribute little (<5%) to ATP resynthesis during moderate-intensity exercise. Metabolism of amino acids may provide up to 20% of energy production after several hours of prolonged exercise in which glucose supply to the muscle is severely limited.

OXYGEN SUPPLY DURING SUSTAINED EXERCISE

The aerobic energy system provides most of the ATP for sustained exercise lasting longer than 3 min and about 20% to 30% of ATP for all-out exercise lasting 30 to 60 s. Oxygen consumption is therefore an important measure of energy expenditure during exercise.

A standard curve of oxygen consumption [$\dot{V}O_2$; the dot over the V indicates a volume rate—that is, volume per unit of time (e.g., L/min)] during exercise has several components (figure 11.5). During the initial few minutes of exercise, oxygen uptake cannot meet the energy demands and the body is said to go into oxygen deficit. During this time, ATP is supplied primarily by the phosphagen system and anaerobic glycolysis. The oxygen deficit occurs primarily because it takes the cardiorespiratory system a few minutes to adjust to the increased energy demand at the onset of exercise. This is felt as discomfort during the first few minutes of exercise, especially in unfit people. After this initial delay, the oxidative system adjusts to provide

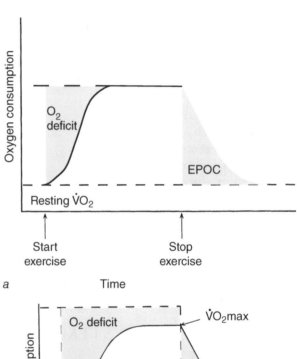

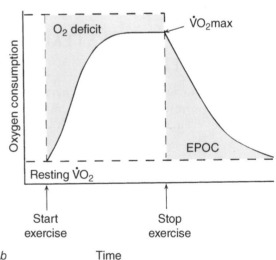

Figure 11.5 Oxygen consumption during exercise. *(a)* During steady-rate, submaximal exercise, $\dot{V}O_2$ reaches a plateau. *(b)* During maximal exercise, oxygen consumption continues to increase until a maximal value ($\dot{V}O_2$max) is achieved. During supramaximal exercise, additional ATP above that produced by oxidative metabolism is generated via anaerobic glycolysis.

the needed energy for the given rate of aerobic exercise.

During submaximal exercise of constant intensity, $\dot{V}O_2$ reaches a steady state. This represents the point at which oxygen uptake is sufficient to provide the ATP needed (figure 11.5a). Theoretically, exercise could proceed at this rate indefinitely. During exercise of continually increasing intensity, $\dot{V}O_2$ continues to increase with work rate until a maximal value ($\dot{V}O_2$max), the maximum aerobic-exercise capacity, is reached (figure 11.5b). This is discussed in more detail later in this chapter. $\dot{V}O_2$max represents the maximum amount of oxygen an individual can consume per minute during exercise. During supramaximal exercise (i.e., above $\dot{V}O_2$max), ATP is generated via anaerobic glycolysis.

At the end of exercise, oxygen uptake does not immediately return to resting levels; rather, it takes some time to return to pre-exercise levels. This slow return of oxygen uptake after the end of exercise is called the **excess postexercise oxygen consumption** (EPOC; figure 11.5, a and b). In the past, EPOC was called the **oxygen debt**, a term implying that the oxygen consumed after exercise is used to repay the deficit occurring at the onset of exercise. We now know that EPOC is more complex than a simple repayment of an anaerobic deficit. A variety of factors beyond the aerobic metabolism of lactate readapt during recovery. Factors such as the restoration of intramuscular PCr and respiratory, circulatory, hormonal, ionic, and thermal adjustments all occur during recovery and require aerobic metabolism. Exercise intensity affects EPOC more than exercise duration does because at higher intensities there is a greater reliance on anaerobic metabolism and a greater depletion of PCr and higher concentrations of lactic acid are produced.

$\dot{V}O_2$MAX AS AN INDICATOR OF ENDURANCE-EXERCISE CAPACITY

The greater the oxygen consumed at maximal levels of aerobic exertion, the higher the capacity for ATP regeneration via oxidative respiration. Thus, $\dot{V}O_2$max gives an indication of endurance-exercise capacity, or the ability to continue exercising for a long time. This is also called **aerobic power**.

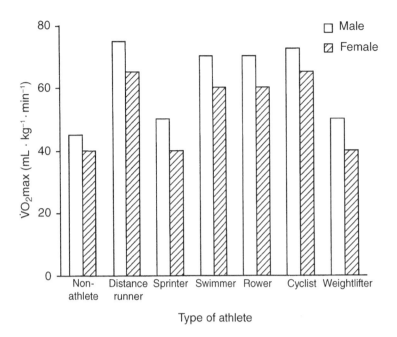

Figure 11.6 Average $\dot{V}O_2$max levels for high-performance young adult male and female athletes in different sports.

Due to a combination of heritability and training, elite endurance athletes generally have high $\dot{V}O_2$max values (figure 11.6). Approximately 40% to 50% of $\dot{V}O_2$max is genetically determined. In addition, $\dot{V}O_2$max may increase by up to 40% with aerobic exercise training. Elite endurance athletes are, to a certain extent, genetically predisposed to excelling in endurance events. However, although a high $\dot{V}O_2$max indicates potential for endurance exercise, $\dot{V}O_2$max is not the best predictor of exercise performance. That is, the individual with the highest $\dot{V}O_2$max will not necessarily be the top performer in endurance events. Other factors such as training, motivation, skill, mechanical efficiency, and the ability to maintain exercise for long periods at a high percentage of maximum capacity contribute to performance. The effects of training are discussed in chapter 12.

MEASUREMENT OF EXERCISE CAPACITY

Exercise capacity can be measured precisely in the laboratory using various exercise machines or ergometers that determine the amount of work performed and energy expended during exercise. Exercise capacity can also be estimated from sport-specific field tests. Standardised testing procedures have been developed to measure different types of exercise capacity (e.g., aerobic power, anaerobic power, anaerobic capacity, muscular strength, and muscular endurance).

AEROBIC POWER

$\dot{V}O_2$max is defined as the highest rate at which an individual can consume oxygen during exercise and is well established as the best single measure of aerobic fitness. Endurance-exercise capacity, on the other hand, is a performance measure, such as the maximum time an individual can exercise at a given speed or the total amount of work that can be accomplished in a given time.

$\dot{V}O_2$max can be measured during most modes of exercise, but the most frequently used modes of measurement are running on a motorised treadmill or cycling on a cycle ergometer (figure 11.7). Other sport-specific ergometers have been developed to simulate activities such as rowing, kayaking, cross-country skiing, and swimming. Measuring $\dot{V}O_2$max on the treadmill is most common for the average person who is generally comfortable with walking and jogging and for athletes in sports involving running. The mode of exercise selected to test $\dot{V}O_2$max should be specific to the athlete's training. For example, a cycling test provides a more accurate measure of $\dot{V}O_2$max for trained cyclists than a treadmill test.

During a $\dot{V}O_2$max test the participant starts to exercise at a comfortable pace. Exercise intensity then increases progressively until the participant can no longer continue to exercise at the given pace. This end point is called volitional exhaustion. Exercise intensity is increased throughout the test by increasing the speed or incline (or both) of the treadmill, increasing pedal resistance on the cycle ergometer, or increasing the speed or resistance on other types of ergometers.

When $\dot{V}O_2$max is assessed in the laboratory, precise equipment can sensitively monitor oxygen consumption throughout the test. The participant normally breathes through a lightweight mouthpiece or mask connected to gas analysers via a sample line or tube. The amounts of oxygen and carbon dioxide and the total volume of air breathed are measured throughout the test and are then used to calculate oxygen consumption.

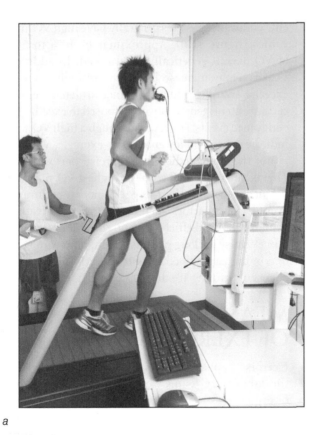

a

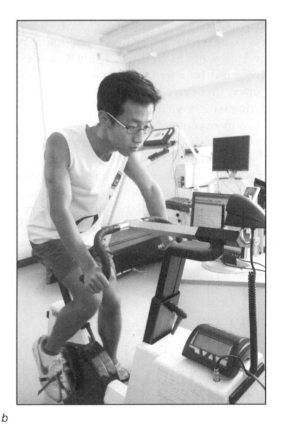

b

Figure 11.7 Exercise tests: (a) treadmill test of $\dot{V}O_2$max; (b) cycle ergometer test for anaerobic power and capacity.

The maximal value is usually achieved during the final minute of exercise, just before volitional exhaustion. The conventional criterion for the attainment of $\dot{V}O_2$max during an incremental exercise test to exhaustion is a leveling off, or plateau, in $\dot{V}O_2$ despite an increase in exercise intensity. However, both the theoretical and methodological bases of the concept of the $\dot{V}O_2$max plateau have been challenged and the validity of this traditional model is a topic of lively debate.

$\dot{V}O_2$max can be expressed as an absolute volume in liters of oxygen consumed per minute (L/min), but because it is strongly related to body size it is usual to adjust for size. Traditionally, this has been achieved by expressing the oxygen consumption in milliliters of oxygen consumed as a ratio of body mass in kilograms ($mL \cdot kg^{-1} \cdot min^{-1}$). Although this is the most common approach, it results in a scaling distortion, penalising the bigger individual and overly advantaging the small.

Theoretically, physiological variables should be scaled according to the general allometric equation $y = ax^b$, where y is the physiological variable of interest, such as oxygen consumption; x is the chosen size denominator, such as body mass; b is the scaling exponent; and a is the constant. When this equation is solved, the resultant power function ratio (y/x^b) is derived. Various studies have shown that this is more appropriate than ratio scaling when comparisons between individuals of differing sizes are being made (e.g., lean versus obese; children versus adults; men versus women).

ANAEROBIC-EXERCISE CAPACITY

Anaerobic power and capacity refer to exercise capacities in activities requiring energy production by the phosphagen and anaerobic glycolytic systems, such as brief, very intense exercise. **Anaerobic power** is the maximum or peak power (expressed in watts), usually occurring in the first 2 to 5 s, that can be achieved in an all-out exercise test. **Anaerobic capacity** represents the total amount of work, expressed as watts or kilojoules, that can be accomplished in a specified time, usually 30 to 60 s.

Anaerobic power and capacity are important in many sports and activities requiring rapid and powerful movement, such as sprinting or jumping. Several procedures—some general (e.g., Wingate

anaerobic test) and others sport specific—are used for measuring anaerobic power and capacity. Among the most commonly used are 10 s and 30 s cycle ergometer tests in which the participant pedals as fast as possible on a special cycle ergometer equipped with a work monitor to measure power and work (figure 11.7b). Other general tests include vertical jumping, sprinting, or stair climbing as measures of explosive power and specific tests for team game sports.

WHY MEASURE EXERCISE CAPACITY?

Precise measurement of aerobic and anaerobic power or anaerobic capacity allows evaluation of an individual's metabolic abilities and is an indication of their current state of fitness as well as the effectiveness of their training program. Such testing can also be used in talent-identification programs for some sports.

$\dot{V}O_2$max provides an accurate and reproducible means of standardising exercise intensity, which is relative to the individual's current ability. This is important for appropriate exercise prescription. Exercise intensity can be set as a percentage of maximum (e.g., a work rate eliciting 70% of $\dot{V}O_2$max). Each individual is therefore working at the rate relative to his or her own ability. Athletes often train at a high percentage of maximum aerobic power (80%-90% $\dot{V}O_2$max), whereas a lower exercise intensity would be recommended for the average healthy nonathlete (see chapter 14).

$\dot{V}O_2$max is a good way to determine initial fitness level from which to develop an individual exercise program and monitor progress. Because it reflects an individual's ability to take up, transport, and utilise oxygen during exercise, it offers the possibility of determining the pathophysiology of exercise limitations and the severity of functional impairment. $\dot{V}O_2$max is therefore routinely used to evaluate the effect of medical, surgical, or rehabilitative treatment on cardiopulmonary function in both adults and children.

Anaerobic power and capacity are important to many sports, especially those requiring explosive movements, such as sprinting and game-type sports. Accurate assessment of these factors is important in evaluating effective training programs and in identifying talent.

COMPONENTS OF MAXIMAL OXYGEN UPTAKE

The Fick equation has established that $\dot{V}O_2$ can be expressed as the product of cardiac output and arteriovenous oxygen difference. Cardiac output is a function of heart rate (HR) and stroke volume (SV). $\dot{V}O_2$max therefore provides a marker of the coordinated response of the cardiorespiratory system, which entails pulmonary, cardiac, and peripheral adjustments to meet the demands in muscular energy (figure 11.8).

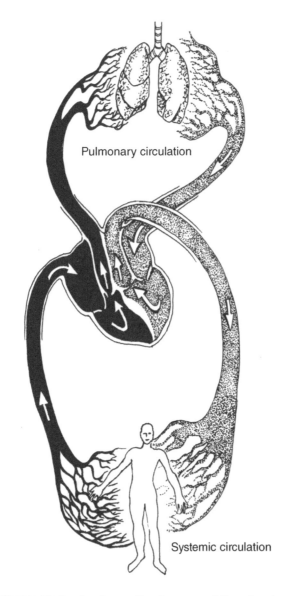

Pulmonary circulation

Systemic circulation

Figure 11.8 A schematic diagram of the circulatory system.

PULMONARY FUNCTION

Exercise pulmonary gas exchange depends on pulmonary ventilation. Pulmonary ventilation is the volume of air brought into the lungs per minute and is a function of both respiratory rate and the depth of each inspiration (i.e., the tidal volume). At maximal exercise, high rates of ventilation are usual and may be 20 to 25 times resting values. The large increase in pulmonary ventilation ensures that blood flowing through the lungs is almost fully saturated with oxygen, even at maximum exercise.

Ventilation at maximal exercise is a less-defined variable because it is dependent on the exercise protocol. During exercise of lower intensity, ventilation reflects increases in oxygen consumption and carbon dioxide production. As exercise approaches maximum, the increase in blood lactate concentration and the subsequent release of carbon dioxide, as a result of bicarbonate buffering of lactic acid, drives the ventilation (figure 11.9e). Protocols that involve large increases in exercise intensity artificially accelerate ventilation.

CARDIOVASCULAR FUNCTION

Cardiac output is the volume of blood pumped through the body per minute. Cardiac output is a function of both HR and SV [cardiac output (L/min) = HR × SV]. Cardiac output increases linearly with increasing work rate and reaches a plateau at maximum exercise capacity. During maximum exercise, cardiac output may reach values approximately four to eight times resting levels (figure 11.9c). Cardiac output represents the ability of blood to circulate and to deliver oxygen to the working muscles and is a major limiting factor for endurance-exercise capacity. Cardiac output and oxygen consumption are linearly related. In other words, a high $\dot{V}O_2$max requires a large cardiac output.

HR increases in an almost linear manner during progressive exercise before tapering off to its maximum (figure 11.9a). Maximum HR is subject to wide individual variations and is dependent on age and sex. Maximum HR is often estimated by the equation 220 minus age. HR is controlled by input from the central nervous system and responds to changes in posture, movement, blood acidity (pH), oxygen and carbon dioxide contents, and temperature.

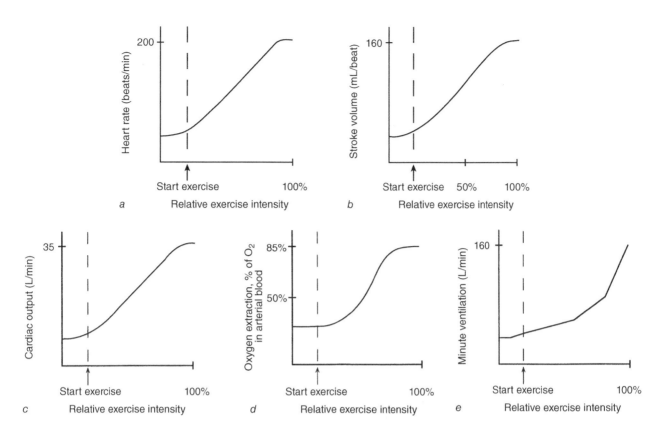

Figure 11.9 Cardiorespiratory system responses to exercise: *(a)* HR, *(b)* SV, *(c)* cardiac output, *(d)* oxygen extraction by skeletal muscle, and *(e)* minute ventilation.

SV is the volume of blood pumped by the heart with each contraction. During exercise, SV increases progressively to values 30% to 40% greater than rest and reach this level at 40% to 60% of $\dot{V}O_2$ max (figure 11.9*b*). SV then plateaus despite further increases in exercise intensity. Subsequent increases in cardiac output rely exclusively on HR.

Oxygen extraction at the muscle also increases during exercise (figure 11.9*d*). At rest tissues extract only about 25% of the oxygen contained in blood, but during maximal exercise the skeletal muscles extract 75% to 85% of the oxygen in blood. Thus, during exercise, increased blood flow to, and oxygen extraction by, the working muscles ensures adequate oxygen delivery.

DISTRIBUTION OF BLOOD FLOW DURING EXERCISE

During exercise the increased cardiac output is not uniformly distributed throughout the body.

Rather, blood is redirected through the circulation so that, with increased exercise, proportionally more blood flow goes to working muscles (figure 11.10). At rest (figure 11.10*a*) only about 20% of blood flow goes to the skeletal muscle; most goes to the brain and internal organs (viscera). At the onset of exercise, blood flow to the viscera is reduced by the narrowing of small arteries (constriction). At the same time, small arteries in the working muscles and skin open (dilate). Circulation increases in proportion to HR and exercise intensity. During submaximal exercise (11.10*b*), about 50% to 60% of blood flow may be directed to working muscles and about 10% to the skin. During maximal exercise (figure 11.10*c*), nearly 80% of blood flow may go to the working skeletal muscles. This redirection of blood flow is important to providing sufficient oxygen, glucose, and fatty acids to the muscles and heart and to removing carbon dioxide, lactic acid, and heat from the working muscles.

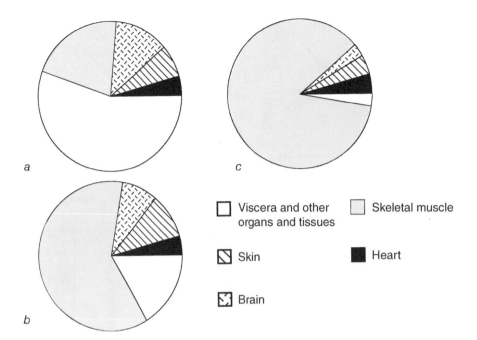

Figure 11.10 Blood flow redistribution during exercise: *(a)* rest, *(b)* submaximal exercise, and *(c)* maximal exercise. Relative proportions are expressed as a percentage of cardiac output.

HUMAN SKELETAL MUSCLE CELLS

Human skeletal muscle cells (called muscle fibres) are not all the same. The different fibre types provide specialisation in muscle function, permitting muscle fibres to adapt to a wide range of demands. It is thought that muscle cells cannot adapt completely to the full range of demands imposed by intense endurance and strength training. The specialisation of fibre types allows certain fibres to adapt optimally to one type of demand (e.g., by increasing in size and glycolytic capacity in response to strength training) and another fibre type to adapt optimally to another set of demands (e.g., by increasing oxidative capacity in response to endurance training).

MUSCLE FIBRE TYPES

As techniques in classifying muscle fibres have improved, we have realised that instead of just two main types of skeletal muscle fibre (Type I or slow-twitch fibres and Type II or fast-twitch fibres), there are now seven recognised types of skeletal muscle fibre. All seven types of fibre are seldom measured; instead, researchers usually place them into three main groups according to their relative speed of contraction and metabolic properties. Slow-twitch (ST) fibres are also known as Type I fibres, and fast-twitch (FT) fibres are also known as Type II fibres. The ST fibres contract within 100 ms, whereas the FT fibres contract within 50 ms. Muscle fibres can also be differentiated by the predominant energy system used to produce ATP. Glycolytic fibres have a high capacity for anaerobic metabolism, whereas oxidative fibres have a greater capacity for oxidative respiration. Combining information about contraction speed and metabolism allows us to classify human skeletal muscle fibres as follows:

▸ Type I, or slow oxidative fibres (ST);

▸ Type IIa, or fast oxidative glycolytic fibres (FTa); and

▸ Type IIb, or fast glycolytic fibres (FTb).

Each muscle fibre is innervated by a single motor neuron (nerve to the muscle cell); the motor neuron determines fibre type. Each motor neuron may innervate up to several hundred muscle fibres. One motor neuron and all the muscle fibres it innervates are called the motor unit (figure 3.12). Because only one motor neuron innervates a motor unit, all muscle fibres in a motor unit have the same fibre type.

In the average human, skeletal muscles comprise approximately 50% ST and 50% FT fibres; about 25% of FT muscle fibres are type IIa and 25% are type IIb. Athletes in particular sports may exhibit different distributions of fibre type compared with the average individual.

ST fibres are well suited to activities requiring low force generation over a long period of time. Muscles used primarily for posture and endurance activities (e.g., the soleus muscle, the deeper muscle in the calf) contain a high proportion of ST fibres. Muscles used for forceful contractions (e.g., gastrocnemius, the surface muscle in the calf) generally contain a higher percentage of FT fibres. The various fibre types are mixed in the muscle so that a microscopic view of a muscle in cross-section reveals a mosaic of the different fibre types, as shown in figure 11.11. The sample depicted was histochemically stained to delineate type IIa and type IIb fibres. Capillaries (C) appear as small, round spots between the skeletal muscle fibres.

As table 11.2 shows, ST fibres are smaller, slower to contract, and not capable of generating as much force as FT fibres. ST fibres are fatigue resistant; that is, they can continue to contract repeatedly without undue fatigue. These fibres contain many mitochondria and are surrounded by several capillaries, ensuring a generous supply of oxygen. Thus, ST fibres have a high capacity for oxidative metabolism and are used primarily for endurance activities. It is also believed that during intense exercise, ST fibres take up and metabolise lactic acid produced by FT fibres.

FTb fibres are the largest, fastest, and most forceful of the three main fibre categories. FTb fibres have a low oxidative capacity but high anaerobic glycolytic capacity and are capable of producing large amounts of lactic acid. These fibres fatigue easily.

FTa fibres exhibit characteristics of both ST and FTb fibres. They resemble FTb fibres in that they are large, fast, capable of forceful contraction, and high in glycolytic capacity. FTa fibres are also similar to ST fibres because they have more mitochondria, a moderate capillary supply, and higher oxidative capacity compared with FTb fibres. FTa fibres are used in activities such as rowing, swimming, sprinting, and moderate-intensity weightlifting.

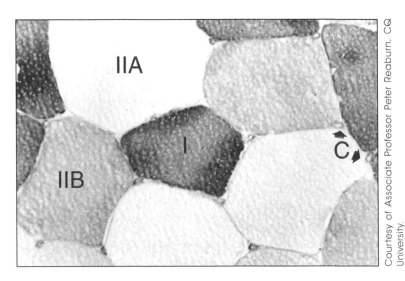

Courtesy of Associate Professor Peter Reaburn, CQ University.

Figure 11.11 Photomicrograph (photograph through a light microscope) of a cross-section of human skeletal muscle from the vastus lateralis (thigh) muscle.

Table 11.2 Characteristics of Human Skeletal Muscle Fibre Types

Characteristic	ST	FTa	FTb
Fibre size	Small	Large	Large
Contraction speed	Slow	Fast	Fast
Force	Low	High	High
Glycolytic capacity	Low	High	High
Oxidative capacity	High	Moderately high	Low
Capillary supply	High	Moderately high	Low
Fatigue resistance	High	Moderate	Low

ST = slow oxidative fibres; FTa = fast oxidative glycolytic fibres; FTb = fast glycolytic fibres.

MUSCLE FIBRE TYPE AND EXERCISE CAPACITY

The different types of muscle fibres allow muscles to respond to a wider variety of exercise demands, from finely controlled movements requiring little force to fast and powerful movements requiring maximum force.

ACTIVATION OF FIBRE TYPES DURING EXERCISE

When a skeletal muscle contracts, not all muscle fibres are activated, or recruited, to produce force. Muscle fibres are activated in proportion to the amount of force required. Muscle fibre recruitment follows a pattern, called the size principle, in which smaller ST fibres are activated first, followed by FTa and then FTb fibres. FTb fibres are activated during forceful contractions requiring 70% of maximum muscular force (figure 11.12). ST fibres are activated during all contractions, and although FT fibres may provide most of the force during near-maximal contractions, ST fibres are still activated and contribute to force production. This principle can be applied to the design of exercise training programs. For example, low-intensity endurance exercise (e.g.,

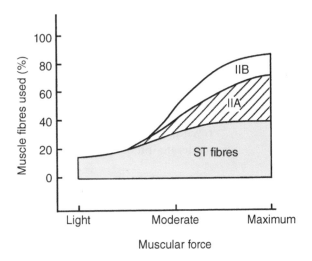

Figure 11.12 Skeletal muscle fibre recruitment during exercise. Smaller ST fibres are activated first, followed by larger FT fibres (FTa and then FTb fibres). FTb fibres are activated during forceful contractions requiring greater than 70% of maximum muscular force.

Adapted, by permission, from J.H. Wilmore and D.L. Costill, 1999, *Physiology of sport and exercise*, 2nd ed. (Champaign, IL: Human Kinetics), 202.

walking, slow jogging) recruits mainly ST fibres; endurance training of higher intensity must be performed in order to recruit and thus induce adaptations in FTa and FTb fibres. Similarly, high-intensity weight training at near-maximum muscular effort will recruit FTb fibres (as well as ST and FTa fibres).

SKELETAL MUSCLE FIBRE TYPING

In humans, muscle fibre typing is performed on a small (20 mg—about the size of a grain of rice) sample of muscle obtained by the muscle biopsy technique. A small (1 cm) incision is made through the skin and underlying tissue down to the muscle layer. A thin muscle-biopsy needle is inserted into the incision and the muscle sample is removed by the needle. The pain is minimal and the procedure is performed under local anesthesia. Once obtained, the muscle sample is frozen to preserve its structure and then cut in cross-section using specialised equipment. The cylindrical muscle fibres appear as circular profiles. Sections are stained with special dyes to visualise muscle fibre properties such as glycogen content or muscle proteins. Muscle biopsy has become a common procedure in adult exercise physiology research. The athlete can usually resume training within 1 d of having a muscle biopsy. However, because a muscle biopsy is an invasive procedure, it is not generally used for routine testing and remains primarily a research tool for understanding the metabolic response to exercise.

IMPORTANCE OF MUSCLE FIBRE TYPES TO SPORT PERFORMANCE

As shown in figure 11.13, the relative proportions of fibre types differ between types of athletes. For example, high-performance distance runners tend to have about 80% ST and 20% FT fibres, whereas elite sprinters and power athletes have 60% to 70% FT and 25% to 40% ST fibres. These differences imply that muscle fibre type is important to elite performance in specific sports. However, as with $\dot{V}O_2$max, top performance is related to many factors. Muscle fibre type gives only a broad indication of potential in sports at the extremes of the energy system continuum, such as distance running, sprinting, or powerlifting. A mixture of fibre types is advantageous in other sports requiring optimisation of all the energy systems and skill, such as soccer, basketball, or tennis. The question of whether skeletal muscle fibre type and number can be changed with exercise training is discussed in chapter 14.

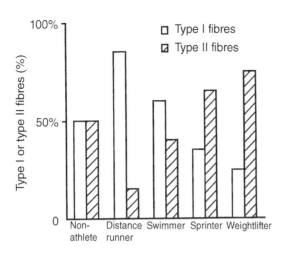

Figure 11.13 Distribution of skeletal muscle fibre types in nonathletes and various types of athletes.

Data from McArdle, Katch, and Katch, 1996.

SUMMARY

Exercise physiologists study exercise capacity and address issues such as what causes fatigue during intense or prolonged exercise and what the limits to human performance are. Exercise capacity might be limited by the rate of energy production in muscle cells. When ATP demand exceeds the rate of production, fatigue occurs and the pace of exercise must slow. Skeletal muscles produce ATP via three key energy systems: the phosphagen system, anaerobic glycolysis, and aerobic metabolism. These three systems ensure that ATP is available for all types of human movement, from short, intense bursts to prolonged exercise. For prolonged exercise, oxygen must be continually supplied to the working muscles. When oxidative metabolism cannot regenerate sufficient ATP, the muscles rely on anaerobic glycolysis. Although $\dot{V}O_2$max is a good general indicator of endurance-exercise capacity, there are better predictors of ability in particular sports that require a high level of aerobic capacity. Human skeletal muscle cells can be grouped into three primary types. ST fibres are small and have high oxidative capacity; FTb fibres are large and have high glycolytic capacity; and FTa fibres are large and have moderately high oxidative capacity and high glycolytic capacity.

FURTHER READING

Baker, J.S., McCormick, M.C., & Robergs, R.A. (2010). Interaction among skeletal muscle metabolic energy systems during intense exercise. *Journal of Nutrition and Metabolism*, 2010:905612; doi:10.1155/2010/905612

Kenney, W.L., Wilmore, J.H., & Costill, D.L. (2012). *Physiology of sport and exercise* (5th ed.). Champaign, IL: Human Kinetics.

Kraemer, W.J., Fleck, S.J., & Deschenes, M.R. (2012). *Exercise physiology: Integrating theory and application*. Baltimore: Lippincott Williams & Wilkins.

Maughan, R., & Gleeson, M. (2010). *The biochemical basis of sports performance* (2nd ed.). Oxford: Oxford University Press.

CHAPTER 12

BASIC CONCEPTS OF NUTRITION AND EXERCISE

The major learning concepts in this chapter relate to

▸ the energy requirements of exercise;

▸ the roles that carbohydrate, fat, and protein play in providing the nutrients needed for exercise;

▸ the fluid requirements for exercise; and

▸ the nutritional and performance benefits of dietary supplements.

Nutrition is of importance to both health and exercise performance. Good nutrition helps with recovery, optimises performance, and is important for preventing problems that might occur if nutritional deficiencies occur. In this chapter we consider the different nutrients that the body utilises to supply energy. We also consider the importance of diet to exercise metabolism and examine how dietary manipulation can affect performance.

ENERGY REQUIREMENTS OF EXERCISE

The energy cost of exercise depends on the type and intensity of exercise, the mechanical efficiency of the person exercising, and the individual's body mass. Environmental factors such as temperature, wind, and terrain may also influence the energy cost of exercise.

As described in chapter 8, mechanical efficiency is the energy cost used to accomplish a specific amount of work. The human body is at most 25% efficient. Mechanical efficiency and the energy cost of exercising at a particular intensity vary across individuals. In other words, a mechanically efficient person uses less energy to exercise at a given intensity than someone who is less mechanically efficient. In some activities (e.g., swimming, rowing), mechanical efficiency or technique is an important determinant of performance. The term *economy of movement* tells us the oxygen cost of that particular exercise [e.g., the oxygen cost ($\dot{V}O_2$) of running at 14 km/h]. Economy of movement and mechanical efficiency are related because oxygen consumption is an indirect measure of energy use.

For activities in which body mass is supported, such as cycling or swimming, energy cost is relatively independent of body mass. However, for activities in which body mass is not supported, such as walking or running, the energy cost increases with body mass. The energy cost is highest in activities that use the entire body or large muscle groups, such as running.

Knowledge about the energy cost of an activity can be used in several ways. For example, because many people exercise to control body mass, knowing how much energy is expended in certain activities is useful in prescribing the appropriate amount of exercise for fat loss. The energy cost of exercise training is also important in the planning of diets for athletes. Athletes, especially endurance athletes, often have difficulty consuming enough energy to meet the high metabolic demands of training. Measuring the energy cost of exercise is therefore important for understanding the physiological responses to exercise and the regulation of metabolism during exercise.

NUTRIENTS FOR EXERCISE

Carbohydrates, fats, and proteins all play important roles in relation to exercise. During low intensity exercise at 50% or less of $\dot{V}O_2$max, a high proportion of energy comes from fat. In contrast, carbohydrates are the dominant fuel source as exercise intensity increases.

CARBOHYDRATE

Dietary carbohydrates such as grains are first broken down into glucose, a simple sugar containing 6 carbon, 12 hydrogen, and 6 oxygen atoms (chemical formula: $C_6H_{12}O_6$). One gram of carbohydrate provides about 4 kcal or 17 kJ of energy. This is about 7% less energy than is produced per gram of fat (see table 12.1). Carbohydrate is the only nutrient or substrate that is utilised in both the anaerobic glycolytic and oxidative metabolic pathways.

Glucose can be stored in the muscles and liver as glycogen, a long chain (polymer) of many glucose molecules linked together. Glycogen is broken down to glucose, which can then be used to resynthesise adenosine triphosphate (ATP).

The amount of carbohydrate an individual needs depends on the type, intensity, and duration of

Table 12.1 Energy Released per Litre of Oxygen Consumed

Substrate	Energy per litre of oxygen, kcal (kJ)
Carbohydrate	5.05 (21.2)
Fat	4.70 (19.7)
Protein	4.82 (20.2)

Compiled from McArdle, Katch, and Katch, 1996; Wilmore and Costill, 1999.

exercise that the individual performs. Carbohydrate ingestion before, during, and after exercise can delay fatigue and improve endurance performance. Exercise capacity is directly related to both pre-exercise muscle glycogen stores and the amount of dietary carbohydrate consumed. A low-carbohydrate diet is associated with low muscle glycogen stores and poor exercise capacity, whereas a high-carbohydrate diet increases muscle glycogen stores and exercise capacity (figure 12.1).

Muscle glycogen stores can be depleted with as little as 40 min of intense continuous exercise and are particularly important in events lasting more than 90 min. Once muscle glycogen stores are depleted, glucose availability is limited and the muscle must rely on fatty acids and, to a limited extent, amino acids for ATP resynthesis. Therefore, glycogen depletion is associated with an inability to maintain the rate of exercise and with the perception of fatigue. In marathon running, this point is called "hitting the wall."

Muscle glycogen stores can also be depleted by prolonged periods of interval training, during which anaerobic glycolysis is the predominant ATP-regeneration system. For example, 30 s of all-out sprinting may deplete 25% of muscle glycogen stores, and 10 sprints (1 min each) may deplete 50% of muscle glycogen stores. Sports in which glycogen depletion may occur include soccer, basketball, and any activity requiring repeated high-intensity sprinting over a prolonged period of time. Swimmers, rowers, and weightlifters training over extended periods of time may also experience fatigue due to glycogen depletion.

Once glycogen is depleted from the muscle it takes 24 to 48 h to fully restore glycogen levels. Athletes who train intensely each day may therefore become chronically glycogen depleted and have difficulty maintaining their training and competitive performance. Appropriate carbohydrate intake is therefore imperative for maximising muscle glycogen stores and, therefore, endurance performance.

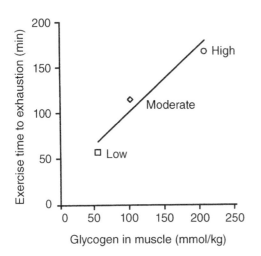

Figure 12.1 Dietary carbohydrate and exercise performance. Increasing dietary carbohydrate enhances muscle glycogen stores. Higher muscle glycogen content increases exercise time to exhaustion by delaying the point of glycogen depletion and, thus, the onset of fatigue.

Reprinted, by permission, from J.H. Wilmore and D.L. Costill, 2004, *Physiology of sport and exercise*, 2nd ed. (Champaign, IL: Human Kinetics), 409.

Normally, athletes training for several hours each day consume a high-carbohydrate diet and would need to eat little and often to achieve the optimal intake. If exercise training is of only moderate duration and low intensity, a consumption of 5 to 7 $g \cdot kg^{-1} \cdot h^{-1}$ is adequate. When training increases to moderate or vigorous intensity, intake should be increased to 7 to 12 $g \cdot kg^{-1} \cdot h^{-1}$. Athletes in extreme training programs, training for at least 4 to 6 h/d, would require 10 to 12 $g \cdot kg^{-1} \cdot h^{-1}$ to achieve their carbohydrate requirements.

During prolonged exercise (at least 90 min), carbohydrate ingestion can help maintain high rates of carbohydrate oxidation and spare muscle glycogen stores. An additional benefit is that some evidence shows that carbohydrate ingestion during long bouts of exercise, such as a triathlon, helps reduce the athlete's perception of fatigue.

Postexercise carbohydrate intake is now known to benefit the athlete, particularly when the athlete must exercise again within 6 to 12 h of the first exercise bout. The recommendation is that the athlete should consume about 1 $g \cdot kg^{-1} \cdot h^{-1}$, beginning within the first 30 min after the first exercise session, and continuing for a couple of hours after the exercise session. This practise is particularly important in multievent competitions in which athletes compete in several events over the course of a few days.

Easily digested foods such as sport drinks, sport bars, fruits, breads, and wheat cereals are effective.

FAT

The important components of fat for metabolism are triglycerides and fatty acids. Fatty acids are long chains of carbon atoms joined together with hydrogen and an acidic component (COOH). Fatty acids can be classified as saturated, unsaturated, monounsaturated, or polyunsaturated. A **saturated** fatty acid contains the maximum number of hydrogen atoms, and no double bonds exist between carbon atoms. **Unsaturated** fatty acids do not contain the maximum number of hydrogen atoms and have at least one double bond between carbon atoms.

Fatty acids are stored as triglycerides. A **triglyceride** is made up of a glycerol molecule plus three fatty acids. Triglycerides are stored mostly in fat cells but are also stored in other tissue such as skeletal muscle. When fat is needed as an energy source, the triglycerides are broken up in a process called **lipolysis**. The fatty acids are then metabolised for ATP regeneration. Glycerol is not metabolised by skeletal muscle, but it can be used by the liver to synthesise glucose. Fat stores in humans are large, and even in the leanest individuals exercise that lasts for hours will not deplete fat stores.

One gram of fat produces about 9 kcal of energy when metabolised. It is recommended that 25% to 30% of daily energy intake should come from dietary fat, although this also depends on the body composition of the individual. If weight loss is required in an athlete, fat intake can be reduced to 15% to 20% of daily energy intake.

PROTEIN

Proteins can be found in both plants and animals. The molecules that make up proteins are called amino acids; there are about 20 of these. Some are **essential amino acids**, which are vital to our diet because the body cannot synthesise them; others are **nonessential amino acids**, which the body can synthesise.

Many athletes, especially sprinters and weightlifters, consume a high-protein diet and supplement their diet with protein or amino acid powders. The belief is that this extra protein will increase muscle growth. Although it is true that athletes need more protein than do nonathletes, a normal, well-balanced diet more than adequately meets the protein needs of virtually all athletes.

The recommended daily protein intake for healthy nonathletes is 12% to 15% of daily energy consumption, or 0.8 g of protein per kg of body mass per day. For a 70 kg individual this would mean a daily intake of about 56 g of protein. For a 55 kg individual this would mean 44 g of protein. The general recommendation is that athletes competing in activities that require strength, power, or speed should consume more protein—about 1.5 to 2 g·kg^{-1}·d^{-1}—whereas endurance athletes should consume 1.5 to 1.6 g·kg^{-1}·d^{-1} (see table 12.2). In developed countries, average protein intake is usually 2 to 3 times higher than the recommendations and a protein deficit is usually present only when severe dietary restriction is coupled with a vegan diet.

Excess dietary protein is not used by skeletal muscle. Rather, it is excreted by the kidneys or used to synthesise fat. Neither is desirable because excess protein excretion places an extra burden on the kidneys and excess fat impairs performance in athletes. High-protein diets and protein supplements are also costly. In addition, a high-protein diet usually does not have sufficient carbohydrate to fuel extended training sessions, and the desired muscle growth may actually be compromised because of inadequate energy to fuel protein synthesis in the muscle cell.

FLUID REQUIREMENTS DURING EXERCISE

Although water is rarely considered a nutrient, it is an essential part of our daily diet. Humans lose heat primarily through the evaporation of sweat; therefore, fluid replacement is of concern in long-duration events in the heat. Sweat rates can be very high—up to 3 L of sweat/h in very hot conditions—but sweating rate is highly dependent on metabolic rate during exercise. Therefore, fluid replacement requirements can vary considerably between individuals. Adequate fluid intake after exercise is important for complete recovery of fluid balance. This is usually achieved by voluntary consumption of both food and fluid over a 24 h period.

When training or competition lasts less than 90 min, consuming fluid offers no physiological or performance benefit except reducing thirst. During prolonged exercise the redistribution of water in the body and the loss of body water via sweating may reduce blood volume. The loss of body water may seriously affect the ability of the body to regulate temperature. The cardiovascular system adjusts to the loss of blood volume by increasing heart rate to offset declines in stroke volume. Therefore, when exercise is performed in hot environments, stress on the cardiovascular system is greater.

Heat illness, which occurs when the athlete's cardiovascular and thermoregulatory systems are seriously impaired, is life threatening if not treated properly. Fluid replacement during events or training sessions of longer duration is important. However, sufficient evidence suggests that fluid replacement does not need to match the mass lost from sweating. Most athletes will drink when they are thirsty. Thirst is stimulated by an increasing serum osmolality and blood pressure. Encouraging excessive intake of water during exercise must be avoided to prevent hyponatremia, a condition that occurs when the kidneys inadequately clear the excess fluid. This results in an expansion of the total body water and a life-threatening change in **osmolality**.

Table 12.2 Meeting Protein Needs Through the Diet

	Nonathlete	Athlete
Total energy consumed (kcal/d)	2,500 (10,500)	3,750 (15,750)
Body mass (kg)	70	70
Percent energy intake as protein	12 to 15%	12 to 15%
Energy intake as protein (kcal/d)	300 (1,260) to 375 (1,575)	450 (1,890) to 562 (2,362)
Grams of protein per day	300/4.2 = 71 to 375/4.2 = 89	450/4.2 = 107 to 562/4.2 = 134
Grams of protein per kg body mass per day	89/70 = 1.3	134/70 = 1.9

When the training session or competition lasts for more than 1 h, a carbohydrate electrolyte (sport) drink is thought to be preferential to water because the drink can double up as a fuel source. In addition, the sport drink replenishes the electrolytes lost through sweating, which may help protect against developing hypernatremia.

Sport drinks are increasingly being used by athletes in an attempt to enhance both recovery and performance. A strong evidence base to support the benefits of this type of nutritional supplementation is still being developed (see "In Focus: Can Sport Supplements Help Athletes Achieve Nutritional Goals?").

IN FOCUS: CAN SPORT SUPPLEMENTS HELP ATHLETES ACHIEVE NUTRITIONAL GOALS?

Billions of dollars are spent globally each year on various supplements—including pills, powders, gels, food bars, and drinks—that are believed to enhance athletic performance. Athletes are usually drawn to these simply because of good marketing rather than a strong evidence base that proves the true value of the product in enhancing performance.

Many recreational and serious athletes drink sport drinks and consume sport gels or sport bars. These products generally have high carbohydrate content and are in an easy-to-consume form. This makes them useful for postexercise carbohydrate-recovery protocols as well as for rehydration. These easy-to-ingest products are effective sources of carbohydrate during long-term exercise training or competitions and provide a useful, low-bulk source of day-to-day carbohydrate for the more extreme endurance athletes who may otherwise struggle to consume the required volume of carbohydrate from a normal diet.

Some of the available sport supplements contain prohormones, which are chemicals that are precursors to certain hormones. Although harmless on their own, prohormones from the steroidal hormonal group can be converted into banned substances such as testosterone or nandrolone once ingested. This is a big problem for serious athletes competing under antidoping legislation because it can lead to unintentional doping.

Caffeine is probably the supplement that is most regularly consumed because of the strong coffee culture in many places across the globe and because so many energy drinks contain caffeine. It is no longer on the World Anti-Doping Agency's list of banned substances, but it does have proven effects on performance. Caffeine is a stimulant that affects the central nervous system and reduces the perception of fatigue. It has been shown to have a positive effect on endurance performance both before and throughout long-duration events, but little evidence shows any effect on short-term, high-intensity exercise.

Creatine monohydrate is probably the most effective legal nutritional supplement for improving performance in high-intensity, intermittent exercise. Creatine is a naturally occurring substance and an important component of the high-energy phosphate system that skeletal muscle utilises to regenerate ATP. Since the early work of Hultman and colleagues (1996), studies have repeatedly shown that creatine supplementation can enhance performance in high-intensity, repeated sprints via increases in muscle creatine and creatine phosphate contents. Evidence also shows that the quality of a resistance-training workout can be improved with creatine supplementation. Although some doubts exist about the long-term safety of creatine supplementation, currently no convincing evidence shows any detrimental effects.

Sources

Burke, L. (2007). *Practical sports nutrition*. Champaign IL: Human Kinetics.

Hultman, E., Söderlund, K., Timmons, J.A., Cederblad, G., & Greenhaff, P.L. (1996). Muscle creatine loading in men. *Journal of Applied Physiology, 81*, 232-237.

Sökmen, B., Armstrong, L.W., Kraemer, W.J., Casa, D.J., Dias, J.C., Judelson, D.A., & Maresh, C.M. (2008). Caffeine use in sports: Considerations for the athlete. *Journal of Strength and Conditioning Research, 22*, 978-986.

SUMMARY

Good nutrition plays an important role in the lives of athletes and influences both training and competition performance. The athlete's diet should supply the body with the right amount of fuel and nutrients to achieve the training goals, to recover properly between training sessions, and to be in optimal condition for competition. Different types of sport and exercise require different dietary strategies. Athletes should seek professional advice to determine the strategy that will best help them achieve their goals.

FURTHER READING

Burke, L.M., & Deakin, V. (2000). *Clinical sports nutrition* (2nd ed.). Sydney: McGraw-Hill.

Burke, L.M., Hawley, J.A., Wong, S.H.S., & Jeukendrup, A.E. (2011). Carbohydrates for training and competition. *Journal of Sports Sciences, 29*(S1), S17-S27.

Montain, S.J., & Cheuvront, S.N. (2008). Fluid electrolyte and carbohydrate requirements for exercise. In N.A.S. Taylor & H. Groeller (Eds.), *Physiological bases of human performance during work and exercise* (pp. 563-576). Philadelphia: Churchill Livingstone Elsevier.

CHAPTER 13

PHYSIOLOGICAL CAPACITY ACROSS THE LIFE SPAN

The major learning concepts in this chapter relate to

- changes in exercise performance over the life span,
- the fact that children are not miniature adults,
- growth versus training in children,
- exercise prescription for children,
- exercise capacity during ageing,
- exercise prescription for older individuals, and
- sex differences in physiological responses and adaptations to exercise.

A great deal of our knowledge of exercise physiology comes from young adult males. Most examples in textbooks default to the standard 70 kg adult male. Many exercise physiology studies from the 1950s to the 1980s were conducted on young men studying to become physical education teachers. Even today a great deal of exercise science research uses university students as subjects, many of whom are pursuing degrees in sport and exercise science. Physiological responses to exercise vary greatly between individuals and are influenced by age and sex. A complete understanding of exercise physiology is therefore possible only if we identify differences in responses to exercise across the life span and between the sexes.

Exercise and physical activity are seen as essential components of a child's life and integral to optimal growth and development. As the child and adolescent age and mature they experience significant growth and improvement in many physiological functions. Continued ageing into older adult life results in a reduction in muscle mass and declines in functional capacity. Many children and older adults compete in sport or exercise regularly for health. The importance of a thorough understanding of the physiological responses to exercise in both the young and the old cannot be overstated.

RESPONSES TO EXERCISE IN CHILDREN

Children are not simply smaller, less-muscular adults. They are growing and maturing, and their physiological responses to exercise are unique and changeable as they progress through childhood into adolescence and on to adult life. The exercise capacity of boys and girls changes accordingly and is perhaps best illustrated by considering running speed over various distances (figure 13.1; data from age-group track and field championships in North America). Figure 13.1, *a* through *c*, shows that

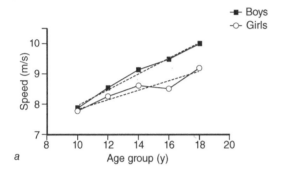

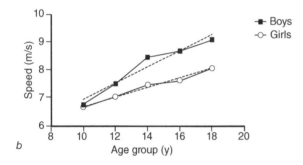

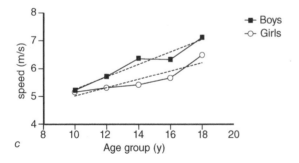

Figure 13.1 Comparison of the running speeds of girls and boys of different ages over distances of *(a)* 100, *(b)* 400, and *(c)* 1500 m.

regardless of the distance, running speed increases with age in both girls and boys. In the primary school years (<12 yr of age) the differences in running speed between boys and girls are minimal, but as girls and boys enter puberty the differences in running speed become more pronounced, with boys achieving running speeds that are about 8% to 15% quicker than girls. This difference persists into adulthood, with men outperforming women by 9% to 15% in distances ranging from 100 m to 200 km.

Running speed in boys improves much more rapidly than in girls across shorter distances (e.g., 100 and 400 m), as shown in figure 13.1, *a* and *b*. The dotted lines on the graphs indicate the line of best fit; greater steepness in the dotted line represents a greater rate of improvement. The rate of improvement over longer distances (e.g., 1500 m; see figure 13.1c) is similar between boys and girls.

Understanding what accounts for the improvements in performance with growth and maturation and why differences that persist into adult life exist between girls and boys leads back to the question of what limits performance. Unfortunately, the physiological determinants of improvements in performance across childhood and adolescence are not yet completely understood.

ENDURANCE EXERCISE IN CHILDREN

Children's aerobic and anaerobic performance is different from that of adults. The physiological responses to exercise alter as children progress into adolescence and on into adult life. This next section provides an overview of these changes and the child–adult differences.

AEROBIC FITNESS IN CHILDREN

The development of aerobic fitness with age has been studied extensively in European and North American children. Peak oxygen uptake (peak $\dot{V}O_2$), the highest rate at which a child or adolescent can consume oxygen during exercise, is recognised as the best single indicator of young people's aerobic fitness. The term *peak*, as opposed to *max*, has become widely recognised as the appropriate term to use with young people because a plateau in $\dot{V}O_2$ is rarely achieved and is not a prerequisite for the valid determination of young people's aerobic fitness.

When expressed in absolute terms (L/min), peak $\dot{V}O_2$ increases almost linearly with age in boys; the average increase is about 150% between 8 and

16 yr of age. An increase in peak V̇O₂ is also apparent in girls, but it is less marked and appears to level off around the age of 14 yr. Girls therefore show a smaller overall increase in peak V̇O₂ of about 80% between the ages of 8 and 16 yr.

When peak V̇O₂ is expressed as a ratio standard with body mass (mL·kg⁻¹·min⁻¹), it shows a progressive decline in girls from 13 yr of age; values decrease from approximately 45 to 35 mL·kg⁻¹·min⁻¹. Mass-related peak V̇O₂ in boys remains almost unchanged; values are 48 to 50 mL·kg⁻¹·min⁻¹ between the ages of 8 and 18 yr.

The consistently lower aerobic fitness of girls from late childhood is, at least in part, attributable to changes in muscle and fat during puberty. If you control for the influence of body fat, the sex difference in aerobic fitness is reduced, although it is still apparent. This difference in aerobic fitness persists into adult life and, even when similarly trained, men have about a 10% greater V̇O₂max than women. Understanding the physiological mechanisms that underlie the sex differences in aerobic fitness requires consideration of the coordinated responses of the pulmonary, cardiac, and peripheral adjustments that are made during exercise. A brief consideration of each is provided here; differences between children and adults and between girls and boys are highlighted.

SEX AND AGE DIFFERENCES IN PULMONARY RESPONSES TO EXERCISE

During puberty, thoracic growth is greater in boys. When coupled with a greater muscle mass, boys have approximately 25% greater lung volumes than girls who are matched for height. As noted in chapter 11, exercise pulmonary responses depend on pulmonary ventilation, which is a function of respiratory rate and tidal volume. Excessive respiratory rate is often indicative of poor conditioning or of abnormalities in breathing mechanics; however, children generally exercise at much greater respiratory rates than adults (commonly in excess of 60 breaths per minute), and this declines with age. Children also have a higher ratio of respiratory rate to tidal volume. In other words, to create the same pulmonary ventilation a child takes more but shallower breaths than an adult. The child's ventilation is therefore not as efficient as the adult's; however, the smaller and less-efficient ventilation in children does not appear to limit aerobic fitness during childhood.

By adult life, in addition to the smaller lung volumes, women have a lower resting diffusion capacity (i.e., capacity for oxygen to diffuse into the blood stream from the alveoli in the lungs). However, if men and women of equal size and the same aerobic fitness are compared, diffusion capacity does not differ during exercise.

AGE AND SEX DIFFERENCES IN CARDIOVASCULAR FUNCTION

There are clear differences in cardiac function at rest and during exercise between girls and boys. Differences in resting heart rate are apparent even before puberty; girls record higher resting heart rates than boys. This is thought to relate to the influence of the sex hormones on the electrical conduction system of the heart. Maximum heart rate is less dependent on age and sex during childhood and remains fairly stable until midadolescence. Usually, healthy children would attain a maximum value around 200 beats per minute running and around 195 beats per minute cycling. We therefore do not use age-predicted maximal values (e.g., 220 minus age) during childhood.

Boys have been shown to have a greater cardiac muscle mass than girls and a higher maximal cardiac output. In the absence of sex differences in maximal heart rate, a higher stroke volume in boys accounts for this greater cardiac output. Reports confirm that boys have a stroke volume that is about 7% to 13% greater than that of girls. This is similar in men, who have a greater cardiac mass than women and greater stroke volume.

There are sex differences in the oxygen-carrying capacity of the blood. By 16 yr of age, the oxygen-carrying protein in the blood, haemoglobin, has increased to about 152 g/L in boys but only 137 g/L in girls. This difference in haemoglobin persists into adult life and is thought to be a factor that limits aerobic fitness levels in women.

AGE AND SEX DIFFERENCES IN THE PERIPHERAL ADJUSTMENTS DURING EXERCISE

Differences in aerobic fitness persist even when haemoglobin levels have been reduced in men so that they are the same as women. This suggests that there are factors other than stroke volume and haemoglobin that limit aerobic fitness in women. There is very little information on the peripheral adjustments to exercise during childhood and adolescence. Maximal arterial venous oxygen difference has been shown to increase during puberty in boys, leading to about a 16% greater oxygen extraction at the muscle in boys compared with girls. There is some evidence of differences in arterial venous oxygen difference between men and women, but these are largely accounted for by the difference in haemoglobin.

It is also possible that differences exist between men and women in the muscle architecture itself, such as muscle fibre composition, metabolic profile, and capillary supply per fibre. Again, information is limited, but these differences are addressed later in this chapter.

HIGH-INTENSITY EXERCISE IN CHILDREN

Children tend not to engage in long-duration activities in their normal, daily lives. Instead, they are more often seen moving very rapidly for very short periods of time. This high-intensity activity requires the energetic support of the phosphagen and anaerobic glycolytic pathways.

The most common anaerobic performance tests used with children are all-out 10 s and 30 s cycle ergometer tests such as the Wingate test and force–velocity tests. These tests provide measures of peak power, which gives an indication of the phosphagen energetic pathway capabilities, as well as mean power, which reflects the capacity of anaerobic glycolytic pathways.

Peak power increases by about 270% in girls from 7 to 16 yr of age. In boys, a greater increase of about 380% occurs over the same period. Mean power also increases with age and is greater in boys than in girls from about 13 yr of age. The differences in peak and mean power are largely a result of the greater increase in muscle mass in boys during puberty. However, even when body size and composition are taken into consideration, differences in peak and mean power remain, suggesting that other factors limit anaerobic power and capacity in girls.

The peak power recorded in an adult elite power athlete is about 6500 W, so the change from childhood represents about a 40-fold increase. When differences in body size are accounted for, the change is reduced somewhat, but a child's power is still only about 30% of that of an adult athlete, again suggesting that other factors, such as muscle fibre recruitment patterns, may differ between children and adults as well as between boys and girls.

Not all child elite athletes in sports requiring high-intensity bursts of activity have unusually high anaerobic power or capacity. It has been suggested that children are metabolic nonspecialists. There is evidence that children have lower glycolytic enzyme activity than adolescents, who in turn have lower glycolytic enzyme activity than adults. The glycolytic:oxidative enzyme activity ratio has been reported to be 60% higher in young adults than in children and 40% higher in adolescents than in children. To compensate for this lower glycolytic support for exercise, children can oxidise pyruvate and fatty acids at much higher rates than adults. This may result in a much higher aerobic energetic contribution to high-intensity exercise.

MUSCLE SIZE AND STRENGTH DURING CHILDHOOD

Even at birth boys tend to have a greater amount of muscle compared with girls. As described in earlier chapters, the overall increases in muscle size influence muscle strength. Even women who are elite track athletes have muscle fibres that are 70% to 80% of the size of those in male athletes. In addition to the size of the muscle, the muscle fibre type distribution is believed to differ between men and women; men have a lower proportion of Type I fibres. There is not a great deal of information on muscle fibre typing in children and adolescents because of the invasive nature of the biopsy technique, but there is evidence to show that differentiation of fibre type occurs during the first few years of life. About 10% of skeletal muscle fibres remain undifferentiated up until puberty, and there is no notable sex difference in the percentage of Type I fibres during childhood. By the end of puberty, young women have more Type I fibres than young men, and the Type II muscle fibres of young men are bigger than their Type I fibres, something that is not evident in young women.

Muscle strength is the ability of muscles to exert force to move one's body, propel an object, or move and resist external forces. Strength increases in an almost linear fashion in both boys and girls from early childhood until about 13 to 14 yr of age. During the pubertal years there is a much greater increase in strength in boys, which slows into adulthood. Strength is an important factor for success in many sports, and the early-maturing boy will gain enormous advantage in sports requiring high levels of strength.

EXERCISE TRAINING DURING CHILDHOOD

The response to training during childhood varies somewhat depending upon the type of exercise that is undertaken. Peak $\dot{V}O_2$ shows adaptation, but to a lesser extent than seen in adults. High intensity training and resistance training are both safe and effective in children providing there is appropriate supervision.

AEROBIC TRAINING

Effective aerobic training programs in children normally consist of a series of sustained-exercise sessions of 40 to 60 min in duration. At least 3 sessions per week would need to be completed for at least 12 wk to achieve a training effect. The biggest difference between aerobic training programs for children and adults is the intensity of the exercise. The relative intensity of exercise required for optimum benefits in children (about 85%-95% of maximum heart rate) is higher than that recommended for adults.

Boys and girls are equally aerobically trainable. Although there was a proposal that children were trainable only once they reached puberty, there is not enough evidence to support this argument. Many studies show that training can be effective before puberty.

HIGH-INTENSITY TRAINING AND RESISTANCE TRAINING

High-intensity training and resistance training are effective ways to improve maximal power output, maximal strength, and athletic performance in children. High-intensity training usually consists of fast, short-duration bouts of exercise that exceed the maximal power of oxidative metabolism. Children can increase their anaerobic power and capacity after high-intensity training; all-out intensities provide the greatest response. This type of training also increases aerobic power in children to the same or a greater extent than endurance-training programs. This is similar in adults, where high-intensity interval training has been shown to result in marked increases in oxidative metabolism.

A wide variety of high-intensity training protocols have been used, from longer, all-out running intervals (approximately 3 min) with short pause durations to short sprint intervals (approximately 30 s) with short or medium pause durations and high-intensity, fixed-distance runs. This makes any specific prescription for high-intensity exercise difficult; however, high-intensity training is a safe and effective training modality in youngsters and is often better tolerated than constant-pace endurance training.

Resistance training was originally believed to be ineffective in children and high in risk because of possible harmful effects on growth. It is now widely accepted that resistance training is safe and viable for youngsters when supervised appropriately.

Improvements in strength of 10% to 40% have been shown in children and adolescents after resistance training for less than 20 wk. Although there is not enough information available to formulate a definitive exercise prescription for resistance training in children, it is known that strength in children can be improved with at least 2 sessions per week. It has also been shown that greater improvements in strength are achieved if the training program consists of high repetitions with moderate loads rather than low repetitions with heavier loads. There are a number of factors that might determine the increase in strength after resistance training in children and adolescents. These may include an increase in muscle size, adaptation of the neural input to the muscle, and more efficient motor unit recruitment.

By late adolescence responses to resistance training are similar to those in adults, and increases in muscle size accompany increases in muscle strength. In contrast, the younger, immature child will increase strength with resistance training, but often in the absence of substantial increases in muscle size. Indeed, it is thought that any change in muscle size in children is about 50% of that observed in adults. This probably led to the original belief that children could not improve strength; however, the increases in strength in the absence of an increase in muscle size suggest alternative mechanisms. It would seem that increases in strength in the prepubertal child are more likely to be the result of increased neural activity, such as increased motor unit activation and changes in muscle recruitment patterns.

Musculoskeletal injuries can arise from strength training in youngsters, but this is usually because of poor technique and the use of unsupervised maximal lifts. A number of guidelines for resistance training in youths have been published to reduce the risk of injury. A summary of these guidelines is provided in table 13.1.

Table 13.1 Guidelines for Resistance Training in Youths

1	To ensure good and safe practice, resistance-training programs should be supervised by appropriately qualified instructors who are familiar with the available guidelines and are experienced in working with children.
2	Resistance training should complement a broader planned sport program and include specific and realistic goals for each child.
3	Training should take place 2-3 times per week on nonconsecutive days.
4	A warm-up period before resistance training is important and should consist of 5-10 min of light aerobic and stretching exercises.
5	The initial few weeks of the program should focus on developing good technique and should include completing only a single set of 13-15 repetitions per exercise using light loads (<50% 1RM).
6	A variety of exercises (e.g., exercises using body weight, rubber bands and tubing, medicine balls, free weights, and child-sized ergometers) can be included in the program. Multijoint exercises should be encouraged because they promote coordinated movements.
7	Progressive increases in the training stimulus are necessary and can be achieved by small increases (5%-10%) in the resistance, the number of repetitions, or the number of sets.
8	Training loads can be individualized using percentages of the 1RM, but assessment of 1RM must be undertaken only after a thorough warm-up and under close supervision.

EXERCISE IN OLDER ADULT LIFE

We are in the midst of a global demographic transition, the result of which is an unprecedented increase in the older population. In 1950 only 3 countries had more than 10 million people over the age of 60 yr. By the year 2000, 12 countries had more than 10 million people over 60 yr of age. It is predicted that by 2050 there will be more than 2 billion people over 60 yr of age and that 33 countries will have more than 10 million people 60 yr of age or older.

Until recently a decline in functional capacity was regarded as simply an inevitable consequence of growing older. However, it has recently been questioned whether the dramatic changes in work capacity are due solely to the ageing process itself or also to the increasingly sedentary lifestyle of many adults. Most of the age-related changes in functional capacity resemble those seen with detraining (cessation of exercise training). It is important to distinguish the inevitable effects of ageing from those attributable to inactivity. If age-related changes are due primarily to lifestyle, then maintaining physical activity patterns throughout adulthood and into old age is imperative, particularly given the rapid global shift in ageing populations.

Ageing in children and adolescents is associated with increases in exercise capacity. However, as we continue to age during adult life, our exercise capabilities and sport performance decrease. Figure 13.2 shows that world-record performance declines after about age 30 in both male and female athletes. The next section considers the possible physiological changes that may be responsible for these declines in exercise capacity in the older adult, classifying those older than 65 yr of age as older or elderly.

VARIABILITY IN RATES OF PHYSIOLOGICAL AGEING

Interindividual variability (i.e., the difference between individuals) increases with ageing. In other words, people age at different rates with regard to exercise capacity. Spirduso (1995) classifies older individuals into five categories based on physical function:

▸ physically dependent (those who are debilitated by disease),

▸ physically frail (those with conditions affecting functional capacity in tasks of daily living),

▸ physically independent (the majority of older adults who are free from disease but do not exercise regularly),

▸ physically fit (those who are sufficiently physically active to achieve health benefits), and

▸ physically elite [a small group of people who train intensely, usually for competitive sport (e.g., masters athletes) or adventure recreation].

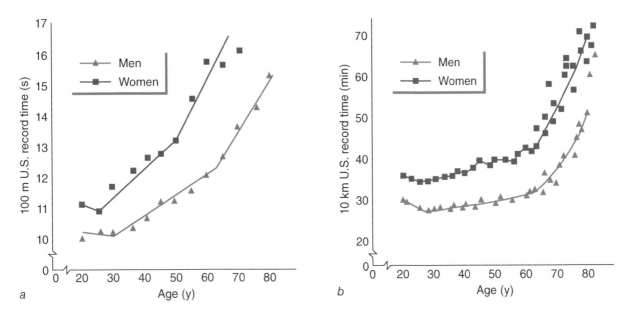

Figure 13.2 Trends for U.S. running records in the (a) 100 m and (b) 10 km by age and sex. For both events, the fastest times are recorded in the age range between approximately 20 and 30+ yr and for males compared with females.

Reprinted, by permission, from W.L. Kenney, J.H. Wilmore, and D.L. Costill, 2012, *Physiology of sport and exercise*, 5th ed. (Champaign, IL: Human Kinetics), 464.

We need to consider this diversity of physical activity patterns and physical capacity when discussing physiological responses to exercise in older people. The following sections on physiological responses and adaptations to exercise provide a general discussion in relation to the average older individual—that is, a person who would be considered physically independent or fit according to Spirduso's classification.

REASONS FOR DECREASES IN EXERCISE CAPACITY DURING AGEING

Both aerobic and anaerobic power and capacity decline with age. Although some degree of change is inevitable with ageing, continued exercise training can slow the rate of decline and help maintain exercise capacity in people who are elderly.

AEROBIC FITNESS IN OLDER ADULTS

Both cross-sectional and longitudinal studies indicate that aerobic fitness, as measured by $\dot{V}O_2$max, declines after about 20 yr of age. The rate of decline averages about 0.5% to 1% per yr, but may vary depending on the group studied (figure 13.3). Between 40 and 50 yr of age, $\dot{V}O_2$max in the average

sedentary person may decline by 5 mL·kg⁻¹·min⁻¹ (up to 10%). In sedentary individuals, the rate of decline accelerates after 75 yr of age.

The decrease in $\dot{V}O_2$max during ageing in a sedentary individual relates to several factors, including loss of muscle mass, decreases in maximum heart rate and cardiac output, and a decline in the oxidative capacity of skeletal muscle. In sedentary individuals, the age-related decline in $\dot{V}O_2$max is generally correlated with changes in body composition. Loss of muscle mass and an increase in fat mass collectively decrease the amount of energetically active muscle tissue and therefore lead to a decrease in oxygen consumption results.

About 50% of the decrease in $\dot{V}O_2$max is attributable to changes in cardiac function. The decrease in maximum heart rate results from changes in the neural input to the heart. Because cardiac output is a function of heart rate and stroke volume, the decrease in maximum heart rate results in a decrease in maximum cardiac output. As discussed in chapter 11, cardiac output is an important determinant of $\dot{V}O_2$max. In contrast, stroke volume does not change much in older life.

The ability of skeletal muscle to extract and use oxygen during exercise also decreases with age in a sedentary population. This decrease is due to a reduction in the number of capillaries in

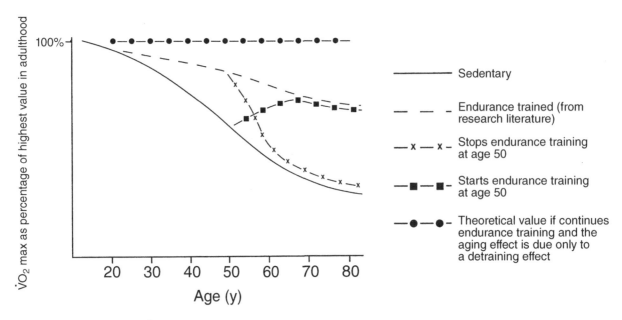

Figure 13.3 Trends in V̇O₂max with age. This graph was drawn from research data, and a proposed outcome of age-related changes due solely to the effects of detraining has been provided in the top line (filled round symbols).

the skeletal muscle and a reduction in the capacity to redirect blood flow to the muscle, which compromises oxygen delivery and extraction at the working muscle during exercise. The oxidative capacity of skeletal muscle also declines in older adults, possibly due to mitochondrial function and protein synthesis in skeletal muscle. Thus, there is both less delivery of oxygen to the working muscles and a lower capacity to produce adenosine triphosphate.

ANAEROBIC CAPACITY IN OLDER ADULTS

Compared with research on aerobic-exercise capacity, there has been far less research on anaerobic capacity in sedentary older adults. As described in chapter 11, tests of anaerobic capacity involve, for instance, a 30 s sprint on a cycle ergometer. It is difficult and possibly unsafe for older individuals who are unaccustomed to intense exercise to exert maximal effort in such a test; it is also unlikely that older, sedentary people would be motivated to perform maximally on such a test.

Anaerobic capacity peaks at about 20 yr of age but may be maintained with continued high-intensity training until the late 30s or early 40s. In older, sedentary people, anaerobic capacity declines by

about 6% per decade. This decline is closely related to loss of muscle mass—especially of the thigh muscles, the main source of power in cycle tests of anaerobic capacity. The decreases in anaerobic capacity and muscle mass are due to decreases in muscle fibre size [especially in the large, stronger fast-twitch (FT) fibres], loss of FT motor units, and changes in coordination. Anaerobic capacity and muscle size decrease with age more in women than in men, although this may occur because older women are less likely to perform intense physical activity.

MUSCULAR STRENGTH

Muscular strength is important to sport performance as well as to basic functions such as posture, balance, and coordination and to simple activities such as walking and climbing stairs. Dramatic changes in muscular strength accompany ageing and may limit an older individual's ability to be physically active and to maintain an independent lifestyle.

For untrained persons, maximum muscular strength is achieved in the early 20s and may be maintained with regular strength training through the late 30s or early 40s. Strength decreases by about 2% to 4% per yr in older, sedentary individuals. Declines in strength are attributable to several factors. Lean body mass decreases gradually between ages 30 and 50 yr, after which the loss of muscle mass accelerates. Between 20 and 80 yr of age, people typically lose 40% of their muscle size. As explained in chapter 5, muscular strength is generally related to muscle mass or muscle cross-sectional area. During increasing older age, muscle mass decreases due to atrophy of the larger, stronger FT muscle fibres and loss of FT motor units. Slow-twitch (ST) fibres do not atrophy as much as FT fibres because ST fibres are recruited during normal activities such as standing and walking. However, as we grow old, we are less likely to perform high-intensity exercise, the type needed to recruit FT fibres.

During older age the amount of connective tissue in skeletal muscle may increase and the size of the muscle fibres may decrease. Because connective tissue does not contribute to active force generation, a muscle's overall size may not decrease as much as its force-generating capacity. That is, the muscle may appear to be the same size, but a relative increase in non-force-producing connective tissue means that the muscle is weaker. In addition, age-related changes in neural input, particularly loss of FT motor units in skeletal muscle, may influence recruitment patterns and reduce force generation. Finally, older, untrained individuals may be reluctant to voluntarily exert maximal force, and this may result in lower strength readings during testing.

PREVENTING OR REVERSING LOSS OF EXERCISE CAPACITY IN OLDER ADULTS

Because people tend to become more sedentary in later adult life, it is unclear how much the age-related changes in exercise capacity are attributable to a reduced level of physical activity and how much to the ageing process itself. In sedentary persons, many of the age-related changes in exercise capacity (e.g., decreases in $\dot{V}O_2$max and muscular strength) resemble the effects of detraining.

Comparisons between masters (veteran) athletes and age-matched sedentary individuals indicate that the athletes have a much slower rate of decline in $\dot{V}O_2$max and muscular strength with age. Moreover, training studies generally indicate that the response to exercise training in older individuals is relatively similar to that seen in younger people.

AEROBIC CAPACITY IN MASTERS ATHLETES

In masters athletes, $\dot{V}O_2$max decreases with age at half the rate observed in an older, sedentary population. For example, whereas $\dot{V}O_2$max may decline with age by 0.5 mL·kg^{-1}·min^{-1} each year in sedentary older individuals, the rate of decline is about 0.25 mL·kg^{-1}·min^{-1} each year in masters endurance athletes who continue to train (figure 13.3). Thus, about half of the decrease in aerobic power appears to be inevitable with continuing old age and half seems to be preventable with continued training. However, it is not clear whether masters athletes continue to train at the same intensity at which they trained when they were younger.

In the masters athlete, the decrease in $\dot{V}O_2$max appears to result from a decrease in maximum heart rate, which cannot be changed with training. Relatively smaller decreases in maximum stroke volume and skeletal muscle oxygen extraction and oxidative capacity may also contribute to the slower rate of decline in $\dot{V}O_2$max in the athlete. In contrast, other factors, such as skeletal muscle capillary number, muscle fibre size (especially in ST fibres), and skeletal muscle oxidative activity may be maintained with continued training. The higher the volume and intensity of training, the more these physiological functions are retained during ageing.

ANAEROBIC CAPACITY AND MUSCULAR STRENGTH IN MASTERS ATHLETES

Few studies have addressed anaerobic capacity and muscular strength in masters athletes. As with endurance capacity, continued training halves the rate of decline of anaerobic capacity and muscular strength in older adults. Much of the age-related decline in muscular strength can be accounted for by loss of muscle mass, even in masters athletes. However, continued training partially prevents the atrophy of skeletal muscle, particularly the atrophy of FT fibres that normally occurs in sedentary individuals. Recent research suggests that strength training induces muscle hypertrophy and improves strength in older people. Despite inevitable changes with age, compared with the average untrained young adult, the masters athlete exhibits high levels of variables such as $\dot{V}O_2$max, muscular strength, and anaerobic capacity.

IN FOCUS: DOES EXERCISE PREVENT BRAIN AGEING?

Exercise physiology has traditionally concentrated on skeletal muscle metabolism; particular emphasis has been placed on the cardiorespiratory system. In contrast, the role of the human brain has been largely ignored. Improved technology, such as functional magnetic resonance imaging, is now providing a wealth of information about brain function and how it changes across the life span and how the brain adapts to exercise. This is an area of particular importance given the degenerative effects elderly life can have on the brain and the very large proportion of the global population that will become elderly in the next 20 yr.

Like the rest of the body, the brain undergoes a number of changes as we enter old age. Contrary to popular belief, not all age-related (>65 yr) changes in brain function are caused by a loss of brain cells or neurons. There is cell damage and cell loss—primarily, loss of the white matter or the insulated nerve fibres in the brain—which can result in an enlargement of the ventricular system (the spaces in the brain that contain cerebrospinal fluid) and a widening of the sulci or grooves on the surface of the brain. This loss also results in a reduction in brain volume. There are also other physiological changes that account for the age-related deterioration in brain function: slowing of neuronal function (the speed at which the neurons can transmit a signal), changes in cerebral blood flow (the blood flow through the brain), and alterations in the chemical and metabolic properties of brain tissue.

The combined effect of all these changes in the brain in older adults is a reduction in the connectivity between the two sides, or hemispheres, of the brain. In other words, the brain responds more slowly. This slow response manifests in many day-to-day activities as slower reflexes and deterioration in memory.

Recently, a number of studies have shown that aerobic exercise and regular physical activity can help reverse some of these age-related losses in brain function. The most compelling information comes from large, well-controlled studies of older adults. In one study, those who followed an aerobic training program for 1 yr showed an increase in the volume of the hippocampus, a region of the brain involved in memory. These same adults also showed improved spatial memory after the training program ended. In another training study, older adults who followed an aerobic training program for 6 mo showed increases in the volume of another region of the brain involved in memory, the cortex. Aerobic fitness has also been associated with a range of other cognitive tasks. For instance, fitter older adults perform better than their less-fit peers in tasks that test reaction time, motor learning, attention, and information processing.

The training prescription for optimal brain health differs little from what would normally be suggested for the older adult: low- to moderate-intensity exercise (50%-60% $\dot{V}O_2$max) for about 150 min/wk, or 20 to 30 min/d for 3 to 4 d/wk. It would appear that staying aerobically fit can enhance the function of the older brain and is an important way of maintaining a good quality of life in older adults.

Sources

Ahlskog, J.E., Geda, Y.E., Graff-Radford, N.R., & Petersen, R.C. (2011). Physical exercise as a preventive or disease-modifying treatment of dementia and brain aging. *Mayo Clinic Proceedings, 86*(9), 876-884.

Lustig, C., Shah, P., Seidler, R., & Reuter-Lorenz, P.A. (2009). Aging, training, and the brain: A review and future directions. *Neuropsychological Reviews, 19*(4), 504-522.

TRAINING EFFECTS IN PREVIOUSLY SEDENTARY INDIVIDUALS

Older people are capable of increasing exercise capacity and performance in both endurance and brief, high-intensity exercise even when they have previously led a sedentary life. In general, however, responses may be less and may take longer than in younger individuals.

AEROBIC TRAINING EFFECTS IN PREVIOUSLY SEDENTARY PERSONS

$\dot{V}O_2$max increases with endurance training in older individuals, although it may take longer to see these increases than in younger adults. As in younger individuals, changes in $\dot{V}O_2$max are closely related to both the total amount of exercise and training intensity. Increases in $\dot{V}O_2$max in the range of 20% to 40% may occur in the older adult after endurance training. Increases in maximum cardiac output (due to increases in stroke volume) and increases in skeletal muscle oxygen extraction and oxidative capacity accompany increases in $\dot{V}O_2$max.

ANAEROBIC TRAINING IN PREVIOUSLY SEDENTARY INDIVIDUALS

The limited data available suggest that high-intensity training at about 85% $\dot{V}O_2$max improves anaerobic capacity as measured by the 30 s Wingate test. The mechanisms are unclear at present but may be related to increases in muscular strength and size, especially of FTa fibres.

STRENGTH TRAINING IN PREVIOUSLY SEDENTARY PERSONS

Increases in isometric and isokinetic strength after resistance training have been well documented in previously sedentary older adults. Early studies suggested that gains in muscular strength occur in the absence of muscle hypertrophy and must therefore be due to neural adaptations. However, more recent work using sensitive imaging techniques (e.g., computer-aided tomography and magnetic resonance imaging) to accurately measure muscle cross-sectional area shows that muscle hypertrophy occurs with resistance training in previously sedentary older adults. Muscle hypertrophy is due to hypertrophy of the FT fibres, especially FTa fibres. However, the increases in strength are greater than can be accounted for by hypertrophy alone, indicating that neural adaptations also contribute to strength gains in older people. It is generally believed that the capacity for hypertrophy is somewhat limited in this population and that neural factors contribute proportionally more to strength gains in older individuals than to strength gains in younger individuals. The limited capacity for muscle hypertrophy in older men compared with younger men may also be due to age-related decreases in levels of the male hormone testosterone.

EXERCISE PRESCRIPTION IN OLDER ADULTS

In older adults, a conservative approach to exercise prescription is needed because older individuals often exhibit a lower functional capacity and slower rate of adaptation to exercise training. As noted earlier, variability in physical capacity widens with age and lifestyle, and there can be large differences in exercise capacity among older individuals.

Table 13.2 lists some of the special concerns one should take into account when prescribing exercise for elderly individuals. The goals of exercise for this group are to improve or maintain functional capacity, muscular strength, muscular endurance, and quality of life and to prevent the onset of disease. The exercise prescription should focus on developing and maintaining aerobic capacity, muscular strength and endurance, flexibility and range of motion around the joints, balance, and coordination.

Low- to moderate-intensity exercise confers many health benefits (discussed further in chapters 14

Table 13.2 Exercise Prescription for the Older Adult

Variable	Consideration for exercise prescription
Low functional capacity	Low-intensity exercise (40%-50% $\dot{V}O_2$max); gradual progression; interval training to avoid early fatigue
Low muscular strength and endurance	Exercises to enhance muscular strength and endurance
Impaired coordination and slower reaction time	Simple movements; supported exercise; exercise machines
Increased risk of disease, especially heart disease, obesity, hypertension, arthritis, and osteoporosis	Health-risk screening and medical examination; modification of exercise prescription according to condition
Prevalence of osteoarthritis	Low-impact, low-intensity exercise; weight-supported or water-based activities
Low heat tolerance	Replacing fluids frequently; avoiding exercise in the heat
Lower capacity for muscle hypertrophy and protein synthesis	Expecting slower progress and improvement in exercise capacity

and 23) and helps prevent possible deleterious effects such as musculoskeletal injury or a cardiac event (e.g., heart attack). Older people are more attracted to low- to moderate-intensity exercise than to exercise at higher intensities, especially if they are in a group. Exercise modes such as walking, low-impact aerobics, water-based exercise, and exercise on machines (e.g., treadmill, cycle ergometer) are appropriate. Intensity is generally recommended in the range of 50% to 60% $\dot{V}O_2$max, although lower intensities (40%-50%) may be appropriate for beginners or those with low functional capacity. Using a self-selected pace may enhance enjoyment and compliance.

Recent research suggests that resistance training can help maintain muscle mass and prevent declines in muscular fitness in elderly exercisers. Current recommendations for healthy adults, including those who are elderly, suggest that resistance training should be performed 2 to 3 times per week and should include 8 to 10 exercises (each performed 8 to 15 times per set) using all the major muscle groups.

There is some evidence that, at least at the onset, resistance training once per week is as effective for increasing muscular fitness as resistance training 2 or 3 times per week. Older people should aim for repetitions of 8 or more per set to avoid maximal lifts and high-intensity resistance exercise because these can cause an excessive increase in blood pressure and musculoskeletal injury.

The risk of falling increases with age. Combined with age-related decreases in bone mass, a fall can lead to a bone fracture (discussed further in chapter 23). In those who are elderly, regular physical activity helps maintain bone mass and muscular strength and prevent falls. A combination of moderate exercises, such as walking, circuit resistance training, calisthenics, and exercise to music, is effective in increasing aerobic fitness, muscle mass, muscular strength, balance, coordination, and agility.

SUMMARY

To effectively prescribe exercise for children or older adults, whether for competition or for general fitness, it is important to understand the effects of growth, maturation, ageing, and sex differences on physiological responses and adaptations to exercise.

Children are not miniature adults and grow and mature at individual rates, which greatly influences exercise capacity. For example, early-maturing boys have greater muscle mass and strength than their average and later maturing peers of the same chronological age. The greater muscle mass and strength of these boys is readily apparent in sport performance. Children do adapt to aerobic, high-intensity, and resistance exercise training, even before puberty, but the magnitude of the training response in children is somewhat lower than in adults.

Girls and boys grow and develop in different ways, resulting in substantial differences in body size and composition. However, the divergent responses to exercise in boys and girls are not explained solely by body size or composition. Many of these differences persist into adulthood and there is evidence of underlying qualitative differences between the sexes in physiological responses to exercise.

Exercise capacity and sport performance begin to decline at about 30 yr of age in both men and women. The rate of decline is slower, by about half, in those who continue with exercise training throughout adulthood and into old age. Some changes are inevitable, such as the age-related decline in maximum heart rate; other changes may be slowed or prevented, such as loss of muscle mass. Older people are capable of improving both aerobic capacity and muscular fitness with appropriate exercise training that takes into account a slower, more gradual rate of improvement. Importantly, resistance training not only improves muscular fitness, it helps prevent falls by enhancing balance and coordination.

FURTHER READING AND REFERENCES

American College of Sports Medicine. (2009). ACSM's position stand: Exercise and physical activity for older adults. *Medicine and Science in Sport and Exercise, 41*(7), 1510-1530.

Armstrong, N., & McManus, A.M. (2011). *The elite child athlete*. Basel, Switzerland: Karger AG.

Rowland, T.W. (2005). *Children's exercise physiology* (2nd ed.). Champaign, IL: Human Kinetics.

Spirduso, W.W. (1995). *Physical dimensions of aging*. Champaign, IL: Human Kinetics.

CHAPTER 14

PHYSIOLOGICAL ADAPTATIONS TO TRAINING

The major learning concepts in this chapter relate to

- ▶ limiting factors in exercise performance,
- ▶ responses of different energy systems to different types of training,
- ▶ adaptation of the muscular system after strength training,
- ▶ effects of exercise training on muscle fibre number and fibre type,
- ▶ basic principles of training,
- ▶ training for cardiovascular endurance,
- ▶ exercise for health-related fitness,
- ▶ methods of strength training, and
- ▶ causes of muscle soreness.

The human body has a tremendous ability to adapt over time in response to repeated bouts of exercise, resulting in an increased capacity for physical work. These long-term adaptations are called chronic or training adaptations.

The purpose of exercise training is to induce metabolic and structural adaptations in muscle that delay the onset of fatigue. Compared with the untrained individual, the trained athlete can perform more work, or exercise at a faster pace or for a longer time, before the onset of fatigue.

Effective training to enhance exercise capacity must take into account the different energy systems used to produce adenosine triphosphate (ATP) in skeletal muscle (discussed in chapter 11). This is true whether the goal of training is to improve performance in an athlete, to enhance health in the average individual, or to treat a patient with a particular illness. As we shall see in this chapter, the outcomes of any training program depend very much on the type of training undertaken.

Muscular fitness is equally important to performance in virtually every sport. Thus, effective training programs must also attempt to induce structural change in the muscles that will result in increased strength, power, and endurance.

TRAINING-INDUCED METABOLIC ADAPTATIONS

One of the challenges for exercise physiologists is to identify the relative contributions of the various energy systems to a particular activity and to use this information to develop training programs that maximize adaptations and, thus, performance. The following are examples of how different types of training result in different types of change.

▶ Muscle glycogen stores increase with endurance training. As discussed in chapter 13, muscles fatigue when depleted of their glycogen stores during endurance events or activities requiring repeated high-intensity exercise over an extended time (e.g., team-game sports). Increasing pre-exercise glycogen stores can delay glycogen depletion and thus the onset of fatigue. In other words, the endurance-trained or team-game athlete will be able to exercise for longer before glycogen depletion causes fatigue.

▶ Muscle phosphocreatine (PCr) stores may be increased by power and short sprint training. As discussed in chapter 11, PCr provides the major means of ATP resynthesis during short, maximal exercise. Because PCr is rapidly depleted, increasing PCr stores via training enables more ATP to be synthesized during short, explosive activities.

▶ An endurance-trained athlete begins to sweat earlier and sweats more during exercise than an untrained person. This allows the athlete to avoid a precipitous increase in body temperature, which could compromise performance, during long-duration events.

FACTORS LIMITING EXERCISE PERFORMANCE

Factors that limit exercise capacity and performance are closely related to the predominant energy system(s) used during a particular activity. For example, factors causing fatigue or limiting sprinting performance are different from those limiting performance in endurance events (see "Summary of Causes of Fatigue During Exercise").

SUMMARY OF CAUSES OF FATIGUE DURING EXERCISE

Brief, high intensity (<1 min)

- Phosphocreatine (PCr) depletion
- Moderate to high lactate levels
- Disturbance of chemical gradients across cell membrane

Longer, high intensity (1-7 min)

- PCr depletion
- High lactate levels
- Disturbance of chemical gradients across cell membrane

Prolonged, moderate to high intensity (10-40 min)

- Moderate lactate accumulation
- Partial glycogen depletion

- Dehydration
- Disturbance of chemical gradients across cell membrane

Very prolonged (>40 min)

- Glycogen depletion
- Dehydration
- Increased body temperature
- Low blood glucose levels
- Disturbance of blood amino acid levels

For power and speed activities lasting only a few seconds, such as sprinting or jumping, performance is limited by the ability to recruit fast-twitch (FT) muscle fibres to generate maximal force and power as well as by the ability to maintain balance and coordination while generating such high muscle force. In brief high-intensity movement in which maximal power is exerted for up to 20 to 30 s, such as 100 to 200 m sprints or 50 m swims, performance is related to the limited amount of ATP and PCr stored in the muscles. PCr is soon depleted during maximal effort. Given that PCr is resynthesized only after exercise, ATP must be supplied by the two other systems (anaerobic glycolysis and oxidative metabolism). Elite sprinters are characterized by an ability to use PCr to resynthesize ATP at a faster rate than nonelite sprinters.

During maximal exercise lasting between 30 s and 2 to 3 min, such as 400 to 800 m runs or 100 to 200 m swims, anaerobic glycolysis is the major source of ATP resynthesis. As discussed in chapter 11, anaerobic glycolysis provides ATP at a relatively fast rate, but when the protons released from the reactions of glycolysis cannot be adequately buffered, metabolic acidosis occurs and fatigue ensues. Fatigue also occurs as a result of disturbance of the chemical and electrical gradients across the muscle cell membrane, due to changes in the intra- and extracellular distribution of electrolytes such as potassium.

Middle-distance events lasting between 3 and 10 min, such as 1,500 to 3,000 m runs, 400 to 800 m swims, and 4,000 m cycles, are limited by a combination of metabolic acidosis, moderate glycogen depletion, and disturbance of electrolyte distribution between the muscle cells and extracellular environment.

Performance in longer events lasting between 10 and 40 min, such as 10 km runs or 1,500 m swims, is limited by a combination of factors, including moderate metabolic acidosis, partial glycogen depletion, dehydration, and disturbance of the chemical and electrical gradient across the muscle cell membrane.

In very long events lasting more than 40 min, such as road cycling or long-distance running, performance is limited by a combination of factors, including glycogen depletion, dehydration, increased body temperature, low blood glucose levels, and changes in the ratios of amino acids in the blood.

An increased body temperature amplifies the demands on the cardiovascular system to direct blood flow to the skin and working muscles. Low blood glucose levels and an altered blood amino acid ratio contribute to fatigue of the central nervous system, especially the sensation of fatigue.

Activities that rely on all the energy systems over an extended time, such as team games (e.g., basketball or soccer), are limited by a combination of factors similar to those described for long-duration events. For example, glycogen depletion may occur after repeated high-intensity sprinting required in many stop–start sports such as tennis or basketball. The extended duration of a match (60-90 min) may also increase body temperature and lower blood glucose levels.

Because training adaptations of a particular energy system lead to improvement in performance in events using that system, it is imperative to understand which energy systems are predominant in which sports and the interplay among those systems. To design an effective training program, then, one must tailor its metabolic demands to appropriately challenge the energy systems used by the sport in question. The following section deals with responses to the demands of various types of training.

IMMEDIATE AND ANAEROBIC-SYSTEM CHANGES AFTER HIGH-INTENSITY SPRINT AND STRENGTH TRAINING

Table 14.1 summarizes the metabolic changes in the anaerobic pathways occurring after high-intensity sprint and strength training. Sprint and strength training increase PCr, ATP, and glycogen stores in muscle fibres, especially FT fibres. These increases facilitate a higher power output in short-duration exercise through an increased capacity for ATP resynthesis via the breakdown of PCr and through anaerobic glycolysis.

The activities of the anaerobic glycolytic enzymes also increase with training, enhancing the amount of ATP generated by anaerobic glycolysis. In addition, sprint training and high-intensity strength training increase muscle glycogen storage, enhancing glycogen availability during repeated sprints and power (e.g., jumping) or strength (e.g., weightlifting) events.

Table 14.1 Metabolic, Physiological, and Structural Adaptations to Sprint and Strength Training

Adaptation	Consequence
Increased muscle ATP and PCr	More ATP at onset of exercise
Increased muscle glycogen	More glycogen available delays onset of fatigue.
Increased anaerobic enzymes	More ATP synthesised via glycolysis
Increased lactic acid buffering	Higher capacity to tolerate high lactic acid levels
Increased muscle fibre size	Increased muscular strength and power
Increased motor unit synchronisation	Greater power generated by muscles
Less disturbance of muscle cell chemical gradient, especially potassium and calcium	Increased force generation and strength

ATP = adenosine triphosphate; PCr = phosphocreatine.

The muscle's capacity to generate and tolerate high levels of lactic acid during maximal exercise also increases with sprint training. The increase in lactic acid production observed after sprint training is attributable to the enhanced rate of conversion of pyruvate to lactic acid via anaerobic glycolysis. Exercise can continue despite high lactic acid levels because lactic acid increases the buffering capacity to neutralize the hydrogen ions (H+) that dissociate from lactic acid.

Muscle fibre size increases, especially in FT fibres, with strength, power, and sprint training; this is called hypertrophy. Because larger muscles contain more cross-bridges capable of generating force, strength and power output increase. High-intensity strength, power, and sprint training also enhance synchronization of motor unit recruitment, increasing the number of active muscle fibres contributing to force generation. Finally, the strength- or power-trained muscle fibres exhibit less disturbance of the internal (intracellular) environment during exercise; that is, cellular levels of ions such as potassium and calcium are better retained during exercise, allowing for continued tension generation by muscle fibres.

CHANGES IN AEROBIC METABOLISM AFTER ENDURANCE TRAINING

Metabolic adaptations resulting from endurance training are summarized in table 14.2. Generally, 6 wk of endurance training will elicit a 20% to 40% increase in $\dot{V}O_2$max. This is largely a result of changes in the cardiovascular system and skeletal muscle cells.

The activity of mitochondrial enzymes involved in oxidative metabolism increases greatly with endurance training, often by more than 100%. Oxygen uptake and oxidative production of ATP increase in slow-twitch (ST) fibres as well as in FTa fibres, provided that training is at a pace that recruits these fibres. Increased capacity for oxidative metabolism means that pyruvic acid can continue through to the Krebs cycle and that higher-intensity endurance exercise can therefore be supported aerobically for longer.

The capacity of skeletal muscle fibres to utilize fatty acids to produce ATP increases after endurance training. This means that at any given submaximal exercise intensity, trained muscle uses more fatty acids and less glycogen to produce ATP. An increased capacity to use fatty acids spares glycogen stores, delaying glycogen depletion and the onset of fatigue during prolonged exercise. In addition, muscle stores of glycogen increase after endurance training, providing more glycogen for prolonged high-intensity aerobic exercise. The combination of extra glycogen and slower rate of its use allows the muscles to work longer before depletion of glycogen and the onset of fatigue.

The capacity of ST muscle fibres and other tissues to buffer lactic acid increases following endurance exercise. This occurs because muscles have a higher oxidative capacity and can use lactic acid to produce ATP. (Remember, lactic acid can be used as a fuel or substrate by the Krebs cycle and electron transport chain.)

Table 14.2 Metabolic Adaptations to Endurance Training

Adapation	Consequence
Increased $\dot{V}O_2$max	Greater endurance performance
Increased muscle glycogen	More work before onset of fatigue
Increased mitochondrial enzymes	Increased oxidative capacity
Increased use of fats as substrate	Less reliance on glycogen; less glycogen depletion
Enhanced lactic acid removal and oxidation	More work before onset of fatigue
Increased lactic acid threshold	More work before onset of fatigue
Increased capillary number	More blood, oxygen, and substrates delivered to muscle and more lactate and carbon dioxide removed from muscle
Increased oxygen extraction by muscle	More oxygen available for ATP production
Increased muscle myoglobin content	More oxygen delivered to mitochondria

ATP = adenosine triphosphate.

Endurance training also increases blood capillary numbers in skeletal muscle, especially the capillary density around the ST and FTa muscle fibres. This increased blood supply enhances the delivery of oxygen to, and removal of carbon dioxide from, working skeletal muscle fibres. Additionally, the increased blood supply facilitates the enhanced removal of lactic acid.

Endurance-trained skeletal muscles contain more myoglobin than do untrained muscles. Myoglobin is an iron-containing protein that transports oxygen through the skeletal muscle cell. Increased myoglobin content enhances oxygen delivery from the cell periphery to the site of oxidative metabolism in the mitochondria.

It is important to note that these metabolic adaptations are specific to the type of training and recruitment pattern of muscle fibre types. Endurance training appears to be optimal for increasing the oxidative capacity of skeletal muscle.

Metabolic changes occur only in those muscle fibres recruited during activity. For example, only ST fibres show changes in oxidative capacity after low-intensity endurance training; training must be of a higher intensity to recruit and train FTa and FTb fibres. Sprinting recruits predominately FTa and FTb fibres, and these fibres exhibit the most profound changes after this type of training. Lower-intensity endurance training does not change muscle fibre size, but higher-intensity endurance training induces hypertrophy of ST and FT fibres, and the increased muscle size results in greater PCr stores.

ENDURANCE TRAINING-INDUCED CHANGES IN THE CARDIORESPIRATORY SYSTEM

The cardiorespiratory system also adapts to endurance training. These adaptations enhance oxygen delivery to skeletal muscle and the muscle's ability to use oxygen to resynthesize ATP during exercise. A summary of these changes is provided in table 14.3.

OXYGEN CONSUMPTION

Resting oxygen consumption is related to body size. There is evidence that long-term effects of exercise training result in increases in resting oxygen consumption because of increases in muscle mass.

Submaximal oxygen consumption at the same absolute exercise pace also remains unchanged after training unless there is a change in mechanical efficiency. Efficiency may change in sports such as swimming in which technique improves with training. In contrast, $\dot{V}O_2$max may increase by 20% to 40% after endurance training.

The extent to which $\dot{V}O_2$max improves during endurance training depends on a person's initial fitness level and previous training, genetics, age, and the type of training program. In general, previously unfit people show large relative gains in $\dot{V}O_2$max with training because they are farther from their genetically determined upper limit of aerobic fitness. Hereditability is also an important factor affecting

Table 14.3 Cardiorespiratory System Responses to Endurance Training

Adaptation	Consequence
Increased $\dot{V}O_2$max	Increased endurance performance
Decreased resting and submaximal heart rate	Less work done by the heart
Increased resting and exercise stroke volume	Increased cardiac output during maximal exercise
Increased maximal cardiac output	Increased blood and oxygen delivery to muscles
Increased blood volume, red blood cell number, and haemoglobin content	Increased oxygen delivery to muscles
Increased oxygen extraction from blood	Increased oxygen delivery to mitochondria
Decreased blood viscosity	Easier movement of blood throughout body
Increased maximal minute ventilation	Increased removal of carbon dioxide

improvement in $\dot{V}O_2$max. Some people respond with larger improvements in $\dot{V}O_2$max than others, even when performing the same training program.

The type of training is also important; training of higher intensity and frequent and longer duration induces larger and faster improvements in $\dot{V}O_2$max. An improved $\dot{V}O_2$max after training is attributable to a combination of metabolic changes, such as increased oxidative capacity and blood supply in muscle, and cardiovascular system changes that enhance delivery of blood and oxygen to working muscles.

HEART RATE

It is not unusual for an endurance athlete to have a resting heart rate of 30 to 40 beats per minute; the normal resting heart rate in an untrained adult is about 60 to 70 beats per minute. This decrease in resting heart rate is called training bradycardia (slowing of heart rate). During submaximal exercise at the same absolute work rate, heart rate is lower in the trained than in the untrained individual. Because of changes in stroke volume (discussed next), the heart is able to pump the same amount of blood with fewer contractions. Thus, the same exercise is less demanding on the endurance-trained heart. The decreases in resting and submaximal heart rate result from changes in the heart's neural control. In contrast to resting and submaximal heart rate, maximal heart rate is more related to age and remains essentially unchanged by training.

STROKE VOLUME

Stroke volume, the volume of blood pumped with each heartbeat, is higher at rest and at all exercise intensities after training. Endurance training over a period of years increases the size and strength of the heart's ventricles, increasing the amount of blood that can be pumped with each contraction. At rest and during submaximal exercise, the higher stroke volume coincides with a lower heart rate, resulting in the same cardiac output. In addition, in the endurance-trained person, stroke volume continues to increase with increasing exercise intensity, whereas it plateaus at a relatively low work rate in untrained individuals (see figure 11.9b). The high stroke volume in well-trained endurance athletes is an important contributor to high cardiac output, $\dot{V}O_2$max, and endurance-exercise capacity.

CARDIAC OUTPUT

Resting cardiac output is related to body size and does not change with endurance training. During submaximal exercise at the same work rate, cardiac output is unchanged or somewhat lower in endurance-trained individuals than in untrained individuals. In contrast, maximal cardiac output may double after endurance training due to the large increase in stroke volume.

OXYGEN EXTRACTION

The ability of skeletal muscle to extract oxygen from the blood depends on blood flow and the muscle's oxidative capacity. Endurance training appears to increase artery size, which increases blood flow to and through the working muscle (see "In Focus: Vascular Adaptations With Training"). Endurance training greatly increases mitochondrial density and myoglobin content, especially in ST and FTa muscle fibres, and the ability of muscles to extract oxygen during submaximal and maximal exercise.

IN FOCUS: VASCULAR ADAPTATIONS WITH TRAINING

We know that anatomical and functional changes occur in the heart after exercise training. The cardiac muscle enlarges, heart chamber volume increases, and cardiac output is improved, largely as a result of an enhanced stroke volume. The vasculature is also an important component of the cardiorespiratory response to exercise. When blood is redistributed to the working muscle during exercise, peripheral resistance is reduced and the flow of blood to and through the muscle is improved. This makes it easier for the heart to push the blood through the circulatory system and therefore allows stroke volume to increase.

Our understanding of how the vasculature responds to exercise training in healthy adults is somewhat limited in comparison with our understanding of cardiac adaptations. However, ultrasound technology has enabled considerable insight into both the structure and function of the vasculature in response to exercise and exercise training.

During exercise, blood flows to the working muscle via the conduit arteries. These large arteries are responsible for supplying blood to both the viscera and limbs and include the aorta, carotid, iliac, femoral, and brachial arteries. Blood flows through the muscle via the smaller resistance arteries. These arteries, with an internal diameter of about 150 μm, control the flow of blood to the capillary bed, allowing optimal gas and nutrient exchange from the blood to the muscle.

Does exercise training in healthy adults result in larger resistance and conduit arteries? It would appear from a variety of studies that resistance arteries are larger in endurance-trained athletes. This increase in size appears to occur only at the site of the working muscle. In other words, the adaptation is localised, resulting in a greater blood flow through the muscle being used. The peripheral conduit arteries have also been found to be bigger in athletes. For example, the femoral arteries have been shown to be larger in cyclists, and the brachial arteries have been shown to larger in elite canoeists, again suggesting localised anatomical adaptations to exercise training in the vasculature.

There is evidence that the artery wall also adapts to exercise training; wall thickness decreases as aerobic fitness increases. In one recent study, handgrip exercises were performed over an 8 wk period in both arms. One arm was cuffed to help to distinguish whether changes in the size of the brachial artery wall resulted from localised hemodynamic responses or from systemic stimuli. Wall thickness in both brachial arteries decreased, confirming that systemic changes were causing the reductions in wall thickness.

In contrast to arterial structure, there is limited evidence to suggest that vascular function is superior in trained, healthy adults. This is in contrast to individuals with cardiovascular disease who show considerable improvement in vascular function after exercise training. It would appear that those with impaired vascular function are most amenable to exercise training adaptations whereas those with normal vascular function are not. There is some evidence that with intense training vascular function is improved; however, it has been proposed that overly strenuous exercise can produce a pro-oxidant environment and lead to long-term impairment of vascular function. Moderate-intensity exercise, on the other hand, promotes an antioxidant state. Moderate-intensity aerobic exercise, or a combined moderate-intensity aerobic and resistance-training program, is probably most useful for preserving rather than changing vascular function in healthy adults.

In summary, exercise training, ideally an aerobic exercise program, can increase the size of both conduit and resistance arteries, reduce arterial wall thickness, and preserve vascular function in healthy adults.

Sources

Green, D.J., Maiorana, A., O'Driscoll, G., & Taylor, R. (2004). Effect of exercise training on endothelium-derived nitric oxide function in humans. *Journal of Physiology, 561*, 1-25.

Green, D.J., Spence, A., Rowley, N., Thijssen, D.H.J., & Naylor, L.H. (2012). Vascular adaptation in athletes: Is there an "athlete's artery"? *Experimental Physiology, 97*, 295-304.

Thijssen, D.H.J., Dawson, E.A., van den Munckhof, I.C.L., Tinken, T.M., den Drijver, E., Hopkins, N., Cable, N.T., and Green, D.J. (2011). Exercise-mediated changes in conduit artery wall thickness in humans: Role of shear stress. *American Journal of Physiology: Heart and Circulatory Physiology, 301*, H241-H246.

BLOOD COMPOSITION

Other adaptations of the cardiovascular system include changes in the composition of blood that enhance blood and oxygen delivery to the working muscles and the removal of byproducts such as lactate and heat from the working muscles. Endurance training increases the total volume of blood in the body, number of erythrocytes (oxygen-carrying red blood cells), and the content of haemoglobin. These changes provide for increased oxygen delivery to the working muscles. In addition, blood becomes less viscous due to a proportionally greater increase in the volume of plasma compared with the increase in erythrocyte number. Lower blood viscosity means less resistance to blood flow, which enhances blood delivery to the working muscles during exercise and reduces the work of the heart. These adaptations also increase the reservoir of fluid available for thermoregulation during exercise in the heat.

Together, these adaptations of the cardiovascular system greatly enhance oxygen delivery to the working muscles. The increase in oxygen delivery, coupled with the higher oxidative capacity of skeletal muscle, means that the body relies more on oxidative metabolism and less on lactic acid-producing anaerobic metabolism during exercise. In addition, these changes improve the capacity to use fat for ATP production, which helps delay glycogen depletion and the onset of fatigue during prolonged exercise. Thus, the capacity for endurance exercise greatly increases in response to training.

ENDURANCE TRAINING-INDUCED RESPIRATORY CHANGES

Due to adaptations in the respiratory muscles, maximum ventilation increases with endurance training. Although ventilation does not usually limit endurance capacity, increased ventilatory capacity means that the respiratory muscles are less likely to fatigue during endurance exercise in the trained person. Moreover, the increased volume of air inspired and expired enhances the lung's ability to rid the body of carbon dioxide and to buffer lactic acid produced during exercise.

ENDURANCE TRAINING-INDUCED CHANGES IN LACTATE THRESHOLD

The amount of lactic acid produced during exercise is a function of exercise intensity and stage of training. As can be seen from a curve of blood lactate concentration plotted against exercise work rate (figure 14.1), blood lactate does not increase linearly with increasing work rate. Rather, the level remains low to a certain exercise intensity and then increases exponentially.

The point at which blood lactic acid level begins to increase has been given many names, including the lactate (or lactic acid) threshold, anaerobic threshold, and onset of blood lactic acid accumulation. The lactate threshold generally occurs at 50% to 65% of $\dot{V}O_2$max in untrained people and 70% to 85% of $\dot{V}O_2$max in endurance-trained individuals. Thus, compared with untrained individuals, endurance-trained athletes can exercise at a higher intensity before lactic acid accumulation exceeds the removal capacity.

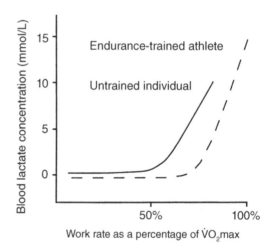

Figure 14.1 Lactate threshold. General trend of blood lactate concentration plotted against relative work rate, expressed as a percentage of $\dot{V}O_2$max. Endurance training shifts the lactate threshold toward a higher relative percentage of aerobic power, so the athlete can exercise at a higher intensity before blood lactic acid accumulates.

The lactate threshold indicates the exercise intensity that can be maintained primarily via oxidative metabolism without significant lactic acid accumulation. This threshold represents the exercise intensity that can be maintained at the upper limit of aerobic metabolic capacity without a major contribution of the anaerobic system. Below this threshold, ATP resynthesis occurs without excessive buildup of lactic acid; above this threshold, ATP production relies on an increasing contribution from anaerobic metabolism. Recall from chapter 11 that the conversion of pyruvate to lactic acid enables a high rate of glycolysis to be maintained via the recycling of protons released from the reactions of glycolysis. These protons need to be removed or metabolic acidosis results. When the capacity of lactic acid as a proton buffer has been exceeded, fatigue ensues and the rate of ATP production via anaerobic glycolysis is slowed. Thus, the lactate threshold represents the intensity of exercise below which an individual can, theoretically, maintain exercise indefinitely without fatigue.

Although $\dot{V}O_2$max is a good general indicator of aerobic fitness, the exercise intensity or pace that coincides with the lactate threshold is a better predictor than $\dot{V}O_2$max of performance among elite endurance athletes. The endurance athlete who can maintain a faster pace of exercise with lower levels of lactic acid will be the better performer. Although the test is complex, regular measurement of the lactate threshold can provide an important means of evaluating an athlete's response to a training program.

CHANGES IN THE MUSCULAR SYSTEM AFTER STRENGTH TRAINING

Elite powerlifters, sprinters, and bodybuilders are excellent examples of the tremendous capacity of skeletal muscle to adapt to strength training. Muscular fitness is important to performance in virtually every sport, and most athletes incorporate some type of muscular fitness exercise into the training regimens.

Muscular fitness refers to the various components of muscle function, including muscular strength, power, and endurance. Muscular strength is defined as the maximum force that can be produced by a muscle or muscle group in a single movement. Muscular power represents the rapid application of force, as that in a quick, explosive movement such as sprinting, throwing, or jumping. Muscular power is a function of both speed and strength, or the product of force multiplied by speed. Muscular endurance, another important aspect of muscular fitness, is defined as the ability of muscle to generate force repeatedly or continuously over time. Muscular endurance is important to performance in many sports, such as swimming, rowing, kayaking, and cycling, and to injury prevention. Muscular strength, power, and endurance are all interrelated because each is a function of the maximum amount of force that can be applied by a particular muscle group or in a specific movement. However, increasing one component does not necessarily lead to increases in another. That is, simply increasing muscular strength will not necessarily enhance muscular power; each component must be specifically trained according to the desired outcome.

MUSCULAR STRENGTH

As noted in previous chapters, muscular strength is closely related to the size or cross-sectional area of muscle as well as to muscle fibre type distribution, neural factors, and hormones such as the male hormone testosterone that stimulate muscle growth after periods of strength training. Muscular strength many increase by 20% to 100% over several months of resistance training.

Strength increases because of a complex interaction of the many factors that are determinants of strength. The relative contribution of each of these adaptations to strength gains varies by individual. For example, in older persons, the capacity for muscle hypertrophy may be limited and strength gains may depend more on neural factors (see also chapter 13).

As with anaerobic and endurance training, muscles adapt to strength training in a way that is specific to the intensity and volume of training. According to the size principle of motor unit recruitment, the largest and strongest FTb fibres are recruited for forceful contractions above 70% of maximum strength. Thus, strength-training programs must include high-intensity lifts to induce adaptations in FTb fibres and to maximise strength gains.

MUSCLE HYPERTROPHY

Muscle hypertrophy begins after approximately 6 to 8 wk of strength training and is the major contributor to continued strength gains after this time. As noted in chapter 5, muscle hypertrophy occurs via increases in the average diameter of muscle fibres and the amount of connective tissue between muscle fibres. The increase in muscle fibre size occurs because of an increased number of myofibrils and contractile filaments, the result of which is more cross-bridges generating force. Protein synthesis and degradation both occur at faster rates in trained than in untrained fibres. However, in trained muscle fibres the rate of synthesis exceeds that for degradation, resulting in a net increase in skeletal muscle protein. The increased connective tissue between muscle fibres is important for preventing injury in the stronger, more forceful muscles.

Muscles also hypertrophy in a manner that is specific to the intensity and volume of strength training and the muscle fibre type used. Low-intensity strength training, which recruits mainly ST and FTa fibres, induces hypertrophy only in these fibres. FTb fibres are the largest and strongest and have the highest capacity for hypertrophy (twice that of ST fibres). Maximum hypertrophy occurs when FTb fibres are recruited through training at high intensity. Maximum hypertrophy also seems to require high-volume strength training.

METABOLIC ADAPTATIONS

Metabolic changes following intense strength training include increases in ATP, PCr, and glycogen content in FT fibres, but these increases are a function of the increase in muscle mass. The activity of several key enzymes involved in PCr breakdown and production of ATP in FT fibres also increase. Together these metabolic changes provide more ATP and PCr at a faster rate so that the muscle can generate more force in brief, maximal muscle actions.

NEURAL ADAPTATIONS

Neural adaptations account for much of the early strength gain, in the first 6 to 8 wk of training, before muscle hypertrophy occurs. Strength training enhances synchronous recruitment and reduces inhibition of motor units. Motor units do not generally fire at the same time; instead, they fire in sequence or asynchronously. Synchronous recruitment of motor units results in a summation of force because more muscle fibres are activated at the same time. Strength training appears to relieve some of the natural inhibition of motor units. Under normal conditions, each motor neuron receives input from several other neurons, some of which are inhibitory. Reduced inhibition by strength training permits more motor units to become active and thus generate greater force. In addition, strength training appears to reduce activation of antagonist muscles (muscles with the opposite action of a specific muscle group) and to increase activation of synergist muscles (muscles that assist the muscles performing a particular movement). The first 4 wk of a strength-training program also improves skill and coordination in lifting heavy weights, both of which contribute to early improvements in strength.

MUSCULAR POWER AND ENDURANCE

Muscular power relates to the ability to generate a large amount of force rapidly; that is, power requires both strength and speed. The amount of muscular force that can be generated depends on several factors, including the speed of contraction. Essentially, force and speed of a contraction are inversely related. For a given muscle, a faster movement reduces the amount of force generated. Thus, maximal power reflects a tradeoff between speed and force. Maximum muscular power does not occur at maximum force but rather is achieved in the midrange of the force–velocity curve, at about 30% to 50% of maximum force. Changes in muscular power are closely related to the movement pattern and velocity of movement during training; these should match, as closely as possible, the demands of the sport. High-speed power training improves muscular power when measured at high speed but does not necessarily improve muscular power or strength at slower speeds of movement. Similarly, training at slower speed may not improve muscular power during fast movements. It appears that the specificity of the type of power training is largely due to neural adaptations.

Muscular endurance, or the ability to maintain force or repeated contractions over time, is related to strength. Increasing strength also improves muscular endurance because, after strength training, a lower proportion of maximum force is required to maintain a given submaximal force. In other words, the same submaximal force can be maintained over time with recruitment of fewer motor units, and thus less fatigue, after training.

TRAINING AND MUSCLE FIBRE NUMBER AND TYPE

It is generally accepted that human skeletal muscle fibre number and fibre type are genetically determined and cannot be changed appreciably through normal activity such as exercise. However, recent research suggests that there is some plasticity (i.e., ability to change) in the structural and metabolic characteristics of muscle fibre types.

Evidence for increased fibre number (hyperplasia) is stronger in experimental animal models than in humans. Hyperplasia has been induced by intense, high-volume strength training in experimental animals. Indirect evidence of hyperplasia has been shown in humans after intense, high-volume strength training. For example, some experienced weightlifters and bodybuilders have very large muscles that contain only average-size muscle fibres. It has been proposed that high-volume, high-intensity strength training over many years may stimulate division of skeletal muscle fibres. This increases the number of fibres, but each fibre remains about average in size. However, even if hyperplasia does occur, its contribution to gains in strength are minimal compared with muscle fibre hypertrophy and changes in neural factors, the two predominant mechanisms responsible for increases in strength after resistance training.

Fibre type is determined by the motor neuron. Fibre type transition, the changing of one fibre type to another, does not occur under most training circumstances. Recent research suggests, however, that some types of training may induce limited changes in the physiological and metabolic characteristics of muscle fibres. Several months of heavy resistance or power training (e.g., weightlifting, sprinting) may cause FTb muscle fibres to more closely resemble FTa fibres. This occurs because of changes in the metabolic characteristics, such as an increase in oxidative capacity and capillary number. The FTa fibres retain the characteristics of FTb fibres, such as size, speed, and strength, but are more fatigue resistant; this change is considered beneficial because it enhances the muscle's ability to generate and maintain force. This change is also reversible if training stops. There is no evidence that ST fibres can become FT fibres, even with very heavy training.

BASIC PRINCIPLES OF TRAINING

Certain basic principles are common to all types of training, from conditioning at the highest (elite) level of sport competition to programming for the average adult who exercises for health benefits. These principles include the specificity of the training; the duration, volume and intensity of the training; and the principle of overload. Training programs need to be tailored to the individual, taking into consideration the individual's goals and competition schedule. This helps to provide the optimal training environment while minimising the risk of injury or over-training.

SPECIFICITY

Skeletal muscle responds specifically to the physiological and metabolic demands of exercise. Thus, training must reflect the specific energy demands of the activity. For example, optimising the oxidative capacity of skeletal muscle is best achieved through endurance-type training. Likewise, only strength training will induce hypertrophy in FT muscle fibres. The concept of specificity also extends to movement patterns and neural adaptations. Training should, as much as possible, simulate the speed, force, and timing of the activity being trained for.

TRAINING VARIABLES

One can manipulate several variables when designing an exercise-training program: type of exercise (mode), duration, intensity, and frequency. The choices regarding each variable depend on the desired outcomes. Someone interested in health has more choice in terms of these variables than does an athlete training for a particular sport. For example, one can achieve health benefits using different modes of exercise (e.g., aerobic dance, swimming, walking), whereas an athlete would spend most of his or her time training using the specific mode of his or her sport. Moreover, one can achieve health-related benefits with much lower intensity, frequency, and volume of training than needed to excel at a high level in sport.

OVERLOAD

The body adapts to the physiological and metabolic demands of training over time. Therefore, the training load must progressively increase to induce continued improvement in exercise capacity. Athletes frequently experience a plateau in performance if training loads are not continually adjusted to accommodate recent adaptations and increased work capacity. Overload is normally achieved by altering the combination of frequency, intensity, time (duration), and type of activity.

INDIVIDUALISATION

Training adaptations are best achieved when the program is individualised to the specific needs of the individual athlete. Although general principles of training apply under most situations, it is important to recognise that the rate or extent of adaptions may vary between individuals. As discussed later in this chapter, individualisation is important for preventing overtraining syndrome.

PERIODISATION

Athletes frequently train in weekly, monthly, or seasonal cycles, varying the intensity and volume of training to achieve their specific goals. Short cycles (1-2 wk) are called microcyles, and longer periods (2 wk-2 mo) are called macrocycles (figure 14.2). In the high-performance athlete, performance plateaus at the end of each microcycle, requiring increased training volume, intensity, or both in the next

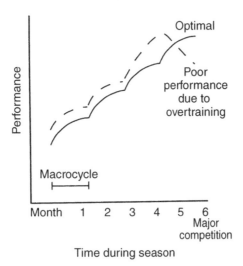

Figure 14.2 Theoretical model of training microcyles, each approximately 6 wk long. Ideally, performance should peak during major competition. Excessive training early in the season may cause overtraining and a decline in performance late in the season and during major competition.

microcycle. Periodisation ensures that the athlete reaches peak form at crucial times, such as major competition. It also allows for variety in training to prevent boredom, injury, and overtraining and permits continued adjustments of the training program in response to the athlete's progress or other factors such as injury or illness.

For team-game athletes, periodisation allows the athlete and team to focus on various aspects of the sport at different times in the season. For example, in sports such as soccer or basketball, early-season training may emphasise conditioning for aerobic fitness and muscular hypertrophy and strength. As the season progresses, training will focus less on endurance capacity and strength and more on speed, power, skill, strategy, and teamwork.

REVERSIBILITY

Training adaptations last only as long as the physiological and metabolic demands of exercise continue (i.e., as long as training is maintained at a certain intensity and volume). Detraining, a reversal of training adaptations, begins within days of cessation of training. Once a certain level of fitness is achieved, a minimum amount of regular exercise is needed to maintain any training adaptations. In general, maintaining training adaptions requires less exercise than is needed to induce the initial changes. Provided that training intensity is maintained, training volume may be decreased by up to 50% with little loss of fitness over the short term (e.g., 2-3 wk).

OVERTRAINING

Overtraining refers to excessive training leading to prolonged fatigue, frequent illness, and poor performance. The athlete is often unable to maintain training loads or perform at the standard expected. Overtraining often results from increasing training volume or intensity too rapidly and not allowing adequate recovery between training sessions. Prolonged rest and a reduction in training loads over several weeks or months are necessary to restore performance in an overtrained athlete.

Optimal training must balance all of the principles discussed previously. Indeed, training of the top athlete is both an art and science, based on scientific principles underlying training adaptations as well as the coach's intuition and observations regarding how an individual athlete adjusts to overload induced by training. Optimal performance depends on maximising the positive training adaptations while minimising the potential negative effects of excessive training loads.

CONTINUOUS AND INTERVAL TRAINING

Training may involve continuous or interval exercise. As their names imply, continuous training is performed without any rest breaks, whereas interval training involves alternating intervals of exercise and rest. Table 14.4 summarises the advantages and disadvantages of each type of training.

Continuous and interval training can be used to develop different aspects of fitness. For example, short-interval training involves repeated high-intensity intervals at maximal pace and induces adaptations in the immediate energy system. In contrast, longer intervals at lower intensity primarily train the aerobic energy system.

CONTINUOUS TRAINING

Continuous training may be further defined by exercise intensity. Lower-intensity continuous training is usually in the range of 70% to 80% $\dot{V}O_2$max for athletes and 50% to 60% $\dot{V}O_2$max for those seeking general health goals. Lower-intensity continuous training can be used in a variety of situations, including development of health-related fitness for the average adult or during early-season aerobic training in many sports. Higher-intensity continuous training, above 85% $\dot{V}O_2$max for the athlete and

60% to 70% $\dot{V}O_2$max for the nonathlete, is generally recommended only for those who are already physically fit. The athlete often exercises at or near race pace in this type of training.

The pace of continuous training need not be constant; it can vary within a given exercise session. Shorter bursts of higher-intensity exercise trains both the anaerobic glycolytic and aerobic energy systems. Longer periods of slower exercise induce adaptations primarily in the aerobic system, enhancing removal of lactic acid produced during higher-intensity exercise. Varying the pace of training adds variety to continuous training sessions, which otherwise may become repetitive and monotonous. Varying the intensity of training within a continuous session can add specificity for team-game athletes in sports such as soccer or basketball in which the athlete may run continuously but at various speeds throughout a match or game.

INTERVAL TRAINING

Internal training involves alternating periods (intervals) of exercise and rest. As shown in table 14.5, the length of both the exercise and rest intervals is manipulated to induce specific training adaptations.

Interval training offers several advantages over continuous training. For competitive athletes, the duration and intensity of exercise intervals can be varied to train specific energy systems, and training may be performed at or even above race pace. Training at or above race pace seems to be important

Table 14.4 Types of Training

Type	Advantages	Disadvantages
Continuous, constant pace	Is time efficient Trains cardiovascular and muscular endurance Easy to follow routine Can be sport specific Less chance of injury because of lower intensity	Athletes may need higher intensity Can be monotonous May not be specific to activities (e.g., team sports)
Continuous, varied pace	Adds variety of pace Train at higher intensity	May not be sport specific Increased risk of injury
Interval	Adds variety of pace and duration Can be very sport specific	Requires more time Increased risk of injury
Circuit	Trains cardiovascular and muscular endurance and strength Can be very sport specific	Improvements in endurance and strength not maximal Requires access to equipment

Table 14.5 Interval-Training Variables

Variable	Short interval	Intermediate interval	Long interval
Work interval	5-30 s	30-120 s	2-5 min
Intensity or pace	At or above race pace	90%-95% race pace	60%-95% $\dot{V}O_2$max
Rest interval	3-6 times work interval	2 times work interval	Equal to work interval
Major energy system	Phosphagens	Anaerobic glycolysis	Oxidative
Major effects on	Speed and power	Speed and power, muscular endurance, lactic acid buffering and tolerance	Cardiorespiratory endurance, muscular endurance, lactic acid removal

for inducing neural adaptations such as increased synchrony of motor unit recruitment. Because of frequent rest intervals, interval training permits the athlete to perform more exercise than in continuous training. Repeated high-intensity exercise intervals also improve lactate buffering and tolerance. Moderate-intensity interval training is an effective way to gradually introduce exercise to the untrained adult interested in health-related outcomes. There is less risk of overuse injury and muscle soreness. In addition, exercise is often perceived as less intense and therefore more comfortable. One disadvantage of interval training is that, because of the rest periods, interval sessions are typically longer than continuous training sessions.

There are three general categories of interval training. Each category corresponds to the predominant energy system used.

▶ Short-interval training, or anaerobic interval training, consists of work intervals lasting 5 to 30 s. Short-interval training relies predominately on the immediate energy system and is used to develop muscular strength, muscular power, and speed; exercise is performed at or above race pace. Rest periods between the intervals should be calculated using a work-to-rest ratio of either 1:3 or 1:6. For example, if a 20 s interval is used, rest can be either 60 or 120 s.

▶ Intermediate-interval training, or anaerobic–aerobic interval training, consists of work intervals lasting 30 s to 2 min and is performed at high intensity (>90% of race pace). Intermediate-interval training relies on PCr breakdown and anaerobic glycolysis for energy production and relies to some

extent on aerobic metabolism. Muscle and blood lactic acid levels are very high at the end of each work interval, and longer rest periods (several minutes) are needed to restore PCr levels and to remove lactic acid. Rest periods should be calculated using a work-to-rest ratio of 1:2. For example, if a 2 min interval is used, rest should be 4 min.

▶ Long-interval training, or aerobic interval training, consists of work intervals lasting 2 to 5 min. This type of training relies primarily on the aerobic system for ATP production. Long intervals induce changes in the oxidative capacity of muscle, especially ST and FTa fibres. If the training is performed at a higher intensity, at 70% to 90% of $\dot{V}O_2$max, FTb fibres are recruited and the anaerobic glycolytic system is taxed, producing moderately high lactic acid levels. Rest periods between intervals should be approximately 1:1.

TRAINING FOR CARDIOVASCULAR ENDURANCE

The optimal type of cardiovascular endurance-training program depends on the individual's objectives and desired outcomes. Obviously, an older person seeking health benefits and a high-performance distance runner require different programs. Thus, exercise programs should be individualised, although certain basic principles apply to all types of endurance training.

There appears to be a minimum threshold in terms of exercise intensity, frequency, and duration for improving aerobic exercise capacity. To improve

$\dot{V}O_2$max, the average healthy young adult needs to exercise for at least 15 min, at a minimum of 60% $\dot{V}O_2$max , at least 3 times per wk. Improvements may occur at a lower exercise intensity (even as low as 30%-50% $\dot{V}O_2$max) in older, less-fit individuals or in persons with disease.

In general, improvements in $\dot{V}O_2$max and endurance exercise capacity are related to the total amount of exercise performed. Exercise intensity, frequency, and duration may be manipulated in many ways to increase aerobic power and endurance-exercise capacity. However, according to the principles of specificity and motor unit recruitment, training intensity and duration for the endurance athlete should approximate those required in competition. There is some debate about the optimal amount, intensity, and frequency of exercise needed to improve physical fitness and health, which is discussed in more detail in chapter 23. Some health benefits, such as the loss of body mass, reduced blood pressure, and lower risk of heart disease, occur without significant changes in physical fitness. Improving physical fitness (i.e., increasing $\dot{V}O_2$max or endurance-exercise capacity) requires slightly more exercise, in the range of 50% to 85% $\dot{V}O_2$max for 20 to 60 min 3 to 5 times per wk. The elite athlete requires a far higher level of exercise—daily training for several hours at an intensity greater than 85% of $\dot{V}O_2$max—to improve physical fitness and performance.

METHODS OF STRENGTH TRAINING

Some types of strength training should be included in all fitness programs, from programs for strength and power athletes to those for individuals who exercise for health benefits. Of course, athletes in sports requiring strength and power, such as weightlifting, bodybuilding, and sprinting, rely heavily on resistance training. However, many other athletes also benefit from strength training, especially those in sports requiring a high level of muscular endurance, such as rowing, swimming, and cycling. Strength training also helps prevent and rehabilitate sport-related injuries.

Muscular-strength exercises can enhance health-related fitness. Moderate-intensity resistance training confers health benefits such as favorable changes in glucose tolerance, body composition, and blood lipids (fats), which are related to heart disease (see

chapter 23). Moderate-resistance training may also help prevent and treat some types of lower-back pain and other conditions such as arthritis and osteoporosis (see chapter 23); maintain lean body mass, muscular strength, and mobility with ageing (see chapter 13); and improve muscular tone and body shape, an important aesthetic consideration for many adults who exercise.

TYPES OF MUSCLE CONTRACTION

As discussed in earlier chapters, the two general types of muscle contractions are static and dynamic (figure 14.3). In a static, or isometric, contraction, the muscle generates force but does not change length because resistance to that force is greater than the force generated by the muscle. Isometric contractions act primarily to stabilise joints during movement. Isometric strength training is primarily used when joint mobility is limited (e.g., after an injury or in a joint affected by arthritis).

In a dynamic contraction, the muscle's length changes while it generates force against a moveable resistance. Dynamic contractions, as noted previously, may be either concentric (the muscle generates force while shortening) or eccentric (the muscle generates force while lengthening). Most force-producing movements (e.g., the propulsive action in running, jumping, or throwing) occur via concentric contractions. Eccentric contractions are used to stabilise or decelerate the body, especially when counteracting gravity. Two common examples are landing after a jump, in which the quadriceps muscles on the front of the thigh lengthen to absorb the impact and prevent falling, and slowing the arm movement at the end of throwing, in which the rotator cuff muscles at the back of the shoulder lengthen to decelerate the arm and prevent loss of balance and shoulder injury.

TYPES OF STRENGTH TRAINING

Strength- or resistance-training programs may include static or dynamic contractions. Given that gains in strength are specific to the type of program, resistance-type activities should match the needs of the sport and desired outcomes. As already indicated, isometric training has limited applications, primarily because the gains in strength

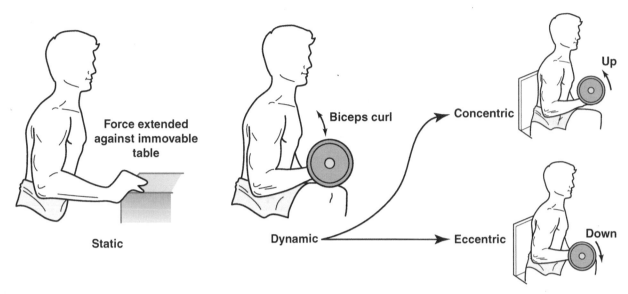

Figure 14.3 Static and dynamic muscle contractions. In a static contraction, tension is generated but muscle length does not change. In a dynamic contraction, the muscle (in this case the biceps brachii) may shorten or lengthen while developing tension; these are referred to as concentric and eccentric contractions, respectively.

are limited to the particular joint angle at which training occurs.

Dynamic strength training may be further classified by the type of resistance against which force is applied. Isoinertial (sometimes called isotonic) training is performed against a constant resistance (e.g., a barbell or fixed-weight machine). Isokinetic training is performed with a constant speed of movement against variable resistance. Specialised equipment, such as the Biodex or ConTrex isokinetic dynamometers, that control the velocity of movement is required for isokinetic training. The machine controls the speed of movement and offers resistance equal to the force being applied; maximal force is measure at each joint throughout the range of motion. The speed may be preset to simulate movement of a particular activity.

The various types of resistance training (isometric, isotonic, and isokinetic) have advantages and disadvantages. Isometric training is low cost and low risk but gives minimal gains in strength. Isoinertial training can be very sport specific and yields excellent improvements in strength but may increase the risk of injury and muscle soreness because of the eccentric component. Isokinetic training gives excellent strength gains with minimal risk of injury but is less sport specific because movement is generally limited to simple flexion and extension at certain joints, and natural movements

rarely occur at constant speed. In addition, many movements in sport occur at speeds faster than an isokinetic dynamometer can provide. Isokinetic training is often used in the clinical setting to assess muscular strength throughout the range of motion, to identify weakness in a certain muscle group or imbalance between muscle groups, and to rehabilitate an injury. Isokinetic training is also used widely in the research laboratory.

TRAINING TO IMPROVE MUSCULAR STRENGTH AND ENDURANCE AND TO INDUCE HYPERTROPHY

As in other forms of training, adaptations to strength training are specific to the training program variables such as the type of training, method of measurement, training intensity, and duration. Improvement in strength also depends on the individual's age, fitness level, and initial strength. In general, training using one type of muscle action improves strength only when measured in a similar manner. That is, isoinertial strength training improves strength only when measured isoinertially, and isokinetic strength training improves strength when measured isokinetically at a speed of movement similar to that used in training. Thus, strength-training programs must be specific to the desired outcomes. Training-program variables that can be manipulated include the following:

- ▶ the number of repetitions or times a weight is lifted;
- ▶ sets or groups of repetitions;
- ▶ training volume or total amount of work performed, which is the product of the number of repetitions and the number of sets; and
- ▶ intensity or resistance.

Intensity can be expressed two ways. Resistance can be prescribed as a percentage of maximal strength, or the weight that can be lifted only once; this is referred to as 1 repetition maximum (1RM). For example, if an athlete's maximum strength, or 1RM, in the bench press is 50 kg, a set might consist of 5 repetitions at 70% of 1RM. In other words, the resistance would be 35 kg. This method requires initial measurement of 1RM for each muscle group. This may be inconvenient or not possible for some groups of individuals. Resistance-training intensity may also be prescribed in terms of the weight or resistance that can be lifted for a specified number of repetitions. For example, a set may consist of 10RM, which is the maximum weight the athlete can lift only 10 times. Although this method does not require the person to perform maximum lifts, it does require trial and error to estimate resistance for each muscle group.

One must consider the pattern of recruitment of skeletal muscle fibre type and specific energy systems in prescribing strength-training programs specific to the desired outcomes. Table 14.6 gives examples of resistance-training programs that improve muscle strength, power, and endurance; improve fitness; and induce hypertrophy. Gains in maximum strength require recruitment of the larger FTb fibres, which occurs above 70% of maximum force. Muscle hypertrophy requires moderate- to high-intensity, high-volume training for several weeks or more. Optimal development of muscular power requires hypertrophy and development of strength first, followed by work emphasising speed and the explosive application of force that simulates the actual movement pattern. Power athletes generally follow a periodised program of resistance training that begins with 6 to 8 wk of training emphasising muscle hypertrophy, followed by a further 6 to 8 wk focusing on muscular strength, and then power training using movement patterns specific to the sport.

ROLE OF ECCENTRIC MUSCLE ACTIONS IN STRENGTH TRAINING

Gains in strength are related to the mode and actions of the muscle(s) used in training. Concentric-only training generally improves strength only when measured concentrically and has little effect on eccentric strength. Most forms of resistance training, such as free weights or pin-weight machines, include both concentric and eccentric muscle actions. Recent research suggests that maximum gains in strength require both types of muscle actions. Although eccentric muscle actions are more likely to induce injury to muscle fibres and muscle soreness (discussed next), it is believed that some degree of cellular damage leads to gains in muscular strength and muscle hypertrophy. Most serious strength and power athletes incorporate both types of muscle actions into their resistance-training programs.

Table 14.6 Prescription for Strength Training

To develop	Repetitions	Sets	Intensity	Rest between sets
Maximal strength	2-6	3-6	High	>3 min
Muscular hypertrophy	8-12	3-6	Moderate	>3 min
Muscular endurance	15-25	1-4	Moderate	<1 min
Muscular power	2-6	3-5	Moderate resistance, fast, explosive	>3 min
Health-related fitness	8-20	1-2	Moderate	<1 min

Data from Wilmore and Costill, 1999.

CAUSES OF MUSCLE SORENESS

All athletes, and most people who exercise, have at some point experienced muscle soreness following heavy training sessions to which they are unaccustomed. There are two forms of postexercise muscle soreness: an immediate form appearing during or soon after exercise and a delayed form that appears 12 to 48 h after exercise. Both forms cause muscle weakness; that is, the muscle is not capable of producing as much force when soreness is present.

The acute or immediate type of soreness feels like a painful or burning sensation in the affected muscles. This sensation and associated weakness is caused by buildup of lactic acid. This form of soreness is temporary, and the sensation disappears after a few minutes or hours.

The second type of muscle soreness is called delayed-onset muscle soreness (DOMS) because it begins the day after exercise and may last for up to several days. DOMS is more likely to occur and be more severe for exercise involving eccentric muscle actions. For example, prolonged downhill walking or running causes DOMS in the quadriceps muscles (on the front of the thigh) because these muscles act eccentrically to counteract the force of gravity. It is generally believed that muscle soreness occurs because fewer muscle fibres are activated during eccentric muscle actions than during concentric muscle actions. The high force generated by each activated muscle fibre is thought to cause damage to the interior structure of that muscle cell as well as to the surrounding connective tissue.

Over the years, several mechanisms have been suggested to explain DOMS. Contrary to popular belief, DOMS is not related to lactic acid accumulation. For example, lactic acid levels are not elevated in the types of exercise most likely to cause DOMS, such as downhill walking. Muscle spasms during and after exercise have also been proposed as a mechanism for DOMS, although there is little experimental evidence to support this concept. The most likely cause of DOMS is damage to both individual muscle fibres and connective tissue surrounding the fibres resulting from intense, prolonged, or eccentric exercise. Damage to the muscle fibres and connective tissue initiates an inflammatory response, causing an influx of inflammatory cells from the circulation, tissue swelling, and the sensation of pain.

The skeletal muscle damage associated with DOMS is temporary, lasting usually less than 1 wk. Muscle fibres are rapidly repaired, and there is no evidence of permanent tissue damage associated with DOMS. As discussed earlier, it is believed that some level of tissue breakdown and repair may be important in initiating training-induced adaptations in skeletal muscle, in particular gains in muscle size and strength. One interesting feature of DOMS is that a single session of eccentric exercise provides protection for up to several months against a subsequent occurrence of DOMS in the same muscle.

SUMMARY

Factors that limit exercise performance vary according to the duration and intensity of exercise and thus the energy systems used. Performance in power and speed activities is limited by depletion of PCr and metabolic acidosis; performance in endurance exercise is limited by muscle glycogen depletion, dehydration, and increased body temperature. The body adapts to exercise training by delaying the onset of fatigue so that the athlete can accomplish more work or exercise at a faster pace or for a longer time before becoming fatigued. Endurance training increases $\dot{V}O_2$max, muscle glycogen stores, and the cardiovascular system's ability to deliver oxygen to, and remove metabolic byproducts from, the working muscles. About half of our endurance-exercise capacity appears to be genetically determined. High-intensity sprint or strength training increases muscle size (hypertrophy); therefore, muscle glycogen, PCr stores, and the enzymes involved in anaerobic metabolism all increase. Muscle fibre type does not change after training, although oxidative capacity and the ability to resist fatigue both increase in FT fibres. DOMS after heavy exercise is more likely to occur in muscles contracting eccentrically (e.g., after downhill running). DOMS reflects minor structural damage to muscle fibres and connective tissue, which is usually repaired within 1 wk.

FURTHER READING

American College of Sports Medicine. (2009). American College of Sports Medicine position stand. Progression models in resistance training for healthy adults. *Medicine and Science in Sports and Exercise, 41*(3), 687-708.

Garber, C.E., Blissmer, B., Deschenes, M.R., Franklin, B.A., Lamonte, M.J., Lee, I.M., Nieman, D.C., Swain, D.P., & American College of Sports Medicine. (2011). American College of Sports Medicine position stand. Quantity and quality of exercise for developing and maintaining cardiorespiratory, musculoskeletal, and neuromotor fitness in apparently healthy adults: Guidance for prescribing exercise. *Medicine and Science in Sports and Exercise, 43*(7), 1334-1359.

National Strength and Conditioning Association (Baechle, T.R., & Earle, R.W., Eds.). (2008). *Essentials of strength training and conditioning* (3rd ed.). Champaign, IL: Human Kinetics.

PART V

NEURAL BASES OF HUMAN MOVEMENT

MOTOR CONTROL

In the preceding sections on functional anatomy, biomechanics, and exercise physiology we gained some understanding of the material structure, design, and energetics of the human machinery for movement. To take our analogy of the human body as a motor vehicle one step further, this section of the book, in introducing the subdiscipline of motor control, examines the control mechanisms we have for movement. Understanding the control of human movement requires an understanding of the functions of the brain and nervous system in a way roughly comparable with studying automotive electronics in order to understand the control systems in a motor vehicle. The analogy of the neural system as an electrical system is a powerful one that has, in a general way, guided much theorising in the motor control field.

DEFINING MOTOR CONTROL

Motor control is the subdiscipline of human movement studies concerned with understanding the processes that are responsible for the acquisition,

performance, and retention of motor skills. Motor development and motor learning are specialised areas of focus in the subdiscipline of motor control. **Motor development** (as discussed in chapter 17) deals with motor control changes throughout the life span, specifically the changes in the acquisition, performance, and retention of motor skills that occur with growth, development, maturation, and ageing. **Motor learning** (as discussed in chapter 18) deals with motor control changes that occur as a consequence of practice (or adaptation), focusing literally on how motor skills are learned and the changes in performance, retention, and control mechanisms that accompany skill acquisition. Over the years a host of other terms such as *motor behaviour, psychology of motor behaviour, motor learning and control*, and *skill acquisition* have also been used to describe the subdiscipline or parts of it. The use of these terms is avoided here because they only confuse rather than clarify the subdiscipline's scope and structure.

The word *motor* in the terms *motor control, motor development, motor learning*, and *motor skills* literally means *movement*. **Motor skills** are goal-directed actions that require movement of the whole body, limb, or muscle in order to be successfully performed. Consequently, the motor control subdiscipline has a broad focus and range of application: in the study of movements as simple as unidirectional finger or eye movements to movements as complex as those involved in fundamental actions such as walking, running, reaching, grasping, and speaking; in workplace tasks such as welding, typing, and driving; in artistic tasks such as dancing and playing musical instruments; and in sporting tasks such as performing a complex gymnastics routine or hitting a fast-moving tennis ball. Despite their obvious diversity, all of the motor skills used in these tasks share in common their purposefulness, their voluntary nature, their dependence to some degree on learning, and the fact that the quality of task performance is directly dependent on the quality of the movement produced.

TYPICAL QUESTIONS POSED AND PROBLEMS ADDRESSED

One way of gaining a feel for the breadth of the motor control subdiscipline is to note some of the many questions currently under examination by motor control researchers. The four chapters in this section discuss some of these questions.

▶ *How are skilled movements remembered?* For example,

- Why is it we can remember how to ride a bicycle or how to swim after many years without practising the skill?

- What elements of movement are stored in memory?

- How can memory for movement be enhanced?

- Do skilled performers have better memories for movement?

▶ *What is the most effective set of conditions for learning a new motor skill?* For example,

- What type of feedback information is best for learning?

- Can we learn when we are fatigued?

- Should practice emphasise consistency or variability?

- Does verbal instruction help or hinder learning?

▶ *How do skilled performers succeed where lesser skilled performers fail?* For example,

- Do experts have faster reaction times than lesser skilled performers?

- Can skilled performers be identified at an early age?

- How do expert game players manage to "read the play"?

- What does it mean when skills become automatic?

▶ *How is movement control affected by fatigue, injury, and disability?* For example,

- Why do stroke sufferers have difficulty with speech and gait?

- What causes clumsiness in some children?

- What is the best way to recover movement control in an injured joint?

- How and why does alcohol affect movement control?

Issues addressed by motor control researchers can clearly range from very basic questions (such as those related to how the nervous system implements the neuromuscular changes needed to make a movement purposefully faster, more forceful, or with altered sequencing or timing) to very applied questions (such as those related to optimising performance or the rate of skill learning or relearning).

LEVELS OF ANALYSIS

The single greatest difficulty for scientists attempting to understand how the control of movement occurs is that the processes in the brain and central nervous system that control movement (and that are modified with maturation and adaptation) are not directly observable. Knowledge about motor control is gained indirectly from inferences about control mechanisms derived from the observation, description, and measurement of observable movement performed under a variety of carefully selected experimental conditions. The more different levels of analysis of the motor system that are undertaken, the greater is the certainty with which inferences about motor control can be made.

Figure 1.1, which schematically represents knowledge organisation in the discipline of human movement studies, indicates that the subdiscipline of motor control draws methods, theories, and paradigms from a range of cognate disciplines, including physiology, mathematics, physics, computer science, psychology, and education. Of these influences the most powerful ones in both a historical and contemporary sense are the influences from physiology, especially neurophysiology, and cognitive science, which is a hybrid of experimental psychology and computer science. Indeed, in the modern field of motor control, it is still possible to identify separate yet complementary neuroscience and cognitive science approaches to the examination of movement control, development, and learning.

The neuroscience approach to motor control focuses on understanding the functioning of the components of the neuromuscular system, especially the functional properties of the movement receptors, the nerve pathways, the spinal cord, and the brain. Its aim, in a crude sense, is to describe the basic wiring, interconnections, and organisation of the neuromuscular system in order to understand the neural architecture—or, to use a computer analogy, the hardware—of the motor system. Physiologists studying the motor system use a variety of methods, including tracing nerve connections (using histological procedures such as staining or labelling procedures where neurotransmitters are traced chemically); measuring metabolic activity in specific areas of the brain (by radioactively labelling a metabolic source such as glucose); examining the functional anatomy of the brain using imaging techniques (such as computer-assisted tomography, **magnetic resonance imaging**, and **positron emission tomography**); measuring electrical activity in the brain (via **electroencephalography** and **magnetoencephalography**) and muscle (via **electromyography**); evaluating the behavioural effects of brain damage or selective lesions to the neural system; and observing the behavioural effects of selective electrical, magnetic, or chemical stimulation of given nerve pathways or regions of the brain. Many of these techniques can be applied only to anaesthetised animals or, in the case of examining the impact of brain damage, to patient populations, although improving technology, especially in brain imaging, allows increasing recordings and observations from the living, undamaged human brain. Basic information about motor control derived from neuroscience is presented in chapter 15.

The cognitive science approach to motor control focuses not so much on the physical structure of the components of the neuromuscular system but rather aims to develop conceptual models to describe and explain the collective behaviour of the some 100,000,000,000,000 (10^{14}) neurons and neural connections that compose the motor system. The favoured approach of psychologists examining movement control is to use experimental tasks with altered movement demands to test their conceptual models. The validity of the conceptual models of the perceiving, deciding, and acting stages involved in movement control are evaluated from measures of both movement outcome and pattern. The usual movement outcome (or product) measures are those of movement speed or accuracy, whereas movement patterns are typically described using one or more of the various measures of kinematics, kinetics, and electromyography as described earlier for the subdisciplines of biomechanics and functional anatomy. In recent times the mixture of the methods of experimental psychology with those of computer science has resulted in the establishment of the field of cognitive science, one of the principal goals of which is to develop powerful computational models and simulations of the functioning of the neuromuscular system. Some of the basic findings concerning motor control derived from a cognitive science perspective are presented in chapter 16.

HISTORICAL PERSPECTIVES

The neuroscience and cognitive science approaches to motor control have quite independent histories. A number of important findings with respect to the neurophysiological basis of movement and movement control were made throughout the 1800s. Foundations for a neurophysiological basis of movement control can be found in Charles Bell's 1830 text *The Nervous System of the Human Body*, in which the motor function of the ventral roots of the spinal cord and the sensory function of the dorsal roots were described, and later in the studies from 1856 to 1866 of Hermann von Helmholtz, in which nerve conduction velocity was estimated and studies of reaction time were commenced. Subsequently, the spring-like characteristics of muscle were described (by Weber in 1846), the electrical excitability of the brain was discovered (by Fritsch and Hitzig in 1870), and studies of the sensory and motor functions of the brain commenced (by Beevar and Horsely in 1887).

Undoubtedly the major historical contributor to the understanding of the neurophysiology of the motor system was Sir Charles Sherrington (1857-1952), whose work on reflexes is still widely credited today. From Sherrington's work, especially his classic 1906 text titled *The Integrative Action of the Nervous System*, arose a number of key concepts such as synaptic transmission, reciprocal inhibition, final common pathway, and proprioception, which are cornerstones of modern neurophysiology. Another influential figure for modern motor control theories was the Russian physiologist Nicolai Bernstein (1897-1966), whose integrative work on phase relations, functional synergies, and distributed control in natural actions such as locomotion appeared only posthumously in the English-language literature. A wealth of knowledge on the neurophysiology of the motor system appeared in the postwar era when improved electrophysiological recording techniques and neurophysiological mapping techniques facilitated new precision and insight into the structure and function of many levels of the neuromotor system. Of particular note in this period was the work of the Canadian neurosurgeon Wilder Penfield, who, along with Rasmussen, used electrical-stimulation studies to map the topographical organisation of the motor and sensory cortices of the brain.

The most noteworthy and influential early studies of movement from a psychological perspective were the basic studies on the control of arm movements by Woodworth in 1899 and the applied studies by Bryan and Harter on the skill acquisition of Morse code operators published in 1897 and 1899. Little motor control research of any kind took place in the first part of the 20th century, and when motor control research reappeared in the 1930s and 1940s it was oriented toward solving practical problems associated with specific motor tasks. Practical concerns that were highlighted in this period included personnel selection and training for war tasks such as flying, steering, tracking, and weapon control. The search for optimal approaches to teaching and coaching in physical education and sport was also to the forefront. An identifiable research field of movement control in psychology, aimed at understanding fundamental processes in movement control, did not emerge until the 1950s. This followed the major conceptual advance in experimental psychology in the late 1940s, triggered by Craik (1947, 1948), Wiener (1948), and Shannon and Weaver (1949), that viewed the brain and nervous system as processors of information in a manner akin to sophisticated, high-speed computers. The 1960s and 1970s in motor control were consequently dominated with information-processing models of motor control and attempts by psychologists such as Paul Fitts to quantify the information-processing capabilities of the motor system. The theoretical contributions of the psychologists Franklin Henry [see "Franklin Henry (1904-1993)" in chapter 2], Jack Adams, Steven Keele, and Richard Schmidt were particularly prominent in this period. Schmidt founded *Journal of Motor Behavior* in 1969, giving the subdiscipline its first specialist journal. With the passage of time the historical distinctions between the neuroscience and psychological (now, more appropriately, cognitive science) approaches to movement control have become less pronounced. The two approaches are complementary—one providing knowledge of the structure and function of the neuromuscular system, the other providing theories and simulations of its collective behaviour and organisation—and synthesis between the two approaches is increasingly sought.

PROFESSIONAL ORGANISATIONS AND TRAINING

People interested in motor control come from a diversity of backgrounds. At any symposium or conference on motor control it is not unusual to find psychologists, neurophysiologists, cognitive scientists, engineers, neurologists, and therapists in addition to researchers with backgrounds in human movement studies. It is perhaps not surprising, therefore, to find no single professional subgroup representing motor control internationally. The major meetings for scientists interested in motor control take place under the umbrellas of international groups such as International Society for Posture and Gait Research (ISPGR) and strong national groups such as, in North America, North American Society for the Psychology of Sport and Physical Activity (NASPSPA), Canadian Society for Psychomotor Learning and Sport Psychology (SCAPPS), Society for Neuroscience, and The Psychonomic Society, and, in Europe, L'association des Chercheurs en Activités Physiques et Sportives (ACAPS) in France. Increasingly, motor control scientists are also attending conferences in biomechanics, such as those hosted by International Society of Biomechanics. In line with its eclectic nature, motor control research is published in a diversity of journals, including specialist journals such as *Human Movement Science, Journal of Motor Behavior,* and *Motor Control;* neuroscience journals such as *Journal of Neurophysiology, Brain Research, Experimental Brain Research, Brain and Behavioral Sciences,* and *Journal of Electromyography and Kinesiology;* and experimental psychology journals such as *Journal of Experimental Psychology: Human Perception and Performance, Quarterly Journal of Experimental Psychology,* and *Acta Psychologica.* Research on motor control also appears regularly in more generalist human movement studies journals such as *Journal of Sports Sciences* and *Research Quarterly for Exercise and Sport.*

FURTHER READING

Latash, M.L., & Zatsiorsky, V.M. (Eds.). (2001). *Classics in movement science.* Champaign, IL: Human Kinetics.

Rosenbaum, D.A. (2010). *Human motor control* (2nd ed.). San Diego: Academic Press/Elsevier.

Schmidt, R.A., & Lee, T.D. (2011). *Motor control and learning: A behavioural emphasis* (5th ed.). Champaign, IL: Human Kinetics.

Thomas, J.R. (1997). Motor behavior. In J.D. Massengale & R.A. Swanson (Eds.), *The history of exercise and sport science* (pp. 203-292). Champaign, IL: Human Kinetics.

SOME RELEVANT WEBSITES

Canadian Society for Psychomotor Learning and Sport Psychology: www.scapps.org

International Society for Posture and Gait Research: www.ispgr.org

North American Society for the Psychology of Sport and Physical Activity: www.naspspa.org

The Psychonomic Society: www.psychonomic.org

Society for Neuroscience: www.sfn.org

CHAPTER 15

BASIC CONCEPTS OF MOTOR CONTROL: NEUROSCIENCE PERSPECTIVES

The major learning concepts in this chapter relate to

- ▶ the nervous system as an elaborate communications network,
- ▶ components of the nervous system,
- ▶ neurons and synapses as the building blocks of the nervous system,
- ▶ sensory receptor systems for movement,
- ▶ effector systems for movement,
- ▶ motor control functions of the spinal cord,
- ▶ motor control functions of the brain,
- ▶ integrative neural mechanisms for movement control, and
- ▶ major disorders of movement and their neurophysiological origins.

Humans are capable of executing movements that are truly incredible in their diversity, complexity, precision, and adaptability. To support such an array of movement capability, we need a neuromuscular system that is highly organised yet that shares the flexible, adaptable, and complex properties of movement itself.

NERVOUS SYSTEM AS AN ELABORATE COMMUNICATIONS NETWORK

The nervous (or neural) system is designed in some ways like a modern telecommunication network. It has receivers (receptors) that pick up important signals; muscles (as effectors) that are able, when instructed, to bring about planned actions; and a vast array of electronic microwiring (neurons) and interconnections that allow for near-infinite linking of receptors to effectors and therefore permit information flow from one region of the network to another. The routine operations of the communication system are achieved at a local level (the spinal level), whereas an overriding authority (the brain) makes executive and policy decisions about what tasks the communication network should attempt to achieve.

The language of the communication system is the coded electronic bursts (nerve impulses) that travel throughout the various links in the network. The communication system as a whole is never static because new connections are constantly being made, damaged connections are replaced and repaired, and frequently used connections are expanded and upgraded to improve their capacity and rate of transmission. In the nervous system, as in the communication network, this capacity for constant change (what is termed *plasticity*) is essential for accommodating growth, development, and adaptation. Just as we are able to use sophisticated communication systems such as the telephone effectively without direct knowledge, or in many cases any understanding, of the structure and function of its component parts, as humans we are able to use our neuromuscular systems to produce all manner of skilled movements with a minimum of awareness of either the neuromuscular system's component parts or its method(s) of operation.

COMPONENTS OF THE NERVOUS SYSTEM

The main physical components of the nervous system are the sensory receptors, the motor units, and the nerves (neurons) and their junctions (the synapses) that permit communication between the sensory receptors, the motor units, and other neurons. There are in all some 10^{12} to 10^{14} neurons in the human nervous system, each of which may have as many as 10^4 synaptic connections with other neurons, receptors, or motor units. As a collective network, the nervous system has two major subdivisions. The **central nervous system** (figure 15.1) consists of the brain and the spinal cord and is responsible for overseeing and monitoring the activation of all sectors of the body, including the muscles. The **peripheral nervous system** carries information from the sensory receptors to the

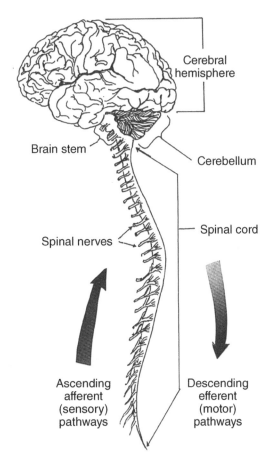

Figure 15.1 The central nervous system consisting of the brain and the spinal cord with its ascending sensory and descending motor pathways.

central nervous system and commands from the central nervous system to the muscles. The sensory information from the receptors reach the brain via ascending **afferent** (from the Latin *afferere*, meaning "to carry toward") pathways, and the commands from the brain to the muscles are carried via descending **efferent** (from the Latin *efferere*, meaning "to carry away") pathways.

The signals from the brain may act to either excite or inhibit the motor neurons that synapse directly with muscle fibres within muscle. Complex connections between neurons in the central nervous system provide for specialised control of movement and for the storage of information that is essential for memory and learning. In the sections that follow, we look in greater detail at the main components of the nervous system and their role in the control of movement.

NEURONS AND SYNAPSES AS THE BUILDING BLOCKS OF THE NERVOUS SYSTEM

The joining together of neurons through synapses—the specialised junctions between nerve cells—provides the essential foundation for the human nervous system and allows the nervous system to provide effective two-way communication between its sensory receptors and its motor units. Without synapses there would be no effective means for information to flow between neurons in different parts of the brain, between neurons in the brain and the spinal cord and muscle, and between the sensory receptors and neurons in the brain and spinal cord. In such a situation, skilled movement as we know it would simply not be possible.

STRUCTURE AND FUNCTION OF NEURONS

The **neuron** or nerve cell is the basic component of the neuromuscular system and provides the means of receiving and sending messages (or information) throughout the entire system. Although neurons vary substantially in both size and shape, depending on their specific function and location in the nervous system, most neurons share a similar structure in terms of having a cell body to which are connected

a single axon and (typically) many dendrites (figure 15.2). The cell body, containing the nucleus, regulates the **homeostasis** of the neuron. The **dendrites**, collectively formed into a dendritic tree, connect with and receive information from other neurons and, in some cases, sensory receptors. The **axon** is responsible for sending information away from the neuron to other neurons. Collateral branches off the main axon permit communication of the nerve impulses from any particular neuron to more than one target neuron. Any single neuron can influence the activity of up to 1,000 other neurons and is itself influenced by the excitatory and inhibitory impacts of some 1,000 to 10,000 other neurons.

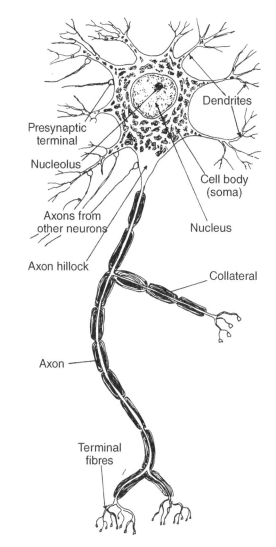

Figure 15.2 A typical neuron or nerve cell.

There are a number of types of neurons. In each case, their structure is dictated by their function (figure 15.3). The sensory or afferent neurons are relatively linear in structure; a single axon connects the sensory receptor ends to the cell body. The motor or efferent neurons vary in structure according to their location. The **alpha motor neurons** of the spinal cord possess many dendritic branches and a relatively long axon, also heavily branched to innervate multiple (100-15,000) skeletal muscle fibres. The **gamma motor neurons**, as discussed in the next section, innervate contractile (intrafusal) fibres located in the muscle receptors. Consequently, gamma motor neurons, which constitute about 40% of the total motor

neurons in the spinal cord, have smaller, considerably less-branched axons than do alpha motor neurons. The cell bodies of both the alpha and gamma motor neurons are located in the spinal cord.

The **pyramidal cells**, located in the motor cortex of the brain, are so named because of their shape, which derives from their branching tree of dendrites, all of which funnel down to a single slender axon. Pyramidal cells send motor commands over the long distances from the brain to the spinal cord and may have axons up to 1 m in length. The **Purkinje cells** in the **cerebellum** also have a single thin axon to which information is sent from an incredibly rich, systematically organised set of

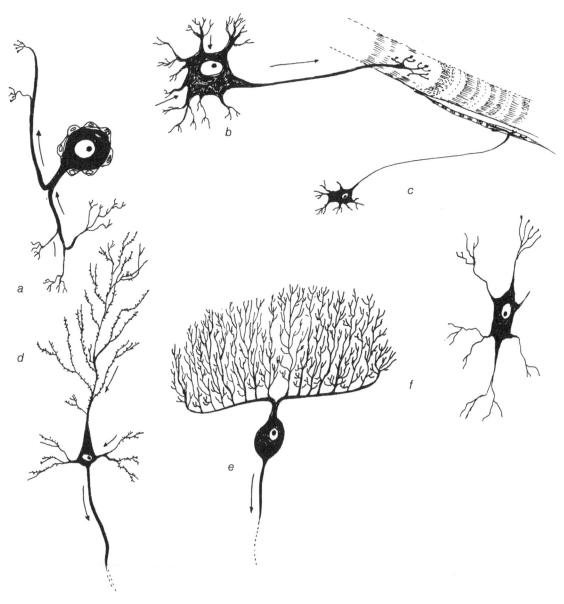

Figure 15.3 Functional types of neurons: (a) sensory neuron; (b) alpha motor neuron; (c) gamma motor neuron; (d) pyramidal cell neuron; (e) Purkinje cell neuron; and (f) interneuron.

dendrites that provide these neurons with a characteristic tree-like appearance.

Interneurons are of a variety of shapes but typically have multiple dendrites and branching axons that permit the connection of multiple neurons with multiple neurons. The structure of interneurons and their connections facilitates both the convergence of multiple input messages onto a single output cell or set of cells and the divergence of a single input message to a number of different motor neurons. Interneurons originate and terminate in either the brain or spinal cord. All neurons in the central nervous system are surrounded by, and outnumbered by, other cells called **glia** or **glial cells**, which provide, among other things, the metabolic and immunological support for the neurons.

Neurons carry messages from their dendrites to the terminal fibres of their axons through a series of electrical pulses that are produced in the axon hillock (see figure 15.2). The electrical pulse produced by the axon hillock is dependent on the spatial and temporal distribution of the pulses impinging on the cell body from its dendritic tree. Signals arriving early and originating from dendrites close to the axon hillock carry more weight than do signals from distant neurons arriving late. If the summed weight of the impulses reaching the neuron exceeds its threshold voltage, the axon hillock triggers a pulse (i.e., the cell fires). This pulse is propagated along the axon to its terminus. The rate at which it is transmitted varies; the rate is greatest in axons of large diameter and in those that are insulated by the fatty substance **myelin**. Loss of myelin in neurodegenerative diseases such as multiple sclerosis results in sensory or motor impairments. Each neuron is therefore more than simply a conductor of electrical signals. Each neuron constantly undertakes complex processing of the input signals it receives from other sources.

STRUCTURE AND FUNCTION OF SYNAPSES

Information passes from one neuron to another via the synapses (which are in many ways the equivalent in the nervous system to the joints in the musculoskeletal system). The term *synapse*, coined by Sir Charles Sherrington, originates from a Greek word meaning "union." At the synapse the axon of one neuron comes in close proximity, but not direct physical contact, to the receptor surfaces of one or more other (postsynaptic) nerve cells.

The electrical activity in the presynaptic neuron is transmitted across the gap (the synaptic cleft or junction) to the postsynaptic neuron via either the direct spread of electrical current or, more frequently, the action of a chemical mediator called a **neurotransmitter**. In the case of chemical transmission, the nerve impulse in the axon of the presynaptic neuron triggers the release of a neurotransmitter from tiny storage sacs (vesicles) in the presynaptic membrane into the synaptic cleft. Specialised receptors on the membrane of the postsynaptic neuron detect the presence of the neurotransmitter triggering either a heightened excitatory or inhibitory response in the postsynaptic neuron (depending on whether the synapse is excitatory and inhibitory). The transmission of information from one neuron to another therefore typically requires a transduction of an electrical signal to a chemical one (at the presynaptic neuron), the diffusion of the chemical transmitter across the synaptic cleft, and then the transduction of the chemical signal back to an electrical one (at the postsynaptic neuron). There are a number of different neurotransmitters, of which **acetylcholine** (ACh), an excitatory neurotransmitter, is the best known. The option of the synaptic connector being either excitatory or inhibitory provides the foundation for more complex functional connections in the nervous system, such as reciprocal inhibition, which, as discussed later, forms the cornerstone of many reflex activities.

SENSORY RECEPTOR SYSTEMS FOR MOVEMENT

The main sensory information to guide the selection and control of movement comes from vision and proprioception. Visual information is derived from the light-sensitive sensory receptors located in the retina of the eye. **Proprioception** (from the Latin *proprius*, meaning "own") is information about the movement and orientation of the body and body parts in space and is provided via **kinesthetic** receptors located in the muscles, tendons, joints, and skin and vestibular receptors for balance located in the inner ear. (The word *kinesthesis* is derived from two Greek words meaning "to move" and "sensation").

Although the sensory receptors for the many facets of vision and proprioception, as well as the receptors for other senses such as hearing, taste, and smell, vary dramatically in their specific structure, all sensory receptors share the common function of transducing physical energy from either beyond the body (e.g., light or sound waves) or in the body (e.g., muscle tension) into coded nerve impulses. These nerve impulses can then be transmitted from one part of the body to another via the nervous system or integrated from one sensory system to another. In this regard, the sensory receptors are very much like transducers in electronics, converting the information they receive into electrical pulses that can be transmitted along the many neural pathways that exist in the human body. Humans, like all other animals, are sensitive to only a limited range of the physical signals in the environment in which we live. Ultraviolet and infrared wavelengths of light, for example, that we know exist in our surrounding environment, are not perceived by us without the assistance of mechanical devices because these signals fall beyond the range of sensitivity of our visual system.

VISUAL SYSTEM

Our rich visual perception of our surrounding environment is achieved through the unique anatomy of the eye and a very complex set of neural processes (figure 15.4). Light reaching the retina (the light-sensitive area at the back of the eyeball) passes through a number of layers of cells to reach the photoreceptors. The photoreceptor cells (the rods and cones) contain chemicals that are sensitive to light, and they send off nerve impulses through their axons to other cells in the retina. This pattern of nerve impulses is specific to the pattern of light falling on the photoreceptors. The rods are most sensitive to light and do not respond to color, and are therefore the primary receptors for night vision. The cones, in contrast, require high levels of illumination to function but enable us to have color vision. The density of both types of photoreceptors is higher around the fovea, giving this area (corresponding to some 2° of the centre of our visual field) the highest level of sensitivity (acuity).

Nerve impulses arising from the photoreceptors are passed through a number of

other layers of interneurons in the retina before being sent to the brain via the optic nerve. The arrangement of nerves in the horizontal, bipolar, amacrine, and ganglion cell layers of the retina allows for early processing of the visual signal, especially in terms of averaging signals over a range of photoreceptors and enhancing contrast between adjacent areas of the visual field.

Visual signals from the retina are carried via the optic nerve along two major pathways, distinct both in structure and function. Some 70% of the connections from the optic nerve link to an area of the midbrain (called the lateral geniculate nucleus) and, from there, to the **visual cortex**, which is located toward the back of the cerebrum. This pathway, contributing to **focal vision**, is specialised for recognising objects, distinguishing detail, and assisting in the direct visual control of fine, precise movements (such as those involved in threading a needle). Most of the remaining nerve fibres from the optic nerve terminate in another section of the midbrain called the superior colliculi. This pathway, contributing to

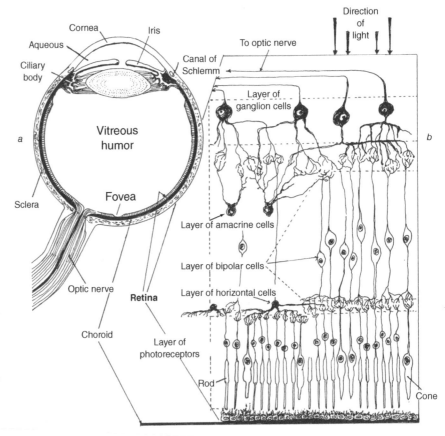

Figure 15.4 Horizontal section of the eyeball (shown on left) with the layered microstructure of the retina (shown on right).

ambient vision, receives information from the whole of the retina, including the peripheral retina, and is concerned with the location of moving objects in the whole visual field. This pathway is especially implicated in providing information about our position in space and our rate of movement through the environment. Damage to the focal vision pathway results in an inability to identify objects but the ability to locate them. The converse is true of damage to the ambient vision pathway.

KINESTHETIC SYSTEM

In addition to information provided through vision, information about the sense of movement is derived from specialised receptors located in the muscles, tendons, joints, and skin. These specialised receptors for movement vary dramatically not only in their location but also in their functioning. Different receptors have evolved to provide different types of information relevant to the perception, monitoring, and regulation of movement.

MUSCLE RECEPTORS

The principal source of sensory information from skeletal muscle is provided by the **muscle spindle**. The muscle spindle is unique as a receptor in that it also contains muscle fibres and hence also has movement capabilities. Muscle spindles are located in all skeletal muscles, although they are particularly abundant in small muscles (such as those in the hands) used to control fine voluntary movements. Muscle spindles provide the central nervous system with information about the absolute amount of stretch plus rate of change of stretch in a particular muscle. This, as discussed in subsequent sections, is invaluable in both the reflex control of movement and the control and monitoring of voluntary movements.

Understanding the control capabilities of the muscle spindle requires an understanding of its unique anatomy (figure 15.5). Under normal circumstances the contraction of any given skeletal muscle is achieved by a burst of neural activity from an alpha motor neuron that causes uniform contraction across the whole length of the large-diameter muscle fibres, called **extrafusal muscle fibres**. Lying in parallel to the extrafusal fibres, and connected to them at their endpoints, are smaller-diameter muscle fibres called **intrafusal muscle fibres** (which form the basis for the muscle spindles). The intrafusal fibres differ from the extrafusal fibres in a number of important ways:

▸ They are smaller and, by themselves, are incapable of directly causing whole muscle contraction.

▸ They are not innervated by alpha motor neurons from the local spinal level but independently by gamma motor neurons whose activity is controlled from descending pathways from the brain.

▸ When stimulated, they contract only at their endpoints and not uniformly across their whole length.

▸ They have sensory receptors located along them and afferent connections back to the spinal cord.

The sensory information from the muscle spindle comes from two sources: primary endings, located in the noncontractile central portion of the spindle and connected to the central nervous system by **type Ia afferent neurons**, and secondary endings, located on the contractile end portions of the spindle and connected to the central nervous system by **type II afferent neurons**. As the primary endings respond to stretch, the Ia afferent neurons send impulses back to the central nervous system under conditions where either the whole muscle is stretched or contraction of the ends of the intrafusal fibres by the gamma motor system is not matched by an equal shortening of the extrafusal fibres, under the control of the alpha motor neuron system.

TENDON RECEPTORS

The sensory receptors located in tendons (the attachments of muscles to bones) are known as **Golgi tendon organs**. These receptors lie close to the surface of the musculotendinous tissue and send their impulses back to the spinal cord by **type Ib afferent fibres** (figure 15.6). The Golgi tendon organs are sensitive to the amount of tension developed in the tendon. Tendon tension increases when a muscle contracts but decreases when a muscle is relaxed; therefore, the Golgi tendon organs act in a manner that counterbalances the action of the muscle spindle. The Golgi tendon organs fire maximally when the muscle spindle is inactive and minimally when the muscle spindle is active.

The Golgi tendon organs appear to serve two major functions with respect to movement control. The first function is a protective one that signals dangerously high tensions in muscle. The Ib afferent neurons are so connected that excessive excitation of the Golgi

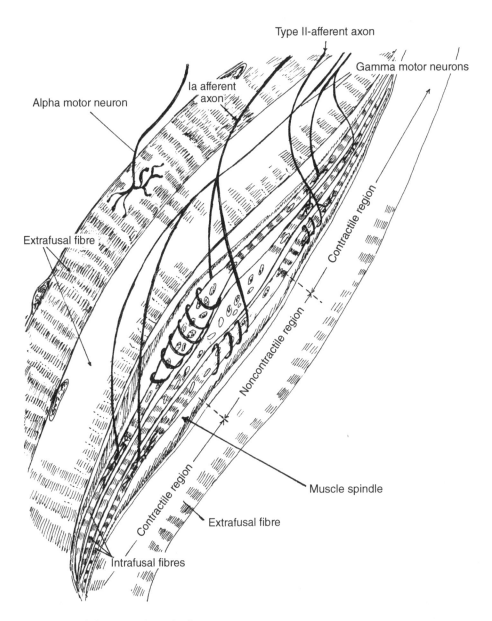

Figure 15.5 Structure of the muscle spindle.

tendon organs acts to inhibit further muscle contraction (by inhibiting the alpha motor neuron innervating the particular muscle), thus preventing damage to the musculotendinous juncture. In this respect the Golgi tendon organ operates somewhat like a fuse in an electrical circuit. The second function of the Golgi tendon organs is to provide sensory feedback to the spinal cord, even at low levels of tension, thereby providing fine-tuned feedback information that can potentially assist in continuous ongoing control throughout a movement. The current line of thought is that the Golgi tendon organs play a particular role in controlling muscle output and tension in response

to fatigue. Another theory is that the Golgi tendon organs, in conjunction with the muscle spindles, help control muscle stiffness (i.e., the force–length relationships in muscle).

Skin Receptors

The skin is an extremely complex and vital organ of the body. It contains a number of types of receptors that can provide useful sensory information for the control of movement. Detection of the deformation of the surface of the skin caused by movement or weight bearing, for example, may be a valuable source of information for monitoring and controlling voluntary

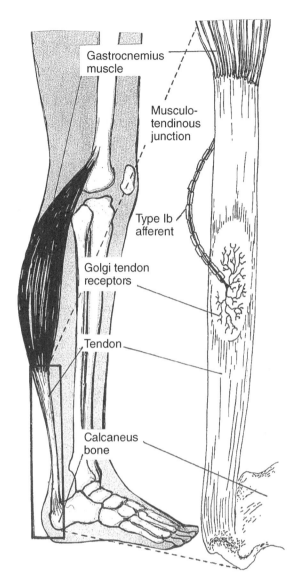

Figure 15.6 Anatomy of the golgi tendon organs.

movements. Receptors such as **Meissner's corpuscles** and **Merkel's discs** (on hairless parts of the skin such as the palms of the hands or the soles of the feet), **Ruffini corpuscles**, and free nerve endings wrapped around hair follicles or other parts of the skin all provide sensory information about light touch or low-frequency vibration. Receptors such as the **Pacinian corpuscles**, located deeper in the skin, respond more to deep compression and high-frequency vibration, especially the onset and offset of such events.

Like other receptors we have examined, the cutaneous receptors are not uniformly distributed throughout the body. They are more densely distributed in regions such as the fingertips that are used for fine, precise movements. Cutaneous sensitivity varies throughout the body in relation to the number of receptors per unit area. Cutaneous receptors clearly have an important role in movement control because it is well known that motor performance deteriorates if these receptors are damaged. Patients with damage to the cutaneous receptors in the soles of their feet, for example, experience difficulty in maintaining balance. Likewise, engineers developing robots to perform movement tasks have discovered that robots lacking touch receptors have great difficulty in performing any tasks requiring fine precision.

JOINT RECEPTORS

There are three types of receptors located in the tissues surrounding and composing joints, and each of these bears similarity to the kinesthetic receptors located elsewhere in the body. There are modified Ruffini corpuscles and modified Pacinian corpuscles (not dissimilar to those found in the skin) located in the joint capsule itself and Golgi organs (not dissimilar to those found in the tendons) located in the ligaments that bind the joint together. Although there is some debate over their function, it appears that the main collective role of the joint receptors is to signal extreme ranges of motion at the joint. The joint receptors are therefore able to play a role in protecting the joint from injury by signalling to the central nervous system when the full range of motion of a joint is being reached.

VESTIBULAR SYSTEM

Whereas the various kinesthetic receptors provide valuable information about the state of individual muscles, joints, and movement segments, the performance of many skilled movements also requires information about the orientation of the whole body in space. This is particularly true of movements in activities such as gymnastics, trampolining, or diving. Some of this information about whole-body orientation can be provided by the visual system, but much of it is provided by a uniquely designed receptor system (the vestibular apparatus) located adjacent to the inner ear (figure 15.7).

The **vestibular apparatus** consists of two types of receptors: the **semicircular canals** (the superior, horizontal, and posterior canals), which respond to angular acceleration in three planes, and the **otolith organs** (the **utricle** and the **saccule**), which respond to linear acceleration. Each of the semi-circular

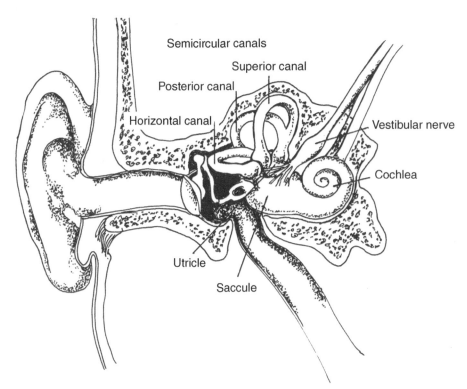

Figure 15.7 Anatomy of the vestibular apparatus.

canals is located at right angles to the other two, allowing for separate information to be sent to the brain on the horizontal, lateral, and vertical angular acceleration of the head. In contrast, the utricle provides sensory information on the linear horizontal acceleration of the head and the saccule information on the linear vertical acceleration of the head. The vestibular apparatus is centrally involved in balance such that any dysfunction of the vestibular apparatus, such as that which occurs with some ear infections, can lead to loss of balance control. The vestibular apparatus is also in close reflex connection with the visual system. Discrepancies between the information provided by the visual and vestibular systems, such as that which occurs in many rides in amusement parks, can give rise to phenomena such as motion sickness.

INTERSENSORY INTEGRATION AND SENSORY DOMINANCE

In many cases common environmental events are experienced by a number of sensory systems in the body, and the challenge for the central nervous system is to integrate these different sources of information. In the maintenance of normal, upright balance, for example, sensory information needs to be integrated from the visual receptors, the many kinesthetic receptors, and the vestibular system in order to ascertain whether balance is being maintained correctly or being lost. That integration is possible at all is a consequence of the sensory receptors transducing their very different sources of physical stimulation into the common language of nerve impulses and the presence of a vast array of neural interconnections and pathways in the brain and central nervous system that allow signals from diverse locations to converge. Although the information coming from the different sensory systems is usually in agreement, in some circumstances the information supplied to the central nervous system may be in conflict (e.g., the visual system may indicate that balance is being lost whereas the kinesthetic system indicates that balance is being retained). The brain and other sections of the central nervous system therefore require some systematic means of resolving this conflict. In humans, any intersensory conflict is almost always resolved in favour of vision, which is referred to as the dominant sensory modality. This can, on occasions, lead to misperception and, in turn, misguided action if the sensations provided by the visual system are inaccurate or misleading (see "Visual Dominance in Balance Control").

VISUAL DOMINANCE IN BALANCE CONTROL

An excellent demonstration of the dominance of vision over information provided from other sensory systems has been provided by the Edinburgh psychologist David Lee and his colleagues. Lee had participants in his study stand upright in (what appeared to be) an enclosed room and gave them the apparently simple task of maintaining their balance and keeping their head and whole body as still as possible. Under control conditions, the information coming from the visual system as well as from the proprioceptive system would all indicate reliably to the participant that they were stationary. What the participants did not know was that the room surrounding them, but not the surface on which they were standing, was able to be moved so that the front wall could be subtly but systematically moved either toward or away from them. In such cases, the visual system senses a loss of balance (overbalancing forward when the room is moved toward the participant and the converse when the room is moved away) even though the independent information from the vestibular apparatus and the various kinesthetic receptors would indicate no loss of balance. In such instances of intersensory conflict, the participant's response is consistently to make compensatory postural adjustments. That such responses are made in the direction opposite the perceived overbalancing clearly indicates that the visual information is the information the central nervous system believes (figure 15.8). Even small room movements of as little as 6 mm can induce marked postural sway in adults and complete loss of balance in young children, demonstrating the dominance of vision (even under situations in which the information provided by vision can be shown to be incorrect).

More recent studies conducted at the University of Virginia in the United States have demonstrated that when vision alone is used to make a conscious judgement of the slope of hills, steps, or ramps, the slope is constantly overestimated. That is, people perceive the slopes to be much steeper than they really are. However, if a matching movement response is needed (e.g., by moving the unseen hand to an angle that matches the same slope), this bias largely disappears.

Sources

Bhalla, A.M., & Proffitt, D.R. (1999). Visual-motor recalibration in geographical slant perception. *Journal of Experimental Psychology: Human Perception and Performance, 25,* 1076-1096.

Lee, D.N., & Aronson, E. (1974). Visual proprioceptive control of standing in human infants. *Perception and Psychophysics, 15,* 529-532.

Lee, D.N., & Lishman, J.R. (1975). Visual proprioceptive control of stance. *Journal of Human Movement Studies, 1,* 87-95.

Lee, D.N., & Thomson, J.A. (1982). Vision in action: The control of locomotion. In D.J. Ingle, M.A. Goodale, & R.J.W. Mansfield (Eds.), *Analysis of visual behavior* (pp. 411-433). Cambridge, MA: MIT Press.

Proffitt, D.R., Bhalla, M., Grossweiler, R., & Midgett, J. (1995). Perceiving geographical slant. *Psychonomic Bulletin and Review, 2,* 409-428.

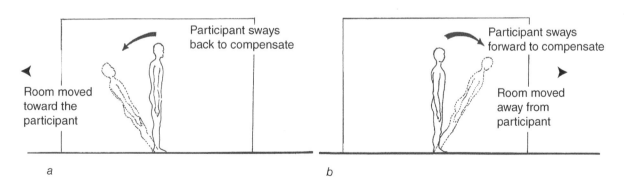

Figure 15.8 Postural adjustments induced by visual movement of room walls *(a)* toward and *(b)* away from the standing subject.

Adapted, by permission, from D.N. Lee and J.A. Thomson, 1982, Vision in action: The control of locomotion. In *Analysis of visual behavior,* edited by D.J. Ingle, M.A. Goodale, and R.J.W. Mansfield (Cambridge, MA: MIT Press), 411-433.

EFFECTOR SYSTEMS FOR MOVEMENT

The motor unit, as the functional unit of interaction between the nervous system and the muscular system, has been considered elsewhere in this text in the context of being the source of neural input to the muscular system (see especially figure 3.12). Consequently, in this section we only briefly reiterate some of the key features of the motor unit, this time in the context of the motor unit's role as the ultimate endpoint for the output of the neural system.

A motor unit consists of a single alpha motor neuron plus all the skeletal muscle fibres (extrafusal fibres) it innervates. This may range from as few as 1 or 2 fibres for the small muscles of the eye that control precise movements to up to a 1,000 in some of the larger postural muscles of the lower leg. As a general rule, the fewer muscle fibres there are in a motor unit, the more precise the control that is possible. Observable contraction of a muscle or motion of a joint crossed by a muscle requires the activation of a number of different motor units. There are three types of motor units, corresponding to the three types of muscle fibre it is possible to innervate (see table 11.2).

With practice and appropriate feedback it is possible to learn to selectively recruit single motor units in a given muscle, but it is not possible to voluntarily activate only some of the muscle fibres in a single motor unit. In natural movements, motor units are typically recruited in order of size; the motor units containing the smaller, less-forceful muscle fibres are recruited first. This order of activation is known as the size principle.

MOTOR CONTROL FUNCTIONS OF THE SPINAL CORD

So far in this chapter we have examined the structures and processes by which the human body is able to receive sensory information of relevance to movement and in turn transmit information through the motor units in order to produce observable movement. We have yet to examine how the central nervous system links (appropriately) this input and output information.

The central nervous system is somewhat hierarchical in structure in that the higher levels of the system, especially the brain, are responsible for higher-order creative and executive mental (cognitive) and motor control functions, whereas the lower levels of the system, especially the spinal cord, are responsible for more routine, repetitive control functions. Because the brain and spinal cord work together in the performance of most skilled movements, understanding the neural control of movement requires understanding the motor control functions of each level of the central nervous system and the way in which these levels interact.

In this section we examine the basic structure of the spinal cord and its motor control capabilities. The spinal cord alone is responsible for the control of reflex movements (rapid movements occurring below the level of consciousness) and for the maintenance of voluntary movements, which are initiated by higher centres in the brain. Much of the knowledge about the motor control capabilities of the spinal cord comes from studies of reflexes in humans and other animals and from studies of the movement capabilities of spinalised animals (animals in whom the nerve pathways from the spinal cord to the brain have been severed).

STRUCTURE OF THE SPINAL CORD

As with virtually all structures in the human body, the anatomical design of the spinal cord can be readily appreciated if its basic functions are first understood. The spinal cord serves two basic functions. Its first role is as a dual-transmission pathway that carries both input information from the sensory receptors to the brain and output information, in the form of motor commands, from the brain to the muscles (see figure 15.1). Its second role is to support reflexes at the local spinal level to provide rapid, essentially automatic responding to noxious (or potentially dangerous) stimuli and to ensure the successful execution of movements already underway.

The spinal cord is about as thick as an adult's little finger and runs from the base of the spine to the point where it joins the brain at the brain stem, located at the base of the skull (figure 15.1). Because of its importance to normal communications functions in the body, the cord, like a well-laid telecommunications cable, is protected throughout its length by the bony structures of the spine. The spinal cord runs throughout the length of the spine in the protection of a canal formed in the vertebral (spinal) column (figure 15.9).

A total of 31 pairs of **spinal nerves** are attached to the spinal cord; each nerve is attached to its side of the cord by two roots. The anterior (or **ventral**) root of each spinal nerve carries the efferent or motor information away from the spinal cord to the muscles, whereas the posterior (or **dorsal**) root carries the afferent or sensory information from the periphery back to the spinal cord. The anterior root consists almost exclusively of the axons of alpha motor neurons, the cell bodies of which are located in the ventral horn of the spinal cord. The posterior root contains both the axons of the sensory neurons and their cell bodies, the latter clustered together to form the dorsal root ganglion. The spinal cord itself, in cross-section, reveals an outer covering of white matter surrounding a central mass of grey matter, roughly approximating the shape of the letter H. The white matter of the spinal cord consists primarily of nerve fibres, and the grey matter consists primarily of the cell bodies of neurons. Nerve fibres in the white matter are frequently bundled together to form tracts that carry impulses up and down the spinal cord.

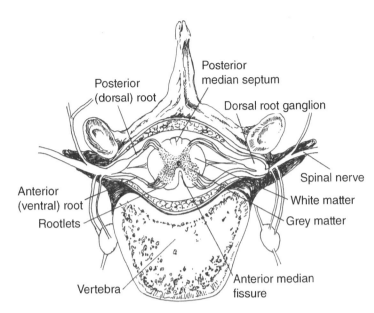

Figure 15.9 Cross-section of the spinal cord and its location in the vertebral column.

SPINAL REFLEXES

A **reflex** (from the Latin word *reflexus* meaning "a bending back") is the simplest functional unit of integrated nervous-system behaviour. The various reflexes scattered throughout the spinal cord provide the foundation on which (directly) all involuntary and (indirectly) all voluntary movement is based.

A minimum of four basic nerve units are needed to form a reflex arc:

▶ a sensory receptor (to detect a pertinent stimulus),

▶ an afferent (or sensory) neuron (to transmit the sensory information to the central nervous system),

▶ an efferent (or motor) neuron (to transmit the output information from the central nervous system), and

▶ an effector, typically a motor unit (to produce a movement response).

The simplest of all reflex systems, which has only two neurons and hence only one synapse, is called, for obvious reasons, a monosynaptic reflex. Most reflex arcs are polysynaptic, containing multiple synapses. The sensory and motor neurons do not synapse together directly; rather, the nerve impulses are passed from one to the other through a series of interneurons. The time it takes for a reflex system to work (its latency, or loop time) is measured from the time of stimulation to the time a response can be recorded in the muscle fibres. Not surprisingly, reflex loop time is longer the more interneurons (and synapses) there are in the reflex arc.

STRETCH REFLEX

The best example of a monosynaptic reflex in the spinal cord is the simple stretch reflex, also known as the **mytotatic reflex** (muscle-stretching reflex). The stimulus for the stretch reflex is excessive stretch on muscle as detected by the muscle spindles (see figure 15.5). This excessive stretch may arise from a number of sources, such as the unexpected addition of a weight, postural sway (figure 15.10*b*), or, as is often the case in a clinical setting, unexpected stretching of the quadriceps muscle when a doctor, using a rubber mallet, delivers a sharp tap to the patellar tendon. In all cases, the excessive stretch detected by the muscle receptors results in a nerve impulse being sent to the dorsal root of the spinal cord via the afferent neuron. This neuron then synapses, within the spinal cord, directly to an alpha motor neuron that transmits its nerve impulse back to the extrafusal fibres of the stretched muscle (figure 15.10a). This typically is sufficient to cause the stretched muscle to contract, thus alleviating the stretch stimulus.

In the case of the patella tap test for nerve function used by general practitioners and neurologists, it is the alpha motor neuron activation of the quadriceps muscle that causes the characteristic and forceful knee extension (kicking) response. If the initial response is insufficient to alleviate the

stretch, two things will occur: The same reflex arc will be activated a second time (because the stimulus still exists), and commands will be sent (via interneuron connections) to other segments of the spinal cord, including higher centres. Typical of most hierarchical organisations, the latter will produce a more powerful but slower response. Whereas the simple monosynaptic stretch reflex may have a loop time as short as 30 ms, reflex responses to alleviate stretch that involve higher segments of the spinal cord, and perhaps even regions of the brain (often referred to as **long-loop reflexes**), may have latencies 2 to 3 times this long.

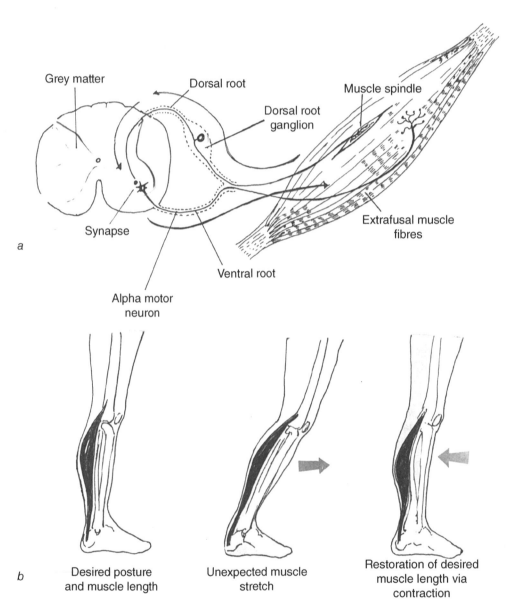

Figure 15.10 The monosynaptic stretch reflex showing in (a) the neural circuitry and (b) a typical situation where the reflex would be activated.

FLEXION REFLEX

A common purpose of many spinal reflexes is to provide rapid protection for the body against potentially injurious stimuli while preserving, at a premium, whole-body balance. The **flexion reflex** is a polysynaptic reflex that causes withdrawal of limbs away from potentially injurious stimuli. In this reflex, stimulation of either a pain or heat receptor in the skin excites, through interneural connections, the motor neurons innervating flexor muscles crossing joints adjacent to the stimulus; this results in the limb being flexed away from the noxious stimuli. Equally important, this rapid limb withdrawal through flexion is facilitated by the extensor musculature being turned off at the same time (figure 15.11). This is achieved through interneuronal inhibitory connections between the afferent neuron and the extensor motor neuron. This action provides a specific example of a general spinal reflex phenomena called **reciprocal inhibition** (the neural control phenomena that ensures that agonist and antagonist muscles do not typically cocontract in opposition to each other).

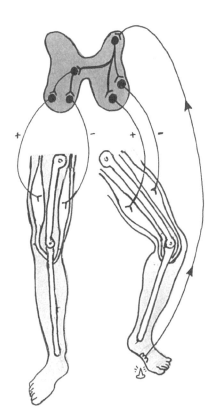

Figure 15.11 The flexion withdrawal and crossed extension reflexes. A painful stretch causes excitation (+) of the ipsilateral flexors and contralateral extensors and inhibition (–) of the ipsilateral extensors and contralateral flexors.

CROSSED EXTENSOR REFLEX

The crossed extensor reflex often functions in conjunction with the flexion reflex to maintain postural stability and, if necessary, help a person push away from a painful stimulus (figure 15.11). Through interneuronal connections that exploit the excitatory–inhibitory potential of different synaptic connections, the crossed extensor reflex ensures the limb closest to a painful stimulus is flexed away from it and that the limb on the opposite side of the body extends (through extensor excitation and flexor inhibition). This reflex provides a good illustration of how nerve impulses pass not only to and from the spinal cord at a given segment level and up and down the spinal cord but also across the spinal cord from one side of the body to the other.

EXTENSOR THRUST REFLEX

The extensor thrust reflex is one of the more complex spinal reflexes; it aids in supporting the body's weight against gravity. Cutaneous receptors in the feet sensitive to pressure, through a vast array of interneural connections, cause reflex contraction of the extensor muscles of the leg. This reflex provides the foundation for standing balance without dependence on brain mechanisms.

SPINAL REFLEXES FOR GAIT CONTROL

All forms of human gait (such as crawling, walking, and running) are characterised by continuous patterns of limb flexion and extension; each limb is one half-cycle different from the other. The flexion and crossed extensor reflexes, with their reciprocal innervation of flexor–extension pairs across matching sets of limbs, and the extensor thrust reflex, with its pathways for balance preservation, provide strong building blocks for basic gait control and maintenance.

Studies conducted on animals, in which spinal connections to the higher centres of the brain have been severed, have demonstrated that the spinal cord has an inherent rhythmicity that plays a major role in gait control. In this respect, the spinal cord is frequently described as a **central pattern generator**. Although the spinal cord, per se, seems incapable of initiating gait (this appears to require either motor commands from the brain or very strong sensory information from the cutaneous receptors of the feet), the spinal cord seems well capable, through its various reflex pathways, of preserving gait once it is initiated, even to the point of controlling a transition from one gait form (e.g., walking) to another (e.g., running). Clearly, spinal reflexes play a major role in involuntary protective actions and in the control of fundamental motor activities such as gait.

ROLE OF REFLEXES IN VOLUNTARY MOVEMENT CONTROL

It is obvious from the preceding section that reflexes clearly play a major role in involuntary movement control (i.e., the control of movements that are below the level of our conscious awareness). Most neurophysiologists and motor control theorists also believe that the spinal cord, through its reflex arcs, plays a major role in ensuring that voluntary movements planned and initiated in the brain are executed as planned. Voluntary movements must, to some degree, use reflexes as their building blocks because the final pathway for all motor commands, regardless of their origin, is through the alpha motor neurons at the spinal cord level to the muscle fibres. Voluntary movements simply involve the spinal reflex pathways being modified or used in ways specified by commands arising from higher centres of the central nervous system.

A good example of the way the reflex structure of the spinal cord can be integrated with the higher-level control provided from the brain is provided through examination, yet again, of the muscle spindle (figure 15.5) and the stretch reflex (figure 15.10). As discussed in the preceding section, the simple stretch reflex provides a means of protecting the muscle against damage from excessive lengthening. In a more functional manner, however, the collaborative activity of the muscle spindle, its sensory neurons, and its alpha and gamma motor neurons can be organised to ensure that voluntary movements are executed as planned.

The progress of any particular movement can be monitored and controlled through a process of **alpha–gamma coactivation**. In this process, actual muscle length is determined by contraction of extrafusal muscle fibres controlled by the alpha motor neurons that originate at spinal level. Intended muscle length is set by contraction of the ends of the intrafusal muscle fibres under the control of the gamma motor neurons that originate from the level of the brain. As the name implies, for any given movement, such as holding a weight in a constant position or maintaining upright stance (figure 15.10), alpha–gamma coactivation results in the simultaneous activation of both the extrafusal

fibres (by the alpha system) and the intrafusal fibres (by the gamma system).

If the movement goes as planned, the change of muscle length of both the extrafusal and intrafusal fibres will be identical and no additional sensory impulses will be sent back from the muscle spindle to the spinal cord. If the movement does not proceed as planned (e.g., there is insufficient extrafusal fibre innervation to shorten the muscle), the sensory receptors on the intrafusal fibres will be placed on stretch and this will evoke, through the usual stretch reflex, additional alpha motor neuron activation to cause the muscle to contract. The alpha–gamma coactivation process therefore provides a good example of how movement plans from the higher centres of the central nervous system can be enacted, using spinal mechanisms to ensure that these movements are executed as planned.

MOTOR CONTROL FUNCTIONS OF THE BRAIN

The human brain possesses a level of complexity and organisation that is beyond comprehension and perhaps unmatched by anything else in the universe. The brain serves many higher-order functions, only some of which are directly related to motor control. In this section we examine the location and function of the main areas of the brain identified as having a significant role in motor control. These areas are the **motor cortex** (located immediately forward of the **central sulcus** in the frontal lobe of the **cerebrum**), the **cerebellum** (located off the brain stem and below the occipital lobe of the cerebrum), the **basal ganglia** (located in the inner layers of cerebrum), and the **brain stem** (located forward of the cerebellum and continuous with the spinal cord and the cerebrum; see figure 15.12). These areas are in constant communication through a rich, interconnecting network of nerve pathways, some of the major ones of which are shown schematically in figure 15.13.

MOTOR CORTEX

The cerebral cortex is the outermost layer of the cerebrum of the brain, is some 2 to 5 mm deep, has an (unfolded) surface area of some 2 to 3 m^2, and contains more than half of the total neurons in the human nervous system. The cerebral cortex is divided into two halves that appear essentially

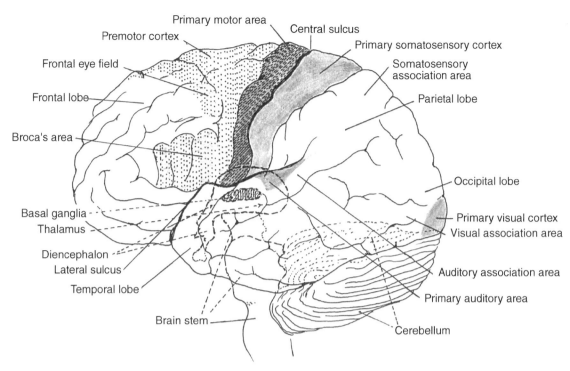

Figure 15.12 Location of the principal motor areas of the brain. Structures located within dashed lines lie underneath or within the external surface of the brain.

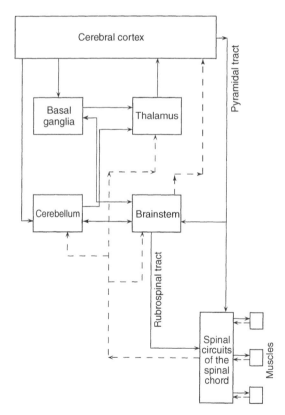

Figure 15.13 Schematic representation of the major motor pathways. Direction of information flow is shown by arrows; afferent input is denoted by dashed lines.

Adapted from *Psychology of human movement*, edited by M.M. Smyth and A.M. Wing, "Physiology of motor control," K. Greer, pg. 33, Copyright 1984, with permissions of Elsevier.

symmetrical, although they are somewhat different in function. These are the left and right cerebral hemispheres; they join at the midline through a thick sheet of interconnecting nerve fibres called the corpus callosum.

Each cerebral hemisphere contains a motor cortex (lying immediately forward of the central sulcus; figure 15.12), a premotor cortex (lying just forward of the motor cortex), and a supplementary motor area (lying on the medial wall of the cerebral hemispheres and forward of the motor cortex). Each of these structures, located in the frontal lobe of the cerebrum, is intimately involved in the production and control of skilled movement. Forward of the premotor cortex are two other areas that also have important, but specialised, motor control functions. The frontal eye fields are involved in the control of voluntary eye movements, and Broca's area (located in the left hemisphere only) has a critical role in the planning of the movements generating speech.

The motor cortex and its associated areas are systematically organised such that each part of the motor cortex controls specific muscles or muscle groups in the body, to the point that all muscles are topographically represented in the brain (figure 15.14). In pioneering studies of the human brain by two Canadian neurosurgeons, Penfield and Rasmussen, muscle maps of the motor cortex were developed by applying weak electrical pulses to

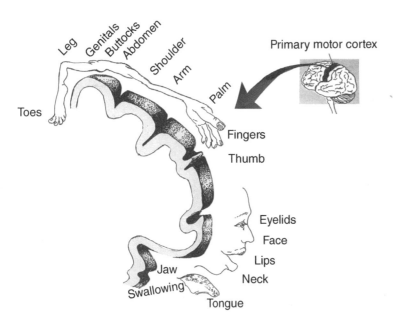

Figure 15.14 Schematic view of the primary motor cortex and the mapping between electrical stimulation of the motor cortex and regions of the body in which movement subsequently occurs. The insert shows the location of the primary motor cortex in the brain.

distinct areas of the motor cortex and observing the muscle contractions that resulted. These electrical mapping studies revealed a number of important things.

- All muscles are not represented proportionally in the motor cortex on the basis of their size; rather, representation is proportional to the precision requirements of different parts of the body. The muscles of the hand and the mouth occupy nearly two thirds of the total area of the motor cortex.

- The representation of distal musculature, such as that crossing the joints in the hands and feet, is entirely contralateral. Consequently, the left motor cortex controls the right hand and vice versa.

- Muscles located more proximally are represented in the motor cortex of both the cerebral hemisphere on the same (ipsilateral) and opposite (contralateral) side of the body.

- As stimulation is moved forward into the premotor cortex, gross movements of muscle groups rather than fine movements of a discrete muscle groups are observed.

Perhaps most importantly, these electrical-mapping studies suggest that the motor cortex acts as something of a relay station and is the final neural station for the organisation and release of the coordinated motor commands to be sent out to the specific muscles or, more correctly, muscle fibres.

The motor cortex has two principal means of relaying its commands to the muscles. The most direct route is via the **pyramidal tract** (or corticospinal tract), which allows neurons from the motor cortex to synapse directly in some cases (and through a minimum of interneurons in most cases) with the alpha motor neurons at the spinal level. This tract carries impulses that are primarily excitatory in nature. Damage to this tract can result in either partial (**hemiparesis**) or complete (**hemiplegia**) paralysis of contralateral movements. Alternative routes, known collectively as the **extrapyramidal tract**, allow nerve impulses from the motor cortex to reach the spinal level through a range of pathways via the cerebellum, basal ganglia, **thalamus**, and brain stem (figure 15.13). Outputs from these pathways are primarily inhibitory in nature. Damage to the extrapyramidal tract can result in **spasticity**.

Damage to the cerebral cortex can result in **apraxia**. Specific damage to the motor cortex results in a loss of fine-movement control, especially in the fingers and toes. Damage to the premotor cortex results in a disruption to movement planning and selection, especially for gross movements involving a number of muscle groups. Damage to the supplementary motor area disrupts the planning of sequential movements and performance of many tasks requiring bimanual coordination. Collectively, the premotor cortex and the supplementary motor area appear to have a particular role to play in movement planning and generation.

CEREBELLUM

As can be seen in figure 15.12, the cerebellum attaches to the brain stem and is located behind and below the cerebral hemispheres. Like the cerebrum, the cerebellum has an outer cortex that is divided into two distinct but interconnected hemispheres. Beneath the cortex are four deep cerebellar nuclei. The cerebellum receives input information from a vast array of areas in the cerebral cortex (including the motor areas), from various areas in the brain stem, from the vestibular apparatus, and, via the spinal cord, from the kinesthetic receptors located on the same (ipsilateral) side of the body.

The cerebellum itself has a very regular anatomical structure based primarily around two types of afferent fibres (called **climbing fibres** and **mossy fibres**) and one output fibre (the Purkinje cell; see figure 15.3*e*). This structure enables the cerebellum to perform a number of very complex signal-processing operations that are fundamental to many aspects of motor coordination. Indeed, the cerebellum has frequently been referred to as "the seat of motor coordination." The cerebellum has two major outputs: one to the thalamus and one to the brain stem (figure 15.13).

A number of major motor control functions have been attributed to the cerebellum, all broadly related to the translation of abstract movement plans into specific spatial and temporal patterns that can be relayed to the muscles via the motor cortex. Principal cerebellar functions appear to be the regulation of muscle tone, the coordinated smoothing of movement, timing, and learning. Patients with cerebellar damage demonstrate one or more of the symptoms of low muscle tone, incoordination or **ataxia** (especially in standing, walking, speaking, or performing precise aiming movements), poor temporal control of muscle recruitment, and difficulty in learning new movements or adapting old ones. Fast, ballistic types of movement appear to be particularly affected.

BASAL GANGLIA

The basal ganglia are a group of five pairs of interconnected nuclei (the globus pallidus, the caudate nucleus, the putamen, the subthalamic nucleus, and the substantia nigra) located deep within each of the cerebral hemispheres and close to the thalamus. The basal ganglia receive input from two major sources (the motor areas of the cerebral cortex and the brain stem) and, similarly, send their output to two locations (the thalamus and the brain stem). Therefore, like the cerebellum, the basal ganglia, although not synapsing directly with spinal neurons, are able to influence alpha motor neuron activity through both the pyramidal tract and the rubrospinal tract (figure 15.13). The basal ganglia work together as a loosely connected unit, although each of the component nuclei is quite different and generally connected in an inhibitory fashion with each other.

Insights into the function of the basal ganglia in motor control have come primarily from studies of patients suffering from two identifiable diseases of the basal ganglia. **Parkinson's disease** is a degenerative disease resulting from deficiency in the natural neurotransmitter substance **dopamine** that assists in carrying nerve impulses from one nucleus in the basal ganglia to another. Parkinsonian patients typically demonstrate a range of motor symptoms including shuffling, uncertain gait, limb tremor, difficulty in initiating movement, and high degrees of muscle stiffness. **Huntington's disease** is a hereditary degenerative disease resulting from damage to the dendrites that produce one of the neurotransmitters used to communicate between selected nuclei in the basal ganglia. Patients with this disease suffer from uncontrollable, involuntary rapid flicking movements of the limbs or facial muscles. Damage to any structure in the basal ganglia may cause slowness of voluntary movement and involuntary postures and movements.

Despite knowledge of the obvious movement problems caused by dysfunction of the basal ganglia, the precise function of the basal ganglia in movement control remains elusive. Some favoured suggestions include the control of slow movements, the retrieval and initiation of movement plans, and the scaling of movement amplitudes, as required in daily tasks such as handwriting. At a general level, the basal ganglia appear to permit selected movements to proceed while inhibiting unwanted movements.

BRAIN STEM

The brain stem contains three major areas that have significant involvement in motor control: the **pons**, the **medulla**, and the **reticular formation**. The brain stem's principal function, as revealed by figure 15.13, is to act as a relay centre, especially for the transmission of information to and from the cerebral cortex.

The pons and medulla, as the main structures in the brain stem, receive input from the cerebral cortex, cerebellum, and basal ganglia as well as all the sensory systems. These structures then integrate this information for output to the spinal cord for use in the control of many involuntary movements, such as those related to posture and cardiorespiratory activity. The brain stem functions in the control of muscle tone and posture and is fundamental to the operation of a number of supraspinal reflexes. Prominent among these are the **righting reflexes**, which maintain the orientation of the body with respect to gravity, and the **tonic reflexes**, such as the tonic neck reflex, which are concerned with the maintenance of the position of one body part (such as the neck) in relation to other body parts (such as the arms and legs). Damage to the pons or medulla disrupts the control of involuntary movements and key orienting reflexes and endangers the control of vital physiological systems.

The reticular formation is a network of neurons that extends throughout the brain stem and, through its ascending connections to the cerebral cortex, has a major role in regulating the activity of the cortex. The ascending reticular formation controls the activation of the cortex in this way and, therefore, the state of arousal experienced by the person. (The issue of arousal is examined in more detail in chapter 19.) The descending fibres of the reticular formation input directly to the spinal reflexes and may modify reflex activity at this level as is necessary to ensure that basic postural needs are met.

INTEGRATIVE BRAIN MECHANISMS FOR MOVEMENT

Given the complexity of both the human nervous system and the movement it produces, it is perhaps not surprising how little is yet known about the neural mechanisms underlying movement control in the brain and how much remains to be discovered. At this point only some speculations can be advanced on the likely flow of neural information through various brain structures and the functional consequences of such information flow. The prefrontal cortex appears to be central to overall movement planning, the basal ganglia and cerebellum to the programming of specific motor commands, and the motor cortex to the release of organised commands to the muscles via the spinal pathways (figure 15.15). Readers should be aware, however, that neuroscientists interested in motor control are still

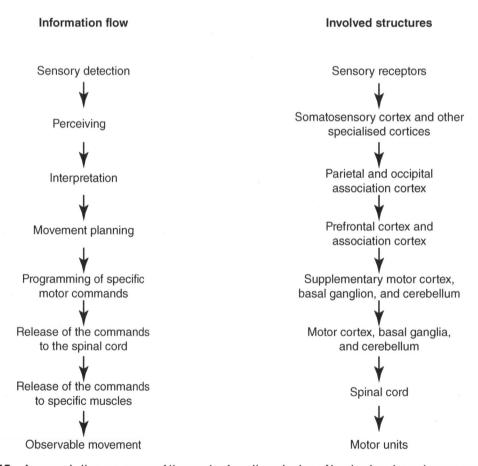

Figure 15.15 A speculation on some of the major functional roles of brain structures in movement control.

many years away from a complete, integrative model of the brain mechanisms for motor control. One approach that may hasten understanding may be to look alternatively or, better still, simultaneously at motor control from a conceptual (psychological) perspective in addition to a neurophysiological one. The next chapter examines basic cognitive science perspectives on motor control.

SUMMARY

The neural control processes underlying skilled human movement are extremely complex. The foundations of the neural control system are built on nerve cells (neurons), sensory receptors, motor units, and the intricate synaptic interconnecting of each of these components. Sensory information for movement comes primarily from vision and proprioception. Proprioceptive information is provided by a host of specialised kinaesthetic receptors located in muscles, tendons, joints, and skin as well as by the vestibular system. Motor units, comprising single motor neurons and the muscle fibres they innervate, provide the work-

ing interface between the nervous and muscular systems. Control of reflex movements and control of the basic flexion–extension pattern generation necessary for locomotion is possible through pathways and connections present at the level of the spinal cord. More sophisticated, voluntary movement requires the involvement of the brain and its specialised pathways and regions responsible for movement planning and initiation. Deficiencies in movement control arising from localised neural damage help provide some insight into the neural basis for movement and postural control.

FURTHER READING

Kandel, E.R., Schwartz, J.H., & Jessell, T.M. (2000). *Principles of neural science* (4th ed.). New York: McGraw-Hill.

Latash, M.L. (2008). *Neurophysiological basis of movement* (2nd ed.). Champaign, IL: Human Kinetics.

Rothwell, J.C. (1994). *Control of voluntary human movement* (2nd ed.). London: Chapman & Hall.

Shumway-Cook, A., & Woollacott, M.H. (2001). *Motor control: Theory and practical applications* (2nd ed.). Philadelphia: Lippincott Williams & Wilkins.

CHAPTER 16

BASIC CONCEPTS OF MOTOR CONTROL: COGNITIVE SCIENCE PERSPECTIVES

The major learning concepts in this chapter relate to

▶ the importance of models of motor control;

▶ key properties that must be explained by models of motor control;

▶ information-processing models of motor control;

▶ processes and limitations related to perceiving, deciding, and acting; and

▶ the possibilities for alternative models of motor control.

In the previous chapter we examined in some detail the structure and function of the main components of the neuromuscular system. This neurophysiological approach to understanding motor control provides us with valuable information about the receptors and effectors for movement and about the major pathways that connect the two. However, the sheer size and complexity of the nervous system (with its some 10^{14} neurons, each with up to 10^4 synaptic interconnections) means that it is impossible to easily or fully appreciate how movement is controlled by studying only nerve pathways. What is needed to complement the knowledge from neuroscience are conceptual theories and models that describe and explain the overall control logic used by the nervous system to collectively acquire, perform, and retain motor skills. Such theories and models have typically originated from the work of experimental psychologists who have focused on the broad, overall functioning of the motor system rather than the more specific, detailed anatomy and physiology of its discrete components. In recent years the methods of experimental psychology have been merged with aspects of computer science as scientists (now referred to as cognitive scientists) attempt to understand the computational capabilities and methods of the human neuromotor system.

In this chapter we examine the role of conceptual models from cognitive science in understanding motor control, outline the key properties of skilled human movement that must be explained by such conceptual models, and describe in some detail one popular model (an information-processing model) of motor control, examining as we do some of its basic assumptions and practical implications.

USING MODELS TO STUDY MOTOR CONTROL

In its structure and function, the human neuromuscular system is clearly incredibly complex. Simplified models of the system provide a valuable means of helping scientists begin to understand how movement skills might be controlled and how this control might change with practice and age. Effective models of movement control help explain the many unique and essential properties of movement control.

ROLE OF MODELS IN SCIENTIFIC STUDY

In all branches of science, models serve the important purpose of aiding in the understanding and advancement of theory. Models enable a theory to be visualised and understood, frequently by drawing comparison with the operation of simpler, everyday systems with which we are familiar. Systems of infinite complexity, such as the physical system of electricity or the biological system of the heart and lungs, can be more easily understood through the use of simplifying models such as those of water flow or the action of a pump. The value of models therefore is their potential to simplify a complex system to a level where understanding can be achieved and experiments for further understanding can be formulated. In a system as complex as the human motor system, there would appear to be great value in developing conceptual models as a means of aiding and advancing our understanding of how the system works.

It needs to be noted, however, that a model is not a theory and therefore should not be taken too literally. A model is also worthwhile only if it accurately captures the key characteristics of the system we are trying to understand. Just as good models can aid understanding, poor models can hamper understanding. As more becomes known about a particular system, the shortcomings of old models are frequently realised and new models are proposed in their place. As shown in this chapter,

one particular model (the information-processing model) has dominated most thinking to date about how movement is controlled, but some of the limitations of this model have become more apparent and alternatives have been suggested.

KEY PROPERTIES TO BE EXPLAINED BY MODELS OF MOTOR CONTROL

A starting point for the development of a model of any system is consideration of the key features of the system that the model must be able to encapsulate. These key features, in a sense, form the constraints for the model. In human motor control there is an impressive array of unique properties that any worthwhile model or theory must be able to be adequately explain. Some of the principal motor control properties that require explanation are the following.

▶ **Degrees of freedom** refer to the capability of the brain and nervous system to simultaneously and continuously consider, and somehow control, the enormous number of independent variables (i.e., motor units, joints, limb angles) that contribute to skilled movement. In theory there is an overwhelmingly enormous computational challenge to the capacity of the brain and nervous system—one that far exceeds the capability of sophisticated computers used to control even simple movements by robots. The nervous system must use clever solutions to overcome this problem and limit control demands to manageable levels.

▶ **Motor equivalence** is the capability of the motor system to perform a particular task, and produce the same movement outcome, in a variety of ways. Even actions as apparently simple, repetitive, and consistent in outcome as writing one's own signature on a piece of paper can be achieved through recruiting different motor units or even different muscle groups. Motor equivalence is a consequence of the many degrees of freedom (i.e., joints, muscles, motor units) we are able to independently control. Any plausible model of motor control must be able to account for how the nervous system rapidly, and apparently effortlessly, selects just one combination of joints, muscles, and motor units from all the options available in order to perform a particular task effectively and efficiently.

▶ **Serial order** is the capability of the motor system to structure movement commands in such a way as to reliably produce movement elements in their desired sequence. Correct sequencing of movement components is fundamental to the performance of virtually all skilled actions, and errors in sequencing inevitably result in errors in performance. Serial-order errors in speech give rise to spoonerisms (e.g., *muman hovement* instead of *human movement*) and in typing to transposition errors (e.g., *cta* instead of *cat*). In gross motor skills, such as throwing, misordering of the recruitment of large proximal muscles (such as those crossing the trunk) and smaller distal muscles (such as those crossing the wrist joint) undermines the effective summation of forces and, through this, limits performance. A useful model of motor control must therefore be able to account for how serial order is generated in movement sequences (and, hence, how errors may arise).

▶ **Perceptual–motor integration** is the capability of the motor system to produce movements closely matched to the current environmental demands (perceived by the performer). Skilled movement is always subtly adjusted to meet changing environmental situations. For example, the skilled tennis player is able to adjust the racquet swing if the ball deviates unexpectedly in flight, and all of us adjust our gait patterns if the surface we are walking over becomes irregular. Such adjustment can be achieved only if there is a tight coupling and integration between perception and action. Such coupling must therefore be a key element of any satisfactory model of motor control. Because the role of particular muscles in either producing or opposing limb movement frequently varies according to contextual factors such as joint position or orientation of the limb with respect to gravity, the central nervous system, in issuing motor commands, must be continuously and accurately informed about the body's posture and position in space by its many perceptual systems.

▶ **Skill acquisition**, as discussed in chapter 18, is the capability of the motor system to learn and improve, given appropriate conditions of practice. Explaining skill acquisition (or motor learning) requires a motor control theory to, in turn, be able to explain how experience is stored and how, once acquired, movements can be modified to meet task conditions never previously encountered. A viable model of motor control must therefore be able to adequately account for the paradoxical capabilities of skilled performers to produce movements that are adaptable yet consistent.

Several models of motor control have been proposed over the years. These models vary in the extent to which they attempt to explain the key properties of movement and incorporate what is known from neuroscience about the structure of the motor system. The models of motor control that have been most widely developed and used can be generally described as information-processing models.

INFORMATION-PROCESSING MODELS OF MOTOR CONTROL

Experimental psychologists have, for a long time now, used the analogy of the nervous system as a computer in order to simplify thinking about the complex neural processes underpinning motor control. By first considering the (relatively) simple operations of a computer we can perhaps attempt to better understand the more complex operations of the human motor system.

MOTOR SYSTEM AS A COMPUTER

Computers are elaborate, engineer-designed devices used for processing information. Information-processing models of movement control are based on the notion that the nervous system, despite being biological in substance, is computer-like in function in that it is capable of sophisticated information processing.

A computer is basically a dedicated electronic device that, through its stored programs, is able to convert input information of one type or another into output of a specific, desired form. The input information may come from data stored on a devise or may come directly from a keyboard (as occurs when we type in letters or numbers) or from some other information-acquisition system. The output information may also be of various forms, such as text (letters and numbers), graphics appearing on the computer screen or sent to a printer (i.e., the hard copy), or electronic commands sent on to other computers or devices controlled by the computer.

The conversion of input information to output information is not a passive process; rather, it is an active reorganisation of information specifically controlled by the commands in the computer program(s). The type of processing the computer is capable of doing, the speed with which it can complete its operations, its capacity to store information in memory, and ultimately the quality and diversity of the output it can produce are limited by two interacting factors: the physical construction of the computer's electrical circuits (i.e., its hardware) and the computer programs that have been written specifically for the computer and that reside in the computer's memory (i.e., its software).

How, then, does the central nervous system act like a computer system? The input information for movement control is the sensory information sent to the central nervous system from (primarily) the visual, kinesthetic, and vestibular receptors (as detailed in chapter 15). The output is the patterns of movement we observe, and can describe biomechanically, that arise as a consequence of the coordinated set of motor commands sent from the central nervous system to selected muscle groups. Input information is converted to output information through a number of information-processing (or computational) stages that take place in the brain and other regions of the central nervous system. The success of the movement that results (the output) depends primarily on the computational programs in the central nervous system that are responsible for selecting and then controlling the movement (figure 16.1). Some of the programs, such as those controlling balance and gait, may be hard-wired into the central nervous system (especially the spinal cord)

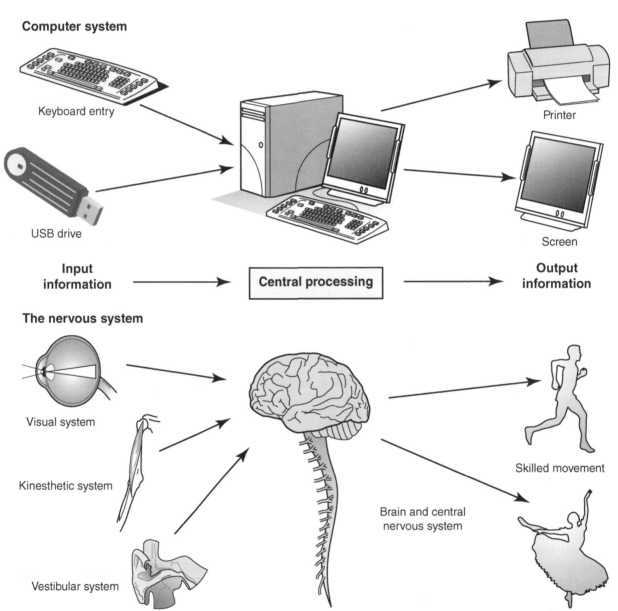

Figure 16.1 Parallels between the information-processing operations of a computer and those of the central nervous system.

from birth and therefore may be considered to be part of the hardware of the central nervous system. Other skills require programs that are not innate and that must be developed, constantly modified, and improved through repeated use and practice. These programs are therefore analogous to specialised software that is written for a particular computer's hardware to fulfil a specific, unique task and stored in the system's memory. In the nervous system this storage may require some changes to hardware as new neural connections are formed.

Just as one cannot understand how a computer works and controls its output simply by inspecting what it produces as output, one cannot expect to understand movement control by simply describing the observable movement patterns produced as output by the motor system. It is important to recognise that movement does not simply occur spontaneously as a consequence of unplanned muscular activity. Rather, movement is the end product of a long series of information-processing stages (or computations) that take place largely beyond observation and within the confines of the central nervous system. To understand movement one must therefore attempt to understand the processes and computations that occur in the central nervous system and that form the link between sensory input and observable movement output. Cognitive scientists have directed much of their energy to attempting to uncover the computational code and programs that the central nervous system has inherited and developed for movement, elaborate on the processing stages used to link input and output information, and determine the capacities and limitations of the various processing stages.

STAGES IN A TYPICAL INFORMATION-PROCESSING MODEL

Most information-processing models of movement control assume that there are distinct and sequential stages through which information must pass (or, more

correctly, be processed) from input to output. Figure 16.2 presents a typical information-processing model. It shows environmental and internal information—present in such forms as light and sound waves and muscle lengths and tensions—being picked up (transduced) through the various sensory receptors described in chapter 15 and then being transmitted along the afferent pathways to the central nervous system. It is this information that then provides the input for central nervous system processes that ultimately produce, as output, motor commands, which are transmitted along the efferent pathways out to the muscle fibres, where they individually cause muscular contraction and collectively generate observable movement patterns. Feedback from the movement itself is monitored via the afferent pathways and can be used to either correct errors in the movement (if the movement is sufficiently slow) or make improvements to the commands for the next time the same or a similar movement is to be produced.

The stages proposed in figure 16.2 are themselves unremarkable and are entirely consistent with the structure and function of the receptors and effectors for movement described in chapter 15. What remains unclear is the nature of the central processing stages, and it is in the conceptualisaton of these central processing stages that most information-processing models differ. Most models, however, accept that at least three sequential processing stages must occur in the brain and other areas of the central nervous system before the initiation of any movement. For simplicity, we refer to these stages as *perceiving*, *deciding*, and *acting*.

Figure 16.3 illustrates these stages with an example from the sport of basketball. Consider in this example the perspective of player 3—the defensive player (without the ball). The player is surrounded by a near-infinite array of physical signals (e.g., light and sound waves, pressure signals, vibrational signals, and chemical signals), only a very limited range of which can be detected by his eyes, ears, vestibular, kinesthetic, and other receptors. Despite this limited range of sensitivity, the brain and spinal cord of the

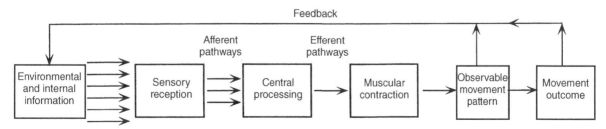

Figure 16.2 A typical information-processing model of motor control.

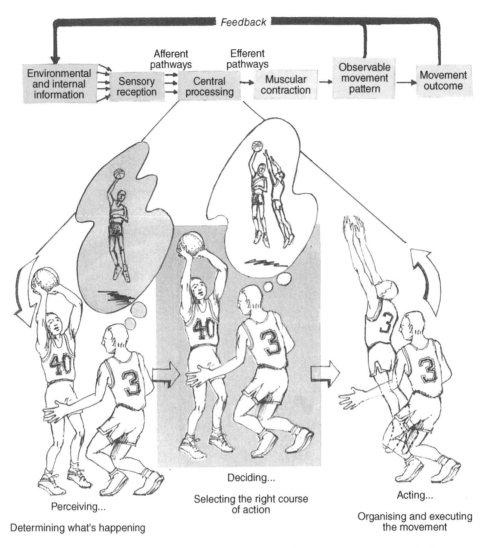

Figure 16.3 Central processing stages in a typical information-processing model of motor control.

player are bombarded every millisecond by an enormous array of input information from the billions of sensory receptors in the body. There is, for example, visual information about the player's own location on the court relative to the basket plus information about the location, velocity, and direction of motion of all four teammates and all five opponents; auditory information from the bouncing ball, the calls of teammates, opponents, coaches, and spectators; kinesthetic information from stretches on muscles, strains on joints, tension on tendons, and pressure on the soles of the feet; tactile information from any contact with the opponent; proprioceptive information from the vestibular apparatus; and smell (**olfactory information**), taste (**gustatory information**), and other information. Some of this information is

highly relevant to the task at hand, whereas much of this information is quite irrelevant.

It is from this enormous array of sensory information that skilled perceiving, deciding, and acting must emanate. In the perceiving stage the player's focus is on determining what is currently happening and what is about to happen in both the external and internal environments. In the deciding stage the focus is on deciding what action, if any, is needed in response to current and future events. In the acting stage the focus is on organising and executing the required movement response in terms of the sequence and timing of the motor commands that have to be sent to the muscles. In the sections that follow we use the example of the basketball player to help examine each of the key central processes in more detail.

PERCEIVING: DETERMINING WHAT IS HAPPENING

The first important process that the brain and central nervous system must perform is that of **perception**. Perception involves selecting only the most relevant information for further processing and then using this information to determine what is currently happening and, implicitly, what is about to happen in the near future. In the case of the defensive player, perception involves making important judgements about external events as well as more internal ones. Externally, the player will need to determine, among other things and primarily through vision, the location and posture of his direct opponent, the structure of the offence as well as the defence, the position and direction of movement of any unguarded offensive players, their position on the court, and especially their proximity to the basket and the relative heights and agilities of teammates and opponents. Internally, the player will need to be aware of his or her own body posture and balance.

UNDERLYING PROCESSES

Perceiving involves determining what is occurring in the outside world (e.g., "Where is each of my teammates located?"), what is occurring in our own bodies (e.g., "Am I in or out of balance? What position are my arms in?"), and the current and ongoing relationship between these internal and external worlds (e.g., "Where am I in relation to my opponent?"). Perception, as noted earlier, is more than simply the passive reception of sensory information by our various receptor systems. It is an active process through which we interpret and apply meaning to the sensory information we receive. Because our prior experiences, accumulated knowledge, expectations, biases, and beliefs all contribute to perception, it is therefore not surprising that two people presented with the same pattern of stimulation (e.g., looking at the same picture or experiencing the same kinesthetic sensations) will often perceive and report different things. This is true for simple visual images typical of the type frequently used by psychologists (figure 16.4) as well as for more complex images typical of natural movement tasks (see chapter 18) and for perception of social environments, such as sport and exercise settings (see chapters 19 and 20).

Perception involves a number of subprocesses. Central processes include

Figure 16.4 A typical simple but ambiguous figure used by psychologists to demonstrate the subjective nature of perception. The pattern may be perceived as either a vase or two faces looking at each other.

- ▸ **detection** (determining whether a particular signal is present),
- ▸ **comparison** (determining whether two stimuli are the same or different),
- ▸ **recognition** (identifying stimuli, objects, or patterns), and
- ▸ **selective attention** (attending to one signal or event in preference to others).

PROCESSING LIMITATIONS

All of the subprocesses that make up perception are limited in their capacity to process information and can therefore potentially limit performance on any particular motor task. In detecting stimuli, humans are limited by the range of physical stimuli to which their various sensory receptors can respond as well as the capability to distinguish the firing of one or more sensory neurons from a background of general neuronal activity. Human ability to detect stimuli consequently varies from situation to situation, from one sensory system to another, and with factors such as arousal (see chapter 19). Humans are also limited in their ability to compare and detect differences between two or more stimuli. For example, in judging the approach velocity of objects (such as when balls are thrown toward a person to catch), the object must increase in velocity from, on average, 4.4 m/s (16 km/h) to 5 m/s (18 km/h) before the change in speed can be reliably detected. The sensitivity for detecting differences varies from one sensory system to the next. The visual and auditory systems are able to reliably detect the smallest change in stimulation.

In regards to recognising stimuli, objects, or events (e.g., recognising different positions of the arm kinesthetically), laboratory experiments have shown that humans are limited to storing seven items before they start to make errors of identification. If more than seven items have to be recognised at any one time, recognition errors occur. In natural settings the number of patterns we are able to recognise (e.g., friends' faces or offensive patterns in basketball) is much greater than seven but nevertheless finite. Recognition in natural settings is enhanced through the use of multiple attributes (e.g., hair colour, length, or style, eye colour, nose size, and ear type in recognising human faces; player location, posture, and size in recognising basketball offensive patterns).

Selective attention is both a limitation and an advantage to human performance. We all know from personal experience that our processing capacity is limited in that we cannot (typically) listen to two separate conversations or attend to two separate visual signals (such as events in the left and right extremes of our field of view) simultaneously. If a movement task requires information from two or more separate locations to be processed at the same time, performance of the task will generally be difficult. However, being able to selectively attend to only one thing at a time can be an advantage in that it provides a means of preventing irrelevant or potentially distracting stimuli from using up some of our valuable processing capacity. A golfer focusing on a putt or a microsurgeon focusing on a suture benefit from being able to apply all their attention to the specific movement and thus effectively block out surrounding noise and other potentially distracting events.

DECIDING: DETERMINING WHAT NEEDS TO BE DONE

Once a person has determined what is currently happening and has predicted future events, the second stage of information processing involves the process of decision making (i.e., determining what, if any, new action or response is required). In many movement tasks, such as sport tasks, the decision-making process equates to picking the correct option from a range of possible response options. Clearly, the quality of the decision that is made will be determined, in part, by the accuracy of the preceding perceptual judgment. In basketball, the

player with the ball has at least six broad response options to choose between on any occasion (e.g., continuing to dribble the ball, shooting the ball at the basket, or passing to one of four possible team-mates). The defending player (number 3) not only must perceive which of these actions the opponent is about to undertake but also decide what action is most appropriate as a response (e.g., jumping to block the shot, attempting to intercept the pass, switching to cover another unguarded opponent). Determining which option to select generally depends on the current perceptual information as well as other situational information such as the state of the game (the score and time remaining) and knowledge about the respective capabilities of matched teammates and opponents.

UNDERLYING PROCESSES

Decision making is essentially the process of response selection—picking the right movement option to match the current circumstances. The quality of the decision that is made depends on the quality of the preceding perceptual judgements as well as the knowledge of the costs and benefits associated with each particular option. The latter is heavily dependent on the extent of the individual's experience. The speed and accuracy of decision making about movement is also influenced by things such as the number of possible options (or response choices) that exist, the costs associated with making incorrect decisions, and the total time that is available to make decisions. Some activities, such as playing golf, offer essentially unlimited periods of time in which to select the correct action, whereas in other activities, such as playing tennis, the time constraints on decision making are severe.

PROCESSING LIMITATIONS

The measure that is used to determine how quickly people can make decisions is called choice reaction time (CRT). In the laboratory, CRT is measured by presenting participants with an array of stimuli (usually lights), each of which has its own associated response (frequently a button press). The participants' task is to view the stimulus array and respond, as soon as possible after a stimulus light is illuminated, by pressing the response button that corresponds to the illuminated light. CRT is then recorded as the time elapsed between the illumination of the stimulus light and the initiation of the button press. Researchers can examine the limitations on rapid decision making by recording CRT as they systematically vary the number of possible stimulus–response pairs.

Studies in which the number of stimulus–response alternatives varied are consistent in their findings. When there is no uncertainty (there is only one possible stimulus and response), the reaction time to the appearance of this stimulus is about 200 ms (0.2 s). This is also the delay in responding that is observed in sprint events between the sound of the gun and the commencement of movement and in hand–eye coordination tasks when there is unexpected movement of the object that is to be intercepted (see "Measuring Reaction Time to Correct Errors"). As the amount of uncertainty is increased by adding more and more possible options, CRT slows substantially (figure 16.5). Each time the number of possible stimulus–response alternatives doubles, CRT slows by a constant amount such that the increase in CRT from a two-choice situation to a four-choice situation approximates the increase in CRT from a one-choice to two-choice situation and from a four-choice to an eight-choice situation.

This relationship between CRT and number of stimulus–response alternatives is frequently exploited in a range of movement tasks. Designers of cars and machinery attempt to minimise the number of options on machine controls in order to reduce the decision-making time of users of the equipment. Skilled sport players attempt to familiarise themselves with the preferred options and patterns of play of their opponents as a means of speeding up their own rates of responding. A basketball player who recognises that his opponent can control the ball only with their right hand might be able to respond more rapidly to his opponent's moves than might a player who considers that both left- and right-hand movements may occur. Conversely, a player who is able to execute, with equal skill, a wide range of options (through having equal shooting, passing, and dribbling skills on both sides of the body) can maximise the amount of information that an opponent has to process and can thus substantially slow the speed of decision making of her opponent.

Close inspection of figure 16.5 reveals two independent components of decision making. One component, given by the intercept of the CRT-alternatives line with the y-axis, corresponds to reaction time when there is only one option and therefore no uncertainty. This component, known as simple **reaction time**, is not influenced by practice and reflects individual differences in the time it takes for the afferent nerve impulses to reach the brain and be registered there and for efferent commands to be sent to the muscles. The second component, given by the slope of the CRT-alternatives line, is a measure of decision-making rate; it estimates the average increase in CRT that occurs for each new additional stimulus–response option. The slope of the line is steep for individuals who are slow decision makers and approaches zero for individuals who are very fast decision makers. Advance knowledge about the probabilities of different events occurring can reduce the amount of information to be processed and make for faster decision making.

ACTING: ORGANISING AND EXECUTING THE DESIRED MOVEMENT

Having selected the desired action (e.g., jumping to attempt to block the opponent's shot), the player must then organise the movement before it can be initiated. This organisation involves sending from the brain motor commands that specify the order and timing of motor unit recruitment. If these efferent commands are not appropriately structured, the resulting movement pattern may lack the force, timing, or coordination necessary to successfully realise the objective of the movement (in the basketball case in figure 16.3, the blocking of the opponent's shot). All three central processes of perception, response selection (decision making), and response organisation and execution (acting) are

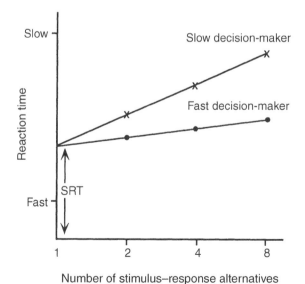

Figure 16.5 Reaction time shown as a function of the number of stimulus–response alternatives for a fast and slow decision maker.

MEASURING REACTION TIME TO CORRECT ERRORS

In a controlled laboratory setting it takes the average person about 200 ms to respond to an unanticipated stimulus. Peter McLeod from the department of psychology at the University of Oxford, England, was interested in ascertaining whether processing delays of this same order exist in natural movement tasks (and therefore act as a limiting factor to skilled performance) or whether, in the natural setting, reaction times might be much faster.

McLeod designed a very innovative experiment to measure the reaction time of cricket batsmen. Cricket batsmen perform the extremely time-constrained task of striking a fast-moving ball as it bounces off a batting surface (the pitch). By placing strips of wooden dowelling under the pitch around the region where the ball would typically bounce, McLeod created a situation in which some balls (landing between the dowelling strips) bounced normally whereas others, landing close by but on the edge of the dowelling strips, bounced and deviated unexpectedly in a lateral direction. By comparing the path of the bat (using standard biomechanical methods) between trials in which the ball bounced normally and those in which it deviated unexpectedly, McLeod was able to measure the minimum time it took the batsman to initiate a corrective

movement response to the unanticipated stimulus (i.e., the deviation of the ball).

The reaction times measured were precisely in the 200-ms range as observed using typical laboratory measures. In natural tasks, skilled performers must therefore develop strategies that enable them to cope with this delay in responding that is inbuilt in their nervous systems.

More recent studies have attempted to determine whether the delay in making simple visuomotor corrections differs between expert and nonexpert athletes. In research conducted at the Université Paris-Sud, expert tennis players were shown to be able to react with a shorter delay (162 ms on average) than nonexperts (221 ms on average) to unanticipated changes in the motion of an object that they were to intercept. Although extensive practice or preselection on the basis of athletic ability may decrease reaction time to a degree, it does not eliminate this delay.

Sources

Le Runigo, C., Benguigui, N., & Bardy, B.G. (2005). Perception-action coupling and expertise in interceptive actions. *Human Movement Science, 24*, 429-445.

McLeod, P.N. (1987). Visual reaction time and high-speed ball games. *Perception, 16*, 49-59.

completed before any observable muscular contraction takes place or whole-body movement occurs.

Feedback during the movement itself may assist in adjusting the motor commands, although the skill of blocking is, in all probability, too short in duration for feedback-based corrections to have time to be effective. Visual feedback derived from the completed action does, however, provide a valuable source of information for the performer to assist in future repetitions of the same or similar actions. If the jump is not high enough, for example, this information can be used to ensure that more motor units are recruited the next time the player opts to attempt a similar block. For the shooter (player 40), comparison of visual information about the outcome of the movement with kinesthetic information from the execution of the movement provides a valuable means of calibrating the force-production system, enabling the player to find his or her range. The relationship between information about movement

execution and that from movement outcome provides essential guidance for future attempts at movement skills of all types, as illustrated in figure 16.6.

UNDERLYING PROCESSES

Once a particular movement response has been selected, the central nervous system is responsible for ensuring that the selected movement response is actually executed as desired. At least three subprocesses are involved at this stage in the processing of information for movement control:

▶ movement organisation (carefully planning out the sequencing and timing of the efferent commands to be sent out to selected motor units),

▶ movement initiation (transmitting the required motor commands to the muscles), and

▶ movement monitoring (adjusting the movement commands on the basis of sensory information about the movement's progress).

Type of evaluation	Was the movement executed as planned?		
	Outcome	Yes	No
Was the goal accomplished?	Yes	Got the idea of the movement	Surprise!
	No	Something's wrong	Everything's wrong

Figure 16.6 The relationship between information about movement execution and information about movement outcome as a basis for guiding future attempts at a skill.

Copyright 1972 from *Quest*, "A working model of skill acquisition with application to teaching," 17:9, by A.M. Gentile, reprinted by permission of Taylor & Francis Ltd., http://www.tandf.co.uk/journals.

PROCESSING LIMITATIONS

The speed and accuracy with which movements can be executed and controlled depends on a number of factors, including the complexity of the movement (the number of joints, muscles, and motor units involved plus the difficulty their coordination may pose for the maintenance of posture and balance), the time constraints imposed on the movement, and the acceptable margins for error in the movement. Movements of relatively long duration (greater than one third of a second) can use feedback generated during the movement itself to assist in their control and precision. Control based on the monitoring of feedback is known as **closed-loop control**. In contrast, very rapid movements require all the efferent commands to be structured in advance. This type of control, known as **open-loop control**, is thought to involve the use of **motor programs**.

For movements controlled in a closed-loop manner, the time taken to complete the movements (**movement time**) is directly dependent on the difficulty of the movement. Movements that involve high precision demands (such as movements to small targets) or traversing a large distance take much longer than movements made over a short distance or to large targets. The relationship between movement time and movement difficulty is a lawful one that is governed by the amount of information that must be processed (figure 16.7). For movements controlled in an open-loop manner,

the time taken to initiate the movements (i.e., reaction time) is directly proportional to the amount of preplanning that must take place. This is greater for more complex movements.

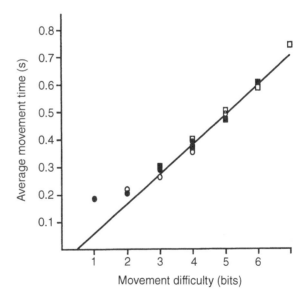

Figure 16.7 Average time to make a movement plotted as a function of the difficulty of the movement. The task involved moving as rapidly as possible for a 10 s period between two targets. Movement difficulty was manipulated by changing the distance between the targets and the size of the targets.

Data from Fitts 1954.

Some Implications of An Information-Processing Model

Now that we have examined the central processing stages proposed by an information-processing model of motor control, we can consider some of the implications and insights such a model may provide. One insight that emerges from consideration of the central processing stages relates to the sources of errors in motor skill performance (see "When Things Go Wrong"). If the ball carrier in our basketball example had attempted to shoot a jump shot but had only succeeded in losing the ball or producing an inaccurate shot, it would be very tempting to simply attribute this error to poor movement-execution technique and attempt to remedy this by having the player practise his or her shooting skills. However, it should be apparent from our preceding discussions that an error in performance could as easily result from poor perception or poor decision making as it could from poor movement execution. Given the sequential nature of central information processing, a perceptual error (e.g., incorrect judgement of the distance from the basket or failure to detect the movement of one of the defensive players) or a decision-making error (e.g., shooting from a position where passing would have been a better option) can just as much cause an ineffective shot as can poor execution of a correctly selected movement. Likewise, a therapist attempting to recover the normal gait of a stroke patient or assist the skills development of a clumsy child needs to be aware that deficiencies in movement control can occur even when the ability to actually execute a selected movement may be essentially normal.

A second, related implication of the information-processing model is that although the three central stages of information processing are not directly observable, they may nevertheless need to be trained just as much as the more observable components of performance, such as technique, strength, speed, endurance, and agility. In many motor skills, the perceptual and decision-making aspects of movement control may act as the limiting factors to performance and therefore warrant systematic training. This is especially true of activities where decisions must be made in a very short time on the basis of limited information, such as fast ball sports. Conditions that slow the speed with which movements can be produced, such as ageing, injury, or disease, also exacerbate the

WHEN THINGS GO WRONG

In a system as complex as the human information-processing system, there is always a high chance of errors occurring. Many errors in movement control (e.g., inadvertently typing the wrong letter on a keyboard or brushing against a door frame as you walk through it) have trivial consequences. However, in other cases, errors in movement control (e.g., a surgeon applying too much force to a scalpel or a bus driver unintentionally applying pressure to the accelerator rather than the brake) can have catastrophic consequences. Because human error rather than machine error is by far the most common cause of major catastrophic accidents in the workplace, ergonomists (scientists who study people in the workplace) are constantly looking for ways to minimise the incidence of errors and to make tasks safer.

The first step toward eliminating errors is to understand how they come about. Understanding the sources of errors can also help provide insight into how movement control is normally achieved. Errors arise from multiple sources, and a number of classification systems (taxonomies) for errors have been developed. Errors in the execution of movement appear to occur most frequently in situations where the ongoing control of movement is not monitored sufficiently, closely, or continuously (using closed-loop control). Under these circumstances, unconscious, automatic control of movement takes over. If these automated movements (ones performed many times in the past) are not the appropriate ones, errors inevitably occur. The car driver who finds him- or herself driving on a familiar route (such as the route home) rather than on their planned

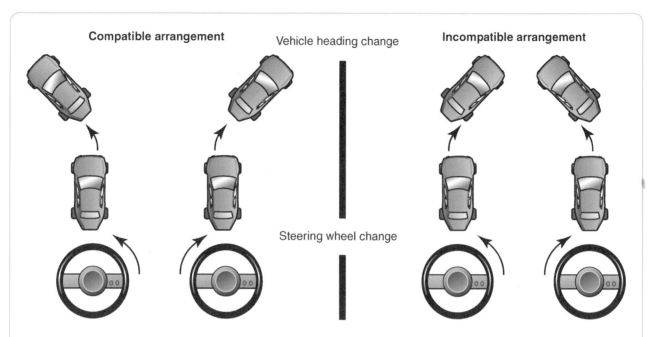

Compatible arrangement

Vehicle heading change

Incompatible arrangement

Steering wheel change

Figure 16.8 Displays of different levels of stimulus–response compatibility. In the compatible display the movement of the steering wheel and the resultant movement of the vehicle are in the same direction, whereas in the incompatible display they are reversed.

pathway to somewhere else has committed an error of this type.

Ergonomists attempt to design and redesign tasks in order to minimise either the potential for processing errors or the consequences of processing errors. One design principle is to ensure that the links between task information and the required movement responses (what is known as **stimulus–response compatibility**) are made as natural as possible. Displays in which the required outcome

and the required action are compatible (figure 16.8) allow for both faster reaction times and fewer errors compared with incompatible displays.

Sources

Proctor, R.W., & Van Zandt, T.V. (1994). *Human factors in simple and complex systems.* Boston: Allyn & Bacon.

Reason, J. (1990). *Human error.* Cambridge, MA: Cambridge University Press.

Schmidt, R.A. (1989). Unintended acceleration: A review of human factors contributions. *Human Factors, 31,* 345-364.

necessity for good perceptual and decision-making skills to offset delays imposed by longer movement times.

SOME ALTERNATIVE MODELS OF MOTOR CONTROL

Modelling the motor control system as an information-processing (or computational) system appears to be a useful way of starting to think about the

neural control of movement; however, it is certainly not the only way. Over the past 30 yr a number of limitations and assumptions in the information-processing model have been highlighted and some alternative models have been suggested. Critics of the information-processing model have been concerned by the implicit assumption that every movement is somehow represented and stored in the central nervous system. They have argued that simply assigning the responsibility of movement organisation to the brain does not explain movement control but rather simply creates the (unacceptable) need for an intelligent little person (a "homunculus") somewhere else in the brain! Critics of the information-processing model have also been concerned about the assumption that the brain and nervous system directly control the specific motor commands for all aspects of a movement, noting that many of the physical properties of the musculoskeletal system (such as the spring-like characteristics of muscle and the natural oscillatory frequencies of limbs) can themselves contribute significantly to movement control without the necessity of any involvement of the central nervous system.

Dynamical models of movement control propose that movement patterns are not represented anywhere in the nervous system by way of a plan or program but rather emerge naturally (or self-organise) out of the physical properties of the musculoskeletal and related systems. The emergence of complex patterns of organisation out of the motor system, such as the complex movement patterns that characterise gaits such as walking and running, are considered to be no more in need of a pattern representation or template than are nonbiological systems, such as chemicals, which, under appropriate environmental conditions, are able to organise and reorganise into complex patterns without the need for a nervous system. For example, under appropriate environmental conditions (level of heat), water may undergo complete pattern reorganisations without the form of the new pattern being anywhere explicitly stored or represented. Because some aspects of movement control share these same characteristics of pattern reorganisation at critical levels of **control parameters** (see figure 16.9), there is considerable current research interest in attempting to further explore this model of movement control by determining the control parameters for a range of human movements and exploring other parallels between movement-pattern formation and pattern formation in physical systems (see "Pattern Transitions in Cyclical Movements"). As more is understood experimentally about movement control, new models that progressively encapsulate more of the many essential and unique characteristics of movement listed earlier in this chapter can and will be developed.

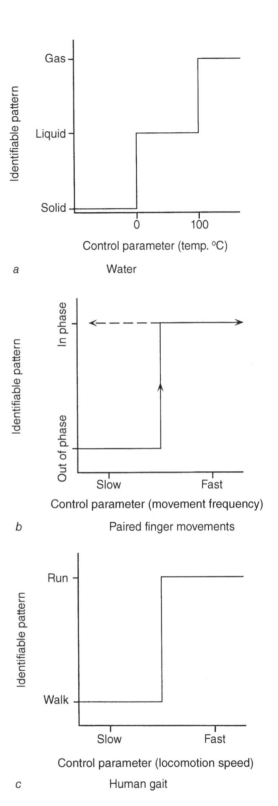

a   Water

b   Paired finger movements

c   Human gait

Figure 16.9 Parallels between the pattern transitions observed in nonbiological systems and those occurring in some human movements.

PATTERN TRANSITIONS IN CYCLICAL MOVEMENTS

In purely physical systems (e.g., chemical and laser systems), spontaneous transitions from one form of organisation or pattern to another occur when key environmental conditions (or thresholds) are reached. For example, we know that the water molecule spontaneously reorganises its structure as it is heated past the critical control temperatures of 0 and 100 °C (figure 16.9). The study of such transitions is called **synergetics**. Synergetic transitions share a number of common characteristics, including sudden, discrete changes in organisation around critical levels of the control parameter, increased variability in structure as the transition point is approached (a property called **critical fluctuations**), and a delay in returning to stability if the system is perturbed in some way when it is near a transition point (a property called **critical slowing down**).

In an extensive series of studies, Scott Kelso from Florida Atlantic University in the United States, Herman Haken from the University of Stuttgart in Germany, and a number of colleagues set out to determine whether transitions in movement patterns also share these synergetic characteristics. They used as their task paired movements of the index fingers. The index fingers were prepared either out of phase or in phase (figure 16.10). When the fingers were commenced out of phase and the participants were required to progressively move the fingers at a faster rate, a critical frequency was reached at which the fingers moved spontaneously from out of phase to in phase. The phase relations between the two fingers were found to increase in variability and be slower to respond to perturbations as the transition point (a point of pattern instability) was approached. Thus, the critical fluctuations and critical slowing-down properties seen in other synergetic systems also exist in some aspects of the human motor system. The in-phase coordination, which was found to be stable, was able to be preserved across all finger-movement frequencies.

These experimental data have made possible the formulation of suitable mathematical models that explain the synergetic phenomena in this task. This model has been the subject of extensive experimentation over the past decade across a range of multijoint coordinative tasks. Attempts have been made to use dynamical models such as this one to explain the transitions that occur between different gait modes

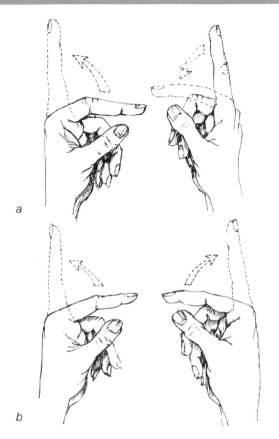

Figure 16.10 When the rate of paired figure movement is progressively increased, a spontaneous shift occurs from *(a)* antiphase (one finger extended while the other is flexed) to *(b)* in phase.

Based on Haken, Kelso, and Bunz 1985.

(e.g., crawling, walking, running) both in response to the need to move at different speeds and as a consequence of development and ageing.

Sources

Diedrich, F., & Warren, W.H., Jr. (1995). Why change gaits? Dynamics of walk-run transition. *Journal of Experimental Psychology: Human Perception and Performance, 21,* 183-202.

Haken, M., Kelso, J.A.S., & Bunz, H. (1985). A theoretical model of phase transitions in human hand movements. *Biological Cybernetics, 51,* 347-357.

Kelso, J.A.S. (1995). *Dynamic patterns: The self-organization of brain and behavior.* Cambridge, MA: MIT Press.

Kelso, J.A.S., & Schöner, G. (1988). Self-organization of coordinative movement patterns. *Human Movement Science, 7,* 27-46.

SUMMARY

Cognitive science has provided a number of conceptual models of how the brain and nervous system control movement. The dominant model has been one that conceptualises the brain and nervous system as an information-processing, or computational, system. Consequently, researchers have been interested in attempting to understand the processing stages, capacities, and limitations of the human computational system. The sequential processes of perceiving, deciding, and acting all have demonstrable processing limitations, the quantification of which helps cognitive scientists understand some of the key characteristics of, and constraints to, human motor behaviour. Because there are parallels between human information processing and the functions of computers, it does not necessarily follow that the human brain and nervous system is a computational system. Alternative models of movement control that are more concerned with the dynamics of the human neuromuscular system are gaining popularity as an explanation of how control is achieved over at least some aspects and types of movement.

FURTHER READING

Kelso, J.A.S. (1995). *Dynamic patterns: The self-organization of brain and behavior.* Cambridge, MA: MIT Press.

Lee, T.D. (2011). *Motor control in everyday actions.* Champaign, IL: Human Kinetics.

Magill, R.A. (2011). *Motor learning: Concepts and applications* (9th ed.). Boston: McGraw-Hill.

Proctor, R.W., & Van Zandt, T.V. (2008). *Human factors in simple and complex systems* (2nd ed.). Boca Raton, FL: CRC Press.

Schmidt, R.A., & Lee, T.D. (2011). *Motor control and learning: A behavioural emphasis* (5th ed.). Champaign, IL: Human Kinetics.

Schmidt, R.A., & Wrisberg, C. (2008). *Motor learning and performance: A situation-based learning approach* (4th ed.). Champaign, IL: Human Kinetics.

CHAPTER 17

MOTOR CONTROL CHANGES THROUGHOUT THE LIFE SPAN

The major learning concepts in this chapter relate to

▶ the development of movement in the first 2 yr of life and the development of fundamental motor patterns in childhood;

▶ the motor performance of people who are elderly;

▶ the major physical changes in the central nervous system, changes in the sensory receptors and sensory systems, changes in the effectors (muscles), and changes in reflex systems that affect neural control of movement across the life span; and

▶ developmental improvements in information-processing capability and declines in information processing with ageing.

In chapters 15 and 16 we examined some basic neuroscience and psychological concepts of motor control. In this chapter these basic concepts are used to examine motor control changes across the life span. The specialised field of study concerned with the description and explanation of changes in motor performance and motor control across the life span is typically referred to as *motor development*. Although the study of motor development has historically focused on the period from conception through adolescence and the changes and stages through which the developing human progresses in attaining adult levels of motor performance, it now increasingly includes the deterioration of motor skills that are apparent in the elderly. Such a broadened focus is important given ageing populations worldwide and given that many changes in the performance and control of motor skills are age related and occur throughout the entire life span.

Studies of motor development have generated knowledge about many topics, including

- the normal rate and sequence of development of fundamental motor skills that arise out of the interaction of biological maturity and environmental stimulation;

- individual differences in the rate of development of specific skills (and motor performance comparisons of early and late developers); and

- deviations from normal development in special populations (such as the intellectually challenged and individuals with conditions such as cerebral **palsy**, **Down syndrome**, or clumsiness).

Such knowledge is important practically in

- assessing the normative development of children,

- screening for neurological and motor disorders as well as identifying and nurturing exceptional talent,

- informing the designers of remedial and therapeutic programs, and

- ascertaining the readiness of individual learners for new challenges.

For both the young and the elderly, knowledge from the field of motor development is valuable in highlighting the role of regular practice and physical activity in the acquisition and retention of movement skills.

Knowledge about motor development is typically derived from two main types of research studies: **cross-sectional studies**, in which motor control and performance are compared between different people of different ages, and **longitudinal studies**, in which the motor control and performance of the same set of people are traced over a number of years as the people mature and grow older. The changes in motor control and performance that occur across the life span may be described, recorded, and explained at a number of levels of observation. In this chapter we first examine observable changes in motor skills across the life span and then, in keeping with the previous two chapters, we examine underlying changes taking place at the neurophysiological level and at the level of information-processing capabilities.

CHANGES IN OBSERVABLE MOTOR PERFORMANCE

One way—the simplest way—of beginning to understand how motor control changes across the life span is to describe and measure motor perfor-

mance at different ages and to determine the specific ages at which key movement skills are first mastered. While individual differences are evident in both the rate of skill learning and the specific ages at which particular movement skills are first mastered, the overall trends in age-related changes in motor performance nevertheless provide an important foundation for the study of motor development. Describing and measuring the changes in motor performance with ageing is an essential first step toward attempting to systematically understand the underlying causes of these changes in movement capability.

MOTOR DEVELOPMENT IN THE FIRST 2 YR OF LIFE

The first 24 mo of life are especially critical for maturation of the neuromuscular system and for the emergence of the basic control skills on which fundamental movement patterns are based. Motor development during this period follows some general principles. Emergence of basic control of posture, locomotion, and movements of the hands can be used to check on normative development.

GENERAL DEVELOPMENTAL PRINCIPLES

Significant advances in voluntary motor control occur over the first 2 yr of life. These advances are generally thought to follow two main principles. The **cephalocaudal principle** highlights that development proceeds in a head-down manner; control develops first in the muscles of neck, then, in order, in the muscles of the arms, trunk, and legs. (Although this principle is generally accepted there have been some challenges to this principle from dynamical theories of movement control. Evidence has been presented to suggest that coordinated kicking movements that are dynamical precursors to walking are present before head and neck control are achieved.) The **proximodistal principle** highlights that development progresses from the axis of the body outward; control is achieved over muscles crossing the trunk (and axial skeleton) before control is achieved over the muscles controlling the limbs (and appendicular skeleton).

These cephalocaudal and proximodistal changes in motor control are paralleled by changes in **muscle tone**. At birth, muscles controlling the movement of the axial skeleton typically have low tone whereas those controlling the movements of the limbs have high tone (or stiffness). As the infant matures this situation reverses; the axial muscles gain tone (to assist in maintaining posture against gravity) and the muscles of the limbs decrease in tone (to facilitate more efficient, voluntary control of limb movements). Reflexes present at or soon after birth disappear, are reduced in strength (attenuated), or are modified during the first 24 mo of life.

MOTOR MILESTONES FOR NORMATIVE DEVELOPMENT

In order to eventually achieve movement independence it is vital in the first years of life that human infants develop control of their general body position (or posture), develop an ability to move (locomote) throughout their environment, and develop an ability to reach, grasp, and manipulate objects using their hands. Because posture, locomotion, and manual control are so vital to infants' development, a number of researchers have studied—and documented in great detail—the ages and stages that infants pass through in the acquisition of these essential motor skills. Knowing the average and range of chronological ages at which key developmental stages in these skills are typically reached provides a set of **motor milestones** that can be used to monitor the motor development of each child and to detect, at an early age, possible movement and neurological difficulties.

The development of the control of posture occurs through three main stages (head control, sitting, and standing), each of which contains a number of identifiable progressions (figure 17.1). Head control

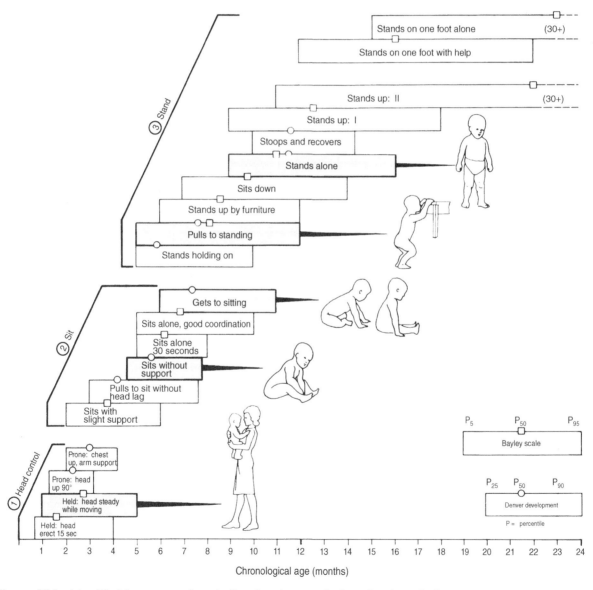

Figure 17.1 Identifiable progressions in the development of postural control.

Reprinted from J. Keogh and D. Sugden, 1985, A general representation of increases in reaction time as information load increases, In *Movement skill development* (Macmillan Publishing Company), 337. By permission of D. Sugden.

is achieved first; a major milestone is the ability to hold the head steady while being carried. This control is typically achieved from 2 to 3 mo after birth. After head control is achieved, sitting without support is then typically achieved at around 5 mo. By 7 mo infants can find their own way to a stable sitting posture without being reliant on being placed in this position. Key milestones in the attainment of postural control for standing occur later. Babies, on average, are able to pull themselves to supported standing by 7 to 8 mo and then stand alone by age 11 mo.

A number of the major motor milestones for the achievement of the locomotion skill of walking occur in parallel with postural control developments (figure 17.2). Upright, unassisted walking appears, on average, at the end of the first year, although walking may first appear anywhere in the range from 9 to 17 mo. A general expectation is that most babies will achieve the basics of walking during the first 6 mo of their second year and then refine this skill considerably by the end of their second year.

The rudiments of mature grasping and reaching control of the hands are in place typically by the end of the first year of a child's life, although the functional use of the hands for self-help activities (e.g., using a spoon) and more complex manual acts

(e.g., writing or tying shoelaces) continues to be refined for many years (figure 17.3). Key milestones in the control of grasping and object manipulation by the hands occur, on average, at 3 to 4 mo (when large objects such as cubes are first picked up), at 6 to 7 mo (when opposition of the thumb is achieved), and at 9 to 10 mo (when the pincer grasp for picking up small objects is first mastered). Reaching skills, controlled more by the proximal muscles of the shoulder and the arm than by the more distal muscles of the wrist and the intrinsic muscles of the hand, are achieved somewhat earlier than manipulation skills. The major motor milestone of reaching to touch a desired object is attained, on average, after 3 to 4 mo. Control of basic reaching and grasping lays the foundation for the acquisition of the complex manual movements that are fundamental to many of our uniquely human actions, gestures, and communications.

MOTOR MILESTONES IN SPECIAL POPULATIONS

Some special populations who suffer movement difficulties as adults are also typically slow in reaching a number of the major milestones for

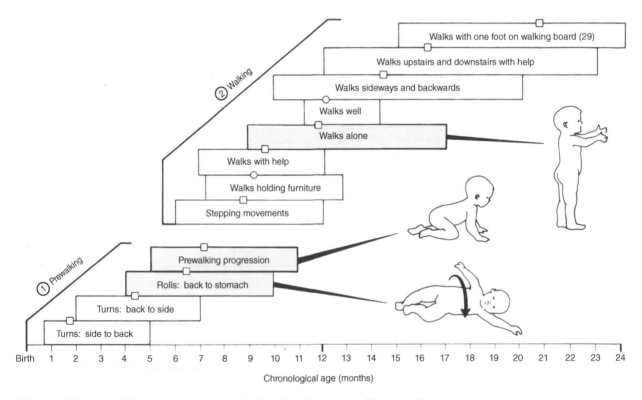

Figure 17.2 Identifiable progressions in the development of locomotion.

Reprinted from J. Keogh and D. Sugden, 1985, A general representation of increases in reaction time as information load increases, In *Movement skill development* (Macmillan Publishing Company), 337. By permission of D. Sugden.

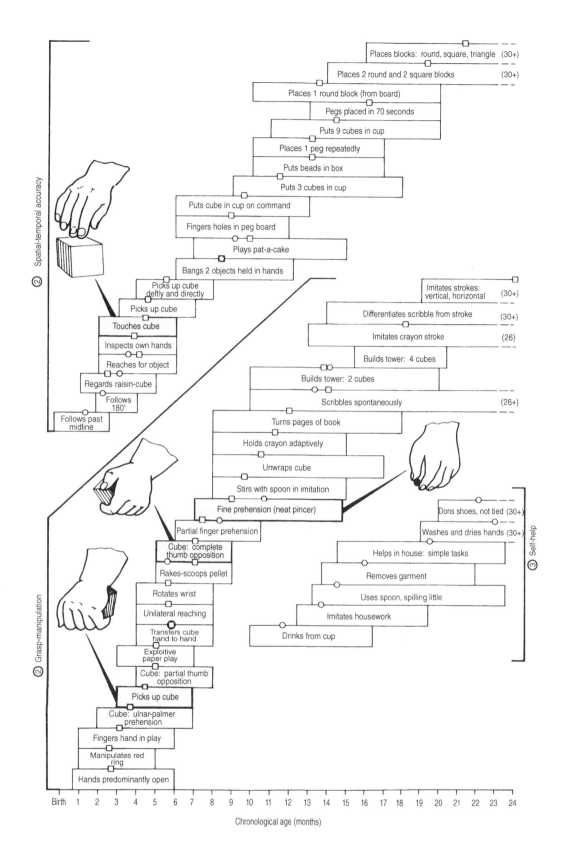

Figure 17.3 Identifiable progressions in the development of reaching (upper panel) and grasping (lower panel).

Reprinted from J. Keogh and D. Sugden, 1985, A general representation of increases in reaction time as information load increases, In *Movement skill development* (Macmillan Publishing Company), 337. By permission of D. Sugden.

motor development. Children with **Down syndrome** (a chromosomal abnormality that affects many aspects of development) are systematically late in achieving virtually all major motor milestones. For example, children with Down syndrome walk unsupported, on average, only after 2 yr of age, whereas, as noted previously, in children without this syndrome walking is generally achieved after 1 yr of age. Mental retardation of all forms also delays the attainment of motor milestones; children with the lowest intelligence quotients are the least likely to stand or walk in the first 12 mo. Motor milestones may act as early markers of potential adult movement problems as well as possible neurological problems and therefore may serve as valuable diagnostic and detection tools. Children described by parents and teachers as clumsy but lacking any apparent neurological problems may also be identified early through assessment of motor milestones and offered appropriate additional training and intervention.

CRITICAL PERIODS

The use and identification of motor milestones is based on the observation that the majority of children pass through the same basic stages in achieving mastery over those motor skills (such as posture, locomotion, and reaching and grasping) that are fundamental to survival. (However, there are notable exceptions. For example, it is not uncommon for children to walk without first learning to crawl.) The basic similarity in stages across children suggests a significant inherited or genetic influence on the emergence of those motor skills that are essential for survival. Obvious questions, therefore, are whether these normal progressions can be altered by environmental factors (e.g., can an enriched environment accelerate the rate of motor development or an impoverished one retard it?) and whether there are **critical periods** during which the child is most sensitive to learning a particular skill. There is clear evidence of critical periods in other aspects of human development. Language, for example, is most easily acquired in childhood. If it is not acquired at this time it is very difficult to pick up subsequently.

The evidence for critical periods in motor development is not clear cut. Some studies of children from cultures that restrain their infants from movement during part of the first year of life have

indicated lower-than-expected levels of adult motor skill, whereas other studies have demonstrated no such detrimental effects; in these studies, early delay in motor milestone achievement is quickly recovered in later years. There is clear evidence available, however, to indicate that enriched environments, in which there is extensive stimulation and many opportunities for motor exploration and play, can speed up the rate at which motor milestones are achieved. What is less apparent is whether the early achievement of milestones translates into improved performance at adult levels. Children who are relatively late in achieving some milestones may attain other skills rapidly, provided that they are exposed to them during a critical period. Programs purported to enhance early motor development (such as educational gymnastics programs, infant swimming programs, and the Suzuki method for learning musical instruments) are all based on the assumption that early attainment of milestones is beneficial to longer term acquisition of motor skills.

PRACTICAL APPLICATIONS

Motor milestone scales (such as those presented in figures 17.1 through 17.3) provide a useful means of comparing the rate of motor development of any specific child against that of children of the same chronological age. As the ranges of ages on each milestone in the figures indicate, there is enormous variability on all these measures, and a child advanced on one milestone may be below the average on another. The normal range of variability also tends to increase as the child gets older and the task becomes more difficult, probably reflecting the cumulative effect of environmental rather than genetic influences. The observation that environment can influence motor development is important because it suggests, among other things, that early intervention for children with movement difficulties offers the best option for effective development. Motor milestones may therefore, within the recognition of the scope of normal variability, be used validly by teachers, therapists, and parents as a screening tool for detecting possible motor and neurological problems early. Further, creating an environment that provides maximal opportunities for movement in the early years would appear to be a concern that should be treated seriously by all parents.

DEVELOPMENT OF FUNDAMENTAL MOTOR PATTERNS IN CHILDHOOD

From 2 yr of age, children develop a range of fundamental motor skills that form the basis for specialised adult motor behaviour. As with the postural, locomotor, and manual control skills from the first 2 yr of life, substantial information is available detailing the ages and stages at which children from 2 through 7 yr of age and older acquire basic locomotor skills (walking, running, jumping, and hopping) as well as nonlocomotor skills (such as throwing, catching, hitting, and kicking). In most cases information is available to outline developmental trends not only in performance (product measures) but also in the movement patterns used (process measures) in performing these skills. In this section we provide two examples of developmental stages in fundamental motor patterns, one

from the locomotor skill of running and the other from the nonlocomotor skill of overarm throwing.

AGES AND STAGES IN THE DEVELOPMENT OF A LOCOMOTION SKILL

A number of children begin to run at around age 18 mo and most run by the time they are 2 yr old. The running pattern of a child goes through a number of developmental stages before becoming mature, on average, by age 4 to 6 yr. Initial running is characterised by short, uneven steps, a wide base of support, and no easily observable flight phase (the phase during which neither foot is in contact with the ground). This first stage of running usually occurs before adult mastery of walking is achieved. In mature running, strides are uniformly long and there is an identifiable flight phase characterised by a full recovery of the nonsupport leg (see "Characteristics of Different Stages in the Development of Running"). The early attainment of a mature running pattern

---

CHARACTERISTICS OF DIFFERENT STAGES IN THE DEVELOPMENT OF RUNNING

A. Initial stage

1. Short, limited leg swing.
2. Stiff, uneven stride.
3. No observable flight phase.
4. Incomplete extension of support leg.
5. Stiff, short swing with varying degrees of elbow flexion.
6. Arms tend to swing outward horizontally.
7. Swinging leg rotates outward from hip.
8. Swinging foot toes outward.
9. Wide base of support.

B. Elementary stage

1. Increase in length of stride, arm swing, and speed.
2. Limited but observable flight phase.
3. More complete extension of support leg at takeoff.
4. Arm swing increases.
5. Horizontal arm swing is reduced on backswing.
6. Swinging foot crosses midline at height of recovery to rear.

C. Mature stage

1. Stride length at maximum; stride speed fast.
2. Definite flight phase.
3. Complete extension of support leg.
4. Recovery thigh is parallel to ground.
5. Arms swing vertically in opposition to legs.
6. Arms are bent at approximate right angles.
7. Minimal rotary action of recovery leg and foot.

Common problems

1. Inhibited or exaggerated arm swing
2. Arms crossing the midline of the body
3. Improper foot placement
4. Exaggerated forward trunk lean
5. Arms flopping at the sides or held out for balance
6. Twisting of the trunk
7. Poor rhythmical action
8. Landing flat-footed
9. Flipping the foot or lower leg in or out

Reprinted, by permission, from D.L. Gallahue, 1989, *Understanding motor development: Infants, children, adolescents*, 2nd ed. (New York: McGraw Hill Companies), 239. © The McGraw-Hill Companies.

is important given the central role running plays in many childhood games and activities. Running speed typically continues to improve throughout childhood and adolescence as strength is gained and subtle refinements in technique produce improvements in efficiency.

AGES AND STAGES IN THE DEVELOPMENT OF A NONLOCOMOTOR SKILL

The overarm throwing pattern is a discrete action used to propel objects and implements, such as balls and javelins. It is a basic movement pattern fundamental to many sport skills such as throwing a baseball or cricket ball, serving a tennis ball, spiking a volleyball, or passing a water polo ball. Overarm throwing performance (regardless of whether the throwing is for distance, accuracy, or form) continues to improve throughout childhood into adolescence. Even though rudimentary throwing patterns are in place by the second year of life, throwing velocity continues to improve steadily throughout the primary school years.

As with running and all other fundamental motor skills, the transition from initial throwing form to mature throwing patterns involves a number of stages. Initial attempts at throwing are characterised by an absence of backswing, weight transference, and upper- and especially lower-body involvement. At intermediate levels there is increased weight transference, trunk rotation, and the involvement of a forward step, albeit from the leg ipsilateral (on the same side) to the throwing arm. Only at the mature stage does the nonthrowing hand lead toward the target, the trunk rotate fully, and the lower body contribute to the action through a forward step of the leg contralateral (opposite) to the throwing arm (see "Characteristics of Different Stages in the Development of Overarm Throwing"). By 12 yr of age most boys have acquired the mature throwing pattern, whereas only a minority of girls have. In overarm throwing, unlike in any of the other fundamental motor skills that have been examined, there appear to be differences in skill performance between boys and girls that cannot be simply accounted for by boys having more practice.

CHARACTERISTICS OF DIFFERENT STAGES IN THE DEVELOPMENT OF OVERARM THROWING

A. Initial stage

1. Action is mainly from elbow.
2. Elbow of throwing arm remains in front of body; action resembles a push.
3. Fingers spread at release.
4. Follow-through is forward and downward.
5. Trunk remains perpendicular to target.
6. Little rotary action occurs during throw.
7. Body weight shifts slightly rearward to maintain balance.
8. Feet remain stationary.
9. Purposeless shifting of feet often occurs during preparation for throw.

B. Elementary stage

1. In preparation, arm is swung up, to the side, and back to a position of elbow flexion.
2. Ball is held behind head.
3. Arm is swung forward, high over shoulder.

4. Trunk rotates toward throwing side during preparatory action.
5. Shoulders rotate toward throwing side.
6. Trunk flexes forward with forward motion of arm.
7. A definite forward shift of body weight occurs.
8. Leg on same side as throwing arm steps forward.

C. Mature stage

1. Arm is swung backward in preparation.
2. Opposite elbow is raised for balance as a preparatory action in the throwing arm.
3. Throwing elbow moves forward horizontally as it extends.
4. Forearm rotates and thumb points down.
5. Trunk markedly rotates to throwing side during preparatory action.
6. Throwing shoulder drops slightly.

7. Definite rotation through hips, legs, spine, and shoulders occurs during throw.

8. Weight during preparatory movement is on rear foot.

9. As weight is shifted, the opposite foot takes a step.

Common problems

1. Forward movement of foot on same side as throwing arm

2. Inhibited backswing

3. Failure to rotate hips as throwing arm is brought forward

4. Failure to step out on leg opposite the throwing arm

5. Poor rhythmical coordination of arm movement with body movement

6. Inability to release ball at desired trajectory

7. Loss of balance while throwing

8. Upward rotation of arm

Based on Gallahue and Ozman, 1989.

PRACTICAL APPLICATIONS

The information that is available on the ages and stages in the development of fundamental motor skills is of potentially great value to teachers of motor skills, especially at the primary school level, for at least three reasons.

▶ First, such information may provide a reasonably objective method for monitoring the motor development of individual children and for detecting any potential movement problems (e.g., most test batteries for movement clumsiness either are based on or include fundamental motor skill items).

▶ Second, the information may provide the teacher with a guide to forthcoming progressions in movement-sequence development and therefore may provide a basis for accelerating acquisition of specific motor skills. For example, given the information provided in the section titled Characteristics of Different Stages in the Development of Overarm Throwing, it may be wise to emphasise an instruction such as "point at the target" to facilitate the transition from the intermediate to the mature stage of throwing. Equally, an instruction such as "step toward the target with your left leg" may be inappropriate for someone at the initial stage of the throwing skill because an intermediate stage involving stepping with the other (right) leg is typically involved as a transition from the initial to the mature pattern of throwing.

▶ Third, assessment of fundamental motor skills can provide an indication of the readiness of children for involvement in more structured activities such as sports in which performance is based around proficiency in one or more fundamental motor skills. There is growing concern in many countries about declining proficiency in fundamental motor skills among school-age children (see "Fundamental Motor Skill Proficiency in Australian Children").

FUNDAMENTAL MOTOR SKILL PROFICIENCY IN AUSTRALIAN CHILDREN

In Australia, as in many industrialised countries, there is growing concern about rising levels of childhood obesity as well as the possibility of a decline in fundamental motor skill proficiency among children. Claims have been made in both the published literature and in debate on public health policy that the latter may be contributing to the former. In one study (Booth et al., 1999), the fundamental motor skills of a sample of 5,518 schoolchildren in the Australian state of New South Wales were assessed. To ensure that the sample tested would be representative, the children tested were drawn from four grades (grades 4, 6, 8, and 10) and from some 90 schools varying in location, size, and structure. The average ages of the children tested in each grade were 9.3, 11.3, 13.3, and 15.3 yr, respectively. Each of the children performed

(continued)

Fundamental Motor Skill Proficiency in Australian Children *(continued)*

six motor skills (run, vertical jump, catch, overarm throw, forehand strike, and instep kick) that are fundamental to performance in a wide range of physical and sport activities. For each child and each skill, the researchers assessed whether a mature motor pattern had been achieved and whether the child had mastery or near mastery of the skill.

The pattern of findings was basically similar for all six skills, and the percentage of children who had achieved mastery of the fundamental movements by each grade level was alarmingly low. For instance, even by grade 10 (approximately 15-16 yr of age) less than half of the boys and less than a quarter of the girls sampled had mastery of overarm throwing. This is telling given that overarm throwing is a movement pattern that forms the foundation of successful participation in many organised sports, including baseball, softball, cricket, and volleyball. It is interesting in this context to note the high numbers of young children who drop out of organised sport because they do not perform basic skills well enough to experience regular (or, in many cases, any) success.

More recent studies have provided evidence of a link between childhood proficiency in object-control skills (kicking, throwing, and catching) and levels of participation in vigorous physical activity as an adolescent. In a study by Barnett and colleagues (2009), object-control proficiency as a child accounted for some 13% to 18% of the variance in adolescent participation in vigorous physical activity. Children who had object-control mastery had some 10% to 20% higher chance of participation in vigorous physical activity as an adolescent. A review of existing studies (Lubans et al., 2010) suggests a positive relationship between proficiency in fundamental motor skills and both physical activity participation and cardiorespiratory fitness. On the basis of these findings, arguments have been advanced for early movement-intervention programs in the preschool years and for increased physical education time and resources in the school curriculum.

Sources

Barnett, L.M., van Beurden, E., Morgan, P.J., Brooks, L.O., & Beard, J.R. (2009). Childhood motor proficiency as a predictor of adolescent physical activity. *Journal of Adolescent Health, 44*, 252-259.

Booth, M.L., Okely, T., McLellan, L., Phongsavan, P., Macaskill, P., Patterson, J., Wright, J., & Holland, B. (1999). Mastery of fundamental motor skills among New South Wales school students: Prevalence and sociodemographic distribution. *Journal of Science and Medicine in Sport, 2*, 93-105.

Hardy, L.L., King, L., Farrell, L., Macniven, R., & Howlett, S. (2010). Fundamental movement skills among Australian preschool children. *Journal of Science and Medicine in Sport, 13*, 503-508.

Lubans, D.R., Morgan, P.J., Cliff, D.P., Barnett, L.M., & Okely, A.D. (2010). Fundamental movement skills in children and adolescents: Review of associated health benefits. *Sports Medicine, 40*, 1019-1035.

Okely, A.D., Booth, M.L., & Patterson, J.W. (2001). Relationship of physical activity to fundamental movement skills in adolescents. *Medicine and Science in Sports and Exercise, 33*, 1899-1904.

The type of early movement and play opportunities experienced by a person, and the context in which these experiences occur, can have a remarkably enduring effect on their ongoing lifetime engagement in regular physical activity and on the success they are likely to achieve in the domain of competitive sport (see "Enduring Effects of Early Experience and Opportunity").

MOTOR PERFORMANCE IN THE ELDERLY

Although the peak performance of motor skills, at least as evidenced by sport performance such as world records, appears to be achieved typically from the late teens through the mid-30s, most movement skills, with regular practice, can be retained at a high level throughout most of adulthood. In older adulthood, however, some declines in motor performance in skills as fundamental as balance and locomotion can seriously impair the movement capabilities of the elderly. Some motor skill deterioration with age appears inevitable, although the extent of the decline can clearly be limited by regular practice.

CHANGES IN BALANCE AND POSTURE

Falls are a major cause of injury and loss of independent-living capability in the elderly. Falls may be particularly incapacitating for older women, as discussed in chapters 5 and 24, because of their susceptibility to osteoporosis and hip fractures.

ENDURING EFFECTS OF EARLY EXPERIENCE AND OPPORTUNITY

In an attempt to understand the factors that contribute to expert performance in sport, a number of researchers have begun doing retrospective analyses of the developmental experiences of outstanding athletes to see whether these differ in any systematic way from that of the general population. A number of surprising findings have emerged that point to the enduring effect that early experiences can have on either facilitating or constraining the chances for future success.

Across a wide range of sports, the time or month of the year in which a person is born has been shown to influence the probability of becoming an elite athlete as an adult. At the elite adult level in sports such as ice hockey and soccer, there is an over-representation of athletes born in the months of the year immediately after the date used for junior age-group cutoffs in the sport (e.g., athletes born in January, February, and March for sports that use a January 1 cutoff date for junior age-division determination). Simply because of where the cutoff date happens to fall, these individuals are relatively older (and likely more mature) than their peers during their initial experience of sport; this provides a favorable situation for both initial success and positive perceptions of ability in the activity. Initial success encourages continued effort, practice, and commitment, which in turn can foster further success. The converse is experienced by those individuals who are relatively younger. They remain under-represented in selected teams, even through to adult level, and are at greater risk of early drop-out from the sport.

In a similar fashion, the physical environment in which one grows up can have a bearing on the chances of acquiring the movement skills necessary for athletic excellence. Both boys and girls who are raised in a rural or small-town environment have an enhanced prospect of becoming an elite athlete in many sports compared with peers raised in large cities. In North America, the number of people born each year in cities with populations greater than 500,000 and less than 500,000 is approximately equal, yet in the professional sports of football, basketball, baseball, ice hockey, soccer, and golf athletes born and raised in areas with populations of less than 500,000 outnumber those born in larger cities almost 3 to 1. As with the relative-age effect, the opportunities created by the small-town environments during the developing years (easier access to safe places for practice; multisport experiences; and the enhanced opportunities to be the best in a school, club, or region) appear to establish benefits in terms of motor skill attainment that endure to adulthood.

Sources

Barnsley, R.H., & Thompson, A.H. (1988). Birthdate and success in minor hockey: The key to the NHL. *Canadian Journal of Behavioral Science, 20*, 167-176.

Cobley, S., Baker, J., Wattie, N., & McKenna, J. (2009). Annual age-grouping and athlete development: A meta-analytical review of relative age effects in sport. *Sports Medicine, 39*, 235-256.

Côté, J., MacDonald, D., Baker, J., & Abernethy, B. (2006). When "where" is more important than "when": Birthplace and birthdate effects on the achievement of sporting expertise. *Journal of Sports Sciences, 24*, 1065-1073.

MacDonald, D.J., Cheung, M., Côté, J., & Abernethy, B. (2009). Place but not date of birth influences the development and emergence of talent in the American football. *Journal of Applied Sport Psychology, 21*, 80-90.

Musch, J., & Grondin, S. (2001). Unequal competition as an impediment to personal development: A review of the relative age effect in sport. *Developmental Review, 21*, 147-167.

Because a major cause of falls is loss of balance, scientists who study ageing (gerontologists) have been particularly interested in studying balance and posture control in the elderly. Such studies have revealed an increase in postural sway as people get older (this effect is most pronounced in women) and slower and less expert recovery of unexpected losses of balance by older people. Poorer vision, increased use of medication, and changes in gait pattern that result in the foot being lifted less off the ground also contribute significantly to increased losses of balance and falls among older people.

CHANGES IN WALKING PATTERNS

There are a number of discernible changes in walking patterns that occur with ageing. In comparison with the gait of younger adults, older people typically lift their feet less, walk more slowly, have shorter stride lengths, have a reduced range of motion at the ankle, and have greater out-toeing in their foot placement.

The net purpose of all these changes is to attempt to increase stability, although this is inevitably to the detriment of mobility. Interestingly, with age, the motor patterns of the elderly regress in many aspects to characteristics seen in the immature gait of children. This is especially true of the out-toeing of the foot, which is a movement pattern used by both young children and the elderly to improve lateral stability.

CHANGES IN MORE COMPLEX MOTOR SKILLS

Like that of fundamental motor skills, the performance of acquired specialised motor skills also declines with age, although the extent of deterioration does not appear to be uniform across all types of skills. Motor tasks that have to be performed under time stress, require complex decision making, or involve psychological pressure (or high anxiety; see chapter 19) typically show the most marked deteriorations with ageing. Importantly, motor skills can still be learned and improved even in late adulthood, so extensive practice of new motor skills can more than offset many of the systematic effects of the ageing process.

PRACTICAL APPLICATIONS

Although it is clear that some decline in motor performance is inevitable with ageing, it is also clear that a number of very positive things can be done to offset these effects. Regular exercise (to offset losses in strength, flexibility, and endurance), regular practice (to consolidate and improve existing movement patterns), and the development and practice of "smart" strategies (such as pacing effort to conserve energy, anticipating to avoid reaction-time delays, and using encoding strategies to improve memory) can all assist in the retention of a high level of motor performance into senescence.

CHANGES AT THE NEUROPHYSIOLOGICAL LEVEL

Examination of the neurophysiological changes in the motor system that accompany development and ageing can offer some useful insight into the mechanisms underpinning the observable changes in motor performance across the life span. Knowing how the nervous system changes with age also provides a foundation for understanding age-related changes in the hardware available for processing that information necessary for the control of movement.

MAJOR PHYSICAL CHANGES IN THE CENTRAL NERVOUS SYSTEM

The growth and development of the nervous system is controlled by a complex interaction between genetic and environmental factors. Early development of the nervous system is primarily genetically regulated and involves in the **prenatal** (before birth) period the formation of nerve cells and in the **postnatal** (after birth) period the branching and insulation (myelination) of the dendrites and axons of these nerve cells. The critical period for the development of the nervous system is from conception through the end of the first year. During this period in particular, the structural development of the nervous system is vulnerable to environmental influences. For example, sedative drugs (barbiturates) in the blood stream of the mother prenatally, or malnutrition postnatally, can dramatically impair the normal development of the nervous system.

At birth the brain weighs some 300 to 350 g, or about 25% of its adult weight. It reaches half its adult weight by around 6 mo, 75% by 2.5 yr, and nearly 100% by 6 yr. Because all the neurons the central nervous system will ever possess are present at birth, the increased brain mass in the early years of life results from increases in size and branching within the neurons, myelination, and growth of the supporting glial cells. Rates of growth vary dramatically in different regions of the central nervous system. In the brain, the cerebral cortex is identifiable from about 8 wk after conception, and the two cerebral hemispheres are formed, but not functional, at birth. The spinal cord, although present, is small and quite short at birth and matures, through myelination, in a top-down (cephalocaudal) manner. The pyramidal tract is myelinated and functional by about 4 to 5 mo, coinciding with the first appearance of voluntary movement control in the infant.

The process of myelination involves surrounding the axons of the nerve cells with a fatty sheath (myelin) that acts to insulate the nerve and substantially increase its speed of impulse conductance, its capability for repetitive firing, and its resistance to fatigue. Myelination of the sensory and motor neurons begins 5 to 6 mo before birth and is completed within the first 6 mo postnatally. Higher centres of the central nervous system, especially the cerebellum and cortex, begin and complete myelination much later than subcortical regions of the system. Damage to the myelin sheath, as occurs with the disease multiple sclerosis, causes tremor, loss of coordination, and, on occasion, paralysis.

A progressive loss of nerve cells occurs throughout life (at a rate of some 10,000 per day) such that by age 65 to 70 yr some 20% of the total neurons present at birth are lost. Glial cells, on the other hand, increase with age. The net effect is a decrease in brain weight in the elderly. Brain activity during the performance of simple movements (as measured using techniques such as electroencephalography and functional magnetic resonance imaging) appears to be fundamentally different for older people (see "Brain Mechanisms for Movement in Older People"). There is also a general slowing of sensory and motor function with ageing, resulting from a reduced impulse (signal) strength relative to background neural activity (noise).

Changes in the Sensory Receptors and Sensory Systems

To begin to understand the changes in motor performance that occur across the life span, it is necessary to understand the developmental and age-related changes that occur in the basic sensory receptors and sensory systems that underpin movement control. Following the structure introduced in chapter 15, we will briefly discuss some of the major changes that occur in the visual, kinesthetic, and vestibular systems across the life span.

Visual System

The eye, like the brain, undergoes most of its growth before birth, even though the size of the eye at birth is only about half of its final size at maturity. The retina is fairly well developed at birth, and the myelination of the optic nerve has commenced at this time and is complete some 1 to 4 mo after birth.

BRAIN MECHANISMS FOR MOVEMENT IN OLDER PEOPLE

In the introduction to the section on the neural bases of human movement we noted that there are now a number of powerful techniques available that neuroscientists can use to examine brain mechanisms of movement control. In recent years a number of research groups have used these techniques to examine whether the brain mechanisms for movement control are any different for older adults compared with young adults.

Alexandra Sailer, Johannes Dichgans, and Christian Gerloff from Tuebingen University in Germany used electroencephalography (EEG) to record the electrical activity of the brain while older (55-76 yr) and younger (18-27 yr) people performed a simple motor task. They found that the electrical activity in different frequency bands in the EEG differed markedly between the two age groups; the older participants recruited greater activation of the primary sensorimotor and premotor regions of both cerebral hemispheres (see figure 15.12). A group of researchers from the National Institute of

Mental Health in the United States used functional magnetic resonance imaging (fMRI) to examine the activation of different regions of the brains of young (<35 yr) and older (>50 yr) people as they performed a reaction-time task. Consistent with the EEG findings of the Tuebingen group, the U.S. team discovered greater activation of a number of motor areas of the brain (contralateral sensorimotor cortex, premotor and supplementary motor areas, and ipsilateral cerebellum) in the older participants. Further, some areas of the brain that were not activated in the younger participants in completing this task (ipsilateral sensorimotor cortex, basal ganglia, and contralateral cerebellum) were activated in the older participants. Increased activation of the ipsilateral sensorimotor cortex in older participants has also been recorded in a number of other fMRI studies of motor tasks.

Collectively, these findings suggest that, even for the performance of simple movement tasks, additional areas of the brain of older participants

(continued)

Brain Mechanisms for Movement in Older People *(continued)*

are recruited. This is possibly an adaptive mechanism to compensate for some of the inevitable loss of neural function that occurs with ageing. There is some recent evidence, however, to indicate that older adults who have maintained high levels of physical activity may be less susceptible to the typical neural correlates of ageing.

Sources

Mattay, V.S., Fera, F., Tessitore, A., Hariri, A.R., Das, S., Callicott, J.H., & Weinberger, D.R. (2002). Neurophysiological correlates of age-related changes in human motor function. *Neurology, 58,* 630-635.

McGregor, K.M., Zlatar, Z., Kleim, E., Sudhyadhom, A., Bauer, A., Phan, S., Seeds, L., Ford, A., Manini, T.M., White, K.D., Kleim, J., & Crosson, B. (2011). Physical activity and neural correlates of aging: A combined TMS/fMRI study. *Behavioural Brain Research, 222,* 158-168.

Riecker, A., Gröschel, K., Ackermann, H., Steinbrink, C., Witte, O., & Kastrup, A. (2006). Functional significance of age-related differences in motor activation patterns. *NeuroImage, 32,* 1345-1354.

Sailer, A., Dichgans, J., & Gerloff, C. (2000). The influence of normal aging on the cortical processing of a simple motor task. *Neurology, 55,* 979-985.

The neural pathways to the visual cortex are functional at birth. **Visual acuity** (sharpness of vision) is poor at birth, and at the first month of life the acuity of human infants for stationary objects is about 5% of that of mature adults. (The infant of 1 mo sees the same degree of visual detail from a distance of 6 m that an adult would see from approximately 250 m!) Acuity improves rapidly during the first years of life. Adult levels are attained by about age 10 yr for the viewing of stationary objects and age 12 yr for the viewing of moving objects. This improved acuity results, to some degree, from an improved capability to adjust the shape of the eye's lens (called **accommodation**) but primarily from improved neuronal differentiation in the retina and connections in the cortex. The visual system matures at a rate sufficient to provide the visual information needed to guide movement at each stage of development.

With ageing there are a number of declines in visual function that can, in turn, adversely affect movement capability. By about age 40 yr there are clinically significant losses in the ability to accommodate to near objects. Material that can be read at 10 cm at age 20 yr must be progressively moved away to distances of 18 cm at 40 yr, 50 cm at 50 yr, and 100 cm at 70 yr of age to retain sharp focus. In addition to acuity losses with ageing, there is reduced sensitivity to glare, declining sensitivity to contrast, narrowing of the visual field size, and increased visual difficulty at low light levels. All of these changes make it increasingly difficult to perform movement tasks (such as driving or catching) that require precise visual information and the accurate judgement of the location and speed of moving objects.

KINESTHETIC AND VESTIBULAR SYSTEM

Because of their central role in many reflex systems essential for the survival of the newborn, the kinesthetic receptors develop early and are functional essentially from birth. Cutaneous receptors in the mouth are functional from as early as 7 to 8 wk after conception and those in the hand from 12 to 13 wk after conception. Muscle spindles are evident in muscles of the upper arm from 12 to 13 wk after conception, although the main development of the muscle spindles, as well as the Golgi tendon organs, joint receptors, and cutaneous receptors, occurs in the period 4 to 6 mo after conception (3-5 mo before birth). The vestibular apparatus is completely formed 2 to 3 mo after conception and may function reflexively from this point onward. The kinesthetic and vestibular systems are thus prepared early in life to support infant activity before the visual system matures.

Although relatively little is known about the effect of ageing on the kinesthetic and vestibular receptors, it is apparent that there are clear functional losses in the elderly with respect to balance, as detected through the vestibular system, and sensitivity to touch, vibration, temperature, and pain, as detected through the cutaneous receptors. It is known that some 40% of the vestibular receptors and nerve cells are lost by age 70. It is also known that the number of Meissner corpuscles in the skin decreases with age and that those that remain undergo changes in size and shape. In addition to a loss of receptors there is a decrease of as much as 30% in the number of sensory neurons innervating the peripheral receptors (a condition known as **peripheral neuropathy**).

CHANGES IN THE EFFECTORS (MUSCLES)

The growth and ageing of muscle is discussed in chapters 5 and 13. The number of muscle fibres increases prenatally and for a short period postnatally, approximately doubling between the last trimester of gestation and 4 mo after birth. Most fibres do not differentiate until the foetus is around 7 mo old. Slow-twitch (or Type I) fibres begin to appear 5 to 7 mo after conception, and these fibres constitute about 40% of all muscle fibres present at birth. Fast-twitch (or Type II) fibres make up about 45% of all fibres present at birth. Slow- and fast-twitch muscle fibres continue to increase in the first year of life, and the relative distribution of fibre types appears to reach steady state by the age of 3 yr.

As children mature, their muscle fibres become wider and longer. This comes about through an increase in both the number and length of the contractile units (the **myofibrils**) in each muscle fibre. Adult-sized myofibrils are attained in adolescence. The reduction of muscle size that occurs in the elderly appears to be a consequence of reduced effectiveness of the nerves that activate the muscles and the selective loss of fast-twitch muscle fibres and motor units primarily. Growing evidence indicates significant changes in the control properties of the motor units with ageing. For instance, discharge rates of motor units become more variable in older people; this contributes to a reduced capability to perform steady, submaximal muscle contractions.

CHANGES IN REFLEX SYSTEMS

Receptors and effectors, as discussed in chapter 15, are the key elements of reflex systems. The development, modification, and, frequently, extinction of reflexes have a complexity greater than that of the maturational changes of the sensory and effector components of the various reflex systems. Reflexes present at birth or soon thereafter can be broadly distinguished in terms of whether their principal function is to assist in the survival of the newborn or to lay the foundations for the development of voluntary movement control.

PRIMITIVE REFLEXES

The human newborn is extremely vulnerable because of its limited mobility and capacity for voluntary movement. Consequently, in the early stages of life, it must depend heavily on adult caretakers and some reflexes for survival and protection. The reflexes present at birth that function predominantly for protection and survival are referred to collectively as **primitive reflexes**. Examples of primitive reflexes are the sucking reflex, which enables the newborn to instinctively gain nutrition from the mother's breast, the searching or rooting reflex, which assists the newborn in locating the nipple, and the Moro reflex, which assists with initial respiration. The primitive reflexes, which dominate movement control at birth, typically weaken with advancing maturity to the point of being either completely inhibited or at least highly localised by 3 to 4 mo after birth. This coincides with the increased maturity of the cortex, suggesting a transition from involuntary control toward greater voluntary control. Persistence of reflexes for extended periods after their expected time of disappearance is used by paediatricians (doctors who specialise in the care of children) as a sign of neurological problems.

POSTURAL AND LOCOMOTOR REFLEXES

Postural reflexes, such as the body righting, neck righting, and parachute reflexes, serve the function of keeping the head upright and the body correctly oriented with respect to gravity. These reflexes are not present at birth but generally appear after 2 mo, around the time early stages of postural control are being established. They disappear after the first year of life and are progressively replaced by more voluntary movement control. The **locomotor reflexes**, such as the walking and swimming reflex, are present from birth or soon thereafter. These reflexes disappear after some 4 to 5 mo and before voluntary walking or swimming are attempted (figure 17.2). Although the exact role of the postural and locomotor reflexes in the development of future voluntary movement control is not entirely clear, it appears that these reflexes collectively play a role in preparing the nervous system and its pathways for the emergence of the voluntary fundamental motor skills discussed earlier in this chapter. Localised reflexes such as the stretch reflex described in chapter 15 persist throughout the life span and appear to alter relatively little in old age.

CHANGES IN INFORMATION-PROCESSING CAPABILITIES

Young people and elderly people both process information through the same basic stages of perception, decision making, and movement organisation and execution outlined in chapter 16. What changes with development, and with ageing, are the speed, efficiency, and sophistication with which information can be processed.

DEVELOPMENTAL IMPROVEMENTS IN INFORMATION-PROCESSING CAPABILITY

The processing of information for perceiving, deciding, and acting improves with maturation and then appears to undergo some decline in people who are elderly. These alterations in information processing are directly responsible for the parallel changes in observable motor performance described in the previous section.

PERCEPTION

At least three general principles can be observed in the development of perceptual skills. One principle is that the maturation of perceptual skills continues well after the sensory system and receptors have matured structurally. We noted earlier that in the visual system the optic nerve is fully myelinated and pathways from the eye to the visual cortex are functional by a few months after birth, yet visual acuity does not reach adult levels until around 10 yr of age. Similarly, the kinesthetic hardware for life is essentially complete at birth, yet improvements in the ability to make precise kinesthetic judgements continue at least until 8 yr of age and typically beyond.

A second principle is that the more complex the perceptual judgement that has to be made, the longer it takes the developing child to reach adult levels of performance. In the visual system, acuity for moving objects matures after acuity for stationary objects. Similarly, relatively simple visual–perceptual judgements (e.g., those involved in comparing object size or depth, differentiating an object from its background, or distinguishing a whole image from its component parts) reach maturity within the first decade of life, whereas more complex judgements (e.g., those involved in anticipating the direction of an opponent's stroke in racquet sports) may continue to improve well into the third decade of life. As with vision, kinesthetic judgments of complex movement patterns mature much later than do simple acuity judgments.

A third principle relates to the integration of information between the different sensory systems. Perhaps surprisingly, the integration of visual and kinesthetic information does not follow the individual maturation of the visual and kinesthetic systems but rather occurs simultaneously with it. For simple tasks, such as shape or pattern recognition, in which the child is permitted active kinesthetic exploration of the shape, the integration of visual and kinesthetic information is mature by around 8 yr of age. By this age, children exploring an object through touch alone are then able to identify the object visually from among other possible shapes or objects. More complex visual–kinesthetic integrations (e.g., those that occur in hitting and catching tasks, where movements must be initiated and guided kinesthetically to coincide with the arrival of a ball perceived visually) take longer to mature, although the attainment of adult levels of performance do not lag behind the separate maturation of the visual and kinesthetic systems.

DECISION MAKING

Reaction time to a single, unanticipated stimulus (simple reaction time) decreases rapidly (i.e., gets faster) until the mid- to late teens, whereupon adult levels are attained (figure 17.4). Being more complex, reaction time for tasks involving decision making (choice reaction time) takes somewhat longer to mature. Children are slower than adults in making decisions for a number of reasons in addition to their slower simple reaction times. Compared with both older children and adults, younger children have a slower information-processing rate (as shown by a steeper slope in figure 17.5), and they frequently process more information in selecting any particular response. This occurs because of their relatively poor ability to separate relevant information from irrelevant information. It is not surprising, therefore, that children's performance in tasks that require rapid decision making show rapid and continued improvements with practice into the mid-20s and beyond.

ORGANISING AND EXECUTING MOVEMENT

Like simple reaction time, the time taken to make simple movements varies across the life span (figure 17.4). Movement time for simple actions reaches its minimum around the mid-teens and remains at that

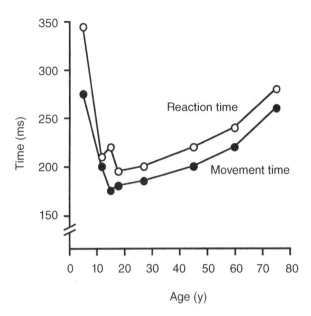

Figure 17.4 Changes in simple reaction time and movement time across the life span.

Data from Hodgkins, 1962.

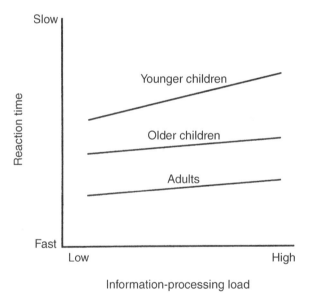

Figure 17.5 Choice reaction time as a function of information-processing load and age.

Reprinted from J. Keogh and D. Sugden, 1985, A general representation of increases in reaction time as information load. In *Movement skill development* (Macmillan Publishing Company), 337. By permission of D. Sugden.

level, typically, until the mid-30s. More complex movements (movements with greater programming or control requirements) are made more slowly by young children than by adults, although it is not clear whether this is due to a slower information-processing rate for movement control for children.

Compared with adults, children show a reduced ability to perform one or more tasks concurrently with a movement task, suggesting that movement is controlled more automatically by adults. A large part (perhaps all) of the information-processing capacity of children is needed for the control of even apparently simple movements, leaving little free capacity to allocate to the performance of other simultaneous tasks.

DECLINES IN INFORMATION PROCESSING WITH AGEING

Much remains to be learned about the changes in information-processing capability that accompany ageing. What is most apparent is the systematic decline in both the speed of reacting and the speed of moving that appears to occur from about the 30s onward (figure 17.4). This reduced speed of responding becomes pronounced in the elderly and is apparent in a wide range of tasks, from walking to driving a car or typing. The loss of speed in responding is more pronounced in complex tasks than in simple tasks. For example, longitudinal studies reveal that slowing with age is much more pronounced for choice reaction time than for simple reaction time; this appears to be especially true for women.

The phenomenon of slowing with age follows a "last in, first out" rule in that simple actions (such as reflexive movements) acquired early in life are more resistant to slowing and loss than complex, coordinated skills learned voluntarily later in life. Although the declines in reaction and movement time with ageing appear to be somewhat inevitable, the rate of decline may be slowed considerably by regular physical activity.

The slowing of simple reaction time with age appears to primarily reflect the neurophysiological changes in the central nervous system, especially nerve cell loss and changes in some of the sensory receptors. In tasks of choice reaction time, slowing reflects a lowered signal:noise ratio in the central nervous system plus a cautious, strategic change toward emphasising accuracy to the detriment of speed. The slowing of movement time with ageing reflects similar central rather than peripheral factors. With ageing comes a reduction in the number of functional neuromuscular units and, from age 20 to 60 yr, a 15% to 35% increase in the amount of neural stimulation needed to excite a muscle to contraction. These changes, along with a loss of nerve cells, effectively lower the signal:noise ratio, slowing the rate at which movements can be initiated and completed. Because new movement skills may possibly be more difficult to learn for older adults, an important preparation for later adulthood is the learning of skills early in life that can be maintained with relative ease through ongoing practice throughout the life span. Chapter 18 examines how skills are learned and acquired through practice.

SUMMARY

Throughout the life span there are continual changes in the capacity to perform various motor skills. In the first 2 yr of life, development occurs in a head-down (cephalocaudal) and centre-out (proximodistal) manner. The emergence of essential postural, locomotor, and manual manipulative skills follows a similar sequence and general time scale for most children. Fundamental movement patterns (for running, jumping, throwing, catching, and so on) develop throughout childhood in a predictable, sequential manner as voluntary control of movement progressively overrides the reflex control that is dominant in the early years of life. Motor performance declines in older people due to a combination of ageing effects and, in many cases, a reduction in activity or training levels. These changes in motor performance that occur across the life span are a consequence of age-related changes in both the underlying functional anatomy and information-processing capacity of the neuromuscular system.

FURTHER READING

Cech, D.J., & Martin, S.T. (2012). *Functional motor development across the lifespan* (3rd ed.). St. Louis: Saunders.

Gallahue, D.L., & Ozmun, J.C. (2006). *Understanding motor development: Infants, children, adolescents, adults* (6th ed.). Boston: McGraw-Hill.

Haibach, P., Reid, G., & Collier, D. (2011). *Motor learning and development*. Champaign, IL: Human Kinetics.

Haywood, K.M., & Getchell, N. (2009). *Life span motor development* (5th ed.). Champaign, IL: Human Kinetics.

Keogh, J.F., & Sugden, D.A. (1985). *Movement skill development*. New York: Macmillan.

Malina, R., Bouchard, C., & Bar-Or, O. (2004). *Growth, maturation, and physical activity* (2nd ed.). Champaign, IL: Human Kinetics.

Piek, J. (2006). *Infant motor development*. Champaign, IL: Human Kinetics.

Shumway-Cook, A., & Woollacott, M.H. (2007). *Motor control: Translating research into clinical practice*. Philadelphia: Lippincott Williams & Wilkins.

Spirduso, W.W., Francis, K., & MacRae, P. (2005). *Physical dimensions of aging* (2nd ed.). Champaign, IL: Human Kinetics.

CHAPTER 18

MOTOR CONTROL ADAPTATIONS TO TRAINING

The major learning concepts in this chapter relate to

▸ training-related changes in observable motor performance, including stages in the acquisition of motor skills and characteristics of skilled performers;

▸ training-related changes in motor control at the neural level;

▸ accounts of learning and plasticity from a neuroscience perspective;

▸ training-related changes in information-processing capabilities, including changes in sensory reception, perception, decision making, movement organisation, and execution;

▸ accounts of skill learning from a cognitive science perspective; and

▸ factors affecting the learning of motor skills, including the quantity and type of practice, instruction, and feedback.

In this chapter we use the basic neuroscience and cognitive science concepts of motor control introduced in chapters 15 and 16 to examine the changes in motor control that occur as a consequence of practice or training. The specialised field of study concerned with the description and explanation of changes in motor performance and motor control that occur with practice is typically referred to as **motor learning** (or, occasionally, skill acquisition).

Learning is the change in the underlying control processes that is responsible for the relatively permanent improvements in performance that accompany practice. Like many aspects of the motor control field, learning is difficult to study because it is a process that cannot be directly observed or measured. Scientists attempting to understand motor learning are dependent on making accurate inferences about learning from observable (and measurable) changes in performance. However, because it is possible to acquire significant skill without necessarily improving performance, changes in the underlying control processes will not always be directly or faithfully reflected in observable performance. The difficulty scientists have in accurately measuring learning is shared with practitioners (e.g., physical educators, music educators, and coaches) charged with the responsibility of attempting to objectively measure skill learning.

Studies of motor learning have generated knowledge about many things, including

- the characteristics of expert performers,
- the effectiveness of different types and schedules of feedback for skill learning,
- the relative merits of different types of practice,
- the transfer of skill from one practice setting to another,
- the retention of skills over time, and
- the relearning of skills following traumas (e.g., joint injury) or damage to the central nervous system (e.g., that which occurs with stroke).

This knowledge is clearly important practically for any profession involved with the assessment, measurement, and improvement of motor skills.

Knowledge about motor learning comes from both cross-sectional and longitudinal studies. In cross-sectional studies, the motor control and performance of people with different levels or types of practice are compared, whereas in longitudinal studies the motor control and performance of the same set of people are examined on a number of occasions throughout the learning or retention of a particular skill. Comparisons of skilled (expert) and less-skilled (novice) performers are examples of the first type of study, whereas training studies are typical of the second type of study. Motor learning can be described and explained on a number of levels. This chapter examines the changes that take place with practice at the level of observable motor performance, at the neurophysiological level, and at the level of information-processing capabilities.

CHANGES IN OBSERVABLE MOTOR PERFORMANCE

Describing the observable characteristics of expert performance and the observable changes in motor performance that occur as a new skill is acquired is a useful starting point for the systematic study of motor learning. In line with the typical scientific approach, a full and detailed description of the phenomenon of interest (in this case the changes in the performance of movement skills with practice) is an essential prerequisite to any attempt at explanation and understanding of the underlying causes of the phenomenon (in this case how learning occurs).

CHARACTERISTICS OF SKILLED PERFORMERS

Skilled performance is the learned ability to achieve a desired outcome with maximum certainty and efficiency, that is, the ability to attain a desired result on a particular task with a minimum outlay of time, energy, or both. Comparing the observable characteristics of expert performers with those of people less skilled on the same task typically reveals a number of differences. Experts, in contrast to the less skilled, are frequently characterised as

- having all the time in the world,
- picking the right options,
- reading the situation well,
- being adaptable,
- moving in a smooth and easy manner, and
- doing things automatically.

Skilled performers are apparently able to balance well the conflicting needs to be

- fast yet accurate,
- consistent yet adaptable, and
- maximally effective while expending minimum attention and effort.

Knowing how skilled performers are able to achieve this balance requires an understanding of the changes in the underlying neurophysiology and information processing that occur with practice.

STAGES IN THE ACQUISITION OF MOTOR SKILLS

Proceeding from being a novice performer to being an expert requires extensive practice over a long period of time. As a very rough rule of thumb, the acquisition of expertise in motor skills, like expertise in cognitive skills such as playing chess, appears to typically take at least 10 yr, 10,000 h of practice, or literally millions of trials of practice. Although skill acquisition is a continuous process, learners pass through at least three identifiable stages during the long transition from novice to expert.

STAGE 1: VERBAL–COGNITIVE PHASE

In the **verbal–cognitive phase**, the movement task to be learned is completely new to the person. Consequently, the learner is preoccupied at this stage with trying to understand the requirements of the task, especially what needs to be done in order to perform the skill successfully. All of the learner's limited information-processing capacity is directed to such issues as where to position the whole body and limbs, where best to gain ongoing feedback about performance, or how the correct movement feels. Consequently, the major activity at this stage of performance is thinking of and planning movement strategies (i.e., cognition). Significant benefits can be gained from good (verbal) instruction and especially demonstrations.

The movements used to make initial attempts at the new task are typically pieces of movement patterns from existing skills that are joined together to meet the challenges of the new task. In other words, old habits are reshaped into new patterns. Instruction that highlights the similarities (and the differences) between the new skill and skills already learned may therefore be beneficial in speeding up the rate at which the new skill is learned. Performance fluctuates dramatically in the early stage of learning new skills as a wide range of movement strategies are tried and many are discarded.

STAGE 2: ASSOCIATIVE PHASE

In the **associative phase** of learning, performance is much more consistent as the learner settles on a single strategy or approach to the task. Learners consequently spend the majority of their time and effort fine-tuning the selected movement pattern rather than constantly switching from one movement pattern to another. Because the basic knowledge of the requirements of the task is established, in the associative phase learners become better able to both produce the movement pattern they had planned and adjust to changes in the conditions in which the movement is to be performed. These developments ensure increased levels of task success.

In contrast to the verbal–cognitive phase, in which the instruction provided by others may be the single most beneficial thing for skill acquisition, in the associative phase there is no substitute for specific practice of the task itself. Progressive increases in task complexity (e.g., adding more complex rhythmic patterns for a pianist, more difficult terrain or traffic conditions for a car driver, or more opponents or time constraints for a basketball player) provide a valuable means of fostering the systematic continuation of skill development.

STAGE 3: AUTONOMOUS PHASE

With sufficient practice of a particular skill, some learners reach the third stage of learning—the **autonomous phase**, so named because at this stage the performance of the skill appears to be largely automatic. The movement apparently can be controlled without the person having to pay attention to it. At this stage of learning, movements are consistently performed with such precision and accuracy that there is no longer a need to constantly monitor feedback to ensure that the movement is correctly performed. Open-loop control therefore largely replaces closed-loop control, and skilled performers have spare attention that can be allocated to other tasks. Consequently, one sign of the expert performer, as discussed later, is the ability to do two or more things at once. For example, typists in the autonomous phase of learning are able to conduct sensible telephone conversations with minimal interference to their concurrent typing speed or accuracy.

The only major drawback with reaching the autonomous stage of learning is that performers at this stage find it difficult, if not impossible, to change their movement pattern if, for example, an error in technique becomes ingrained. Experts also often cannot verbally describe how they perform the skilled movements they execute; in the autonomous phase, movement control operates below the level of consciousness. Even at the autonomous phase of learning, there exists room for improvement in both motor performance and control. As discussed later, there is no reason that learning of movement skills cannot occur continuously even after the learner has reached the autonomous phase.

SPECIFICITY OF MOTOR SKILL

Skill acquisition is highly specific, and transfer of training from one motor skill to another appears to be quite limited. Level of performance on any one particular motor skill is typically of limited use in predicting the rate of learning or the ultimate level of performance of an individual on any other motor skill. An individual must therefore pass through the verbal–cognitive, associative, and autonomous stages of learning for each skill for which he or she wishes to become highly proficient.

CHANGES AT THE NEUROPHYSIOLOGICAL LEVEL

Learning, as noted earlier, is reflected in a relatively permanent change in observable performance. This change must be underpinned by biological changes at the level of the nervous system. Consequently, extensive efforts have been made by neuroscientists in searching for neural correlates of learning and memory.

CHALLENGES FOR A NEUROPHYSIOLOGICAL ACCOUNT OF LEARNING

Understanding the physiology of learning and memory is perhaps the ultimate challenge to researchers in the neurosciences. The task is difficult for a number of reasons. First and foremost, both learning (and memory) and the brain (and the rest of the central nervous system) are incredibly complex, so matching the complex phenomena of learning and remembering with the complex structure of the brain and central nervous system is always going to be an incredibly difficult, if not impossible, task. Any attempt to locate neural changes underpinning learning is complicated by the possibility that learning a particular motor skill may be associated with structural and functional changes in only a few of the brain's many billions of neurons or, more likely, associated with relatively subtle changes in the relationships between a large number of neurons distributed at diverse locations throughout the brain. The neural changes that underpin learning may take place over very short or quite lengthy time periods, and techniques are only now being developed that may be sufficiently sensitive to record some of the many subtle neural changes that accompany the acquisition of skill.

Given the complexities in both the learning phenomena and the structure of the brain and the technical difficulties in observing and recording possible neural events associated with learning, most existing neurophysiological studies of learning have, by necessity, used animals (other than humans) and very simple learning tasks (rather than the complex ones that are typical of human movement). Many types of learning exist, yet, to date, the majority of learning studies from a neurophysiological perspective have focused on simple stimulus-response learning or conditioning. Such studies have typically sought to find a neural mechanism for memory because memory for events and experiences past is seen as fundamental for learning. Learning would be impossible if the organism is not changed by past experience so that the experience can influence future behaviour. Although the inferences that can be made about the neurophysiological bases of human motor learning at this stage are only very preliminary and somewhat speculative, it does appear that the search for a single neural location for memory, in particular, is likely to be fruitless. Increasingly, it appears that memory is a phenomenon that is distributed throughout networks of neurons and not confined to a single site.

PLASTICITY AS THE BASIS FOR LEARNING

Although psychologists frequently conceptualise memories as records stored in filing cabinets or other similar archives, this is not the way experience is retained in the nervous system. Experience is not stored literally. Rather, it influences the way we perceive, decide, and organise and execute movement by physically modifying the neural pathways and circuits responsible for perceiving, deciding, and acting. Remembering and learning are therefore possible only because the nervous system is highly plastic; that is, it is able to dynamically alter its structure to accommodate the new functions it is required to perform. **Plasticity** exists on a continuum from short-term changes in synaptic function to long-term structural changes in nerve connections and the organisation of neural networks.

Whereas the nervous system retains some degree of plasticity throughout life, the brain is especially plastic during the early years and during the critical periods for skill acquisition described in chapter 17. If, for any reason, a region of the brain is injured or damaged during the first few years of life, other regions of the brain are often able to take over the functions normally performed by the injured region. The level of plasticity is reduced later in life, making full recovery of function following traumas, such as stroke (which damages the brain, commonly resulting in deficits in motor skills such as speech and gait), difficult in the elderly. Significant neurological reorganisation in adults is nevertheless possible.

Injury to the central nervous system can cause complete loss of some neurons, disruption to some of the axonal projections of others, and denervation of other neurons innervated by the injured neuron. Full or partial recovery of function is achieved through a number of means, both local and global. Locally, postsynaptic neurons connected to the damaged neuron(s) may become hypersensitive (and hence easier to activate), previously silent synapses may become functional, injured axons may regenerate, and collateral axons may sprout from neighbouring nerve cells to create alternatives to the damaged connections. More global means include the reorganisation of the connections between the motor cortex and the affected limb or region and increased contributions of parallel nerve pathways on the other side of the body.

As with the response to injury, the nervous system may undergo a number of structural and functional changes as either a cause or a consequence of learning. Learning may result in a change in the response characteristics of neurons, the establishment of new synaptic connections (as well as the atrophy and eventual disappearance of old, unused connections), and changes in the response characteristics of synapses. All of these changes are interrelated and contribute to synaptic plasticity. Collectively, changes in the nature of synaptic structure and function across one or more synapses modify neural circuitry that, in turn, can modify perceiving, deciding, and acting.

These neural changes are underpinned by biochemical and molecular changes. For example, the reinforcement or repetition of successful responses is associated with increased receptivity to the chemical neurotransmitter dopamine. (We saw in chapter 15 that this chemical is important in transmitting messages in the basal ganglia, among other places, and that deficits in this chemical in specific regions of the basal ganglia appear to be responsible for Parkinson's disease.) Recent evidence also suggests that long-term memory may require the synthesis and growth of new proteins, whereas short-term memory may not.

SYNAPTIC PLASTICITY AND LONG-TERM POTENTIATION

It has long been recognised that synaptic changes must be fundamental to learning and memory. In the 1940s the Canadian neuropsychologist Donald Hebb proposed that the strengthening of synapses provides the neural foundation for learning and that a synapse is strengthened by simultaneous activity in both pre- and postsynaptic neurons. More specifically, Hebb hypothesised that if a synapse repeatedly becomes active at around the same time that the postsynaptic neuron fires, structural or chemical changes will take place in the synapse to strengthen it. Techniques were not available to experimentally test these proposals at the time Hebb proposed these effects.

More recent neurophysiological studies have demonstrated one mechanism through which the type of synaptic strengthening hypothesised by Hebb could be achieved. When intense, high-frequency electrical stimulation is applied to afferent neurons in some regions of the brain, a subsequent increase in the magnitude of the neural response in this region of the brain is frequently seen when a standard (test) stimulus of known electrical strength is subsequently given. This increased response may last for up to several months and is referred to as **long-term potentiation**.

Long-term potentiation provides a measure of increased synaptic efficiency in that a test stimulus of the same magnitude elicits an increased response. The exact cellular and biochemical mechanisms underpinning long-term potentiation are still not clear but are thought to involve the combined effects of depolarisation of the postsynaptic membrane and the activation of unique receptors sensitive to a neurotransmitter called glutamate. Long-term potentiation has been observed primarily in one specific area of the brain (the **hippocampus**) but has also been observed at other sites. Because disruption to long-term potentiation impairs learning of the type usually attributed to the hippocampus, it seems reasonable to conclude that improved synaptic efficiency through long-term potentiation is an important neural foundation for skill acquisition.

Learning appears to be associated with long-term changes in both synaptic efficiency and synaptic connections. Recent studies, using a procedure known as **transcranial magnetic stimulation**, have demonstrated significant changes in the topographical organisation of the motor cortex and surrounding regions (see figure 15.14) as the learning of skills progresses from a cognitive stage, in which conscious control is exerted over movement, to a more automatic stage, where movement is controlled without explicit knowledge.

CHANGES IN INFORMATION-PROCESSING CAPABILITIES

Comparative studies of expert and novice performers on different motor skills reveal a number of changes in information-processing capability with the acquisition of skill. Changes are apparent in each of the central processing stages of perception, decision making, and movement organisation and execution described in chapter 16 as well as in the observable movement patterns and outcomes produced (see figure 16.3). In all cases, expert–novice differences are systematically present only for the processing of information specific to the motor task for which expertise has been developed.

SENSORY RECEPTION

The sensitivity of the key sensory systems for movement—the visual, kinesthetic, and vestibular systems—appears to change relatively little as a consequence of learning, practising, and improving specific motor skills. Although it may be possible to improve characteristics such as visual or kinesthetic acuity by specifically training these attributes, as a general rule, sensitivity to the sensory information needed to control movement is not the limiting factor in motor performance. Perhaps surprisingly, expert performers from a range of motor skills are not characterised by above-average levels of visual acuity or kinesthetic sensitivity, at least when these characteristics are measured using standardised tests (figure 18.1). What appears to be more related to skilled performance is how the basic information provided by the various sensory receptors is subsequently processed, interpreted, and used by the central nervous system.

PERCEPTION

With practice and the acquisition of expertise, a number of systematic changes take place in the way performers perceive environmental and internal

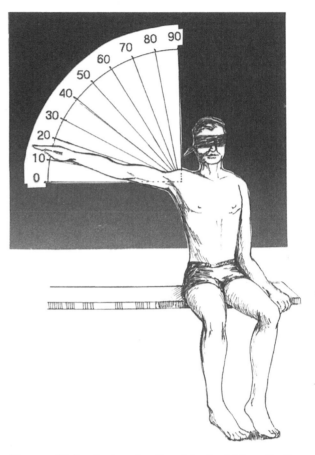

Figure 18.1 A standardised test of kinesthetic acuity. In this test acuity is determined by the subject's accuracy in reproducing the joint angle of a previously experienced movement.

events. Among other things, experts are superior to novices in recognising patterns and predicting (anticipating) forthcoming events. The experts' superiority is not general, however, and holds only for patterns and events drawn specifically from the experts' domain of expertise. Expert pianists, for example, are better at recognising patterns in pieces of music than are novice pianists but are typically no better than novices at recognising patterns present in skills other than music.

The improved pattern recognition that accompanies skill acquisition is especially evident in team sports, such as hockey, basketball, volleyball, and football, where selecting the correct movement response is dependent on being able to quickly and accurately recognise the defensive and offensive patterns of play of the opposing team. In a number of studies, photographs were taken of an opposing team during an actual game. These images were shown to the study participants for brief periods

Figure 18.2 A typical structured-pattern recognition situation from the sport of field hockey. Displays such as this are typically shown to players of different skill levels for durations of about 5 s, and then the players are required to remember the position of as many players in the display as possible.

Photo copyright by the National Sports Information Centre, Australian Sports Commission. Reprinted with permission.

(usually 5 s). Participants were then asked to recall the position of all the players (both attacking and defensive) shown in the photograph. When the images that were shown depicted structured patterns of play (figure 18.2), the recall of expert team-sport players was superior to that of less-skilled players. This perceptual advantage disappeared when the players shown in the photographs were in random position (i.e., no familiar pattern was present). This systematic finding, illustrated in figure 18.3, demonstrates that the experts' superior perception is due to skill-specific experience and is not a consequence of expert performers having generically superior perceptual skills that hold across all situations. To use the computer analogy introduced in chapter 16, the main change with the acquisition of expertise on a task is improvement in the specific programs (or software) used to process the input information provided by the sensory receptors.

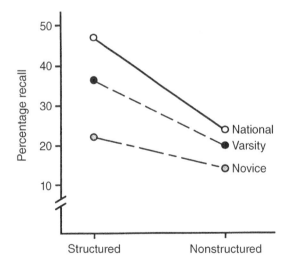

Figure 18.3 Pattern recall accuracy of field hockey players of different skill levels for displays with and without structure.

Reprinted, by permission, from J.L. Starkes, 1987, "Skill in field hockey: The nature of the cognitive advantage," *Journal of Sport and Exercise Psychology* 9(2): 152.

Many motor skills, especially skills such as those involved in fast ball sports, involve substantial time constraints that pose major demands on human information-processing capabilities. However, as noted earlier, one of the distinguishing characteristics of expert performers is their ability to give the impression of having all the time in the world. One important means by which expert performers overcome the time constraints placed on them is anticipating likely events from any information available to them. Experts are better able than less-skilled players to predict actions in advance based on information from the actions and postures of their opponents. For example, compared with novice players, expert badminton players are able to make more accurate predictions of where an opponent's stroke will land from information available before the opponent strikes the shuttlecock (figure 18.4).

Experts are also able to pick up information from cues other than those used by novices. For example, badminton experts are able to pick up advance information from the motion of the racket and the arm holding it, whereas novices can use only the racket as a cue (figure 18.5). Recent studies show that the expert advantage in perception that is evident from behavioural measures is also evident at the neural level (see "Brain-Function Differences Between Experts and Novices"). Expert–novice differences in the ability to anticipate occur even when the two skill groups may be looking at the same features of their opponent. This demonstrates that the limiting factor in the perceptual performance of untrained people is not the ability to pick up the necessary sensory information but rather the ability to interpret, understand, and use it to guide decision making and movement execution.

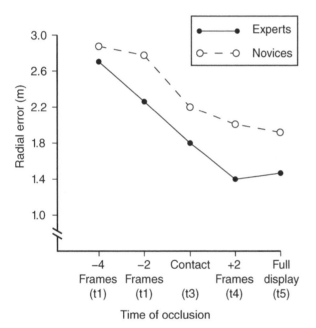

Figure 18.4 Error in predicting the landing position of a badminton stroke made from watching a film of an opposing player. Error is shown as a function of both skill level of the participants and when the display was occluded. Time of occlusion is expressed in relation to the contact of the opponent's racket with the shuttle (each frame is approximately 40 ms).

Reprinted, by permission, from B. Abernethy and D.G. Russell, 1987, "Expert-novice differences in an applied selective attention task," *Journal of Sport and Exercise Psychology* 9: 331.

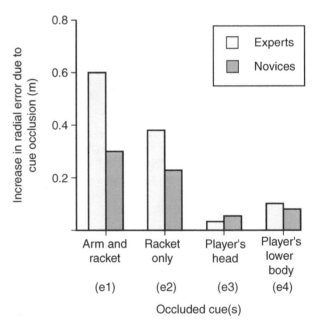

Figure 18.5 Error in predicting the landing position of a badminton stroke when visibility is occluded to different regions of the opposing player's body. An increase in error indicates that the region occluded contains cues that are of use in anticipating the stroke direction and force.

Reprinted, by permission, from B. Abernethy and D.G. Russell, 1987, "Expert-novice differences in an applied selective attention task," *Journal of Sport and Exercise Psychology* 9: 338.

BRAIN-FUNCTION DIFFERENCES BETWEEN EXPERTS AND NOVICES

Over the past decade, improvements in both the quality and the accessibility of imaging techniques such as functional magnetic resonance (fMRI) have made it possible for movement scientists to measure the brain activity of experts and novices as they undertake perceptual tasks that require anticipation. Regions of the brain that are active during a particular task require increased blood flow. With fMRI, it is possible to record changes in blood flow to different regions of the brain with a high degree of spatial resolution.

In an early study from Brunel University, psychologist Michael Wright and sport scientist Robin Jackson collaborated to examine which regions of the brain are active when novice tennis players are required to predict the direction of an opposing player's serve when this is presented via a video (occluded at the point of racquet–ball contact). Wright and Jackson found that areas in the parietal lobe and frontal cortex were particularly active (see figure 15.12 for reference). These areas form part of a network known as the mirror neuron network, which is a set of connected structures in the brain that appears to be equally activated regardless of whether a person is perceiving a particular movement or producing the same movement themselves.

In a subsequent study in which both expert and novice racquet sport players were examined, it was found that although activation was evident in the mirror neurons system for both expert and novice players, the level of activation was greater for the more skilled players. This finding was consistent with the findings of a study of professional and novice dancers viewing videos of different ballet moves. In this study, experts showed greater activation of the mirror neuron network when viewing movements they had been trained to perform than when viewing other movements. Collectively, these findings suggest that the human brain may attempt to perceive, predict, and understand movement by simulating the production of the same movement and that experts may be able to anticipate movements from their domain of expertise better than novices due to their own experience in producing the same movements. More research is needed to examine this notion further.

Sources

Calvo-Merino, B., Glaser, D.E., Grèzes, J., Passingham, R.E., & Haggard, P. (2005). Action observation and acquired motor skills: An fMRI study with expert dancers. *Cerebral Cortex, 15,* 1243-1249.

Wright, M.J., Bishop, D.T., Jackson, R.C., & Abernethy, B. (2010). Functional MRI reveals expert-novice differences during sport-related anticipation. *NeuroReport, 21,* 94-98.

Wright, M.J., & Jackson, R.C. (2007). Brain regions concerned with perceptual skills in tennis: An fMRI study. *International Journal of Psychophysiology, 63,* 214-220.

DECISION MAKING

A capability to make decisions both quickly and accurately is clearly beneficial for the performance of many motor skills. As discussed in chapter 16, a faster choice reaction time on any particular occasion may be achieved with practice in one of two ways: facilitating the overall decision-making rate (decreasing the slope of figure 16.5) or decreasing the total amount of information that has to be processed before making a decision. Choice reaction-time tasks that use stimuli and responses that are skill specific provide some support for the proposition that expert performers have faster decision-making rates than novice performers. Nevertheless, the more potent strategy used by skilled performers to decrease decision-making time is to reduce the absolute amount of information that they have to process. Expert performers do this by using knowledge, acquired through experience, of options and event sequences that are either not possible or improbable. Experts are more accurate than novices in predicting the probability of particular events occurring. This knowledge is invaluable in reducing the amount of information to be processed and, in turn, allowing decisions to be made more rapidly (see "Response Speed in Soccer Players").

RESPONSE SPEED IN SOCCER PLAYERS

In order to examine the decision-making ability of soccer players of different skill levels, Werner Helsen and J.M. Pauwels of Leuven University in Belgium developed an experimental setting in which players were required to respond to life-size dynamic simulations of situations drawn from typical soccer games. The players would view a scenario from a soccer match unfold on a large screen in front of them and then, at a critical moment, the ball would be played by one of the attackers (shown on the screen) in the direction of the player (located in the laboratory). The viewer's task was to move as quickly and accurately as possible to execute the movement that was most appropriate for the situation. The options that were available to the player were shooting a goal, dribbling around the goalkeeper or defender, or passing to a free teammate. Time taken to respond and the accuracy of the responses were recorded. Where the players were looking was also measured using an eye-movement recorder.

It was found that the expert soccer players had total response times that were, on average, nearly 260 ms faster than those of nonexperts and that experts were more accurate in their responses. Experts chose the correct option 92% of the time, compared with 82% by the nonexperts. These differences occurred in the absence of any major differences in the eye-movement patterns of the different skill groups. These general observations have since been replicated by Mark Williams and colleagues from Liverpool John Moores University on a number of subgame situations (e.g., 3v3; 1v1).

Sources

Helsen, W., & Pauwels, J.M. (1993). The relationship between expertise and visual information processing in sport. In J.L. Starkes & F. Allard (Eds.), *Cognitive issues in motor expertise* (pp. 109-134). Amsterdam: North-Holland.

Williams, A.M., & Davids, K. (1998). Visual search strategy, selective attention, and expertise in soccer. *Research Quarterly for Exercise and Sport, 69,* 111-128.

MOVEMENT ORGANISATION AND EXECUTION

A number of changes in the way movement is organised and executed are apparent with training. The rate of processing movement information is improved with practice. This is due, at least in part, to transition early in learning away from an exclusively feedback-dependent, closed-loop type of motor control toward greater use of open-loop control. Open-loop control occurs via motor programs that are prestructured sets of commands that allow a movement sequence to be executed without reliance on feedback. As skills are learned, it appears that bigger and better motor programs are formed, bringing progressively more movement elements under the control of a single program. The collective effect of a reduced need to constantly monitor feedback, as well as a reduction in the number of separate programs needed to produce a particular action, is a reduction of the demands that the organisation and execution of movement place on the information-processing resources available in the central nervous system. Consequently, highly skilled performers have significant amounts of spare information-processing capacity that can be allocated to the performance of other tasks.

Expert performers are therefore markedly superior to less-skilled performers in the performance of a second task concurrent with their usual movement-control skills. Figure 18.6 illustrates the superior performance of experienced pilots undertaking a reaction-time task while performing a landing procedure on a flight simulator. Having to do a second task can sometimes disrupt the ability of less-skilled performers to complete the primary task effectively. Figure 18.7 illustrates the performance of rugby league players of different skill levels who were required to perform a 2v1 offensive maneuver with or without a secondary task of recognizing an auditory tone. Doing a second task did not disrupt the performance of the high-skilled players; however, it caused a significant disruption to the performance of the less-skilled performers.

OBSERVABLE MOVEMENT PATTERN AND MOVEMENT OUTCOME

By definition, the movements produced by skilled performers more consistently and precisely match the requirements of the movement task than those produced by less-skilled performers. In other words, expert performers develop, with practice, a greater

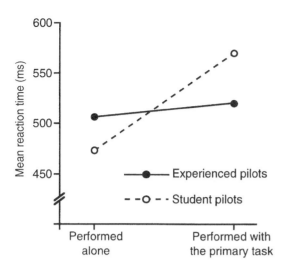

Figure 18.6 Mean reaction time on the secondary (memory) task for experienced and student aircraft pilots.

Reprinted from J.V. Crosby & S.R. Parkinson, "A dual task investigation of pilots' skill level," *Ergonomics* 22: 338, reprinted by permission of Taylor & Francis Ltd, http://tandf.co.uk/journals.

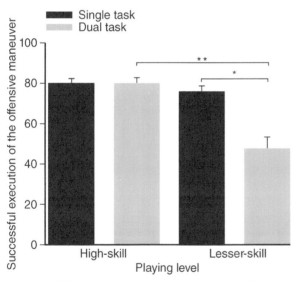

Figure 18.7 Primary task performance of rugby league players of different skill levels when performed alone or in conjunction with a secondary (auditory-tone counting) task.

Reprinted from T. Gabbett, M. Wake, and B. Abernethy, 2011, "Use of dual-task methodology for skill assessment and development: Examples from rugby league" *Journal of Sports Sciences* 29: 5. Reprinted by permission of Taylor & Francis Ltd, http://tandf.co.uk/journals.

capability to produce exactly the movement outcomes needed for successful performance. These successful movement outcomes are achieved through movement patterns that also show some reliable differences between expert and novice performers. Movement patterns, as discussed in chapter 7, can be best described in terms of their kinematics, kinetics, and underlying neuromuscular patterns using the techniques of biomechanics.

Studies drawing on both biomechanics and motor control have revealed a number of important changes in observable movement patterns with growing expertise. In terms of kinematics, the movement patterns of expert performers are characterised by greater consistency with respect to overall movement duration as well as movement trajectories and displacement-time characteristics. This greater consistency in the observable movement pattern is necessarily also reproduced in the underlying movement kinetics. Typically, the force–time curves are more consistent on a trial-to-trial basis for expert performers and show a clearer, more distinct pattern of force pulses. Expert performers make greater use of the external forces (such as gravity and reactional forces) available in movements and restrict the injection of muscular force generated by the body to only those points in the movement where it is needed and can act most effectively. The time course of power generated in movements consequently varies significantly between experts and novices. Novices tend to supply muscular force more often throughout a movement, often either inefficiently in opposition to external body forces or as an unnecessary supplement to external forces. It is therefore not surprising that neuromuscular recruitment patterns (as revealed from electromyography) become more discrete with practice and that there is a general reduction in recruitment as muscular contraction extraneous to the movement of interest is eliminated.

IMPLICATIONS FOR TRAINING

The value of knowledge about expert–novice differences in information-processing capability is that it provides a guide to where energy and attention should be directed in practice and training. Training based on improving the information-processing factors known to be related to the expert's advantage on a task appears to be more sensible than training focusing on factors that provide little or no discrimination between experts and novices. This logic is unfortunately not always fully appreciated or considered in the design and recommendation of practice. For example, the generalised visual and kinesthetic training programs that become popular from time to time and that are designed to improve motor skills through improving the general sensitivity of the sensory systems for movement are most unlikely to be beneficial for skill learning. The reason is that they do not, under most circumstances, train any of the limiting information-processing factors for skilled performance. The available evidence on motor expertise also clearly suggests that the training of perceptual and decision-making skills is, in many cases, just as important as the training of movement-execution skills, if not more so. However, this is also not frequently reflected in current training practices. Given that good practice is clearly fundamental to motor skill learning, the next section focuses on some of the major factors known to affect the learning of motor skills.

FACTORS AFFECTING THE LEARNING OF MOTOR SKILLS

An age-old adage about skill learning is that practice makes perfect. Studies of experts from a range of motor domains demonstrate the extraordinary amounts of practice (typically >10,000 h) undertaken in the acquisition of expertise and how the sheer volume of practice is one of the key discriminators between experts and less-skilled performers (figure 18.8). This practice may be accumulated in a deliberate way where the principal purpose is skill enhancement or may be undertaken less instrumentally through engagement in more play-like activities where the principal purpose is fun and social engagement.

Although it is certainly true that extensive amounts of practice are necessary for high levels of skill to be developed, practice is a necessary but not itself a sufficient condition for learning. Therefore, the adage that practice makes perfect, although essentially true, needs to be qualified in a number of ways. These qualifications reveal much about the factors that affect the acquisition of movement skills.

IMPERFECTABILITY OF SKILLS

One important qualification to the "practice makes perfect" adage involves the recognition that although motor skills improve with practice, there is no reason to suggest that they ever become perfect or that there is no room for further improvement

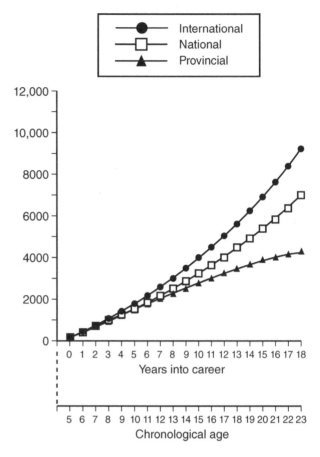

Figure 18.8 Accumulated hours of practice as a function of years of playing and chronological age for soccer players of different skill levels.

Reprinted, by permission, from W.F. Helsen, F.L. Starkes, and N.J. Hodges, 1998, "Team sports and the theory of deliberate practice," *Journal of Sport and Exercise Psychology* 20: 12-34.

through learning. Even in extremely simple tasks, such as hand-rolling cigars in a factory, improvements in performance have been shown to continue even after as many as 100 million trials of practice! In more complex motor skills that involve many more components that can be potentially improved with practice, improvements are likely to extend over an even greater time scale. There is no evidence that skill learning ever ceases, provided that practice is ongoing. The levelling out of performance observed after a number of years of practice or performance in various motor skills is more likely attributable to either psychological or physiological factors or to measurement difficulties. In light of this, a more appropriate adage may be "practice makes better."

NECESSITY OF FEEDBACK FOR LEARNING

Although practice is necessary for learning, practice alone does not guarantee learning. In particular, learners must be able to regularly derive feedback information about their performance for practice to be effective in improving learning. If learners are not able to gain information about the success of each attempt they make at a new task, learning will be impaired and indeed may not occur at all. For this reason, the adage "practice, the results of which are known, makes better" may more accurately encapsulate the nature of the relationship between practice, feedback, and learning.

Not all feedback is, of course, equally effective. As a general rule, feedback information is more effective the more specific it is and the more it provides information from the learner's perspective rather than from a third-person perspective. This is true regardless of whether the feedback information is derived by the learners themselves, gained from media such as video, or provided by a teacher, coach, or instructor. Feedback information must be limited to key features to avoid overloading the information-processing capacity of the learner. For this reason, summary feedback presented at the end of a block of trials rather than after each individual trial may be advantageous. This approach also encourages learners to develop the ability to extract their own feedback information (rather than relying on feedback from external sources). This ability is essential for learners' progression toward the autonomous phase of learning.

IMPORTANCE OF THE TYPE OF PRACTICE

Just as different types of feedback vary in their effectiveness for learning, so to do different types of practice. The teacher or coach of motor skills charged with the responsibility of designing training or practice regimes for maximal effectiveness must consider a range of issues such as the length of the rest intervals between practice attempts, the extent to which fatigue should be included or avoided in the practice sessions, and the degree to which the practice should be repetitive as opposed to variable.

A common guiding principle, which holds across all motor skills, is specificity. This is the notion that skills should be practised under conditions that most closely replicate the information-processing demands of the situation in which the skills must ultimately be performed. Consequently, the best type of practice differs for different motor skills. In some cases, the best type of practice may differ from commonly and traditionally accepted methods of practice for a particular skill or set of skills. Consideration of practice for the motor skills involved in the sport of golf may well illustrate this point.

Playing golf, like undertaking many other human activities, requires mastery over a number of motor skills. To play golf successfully a player must be able to drive the ball a long way with wooden and iron clubs and must be able to accurately pitch and chip the ball, play from sand bunkers, and putt with precision. The golfer must be able to adapt these skills on a shot-by-shot basis to accommodate such things as the lie of the ball, the force and direction of the wind, and the position of the hazards and obstacles. A key issue for the player and coach is how these skills might be best practised.

The traditional form of structured practice for golfers is to take a bucket of balls to the practice range and hit the same club over and over again until the skill is executed effectively and is ingrained. This form of practice, where each component skill is practised repetitively, is referred to as **blocked practice**. Most practice that involves drills is of this type. This type of practice can be contrasted with **random practice**, in which clubs and shots are practised in essentially random order, in a manner not dissimilar to what occurs in actually playing a round of golf. A random-practice schedule might involve, for example, hitting in order a driver, a seven iron, a sand wedge, and a putter. With such a practice method the same specific task is never repeated on two successive practice trials. Normally, different clubs are used on successive trials, but if the same club (e.g., the putter) were to be used twice in a row it would be from a different position or distance. The critical question is which type of practice is most effective for learning and performing golf skills.

We may gain some insight into this question from the results of a study on blocked and random practice, shown in figure 18.9. In this study participants had to learn three rapid hand- and arm-movement tasks. One group of participants learned the tasks through blocked practice and the other through random practice. The total amount of practice was the same for the two groups of participants; all that

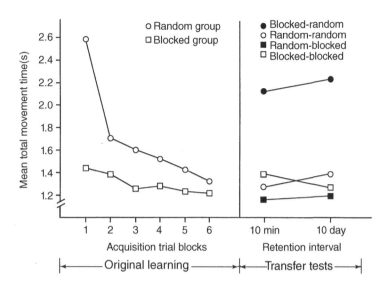

Figure 18.9 Performance during the acquisition and retention of a complex movement task for groups practising under either random or blocked conditions.

Reprinted, by permission, from J.B. Shea and R.L. Morgan, 1979, "Contextual interference effects on the acquisition, retention, and transfer of a motor skill," *Journal of Experimental Psychology: Human Learning and Memory* 5: 183.

varied was the type of practice. During practice, the blocked group performed best. This is not surprising given that they were exposed to the same task requirements trial after trial.

What is important, however, is how well the two groups were able to retain what they had practised (remembering that learning is a relatively permanent change in the ability to perform skills). The skills were retained best (and hence, by inference, learned best) by the group who had experienced random practice. Blocked practice was particularly ineffective when subsequent performance of the practised skill took place under random conditions. This is a crucial observation because this is precisely the situation that exists for the golfer who does repetitive practice at the driving range and then attempts to use these skills under playing conditions. Thus, while blocked practice may be valuable in the very early stages of learning, blocked practice may give an inflated view of actual skill level and may be a less-than-optimal preparation for a task that ultimately requires component skills to be executed in essentially random order. This example illustrates the importance of designing practice in a way that best simulates the demands of the actual performance setting.

LIMITATIONS OF VERBAL INSTRUCTION AND CONSCIOUS ATTENTION

The conventional approach to skill acquisition relies heavily on the premises that detailed verbal instruction about how to perform the skill and conscious attention to this information by the learner are beneficial to learning. Learning achieved under these circumstances is referred to as **explicit learning**. One of the defining characteristics of this type of learning is the concurrent acquisition of both the movement skill and knowledge about the performance of the skill.

Although explicit learning may be beneficial in the early stages of skill acquisition, there is a danger with becoming bogged down with too much knowledge. An excess focus on explicit knowledge can prevent desirable progression to the autonomous stage of learning, and too much conscious attention to performance can interfere with the automatic production of movement and lead to errors of "paralysis by analysis." Making movement control conscious can leave skill performance open to disruption from other conscious processes, such as those that are discussed in chapter 19 in relation to anxiety. It is for this reason that there is now growing interest in **implicit learning** of motor skills. In implicit learning, conscious attention is directed away from the task at hand (e.g., through the use of a concurrent secondary task), verbal instruction is minimal or absent, and the participants simply practise and acquire the skill without the concurrent acquisition of explicit knowledge about the performance of the skill. In cognitive tasks, such as pattern recognition, the performance of people who have learned these skills implicitly has been shown to be both more resistant to forgetting and more resistant to stress and anxiety than the performance of people who have learned the comparable skills explicitly. A growing body of evidence suggests the same is also likely true for the learning of motor skills (see "Implicit Learning of Motor Skills").

IMPLICIT LEARNING OF MOTOR SKILLS

Rich Masters from the University of Hong Kong has long been interested in the question of the relative merits of implicit and explicit approaches to learning movement skills such as putting in golf. In an early study he randomly allocated novice golfers to one of five experimental groups. Participants in the implicit learning (IL) group and the implicit learning control (ILC) group received no instruction on how to putt and were required to do a random letter-generation task while they practised putting in order to divert their conscious attention away from the putting skill. Participants in the explicit learning (EL) group were given very specific instructions on how to putt and were required to follow these instructions throughout practice. Participants in the stressed control (SC) and the nonstressed control (N-SC) groups were given no instructions or any secondary task and were simply instructed to improve as much as possible. After 4 practice sessions of 100 putts, the groups were tested at a final session of 100 putts. In this test condition, three of the groups (IL, EL, and SC) were subjected to performance stress (through monetary rewards for performance and evaluation by an expert assessor), whereas the others (ILC and N-SC) were not. Analysis of putting performance showed that those who learned implicitly were less likely to suffer performance decrements under stress. In later studies, Masters and colleagues demonstrated that implicit approaches can provide learning benefits in other domains as well, including the learning of the movement skills needed for laparoscopic surgery.

In 2002, Sian Beilock (then from Michigan State University and now at the University of Chicago in the United States) and colleagues examined the performance of experienced golfers under conditions where they undertook a dual task to distract attention from putting and a condition where they gave conscious attention to step-by-step monitoring of movement execution. In both this experiment and a second experiment on soccer dribbling, the findings indicated that deliberate attention to the conscious, step-by-step monitoring of performance can actually impede the performance of experts whereas it benefits the performance of novices and less-skilled performers.

Sources

Beilock, S.L., Carr, T.H., MacMahon, C., & Starkes, J.L. (2002). When paying attention becomes counterproductive: Impact of divided versus skill-focussed attention on novice and experienced performance of sensorimotor skills. *Journal of Experimental Psychology: Applied, 8,* 6-16.

Masters, R.S.W. (1992). Knowledge, knerves and know-how: The role of explicit versus implicit knowledge in the breakdown of complex motor skill under pressure. *British Journal of Psychology, 83,* 343-358.

Masters, R.S.W., & Maxwell, J.P. (2004). Implicit motor learning, reinvestment and movement disruption: What you don't know won't hurt you? In A.M. Williams & N.J. Hodges (Eds.), *Skill acquisition in sport: Research, theory and practice* (pp. 207-228). London: Routledge.

Zhu, F.F., Poolton, J.M, Wilson, M.R., Hu, Y., Maxwell, J.P., & Masters, R.S.W. (2011). Implicit motor learning promotes neural efficiency during laparoscopy. *Surgical Endoscopy, 25,* 2950-2955.

IMPORTANCE OF SLEEP

Sleep offers an essential opportunity for the consolidation of newly learned information into memory. Disruptions to sleep are known to interfere with the learning of skills, including motor skills. During the learning of a new skill, improvements are typically observed across the course of a training session. These improvements from the start to the end of the training session provide a measure of **practice-dependent learning**. Further improvements in performance are also generally seen for at least 24 h after training (such that the starting performance level at the next practice session may exceed the performance level attained at the end of the preceding practice session without any additional trials of physical practice having taken place). Growing evidence demonstrates that these improvements occur only if sleep occurs during the intervening period. Indeed, such improvement is now referred to as **sleep-dependent learning**. As is illustrated in figure 18.10, sleep, rather than simply the passage of time, is responsible for the consolidation of learning gains. Some 20% improvement in motor-performance

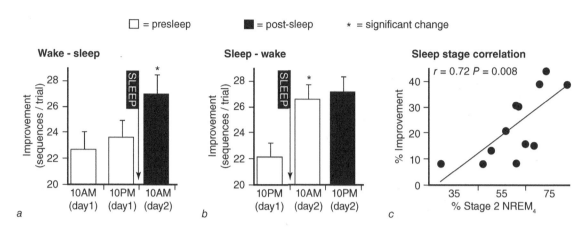

Figure 18.10 An example of sleep-dependent motor learning. No improvements in performance were evident when testing occurred 12 h later without a sleep (a), but improvements were evident after sleep (a and b). Improvement correlates with the amount of stage 2 (non-REM) sleep in the last quarter of the evening.

scores occurs if training is followed by sleep, and minimal change occurs if sleep is not provided. Stage 2 (non-REM sleep) seems particularly important to sleep-dependent learning; improvements are correlated with the duration of this sleep stage.

DEPENDENCY OF LEARNING ON READINESS

As noted in chapter 17 in the context of the notion of critical periods, practice may be spectacularly ineffective for improving skill if the learner is not developmentally ready, yet skills may be acquired with surprisingly few trials of practice when the learner reaches an appropriate developmental stage. Although developmental factors have the most pronounced effects in moderating the effectiveness of practice, other psychological factors present throughout the life span are also important in determining the extent to which practice translates into actual learning. Foremost among these are the motivation of the learner and their level of arousal and anxiety. These concepts are considered in part VI, which introduces the subdiscipline of sport and exercise psychology.

SUMMARY

Practice (training) results in changes in information-processing capabilities and in the underlying structure and function of the brain and neuromuscular system. These changes collectively produce significant observable changes in motor performance. With practice, skills become less consciously controlled and more automatic. This provides highly skilled performers with both the capacity to concurrently perform multiple tasks and improvements in efficiency that help delay the onset of fatigue. Clear expert–novice differences are evident in all three aspects of central information processing (perceiving, deciding, and acting), indicating that the nervous system responds to training through functional adaptations in much the same manner as do other key systems of human movement. Although the precise neural mechanisms for learning are not yet well understood, it is apparent that the nervous system possesses considerable plasticity. Short-term changes in synaptic efficiency and long-term changes in synaptic connectivity appear to be fundamental neural foundations for learning. Under the appropriate set of practice conditions, and in the presence of suitable feedback, continuous refinement and improvement of all motor skills seems possible (even for simple tasks performed by acknowledged experts). The challenge for researchers and practitioners alike is to understand more fully the optimal practice conditions for the continuous learning of different types of motor skills.

FURTHER READING

Abernethy, B., Wann, J., & Parks, S. (1998). Training perceptual-motor skills for sport. In B.C. Elliott (Ed.), *Training in sport: Applying sport science* (pp. 1-68). Chichester, U.K.: Wiley.

Ericsson, K.A. (Ed.). (1996). *The road to excellence: The acquisition of expert performance in the arts and sciences, sports and games.* Mahwah, NJ: Erlbaum.

Farrow, D., Baker, J., & MacMahon, C. (2008). *Developing sport expertise: Researchers and coaches put theory into practice.* London: Routledge.

Hodges, N.J. & Williams, A.M. (Eds.) (2012). *Skill acquisition in sport: Research, theory and practice.* Abingdon, Oxon: Routledge.

Magill, R.A. (2011). *Motor learning: Concepts and applications* (9th ed.). Boston: McGraw-Hill.

Schmidt, R.A., & Lee, T.D. (2011). *Motor control and learning: A behavioural emphasis* (5th ed.). Champaign, IL: Human Kinetics.

Schmidt, R.A., & Wrisberg, C. (2008). *Motor learning and performance: A situation-based learning approach* (4th ed.). Champaign, IL: Human Kinetics.

Starkes, J.L., & Ericsson, K.A. (Eds.). (2003). *Expert performance in sports: Advances in research on sport expertise.* Champaign, IL: Human Kinetics.

PART VI

PSYCHOLOGICAL BASES OF HUMAN MOVEMENT

SPORT AND EXERCISE PSYCHOLOGY

Sport and exercise psychology is the subdiscipline of human movement involving the scientific study of human behaviour and cognition (thought processes) in the context of physical activity. Within the subdiscipline, the distinction between sport psychology and exercise psychology is not always crystal clear. Sport psychology is concerned with human behaviour in the sport environment, whereas exercise psychology focuses on human behaviour in the exercise environment. Issues relating to competition and sporting performance traditionally fall under the jurisdiction of sport psychology. Exercise psychology typically involves the study of why people do or do not exercise and the psychological effects of exercise on people. Some overlap between sport psychology and exercise psychology is inevitable. Many competitive athletes participate in exercise that is not sport specific to improve various aspects of fitness. Similarly, some outdoor recreation activities could be classified as either sport or exercise.

There are two major questions that sport and exercise psychologists address:

▸ What effect does participation in physical activity have on the psychological makeup of the participant?

▸ What effect do psychological factors have on physical activity participation and performance?

Areas of interest related to the first question include the effects of exercise on psychological well-being and the effect of participating in youth sport on the development of character. Examples of themes related to the second question are the effects of anxiety on competitive performance and the effects of self-confidence on participation in physical activity.

TYPICAL ISSUES POSED AND PROBLEMS ADDRESSED

A few examples of topics dealt with in sport and exercise psychology are given in the description of the two main questions addressed by the subdiscipline. Additional issues in the field include:

▶ the effect of personality on participation or performance;

▶ reinforcement, feedback, and performance;

▶ leadership effectiveness;

▶ the enhancement of sporting performance or exercise adherence through the development of psychological skills;

▶ techniques for developing self-confidence;

▶ the effects of external rewards on motivation;

▶ the influence of an audience on performance;

▶ determinants and consequences of team cohesion;

▶ psychological predictors of athletic injuries;

▶ goal setting;

▶ concentration and attention;

▶ drug abuse in sport; and

▶ appropriate supervision of applied practice.

LEVELS OF ANALYSIS

Because sport and exercise psychologists do not have the technology to directly record individuals' thoughts or feelings at a specific moment in time, they rely on the use of multiple levels of analysis. The most commonly used levels of analysis are the behavioural or observational level, the cognitive level, and the physiological level. The behavioural or observational level involves watching individuals and recording what they do. For example, when studying anxiety during competition, psychologists might use the behavioural or observational level of analysis for indications of anxiety such as irritability, the return to old habits, yawning, the trembling of muscles, the inability to make decisions, or the inability to concentrate. The study of competitive anxiety using the cognitive level of analysis typically entails the use of questionnaires or inventories. The physiological level of analysis involves the direct measurement of physiological variables. For example, studying competitive anxiety might consist of measuring heart rate, respiration rate, or muscle tension. There is frequently a limited coherence across different levels of analysis in sport and exercise psychology. Therefore, making generalisations on the basis of a single level of analysis may be misleading. For this reason it is important to use multiple levels of analysis within this subdiscipline.

HISTORICAL PERSPECTIVES

Sport and exercise psychology is a much newer area of study than some of the other biophysical subdisciplines of human movement studies. The first recognised studies in sport psychology appeared in the late 1890s. These studies were isolated and included the investigation of topics such as reaction time, audience effects, mental practice, and the personality of athletes. In North America the first systematic research in sport psychology is attributed to Coleman Griffith, who wrote books on the topic, worked as a practitioner with athletes, and established a sport psychology laboratory at the University of Illinois in the 1920s. At the same time, organised sport psychology was beginning in Eastern Europe with the establishment of The Institutes for Physical Culture in Moscow and Leningrad. Extensive research began in the Soviet Union in the 1950s to help cosmonauts control bodily functions and emotional reactions while in space. Later these methods were applied to Soviet and East German elite athletes. Although little of the sport psychology research and practices from within Eastern Europe were available outside the region, it is apparent that there was strong government support (and control) of applied research on the enhancement of the performance of elite athletes.

In North America, little was accomplished in the area of sport psychology after Griffith until the 1960s when numerous research scientists became actively involved in the area and professional organisations began to form. Many of these research workers had been trained in motor control but saw the need to develop sport psychology independently of the study of motor control. In 1965, the International Society of Sport Psychology was formed in Rome, and professional societies for sport psychology and motor control in the United States and Canada formed soon after (1967 and 1969, respectively). Although the emphasis in Eastern Europe remained on field research, North America emphasised laboratory research. It was not until the 1980s that greater importance was placed on applied and field-based research outside Eastern Europe. As both the research and practice of sport psychology burgeon in most countries throughout the world, financial constraints and government changes may, at least temporarily, limit the pursuit of sport psychology in areas that were previously part of the Soviet Union and were historically strong in applied sport psychology.

Although studies on exercise and psychological factors have been intermingled in sport psychology studies during the past 30 or 35 yr, and the International Society of Sport Psychology helped identify exercise psychology as a separate area of study in 1968, exercise psychology emerged as a specialist area only recently. In 1988, *Journal of Sport Psychology* became *Journal of Sport & Exercise Psychology*. The first textbooks focusing solely on exercise psychology were not published until the 1990s. In most professional organisations exercise psychology is seen as a subdivision of sport psychology. However, just as sport psychology and motor control have split into two distinct fields, the future may see the professional division of sport psychology and exercise psychology.

PROFESSIONAL ORGANISATIONS

The International Society of Sport Psychology (ISSP) was organised in 1965 with the stated purpose of promoting and disseminating information throughout the world. In addition to publishing *International Journal of Sport Psychology*, ISSP holds a world congress in sport psychology once every 4 yr. The location of the congress varies. For example, the 10th, 11th, and 12th world congresses were held in Greece, Australia, and Morocco, respectively. The 13th world congress in 2013 will be held in Beijing, China.

Numerous regional organisations of sport psychology have also been formed. For example, in Europe there is European Federation of Sport Psychology, which is officially called Fédération Européenne de Psychologie des Sports et des Activités Corporelles (FEPSAC). There are also several national sport psychology organisations in Europe, such as Associazione Italiana Psicologia dello Sport (Italy), Sociedade Portuguesa de Psicologia do Desporto (Portugal), and Société Française de Psychologie du Sport (France).

In North America, two major associations have emerged. The primary goal of the North American Society for the Psychology of Sport and Physical Activity (NASPSPA) is the advancement of the knowledge base of sport psychology through experimental research. Perhaps because of its relatively early founding in 1967, NASPSPA has a comparatively heavy emphasis on motor learning and control and motor development. Founded in 1985, the Association for Applied Sport Psychology (AASP) was formed to promote the field of applied sport psychology. AASP is concerned with ethical and professional issues related to the development of sport psychology and to the provision of psychological services in sport and exercise settings. AASP promotes the development of research and theory but also focuses on intervention strategies in sport psychology.

This list of specific organisations is not intended to be representative of all sport psychology associations, societies, and organisations. Sport psychology is growing rapidly in many countries in addition to those in Europe and North America, such as Australia, Brazil, China, India, Korea, and Nigeria.

FURTHER READING

Weinberg, R.S., & Gould, D. (2011). *Foundations of sport and exercise psychology* (5th ed.). Champaign, IL: Human Kinetics.

SOME RELEVANT WEBSITES

Association for Applied Sport Psychology: http:// appliedsportpsych.org/

College of Sport Psychologists (Australian Psychological Society): www.psychsociety.com.au/units/ fr_aps_units.htm

Exercise and Sport Psychology (Division 47 of the American Psychological Association): www.psyc. unt.edu/apadiv47

International Society of Sport Psychology: www. issponline.org

Journal of Sport and Exercise Psychology: www. humankinetics.com/products/journals/journal. cfm?id=JSEP]

North American Society for the Psychology of Sport and Physical Activity: www.naspspa.org

The Sport Psychologist: www.humankinetics.com/ products/journals/journal.cfm?id=TSP

CHAPTER 19

BASIC CONCEPTS IN SPORT PSYCHOLOGY

The major learning concepts in this chapter relate to

- ▶ the difference between the trait and interaction frameworks of personality;
- ▶ limitations of personality research in sport;
- ▶ the three components of motivation: direction, intensity, and persistence;
- ▶ different definitions of success and their influence on motivation;
- ▶ the role coaches, parents, and teachers can play in establishing the motivational climate;
- ▶ the difference between arousal and anxiety and how they relate to performance; and
- ▶ what imagery is, how it works, and why it should be used.

Sport psychology covers many topics, and an introductory text such as this cannot provide comprehensive coverage of the field. Instead, four of the major domains of the field are introduced in this chapter. In terms of the analogy of the car as the human, first introduced in the preface, sport and exercise psychology, with its focus on mental processes and behaviour, is about investigating the driver of the car. What assumptions might be made about an individual who owns a brand-new Jaguar versus someone who drives an old, beat-up Ford? It might be presumed that these individuals have different personalities. In addition to presenting information about personality and sport, this chapter considers a few of the factors that might influence the performance of the driver. Specifically, motivation, anxiety, and imagery are briefly introduced and considered in this chapter. There is also a short section on the practise of applied sport psychology, that is, what sport psychologists actually do.

PERSONALITY

Personality has been defined in various ways, but certain elements are common to all definitions. In its simplest form, personality is the composite of the characteristic individual differences that make each of us unique. Think about two people you know who act very differently. What about them makes them different? For example, is one outgoing and the other shy? These differences are what make each of us unique. In portraying the differences between the two people you have considered, you have depicted aspects of their personalities.

TRAIT FRAMEWORK OF PERSONALITY

One framework for studying personality, the trait framework, suggests that everything we do is the result of our personalities. Our behaviours are determined completely by our personalities. According to the trait framework, each individual has stable and enduring predispositions to act in a certain way across a number of different situations. These predispositions, or **traits**, predict how we will respond. For example, if individuals have the trait of shyness they would be expected to be reserved or timid when joining new teams. On the other hand, athletes with the trait of being outgoing would be expected to be extroverted and sociable when first meeting with new teammates.

The trait framework of personality suggests that personality traits can be objectively measured. Because traits are considered to be enduring and stable, inventories are used in an attempt to measure personality characteristics.

ARE ATHLETES DIFFERENT FROM NONATHLETES?

Sport personality researchers traditionally have administered personality inventories to groups of athletes and nonathletes and then examined the results for any differences. Some studies have found differences that indicate that nonathletes are more anxious than athletes and that athletes are more independent and extroverted than nonathletes.

There have, however, been problems with some of the traditional sport-personality research. Most personality inventories were designed for a specific purpose and a specific population. Several of these inventories were created for use with clinical populations and therefore are inappropriate for use with nonclinical populations such as athletes. Additionally, a number of the questionnaires used in the research have not been shown to be valid and reliable. Certain questionnaires have not been demonstrated to actually measure what they say they measure (poor validity), and some questionnaires provide different results when administered to the same person, even over short periods of time (poor reliability).

Aside from possible difficulties with the inventories used in the research, other complications arise when the definitions of athletes and nonathletes are considered. Trying to generalise findings across studies is difficult because the terms have not been defined in the same manner in the different studies. How does one define an athlete? In some studies athletes have been defined as those competing in intercollegiate sport, suggesting that individuals participating in club sport are not athletes. Other studies consider individuals to be athletes only if they have achieved a particular level of performance and not if they train and compete regularly at a lower level of proficiency. Would people who run on their own many times per week be considered athletes if they do not participate in competitions? What about someone who competes in social tennis matches only once or twice per year? As you can see, even if valid, reliable, and relevant personality inventories can be obtained, actually determining who is and who is not an athlete makes comparing the personalities of athletes and nonathletes difficult.

INTERACTION FRAMEWORK OF PERSONALITY

Another model for studying personality is the **interaction framework**. The interaction framework still recognises that personality traits influence behaviour; however, the situation or the environment is also acknowledged as influencing how we act. In the interaction framework, the environment refers to all aspects of the situation that are external to the individual, including the physical surroundings as well as the social milieu (other people). It is important to note that the interaction framework is just that—an interaction. It is not only that personality traits and the environment both influence behaviour, but that the two interact and affect each other as well. For example, when shy athletes attend the first training sessions of new teams, there are aspects of the environment that could influence their behaviours. If the other team members are inseparable and suspicious of newcomers, the shy newcomers' behaviours would probably be different than if the team members went out of their way to welcome new additions to the teams. Similarly, if extroverted athletes joined new teams, their outgoing personalities would influence the behaviours of the people around them, thus influencing the team environment.

The interaction framework considers not only traits but also states. Traits, as previously described, are stable, enduring personality predispositions. **States**, on the other hand, refer to how someone feels at a particular point in time. Traits may influence states, but they do not directly determine states. For example, certain athletes may have low levels of trait anxiety. Generally speaking, they are relaxed and calm people who do not get anxious easily. In most situations they are tranquil and unruffled. These athletes could be in the finals of major competitions and still remain calm. Certain situations, however, may cause them to react with high levels of state anxiety. An illustration might be if they found themselves hurtling down the ramp of a ski jump (assuming these athletes are not ski jumpers) or, more realistically, shooting free throws with one second left on the clock when the team is down by one point (or similar time-pressured moments common to many team sports).

The main difference of the interaction framework is that it acknowledges that the situation or the environment can influence how we react. Behaviour is not solely determined by personality traits. Because the situation can influence behaviour, the use of personality inventories as a method of establishing how individuals will perform in all situations is not viable. Although these inventories may give some indication of the personality traits of individuals, measures that take into account specific situational factors are needed to determine states.

PRACTICAL IMPLICATIONS OF PERSONALITY IN SPORT

If it could be determined that people with certain characteristics or traits perform better in particular sports, then personality inventories could be given to individuals to determine which sports suit them best. This process should also incorporate anthropometric, biomechanical, physiological, and motor control facets of the individuals. The challenge, of course, is to base any decisions of this type on factors that will continue to be predictive of performance over time.

MOTIVATION IN SPORT

Many coaches complain that certain individuals would be great athletes if only they were motivated. The athletes are seen to have all of the anthropometrical, biomechanical, physiological, and skill components necessary for performing at a high level except that they just do not seem to care. They might show up late to practice, fail to try hard during drills, or just not show up at all. People interested in community health may not be so concerned about the performance of specific individuals, but are instead troubled by the large number of people who no longer participate in sport at all and who lead predominantly sedentary lives. Both concerns relate to the concept of motivation.

WHAT IS MOTIVATION?

Motivation is made up of three components: direction, intensity, and persistence.

▸ **Direction** refers to where people choose to invest their energy. There are few unmotivated people in the world; it just may be that some are motivated in a different direction than the one in which we would like them to be motivated. For example, a roommate may be motivated to go to the movies but not motivated to clean the house. Similarly, an athlete with great performance potential may be motivated to go to a party instead of to practice. Neither of these people lacks motivation; they have just chosen a different direction in which to invest their energy.

▸ **Intensity** refers to how much energy is invested in a particular task once the direction has been chosen. Two athletes may both choose the direction of practice, but where one invests minimal effort, the other tries hard and works at a high level of intensity. It is worth noting, however, that the same exercise or drill will require different amounts of intensity from different individuals depending on their levels of fitness and skill.

▸ **Persistence** alludes to the long-term component of motivation. It is not enough to have an athlete choose the direction of practice and work out at a high level of intensity during one or two sessions. Athletes need to continue to practise and participate over time. It may be preferable to have athletes work out at a moderate level of intensity over an entire season rather than have them work out at an extreme level of intensity at the beginning of the season and then drop out.

Individuals interested in enhancing the motivation of sport participants should keep in mind all three components. Direction can be influenced by making the path of participation enjoyable. Intensity can be increased by stipulating a reason for doing any particular activity. Workers are likely to put minimal (if any) effort into a job if they think there is no reason for doing it. The same principle holds true for sport. Athletes will be more likely to put effort into a sporting drill or activity if they perceive that there is a reason for doing so. Persistence can be improved by providing positive feedback to participants. Athletes are more likely to continue participating if they are convinced of the long-term benefits. Similarly, students are unlikely to continue to study if they honestly feel that it does not accomplish anything. If, however, they feel that studying gives them greater understanding of the subject, better prepares them for advanced courses, leads to higher grades, or causes them to get positive recognition from someone they care about, they will be more likely to persist. Believing that continued participation leads to greater success increases persistence.

WHAT IS SUCCESS?

Traditionally, success in sport has been considered in terms of winning and losing; winning equals success and losing equals failure. If outcome is the only criterion for success, anyone who does not win fails. Using this definition, only one team in a given league can be considered successful at the end of the season, only one swimmer in a particular event is successful, and only one runner in a marathon is successful. Everyone else is a failure. Given that experiencing success enhances motivation, the outcome definition of success leaves the majority of participants with limited motivation because they experience only failure.

Success, however, can mean different things to different people. Some people do define success in terms of outcome. A performance is considered to be successful only if the athletes were able to demonstrate that they were better than everyone else. Nevertheless, many athletes feel successful if they improved their own performance, regardless of outcome. If technique improved, times decreased, or some other aspect of performance enhanced, then

success was experienced. In this case, comparisons are made with one's own previous performances rather than with those of others.

Some people do not use performance as the basis of their definition of success. Success for these people is achieved when they get social approval or recognition from others. For example, a field hockey player may consider the season to be successful because she made a lot of friends or was recognised by the media. Similarly, a basketball player may consider a game to be a success because he received a compliment from the coach for his efforts.

These different definitions of success are called **achievement goal orientations**. The two most frequently studied achievement goal orientations are task and ego. A person with a strong **task orientation** has a self-referenced definition of success, whereas someone with a strong **ego orientation** defines success as being better than others. Athletes often have a combination of task and ego achievement goal orientations. They may want to improve their skills as well as win. In addition, an individual may have a stronger ego orientation for one sport but a stronger task orientation for another.

ACHIEVEMENT GOAL ORIENTATIONS AND MOTIVATION

These different definitions of success influence motivation. For example, if a swimmer is in a race and is predominantly ego oriented, effort may be decreased if the swimmer is well ahead of the others or if the swimmer is well behind and perceives that there is no chance of winning. If, however, the swimmer has a strong task orientation, effort will be maintained no matter where the swimmer is placed in the race because success is related to improving one's own time or technique.

Coaches should be aware that athletes often vary in their achievement goal orientations. Unfortunately, many coaches assume that all their athletes define success the same way they do and that their reasons for participating are also identical. This assumption can be problematic if the goals of the coach and the athletes are in fact different. There is a greater likelihood of athletes dropping out of a situation if they perceive that their needs are not being met. Therefore, if a coach is interested only in winning but the athletes are interested in making friends or learning new skills, the athletes may feel that they are not getting what they want and abandon the sport altogether. Similarly, if a coach stresses individual growth and improvement by ensuring that every athlete gets similar playing time and some of the athletes are solely interested in winning, friction and unhappiness may result.

MOTIVATIONAL CLIMATE

Although there are dispositional tendencies that predispose people to be high or low in task and ego orientations, the environment also has an effect. Perceived motivational climate is determined by the emphasis of situational goal perspectives. Coaches, parents, and physical education teachers create environments that differ in how much emphasis is placed on task or ego. The majority of research on motivational climate in physical activity has taken place in physical education classes. A task or mastery climate is linked to enjoyment, perceived ability, and effort. An ego environment increases the tension and pressure felt by participants and negatively affects their interest and enjoyment.

SELF-DETERMINATION THEORY

A lot of the research and practice in the area of motivation in sport (and exercise) is currently based on Deci's self-determination theory. According to theory, people are inherently and proactively motivated to master their environment. The theory focuses on three personal needs: autonomy, competence, and relatedness.

▶ **Autonomy** (i.e., self-determination) involves a combination of personal control and perceived choice. Control on its own is not enough. For example, an athlete who is forced to return from competing overseas to compete in national Olympic selection trials even when she or he is clearly the best athlete in the country can easily control performance, but because the perception of choice is not present, there is no sense of autonomy or self-determination.

▶ **Competence** involves the need to have optimally challenging activities to promote a feeling of competence. If a task is too hard, individuals will not feel competent; similarly, if the task is too easy, there will be no sense of accomplishment. Positive feedback from others can promote competence, but only if it is perceived to be sincere.

▶ **Relatedness** is the feeling of being connected to others. According to self-determination theory, we all have the need to be accepted by others and belong to groups. Team sport can be a source of relatedness.

Coaches play an important role in facilitating adaptive forms of motivation. See "Practical Application of Self-Determination Theory in Elite Sport" for an example of how a coach promoted athlete autonomy in an Olympic relay team.

PRACTICAL APPLICATION OF SELF-DETERMINATION THEORY IN ELITE SPORT

Self-determination theory indicates that there is a motivational continuum ranging from intrinsic motivation to amotivation. Between the two extremes are two forms of extrinsic motivation: self-determined extrinsic motivation (SDEM) and non-self-determined extrinsic motivation (non-SDEM). Non-SDEM does not involve the perception of choice (e.g., athletes are coerced by their coaches to train or they train because they feel guilty if they do not). SDEM involves the perception of choice or autonomy (e.g., athletes value training because they realise it will help them achieve their performance goals). Athletes with higher levels of SDEM (and intrinsic motivation) tend to perform better, persist longer, try harder, and cope with stress better than those with lower levels.

Mallett, a researcher, sport psychologist, and elite coach, created an autonomy-supportive motivational climate in his 2 yr preparation of the Australian men's Olympic relay teams. The athletes were given choices regarding training content, training venues, training times, and uniforms. Another example of

(continued)

the creation of the perception of choice was the 4 × 400 m team's running order for the Olympic final. As the coach, Mallett decided on the composition of the team (i.e., which four of six athletes would compete) but then outlined the pros and cons of two preferred running orders. The athletes made the decision on the running order, meaning that they were then fully committed to the decision. Having the power to make the decision provided autonomy and promoted perceptions of competence (another of the three basic needs of self-determination theory).

Mallett also promoted autonomy (and competence and relatedness) through the use of a problem-solving approach to instruction and learning. For example, when working on baton exchanges, athletes would provide strengths and weaknesses of their own performance, provide similar feedback to teammates, and then compare that feedback with feedback available from video. As the coach, Mallett asked questions (e.g., "What was another option in that scenario?") to encourage athletes to think critically about their event.

Although there was no control group as needed for strong experimental research, both objective and subjective assessment of the athletes' performances suggest that the autonomy-supportive approach was effective. The men's 4 × 100 m team ran faster than their previous season best time and finished 6th after being ranked 14th coming into the competition. The men's 4 × 400 m team also ran faster in the final than their previous season best time and finished 2nd after being ranked 13th coming into the Olympics. As the coach, Mallett believed that his approach resulted in teams that were enthusiastic, positive, confident, and willing to work together. Although additional research is needed, Mallett's results indicate that by developing autonomy-supportive motivational climates, coaches can help satisfy the three basic human needs of autonomy, competence, and relatedness.

Source

Mallett, C.J. (2005). Self-determination theory: A case study of evidence-based coaching. *The Sport Psychologist, 19,* 417-429.

AROUSAL, ANXIETY, AND SPORT PERFORMANCE

Motivation is not the only psychological factor that influences participation and performance in sport. Arousal and anxiety levels have a profound effect on the performance of athletes.

AROUSAL

Arousal is traditionally considered to be physiological activation. When individuals are highly aroused or activated they often have tense muscles and higher blood pressures, heart rates, and respiration rates than they do when they are calm. Arousal can involve mental activation in addition to physiological activation. Therefore, arousal can be considered to be the degree of mental and physical activation or intensity. An example of a low level of arousal would be the grogginess experienced when first waking in the morning. An example of high arousal would be the heightened mental alertness and physiological activation experienced before a major competition.

ANXIETY

Confusion is often caused when people use the terms arousal and anxiety interchangeably. **Anxiety** is the subjective feeling of apprehension and is usually accompanied by increased arousal levels. High levels of arousal, however, are not always accompanied by anxiety. If a team just beat the defending league champions, they would probably be fairly aroused, yet they would almost certainly interpret this arousal as excitement rather than anxiety.

Anxiety is usually experienced when a situation is perceived to be threatening. This commonly occurs when there is a perceived imbalance between the demands of the situation and individual ability to meet those demands. For example, if athletes believe that they must perform a skill (e.g., for the team to win) yet do not believe that they are good at that particular skill, they are likely to experience anxiety. This anxiety will increase the more the athletes believe that there may be negative repercussions if they perform poorly. If, for example, individuals do not care about the outcome of the game or their own performances and are playing only to have fun with their friends, they will probably experience only mild anxiety, if any. On the other hand, if they believe the outcome of the performance to be extremely important, they will likely experience a relatively high level of anxiety. Anxiety, therefore, is determined by the perceptions of the individual. Two individuals may be in the same situation and one may perceive it to be slightly challenging whereas the other perceives it to be extremely threatening.

TRAIT AND STATE ANXIETY

This sensation of anxiety is a good example of the interaction framework mentioned in the personality section of this chapter. The level of anxiety experienced is a result of the interaction of personal factors (e.g., personality, needs, and capabilities) and situation factors (e.g., opponent, task difficulty, and the presence of other people). An athlete with a predisposition to perceive situations as stressful or threatening would be considered as having a high level of **trait anxiety**. **State anxiety**, on the other hand, is the experience of apprehension at a particular point in time.

COGNITIVE ANXIETY AND SOMATIC ANXIETY

State anxiety is a multidimensional concept. **Cognitive anxiety** is the mental facet of state anxiety. Worry, perceived threat, and self-defeating thoughts are possible aspects of cognitive state anxiety. Somatic anxiety refers to physiological anxiety responses such as butterflies in the stomach and sweaty palms. Although somatic and cognitive state anxiety are related, it is possible to experience a higher degree of one than the other. For example, individual students experience the anxiety of examinations differently. Some get tense physically, whereas others drive themselves (and everyone else) crazy with their worrying. The same is true for the experience of anxiety in sport. Some athletes have more signs of somatic anxiety, and others experience greater cognitive anxiety.

AROUSAL–PERFORMANCE RELATIONSHIP

If individuals had just awakened and were feeling sluggish and tired, they probably would not perform well if asked to engage in competitive sport. Their

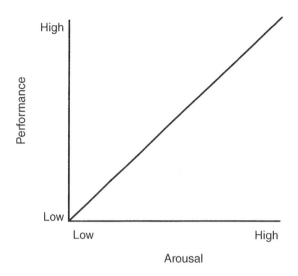

Figure 19.1 Drive theory.

performances would probably improve as they became more alert and awake. This proposed linear relationship between arousal and performance is called **drive theory** and is illustrated in figure 19.1. According to drive theory, performance will continue to improve with further increases in arousal or activation. Therefore, if drive theory is to be believed, one way to ensure optimum performance is to arouse athletes as much as possible.

Some coaches have done some strange things because of their belief in drive theory. Coaches have had their players bite the heads off of live chickens, castrate a bull, stage a mock gun battle in a school cafeteria (complete with fake blood), and watch films of prisoners being murdered in concentration camps. Fear, anger, and horror were seen as emotions that could raise arousal levels. All of these activities were done to arouse the athletes as part of precompetition preparation in the belief that drive theory is correct.

Unfortunately, there are still some coaches who continue to base precompetition preparation on drive theory. A more useful method of considering the relationship between arousal and performance is to follow the **inverted-U hypothesis**. As can be seen in figure 19.2, when the relationship between arousal and performance is plotted according to this hypothesis, the graph forms an upside-down U. As with drive theory, there are increases in performance with increases in arousal. These increases in performance, however, occur only up to a certain point of arousal. If arousal continues

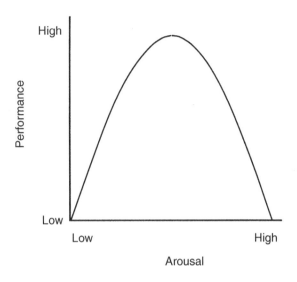

Figure 19.2 Inverted-U hypothesis.

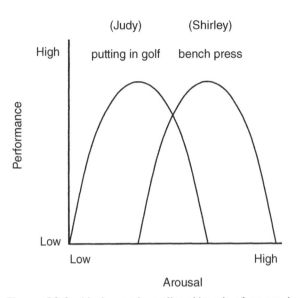

Figure 19.3 Variance in optimal levels of arousal.

to increase past that point, performance begins to deteriorate. The point at the top of the inverted U is called the point of optimal arousal. It is at this point that performance is best. If arousal is below that point (i.e., underarousal), performance will be less than optimal. If arousal is past the point of optimal arousal (i.e., overarousal), performance will also be less than optimal. Therefore, continually increasing arousal can be problematic because it can cause athletes to become overaroused and therefore impair performance.

The challenging aspect of this relationship for coaches is that the **optimal level of arousal** varies across individuals, even in the same sport (figure 19.3). Some athletes will perform their best when feeling relaxed, but others will perform their best when feeling pumped up and activated. Differences in optimal arousal also can be seen when considering different sports. For example, putting in golf generally requires a lower level of activation or arousal than attempting to lift a personal best in bench press.

In practical terms, these individual differences in optimal arousal mean that the identical precompetition buildup will not be equally effective for everyone. Athletes who arrive at the competition feeling sluggish and underaroused may benefit from a big motivational speech from the coach. If, however, other athletes on the team are already at their optimal levels of arousal or are already overaroused, a motivational pep talk would probably have a negative effect on their performances by causing them to become overaroused. Some athletes may need to focus on relaxing and lowering their levels of arousal before competing.

ANXIETY–PERFORMANCE RELATIONSHIP

As mentioned previously, state anxiety is considered to be multidimensional. Somatic state anxiety and cognitive state anxiety have different effects on performance. The relationship between somatic state anxiety and performance is virtually identical to that between arousal and performance: The relationship is curvilinear. As state somatic anxiety increases, there are initial increases in athletic performance. If somatic anxiety continues to increase, performance gradually decreases.

Cognitive state anxiety, however, is believed to have a negative linear relationship with performance. As cognitive anxiety increases, performance decreases (figure 19.4). In simple terms, as people begin to worry, their athletic performances begin to suffer. The more worry and distress, the worse the performance.

Things become complicated when we try to consider the simultaneous effects and interaction of cognitive anxiety and somatic anxiety (or arousal) on performance. When cognitive anxiety is low, the relationship between arousal and performance takes the form of the inverted U. When cognitive anxiety is high, however, the relationship between arousal and performance takes the form of the **catastrophe model**. According to the catastrophe model, if arousal continues to increase after the point of optimal arousal, there is a sharp and rapid decline in performance rather than a gradual decrease in performance. Basically speaking, a catastrophe occurs.

If cognitive anxiety is high, an increase in arousal can cause a terrible performance. For example, if

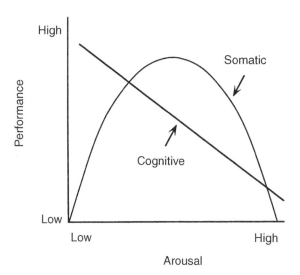

Figure 19.4 Relationship between multidimensional state anxiety and performance.

athletes are playing basketball and have high levels of cognitive state anxiety, an increase in arousal past the optimal point could cause them to miss easy shots and make multiple errors in shot selection, defense, and passing. According to the catastrophe model, for these players to return to their optimal levels of performance while experiencing high cognitive anxiety, they would need to allow their physiological arousal to return to relatively low levels before gradually building their arousal and performance back to optimal levels. Although a relatively small increase in arousal can cause a catastrophe in performance, a minor decrease in arousal will not allow performance to return to its previous level. Instead, a significant decrease in arousal is needed before performance will return to its previous standard.

This catastrophe model, which explains the relationship between arousal and performance under high levels of cognitive anxiety, can be thought of as being like a wave crashing on the beach. Once the wave gets to its highest point, it does not decrease in height gradually, but rather decreases in height suddenly and dramatically. The water is unable to just wash down the beach to where it broke to build up its height again. Instead, it washes back into the ocean before once again gradually building up to its optimal height.

MEASURING ANXIETY

Most researchers and practitioners use self-report measures of anxiety. Although physiological measures such as respiration or heart rates, biochemical changes, and electrophysiological factors can pro-

vide information about anxiety, they are problematic because they do not tend to agree with each other, and individuals differ in how they physiologically respond to anxiety. In addition, most sports involve physical activation that affects these measures. Generally, only stationary sports such as rifle shooting are able to use these measures without confounding anxiety and physical activation.

Self-report measures of anxiety use questionnaires that measure either trait or state anxiety. The trait measures indicate individuals' typical levels of anxiety. State measures determine the levels of anxiety experienced in particular situations. Most questionnaires contain separate scales for cognitive and somatic anxiety. Traditionally, the scales have measured the intensity of anxiety experienced. In the 1990s some researchers added a directional component to scales because two people may experience the same intensity of a symptom of anxiety but interpret its effect differently. For example, athletes might interpret butterflies in the stomach as a sign that the adrenaline is pumping and that they are ready to perform (facilitative), but others might interpret the same butterflies as meaning they are getting nauseous and will perform poorly (debilitative).

THE PRACTICE OF APPLIED SPORT PSYCHOLOGY

What sport psychologists do varies across practitioners. Those who are trained through the sport and exercise sciences tend to focus on performance enhancement and help athletes improve the quality and consistency of their performances by teaching them psychological skills such as goal setting, imagery, self-talk, and relaxation. This approach to applied practice is often termed *educational sport psychology*. Sport psychologists who are educated in the areas of clinical or counseling psychology tend to have broader views of applied practice and treat and care for athletes both in and out of sport. Not surprisingly, this approach to applied practice is usually termed *clinical sport psychology*. Some organisations have created a third category, termed *research sport psychology*, for those who conduct investigations to increase knowledge in the field but rarely work directly with athletes. Ideally, people entering the field will obtain training in both psychology and the sport and exercise sciences. In addition, individuals who will be working with athletes need to complete supervised practica. The specific courses and the number of hours of supervised experience required to be a sport psychologist vary across countries.

IMAGERY: AN EXAMPLE OF PSYCHOLOGICAL SKILL

Both educational sport psychologists and clinical sport psychologists use cognitive–behavioural interventions. Interventions designed to enhance performance are typically based on techniques that allow athletes to learn about the relationships between cognitions, feelings, and behaviours and to control factors that prompt or reinforce behaviours. There are entire textbooks devoted to these psychological skills; what follows is a brief introduction to one of the skills: imagery.

Imagery is a skill that involves using all the senses to create or recreate an experience in the mind. Terms such as *mental practice*, *mental rehearsal*, and *visualisation* have been used interchangeably with the term *imagery*. Mental practice and mental rehearsal could be considered as specific forms of imagery (i.e., using imagery to mentally practise or rehearse particular skills or techniques). Visualisation and imagery are often considered to be synonymous; however, the term *imagery* is preferable because the term *visualisation* implies a restriction to the sense of vision, whereas the term *imagery* more easily incorporates sound, smell, taste, and feel as well as sight.

HOW DOES IMAGERY WORK?

Although sport psychologists, athletes, and coaches generally agree that imagery helps individuals learn new skills and improve performance, there is less agreement about how it works. Three of the existing theories proposed as possible explanations of how imagery works are briefly described; however, no single theory has been proven to be the definitive answer. Nevertheless, each of the theories presented here should help create an understanding of imagery.

▶ **Psychoneuromuscular theory** states that when we image ourselves moving, our brain is sending subliminal electrical signals to our muscles in the same order as when we physically move. Because all of the coordination and organisation of movement takes place in our brains, all of that organisa-

tion and coordination takes place when we image movement. Our brain sends electrical signals to our muscles concerning which muscles should contract when and with what intensity. Therefore, whether an athlete is actually performing a movement or vividly imaging the performance of the same movement, similar neural pathways to the muscles are used. The psychoneuromuscular theory claims that it is through this mechanism that imagery may aid skill learning.

▶ **Symbolic learning theory** is similar to psychoneuromuscular theory except that electrical activity in the muscles is not required. According to the symbolic learning theory, imagery helps the brain make a blueprint (or plan) of the movement sequence without actually sending any messages to the muscles. Imagery helps develop a blueprint that can be followed when action is required.

▶ According to the **attention–arousal set theory**, imagery works because it helps the athlete reach the optimal level of arousal and focus attention on what is relevant. According to this theory, imagery does not send messages to the muscles or help develop a blueprint, but instead primes the athlete for performance. Physiologically, imagery helps the athlete raise or lower arousal levels to the appropriate point. Cognitively, imagery helps the athlete attend to what is important for good performance, decreasing the chance of distraction.

WHY USE IMAGERY?

Imagery has many advantages. It is not physically fatiguing, so practising with imagery just before competition does not cause a decrease in the energy available. Imagery can be practised anywhere and anytime: when sitting on a bus, waiting in the marshalling area, or taking a shower. Imagery can also be a welcome change of pace during a practice session. When an athlete is physically working hard, it can be useful to take a time-out and image specific aspects of technique. Imagery also uses a language that is understood by the body. Sometimes technical corrections of a skill can be understood by the individual athlete yet still fail to translate to the actual movement. Using imagery can enhance the translation. Additionally, in these times of expensive equipment, shoes, and gadgets, imagery is free!

Imagery can be used in different ways. Athletes can use imagery to learn to control emotions. For example, through imagery athletes can imagine themselves shrugging off poor calls made by officials and focusing on what they need to do to perform well instead of losing their cool and receiving penalties or fouls. If athletes practise appropriate reactions through imagery, the desired reactions will become natural when similar situations occur in real life.

Imagery can also be used to improve confidence. If athletes can see and feel themselves performing the way they want to immediately before they go out to perform, they will feel better about themselves. Another way of thinking about this notion is to consider how previous performances relate to confidence. For example, if athletes were about to perform skills in a pressure situation (e.g., in the last minutes or seconds of a game), they would probably feel more confident if they had successfully executed the skill in their last 20 attempts than if they had made errors in their last 20 attempts. Basically, the more times something is done, the easier it is to do. This situation is where imagery can come in handy. Athletes can create that feeling of confidence by imaging themselves performing the way they want to immediately before the physical performance.

Imagery can be used in combination with physical practice to enhance the learning of new skills and the performance of known skills. Although there is some disagreement in the literature, most of the research supports that combining physical practice and mental practice is more effective than either alone. For example, novices learning how to serve in tennis will learn the skill more quickly if they both image themselves serving correctly and physically practise serving. The combination of the two techniques works better than either in isolation.

Although there are many other uses of imagery, only one more is discussed here. Imagery has been shown to help with the control of pain. The most common form of pain-control imagery is removing one's self mentally from the situation. By conjuring up scenes that are unrelated or incompatible with pain, people can learn to distract themselves from pain. For example, instead of thinking about how much a reconstructed knee hurts, an athlete can image relaxing at the beach with the waves gently breaking, the sun warming the skin, and the breeze wafting across the face.

DEVELOPING VIVIDNESS AND CONTROL

If imagery is to be effective, two factors need to be developed: vividness and control. The more vivid an image is, the more likely the brain is going to be convinced that the image is real. When physically participating in sport, most athletes are aware of the feel of the movement as well as sounds and sights. Some sports, such as swimming in a chlorinated pool, are also associated with smells and tastes. The more senses that can be included in an image, the more vivid and therefore the more effective the image will be. Athletes will get more out of imagery if they image in colour and surround sound than if they image in fuzzy, silent black and white. See "Imaging With All the Senses" for an example of guided practice in these techniques.

IMAGING WITH ALL THE SENSES

Try to image the following sensory experiences. Are some images more vivid than others?

See	A colorful sunset The face of a friend
Hear	The roar of a crowd after the home team has scored A door slamming shut
Feel	Stepping into a cold shower A bear hug from a close friend or relative
Smell	Freshly mown grass Cigarette smoke
Taste	Sour milk Chocolate

Vividness alone, however, is not enough. Athletes also need to be able to control imagery for it to be effective. If we try to image a good performance and instead image only mistakes and errors, anxiety will increase and self-confidence will waver. Following prewritten or prerecorded imagery scripts can often help athletes image the desired performance. Following a guided imagery script is often easier than trying to develop an image from scratch. If both vividness and control can be developed, then imagery can be a productive technique for athletes.

SUMMARY

Personality, or the composite of the characteristics that make each of us unique, has been studied using trait and interaction frameworks. Due to problems with some of the questionnaires used as well as inconsistent operational definitions, traditional sport personality research has been inconclusive when trying to determine whether athletes and nonathletes have different personalities.

Motivation is made up of direction, intensity, and persistence. Making the direction of participation enjoyable, providing reasons for particular activities, and providing positive feedback can enhance the motivation of athletes. The motivation of athletes can also be increased by meeting their basic needs for autonomy, competence, and relatedness.

Success in sport is not always equivalent to winning. Individual definitions of success are called achievement goal orientations. The two most studied achievement goal orientations are task and ego. People with a high ego orientation define success as being able to demonstrate that they are better than everyone else. People with a strong task orientation define success as improving their own performances. Coaches, parents, and physical education teachers create motivational climates that differ in how much emphasis they place on task or ego. The majority of research reports that creating a task environment is preferable because it is linked to enjoyment, perceived ability, and effort.

Arousal is traditionally considered to be physiological activation. Anxiety is the subjective feeling of apprehension usually experienced when a situation is perceived to be threatening or when there is a perceived imbalance between the demands of the situation and the ability to meet those demands. Cognitive anxiety, or worry, is the mental side of anxiety. Somatic anxiety, similar to arousal, is the physical facet of anxiety.

Although some coaches continue to subscribe to drive theory, most people involved in sport now recognise that the inverted-U hypothesis more accurately reflects the relationship between arousal and performance. Athletes can have poor performances due to either over- or underarousal. Good coaches realise that the optimal level of arousal varies across sports and individual athletes.

The relationship between somatic anxiety and performance is virtually identical to the relationship between arousal and performance. Cognitive anxiety, however, has a negative linear relationship with performance. When cognitive anxiety is low, the relationship between somatic anxiety and performance takes the form of the inverted U. When cognitive anxiety is high, however, the relationship between somatic anxiety (or arousal) and performance takes the form of the catastrophe model.

Educational sport psychologists and clinical sport psychologists have different training and different views regarding what is involved in the practice of applied sport psychology. Both types of sport psychologists, however, help athletes develop and use psychological skills. Imagery is one example of a commonly taught psychological skill.

Imagery is a skill that involves using all the senses to create or recreate an experience in the mind. The psychoneuromuscular theory, the symbolic learning theory, and the attention–arousal set theory all attempt to explain how imagery works. Although there is not agreement in terms of how imagery works, most people in the field agree that imagery can be used to learn to control emotions, improve confidence, enhance the learning of new skills and the performance of known skills, and control pain. For imagery to be effective, however, individuals need to develop vividness and control of their images.

FURTHER READING

Andersen, M.B. (2000). *Doing sport psychology*. Champaign, IL: Human Kinetics.

Cox, R.H. (2012). Alternatives to inverted-U theory. In *Sport psychology concepts and applications* (7th ed.) (pp. 183-208). Boston: McGraw Hill.

Vealey, R.S., & Greenleaf, C.A. (2010). Seeing is believing: Understanding and using imagery in sport. In J.M. Williams (Ed.), *Applied sport psychology: Personal growth to peak performance* (6th ed.) (pp. 267-304). Boston: McGraw-Hill.

Weinberg, R.S., & Gould, D. (2007). Personality and sport. In *Foundations of sport and exercise psychology* (4th ed.) (pp. 27-50). Champaign, IL: Human Kinetics.

CHAPTER 20

BASIC CONCEPTS IN EXERCISE PSYCHOLOGY

The major learning concepts in this chapter relate to

- ▶ the difference between exercise participation motivation and exercise adherence motivation;
- ▶ the biological, psychological, sensory, and situational factors that interact to influence exercise adherence motivation;
- ▶ the role of goal setting in enhancing motivation;
- ▶ setting goals that are challenging but realistic, positive, controllable, specific, and measurable;
- ▶ the different stages of the transtheoretical model;
- ▶ the possibility of exercise addiction;
- ▶ possible explanations for why exercise enhances psychological well-being; and
- ▶ the influence of exercise on negative mood states and cognitive performance.

The purpose of this chapter is to examine the reciprocal links between psychology and exercise—namely, the effects of psychological functions, such as motivation, on exercise—and the effects of exercise on psychological factors such as feelings of well-being, mood states, and mental performance.

EFFECTS OF PSYCHOLOGICAL FACTORS ON EXERCISE

The three components of motivation—direction, intensity, and persistence—mentioned in chapter 19 in relation to sport apply to exercise motivation as well. Direction clearly relates to whether an individual chooses the direction of the gym or the couch, the stairs or the elevator, or the pool or the bath. Once the direction of exercise has been chosen, intensity and persistence become important. For example, as a New Year's resolution two individuals may have both decided to join a gym. They have both chosen the direction of exercise. During their first day at the gym, one rides the bicycle for 30 min, tries the stepper and the rowing machine, completes multiple sets on the different weights available, and finishes off with an aerobics class. The other individual begins with just a few minutes on the bicycle and then completes just one set of each of the different weights exercises using light weights. The first person in this example exhibits high intensity and the second person exhibits low intensity. If we stop at this point in the example, you may conclude that the first person is more motivated than the second. Your opinion may change when the third component of motivation—persistence—is considered. Two days later the first individual is home in bed, too sore to move (thinking that exercise is painful and should be avoided whenever possible). The second individual, however, is once again at the gym adding a couple of minutes to his time on the bicycle and sticking to his light-weights workout.

EXERCISE PARTICIPATION MOTIVATION

Exercise participation motivation refers primarily to the direction component of motivation. Exercise participation motivation is the initiation of exercise. A variety of factors influence whether individuals initiate an exercise program.

Knowledge, attitudes, and beliefs about exercise influence motivation toward exercise participation. Individuals who understand the importance and value of regular exercise are more likely to initiate an exercise program than those who do not. Similarly, if people have positive attitudes about the value and importance of regular exercise they will have greater motivation to participate in exercise than will people with negative attitudes.

Valuing the importance of exercise, however, is not the only determinant of exercise participation. Beliefs about ourselves influence motivation as well. Even if individuals understand that exercise is important, they will be unlikely to begin an exercise program if they believe that they cannot succeed at it. If they believe that the exercise program is too difficult or that it requires more fitness, strength, coordination, or time than they have, it is doubtful that they will join the program. This confidence in one's ability to succeed at an exercise program is called **exercise self-efficacy**. It is logical that self-efficacy would influence behaviour. How likely would people be to do something they were convinced they could not do, particularly if they had to pay to do it? How likely would they be to invest any energy in pursuing that activity? Most people in this situation would not attempt the activity. People who have such feelings are described as having low self-efficacy. As **self-efficacy**, or one's belief in one's ability to succeed at a particular task, increases, so does the likelihood of undertaking that task.

How can we enhance exercise participation motivation? Educating people about the importance and value of exercise can be a valuable first step because individuals who understand the merit of exercise are more likely to adopt an exercise program. Unfortunately, imparting knowledge is not enough. A lot of us do things that we know are not good for us and, similarly, do not do things that we know are good for us. Enhancing the exercise self-efficacy of individuals will increase their motivation. Demonstrating how individuals can control their own activity is useful. Some people have low self-efficacy about exercise because they believe that they are too unfit to begin exercising. They may equate exercise with young, thin, Lycra-clad gym enthusiasts whom they see in the media. Programs that emphasise choice of activities, illustrate exercisers similar in age and fitness level to the potential exercisers, and reveal that exercising can be enjoyable may increase exercise participation motivation. Exercise programs that begin with activities that the individuals already know they are capable of doing, such as walking and climbing stairs, may also increase self-efficacy and thereby increase motivation. Imagery may also help individuals enhance their self-efficacy beliefs (see "Imagery, Exercise, and Self-Efficacy").

EXERCISE ADHERENCE MOTIVATION

Although many people get motivated to begin an exercise program, many of the people who begin fail to continue. Approximately 50% of individuals who begin a regular physical-activity program drop out within the first 6 mo. These people had exercise participation motivation, but they lack exercise adherence motivation, the persistence component of motivation.

Biological, psychological, sensory, and situational factors all interact to influence exercise adherence. Biologically, body composition, aerobic fitness, and the presence of disease influence adherence. Unfortunately, it is usually the people who could gain the most from exercise who are the least likely to adhere. Overweight or obese, unfit, or chronically ill people are less likely to adhere to an exercise program than are thinner, fitter, and healthier people.

As is the case with exercise participation motivation, attitudes and beliefs influence exercise adherence motivation. Attitudes and beliefs about the importance of exercise play a role in adherence, but so too do individuals' expectations about the effects that exercise is having on them personally. For example, if individuals believe that major changes in fitness and body composition should occur after 6 wk of regular exercise and they do not perceive major improvement in their own bodies after 6 wk,

IMAGERY, EXERCISE, AND SELF-EFFICACY

Individuals with high levels of efficacy tend to be more physically active than those with low levels of efficacy. Different types of efficacy, however, may affect different types of behaviours. Task efficacy is confidence in one's ability to perform basic aspects of exercise. Coping efficacy is confidence in one's ability to deal with difficulties and challenges related to exercise (e.g., exercising when in a bad mood). Scheduling efficacy is confidence in one's ability to effectively find time to exercise on a regular basis.

Imagery has been proposed as an important determinant of exercise behaviour. Nevertheless, there are different types of exercise imagery: appearance health (e.g., imaging one's self as fitter), exercise technique (e.g., imaging good lifting technique), exercise self-efficacy (e.g., imaging completing workouts), and exercise feelings (e.g., imaging being relaxed from exercising). Questions remain about which types of imagery affect which types of behaviour or efficacy beliefs.

In a study by J. Cumming, 162 exercisers at different fitness clubs in England completed questionnaires that measured exercise imagery, self-reported exercise behaviour, and the three types of exercise efficacy. Individuals who indicated frequent images related to appearance and health tended to report greater exercise behaviour and greater coping efficacy compared with those with less-frequent images. Cumming suggested that these results indicate that appearance- and health-related images may help increase individuals' intentions to exercise and help sustain or increase their exercise behaviours.

Another significant finding in the study was that technique imagery was related to task efficacy. Individuals who engaged in frequent technique imagery reported greater confidence in their abilities to perform exercise activities.

Nevertheless, before hailing imagery as the cure for couch potatoes, additional research is needed. This study was cross-sectional, meaning that conclusions cannot be drawn about the causal relationships between exercise imagery and efficacy or behaviour. Intervention studies are needed to determine whether exercise imagery leads to changes in efficacy or behaviour. In addition, some individuals may have limited imagery ability and need to first develop imagery skills such as vividness and control before imagery might effectively lead to increases in self-efficacy beliefs and exercise behaviour.

Source

Cumming, J. (2008). Investigating the relationship between exercise imagery, leisure-time exercise behavior, and self-efficacy. *Journal of Applied Sport Psychology, 20,* 184-198.

they may believe that exercise does not do what it should and therefore quit. Even though it is unrealistic to believe that 6 wk of exercise can make up for 6 yr of inactivity, it is individuals' beliefs, not reality, that influence behaviour. Therefore, when introducing newcomers to exercise it is important that they have realistic expectations about the time and effort required and the anticipated effects of the proposed program.

In addition to attitudes and beliefs, other psychological factors influence exercise adherence motivation. Extroverts (people who are social and outgoing) tend to adhere to exercise programs better than do introverts. Exercise programs that are executed in the presence of other people are probably more comfortable for extroverts than for introverts. Extroverts tend to enjoy the interaction with class members and exercise partners, possibly encouraging their adherence. Introverts, on the other hand, may adhere better to individual, home-based exercise programs. The bulk of the research on exercise adherence has involved programs that take place on site at fitness facilities with other people. This setting may have led to the conclusion that extroverts are better adherers than introverts. If the research had been done on independent, home-based exercise programs, it might have been found that introverts were better adherers. Clearly efforts need to be made to match the social environment of the exercise program to the personality of the exerciser.

Individuals with high levels of self-motivation are more likely to adhere to exercise programs than are individuals with low levels of self-motivation. It is logical that highly self-motivated people have better adherence rates. The challenge is to help those individuals with low levels of self-motivation. One of the most effective methods of helping these individuals is to encourage their involvement in the goal-setting process.

GOAL SETTING

Setting goals can help enhance motivation for a number of reasons. Goal setting addresses all three components of motivation. Goals give direction by providing a target. Intensity and effort also can be enhanced because goals provide reasons for participating in the activity. We are all more likely to put in effort when we feel there is a reason for doing so. If individuals are given two jobs to do at work, one that has a particular target and objective and one that seems vague and purposeless, into which job are they more likely to put their effort? Having goals helps focus attention and effort. In addition, goals can augment persistence by fostering new strategies. If individuals have a goal to which they are committed and initial tactics appear unsuccessful, they will search for alternative strategies to achieve their aim. If the goal had not been set in the first place, instead of persisting with different plans of action, they would likely give up.

Goals are also beneficial because they reflect improvement. Too often people make short-term comparisons regarding their strength, fitness, flexibility, or weight. Because the positive effects of exercise take time to emerge, improvements being made are often not noticed when people use a short timeframe for comparison. If a goal is achieved, evidence of improvement exists.

Goal setting involves a number of steps:

▸ setting the goal,

▸ setting a target date by which to achieve the goal,

▸ determining strategies to achieve the goal, and

▸ evaluating the goal on a regular basis.

If a target date is not set (e.g., "One of these days I'll ride the exercise bike continuously for an hour"), the goal is really just a dream. For this idea to be a goal, the individual needs to set a specific date by which to achieve the behaviour. Goal setting usually involves both long-term and short-term goals. The long-term goal provides direction; the short-term goals provide the increase in intensity and effort. People often err by setting only long-term goals. They begin to work toward achieving the goal, but success seems so far away that they give up before they get there. If someone had decided to ride a bike for 1 h and was currently having trouble lasting 10 min, 1 h would seem virtually impossible. Achieving short-term goals along the way to the long-term goal boosts confidence and motivation because it is obvious that the effort is worthwhile because improvement is being made. Target dates are set for each short-term goal in a progressive order until the long-term goal is achieved. This pattern of goal setting can be considered as a staircase, where each short-term goal is a step on the way to the long-term goal (see figure 20.1).

For goals to be effective, however, they need to be properly set. Goals can be considered to be good

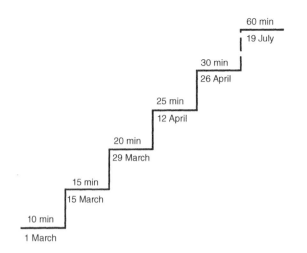

Figure 20.1 The goal-setting staircase.

if they meet certain criteria (table 20.1). Goals need to be challenging but realistic. If goals are not challenging, they probably are not requiring any real change in behaviour and therefore will have little effect. However, if goals are so challenging that they are unrealistic, people are setting themselves up for failure. Continued failure leads to lowered confidence and less motivation.

Goals also need to be specific and measurable. Saying "I want to be fitter" or "My goal is to be stronger" does not provide any way of knowing when success has been achieved. What is "fitter"? How strong is "stronger"? There needs to be some way of knowing whether the goal has been achieved when the target date arrives. The easiest way of making goals specific is to make goals numerical. Numbers can easily be used to quantify time spent exercising, distance travelled, repetitions accomplished, weight lifted, or exercise sessions attended.

In addition to being specific and measurable, goals need to be positive. If an exerciser makes it her goal not to recline and rest during the abdominals section of her aerobics class, she will be thinking about reclining and resting. If instead she makes it her goal to complete first 1 min and then 2 min of the abdominal exercises, she will be thinking about doing the exercises, thus increasing her chances of doing them. If we tell you not to think about pink elephants, what is the first thing you think about? Similarly, if you set a goal of not letting your back arch off the bench when you do bench presses, you will be thinking about your back arching. When you think about your back arching, your brain may be sending messages to the muscles that make your back arch (just like in the psychoneuromuscular theory of imagery discussed in chapter 19). By making your goal "not arching," you may actually be increasing the likelihood of arching your back. A goal of keeping your back pushed flat against the bench would be much more effective. Goals should stipulate the desired behaviour. Positively worded goals help you think about, plan for, and prepare to do what it is you want to do.

For goals to be effective, it is also important that the person setting the goal has control over the activity. Goals should be related to the performance, technique, or behaviour of the goal setter. Goals are ineffectual when they rely on the behaviour of others. Ultimately, people have control only over what they do. Therefore, goals such as being the strongest person in the gym or getting everyone in one's family involved in regular exercise are not completely under personal control. Best efforts might result in failure because individuals cannot control others. If someone really does want to be the strongest person in the gym, goals that reflect what

Table 20.1 Goals Need to Be . . .

Principle	Question to determine whether the principle has been met
Challenging	Will the goal require effort?
Realistic	Is the goal reasonable?
Specific	Is it obvious what the precise objective is?
Measurable	Is there an exact method of determining whether the goal has been achieved?
Positive	Does the goal stipulate the desired behaviour?
Controllable	Does the goal relate to the performance, technique, or behaviour of the goal setter?

he thinks he needs to do to achieve that should be set and he should focus on his own behaviour. No one can help it if the defending world powerlifting champion suddenly moves to the neighborhood and joins the local gym. We can, however, set goals about how much weight we will lift, how many sets or repetitions of what specific exercises we will do, or how many times each week we will train. Similarly, if an individual really wants everyone in her family to be involved in regular exercise, goals related to what she will do to try to achieve that should be set. For example, she might set a goal of organising weekly family fun days that involve physical activity, creating a package of exercise options that fit their schedules, or babysitting the young kids for her brother and his wife so they have the opportunity to go to the gym together. The focus is on what individuals can do themselves rather than on what other people may or may not do.

In summary, goal setting can be an effective method of improving exercise adherence motivation. The goals need to be challenging but realistic, specific and measurable, positive, and controllable. If the goals require a lot of work or a major change in behaviour, a series of short-term goals that lead to the long-term goals should be set. For each goal there needs to be a target date and a list of strategies that can be used to achieve the goal. Finally, the goals need to be evaluated on a regular basis. If a goal has been achieved, great! People can reward themselves by giving themselves a pat on the back, buying themselves that CD they have been wanting, or just feeling good because of their achievement. If, however, the goal has not been achieved, they need to think about what may have happened. Possibly the goal was too big of a step on the staircase. In that case, maybe a smaller step would be more appropriate. Maybe they became sick or injured and could not work toward achieving their goal. In that case, a new target date should be set. Perhaps the goal is realistic but the strategies selected to achieve it were not suitable. In that case, a new target date should be set and different strategies implemented. Goals can strengthen self-motivation, which, in turn, can enhance exercise adherence motivation.

SENSORY FACTORS

The same workload is going to be perceived differently by various individuals. If someone perceives an exercise program as being excessively stressful, that person is apt to drop out. Therefore, when exercise is prescribed, it is important to take into account individuals' perceptions of the difficulty of the exercise and how much distress they experience when participating.

Problems sometimes arise when fit individuals who are used to training with their peers invite less-fit (or unfit) friends to join them while exercising. Although they genuinely may have intended to help their friends by encouraging them to exercise, the opposite may occur. The usually sedentary friends perceive the exercise sessions as anything but enjoyable. Because of the negative experiences, they are even less likely to exercise in the future.

Exercise professionals who are qualified to prescribe exercise take into account the initial fitness levels of clients. This process decreases the chances of novice exercisers experiencing excessive stress when exercising. Nevertheless, when working with a large number of clients, exercise-management professionals sometimes are tempted to generalise. People with certain fitness levels tend to be given particular exercise programs. Although individual variations in fitness are accounted for, individual differences in perceived exertion and perceived effort are sometimes ignored. People exercising at the same relative work intensity may have diverse experiences of that exercise. It is not only the fitness levels that need to be considered, but also the individuals' perceptions of the exercise. One of the challenges of the exercise-management profession is to cater to a large number of people while taking into account individual differences in physiological fitness and subjective individual differences in perceptions of the exercise experience.

One determinant of how exercise is perceived is whether we dissociate or associate while exercising. Associating is attending to the body while exercising and being aware of what our muscles are feeling, how we are breathing, and even our heart rates. Dissociating is using attentional strategies that distract ourselves from the fatigue-producing effects of exercise. Dissociation can involve listening to music, daydreaming, planning our day, checking out the bodies of the other exercisers, or focusing on anything else that keeps our minds off the actual exertion of our own bodies. People who dissociate while exercising tend to have better rates of adherence than do those who associate. This phenomenon may be one reason why aerobics classes have become so popular. It is doubtful that many of the people currently participating in aerobics classes would continue to do the same exercises if they were done in silence. Focusing on moving the right way at the right time and attending to the music, the instructor, or the other people in the class keeps the mind off of how the heart is beating, the muscles are straining, or the breath is gasping.

SITUATIONAL FACTORS

A number of situational factors also influence exercise adherence motivation. The size of the exercise group has been shown to influence adherence. Although the ideal group size has not been determined precisely, if the group is too large adherence may decrease because the individual may feel lost and unimportant. The larger the group, the less individual feedback each participant receives from the exercise leader. Additionally, many people find it difficult to get to know people in a large group. With a great number of people, there is less chance that anyone will notice the absence of a single individual. Some exercisers are motivated to adhere to a program because they do not want others to think they could not cope. Smaller groups allow for the development of social relationships and more attention and recognition from the instructor. It should be noted, however, that some exercisers, particularly those with low self-confidence, prefer the anonymity of a large class.

Ease of access to the exercise venue also affects adherence. Convenience of the program influences how often people are likely to attend. Convenience can involve a number of components. If the program is in a location close to home or to work, there is greater chance of attendance than if an additional long journey to the venue is required. Similarly, even if we drive straight past the venue when returning home from work, difficulty in finding a parking place could make us think that the hassle is not worth the bother. Exercise programs are more accessible if child-care facilities are available to parents of young children. Additionally, how pleasant the exercise environment is may sway individuals to attend more or less frequently. Although some gym junkies may think that a smelly, confined space in need of a coat of paint is perfectly fine for lifting weights, others may be immediately dissuaded by these conditions.

Having the social support of others also can influence adherence. If a spouse, friend, or significant other is supportive of our exercise behaviour, many advantages may ensue. First, it is likely that we will receive positive reinforcement from the person, increasing our self-worth and feelings of competence. In addition to encouraging our exer-

cise behaviour, others may provide informational support by giving advice or suggestions that may decrease the chance of injury or increase the benefits gained through exercise. Tangible or instrumental support can also be provided by others. For example, they may lend exercise equipment, provide transportation, or share child care. Social support can be a definite advantage.

A lack of social support would be better, however, than the presence of social disapproval. Ridicule of exercise attempts indisputably can have a negative effect on adherence. Outright resistance to exercise involvement can create additional hurdles. For example, an individual may try to make a positive health change by stopping by the local gym for a 1 h exercise class after work 3 times per week. A spouse who condemns this activity can become an insurmountable barrier to adherence. This disapproval may be exhibited by the silent treatment, sarcasm, or outright rage. When one person makes a behaviour change, it may have an effect on others. Ride-sharing arrangements may have to be changed, dinner may be an hour later, or child-care responsibilities may vary. If the partner is opposed to the lifestyle alterations, the exerciser may experience the exact opposite of social support: social disapproval.

Many factors influence exercise adherence motivation. Understanding the benefits of exercise is only a preliminary step in encouraging exercise behaviour. Biological factors, attitudes, beliefs, personality characteristics, goal setting, sensory perceptions, group size, program convenience, and social support all influence exercise adherence. If people want to enhance the exercise adherence of themselves or others, they should consider all of these factors.

TRANSTHEORETICAL MODEL

When developing interventions to increase the exercise behaviour of individuals, it is important to note that no single intervention will be effective for everyone. Some of these individual differences are due to people being at different stages of change. Changing behaviour is a process that occurs over time. Individuals do not instantaneously change from smoker to nonsmoker, from junk-food junkie to health nut, or from couch potato to exerciser. The **transtheoretical model** suggests that individuals progress through a series of stages of change: precontemplation, contemplation, preparation, action, and maintenance.

In the **precontemplation** stage, individuals are not even thinking about changing their unhealthy behaviours. In terms of exercise, precontemplators are not exercising and have no intention of beginning an exercise program in the next 6 mo. Individuals at this stage gain more from increasing their awareness of the positive effects of exercise and recognising the barriers that may be preventing them from exercising than from specific information about how to start an exercise program.

In the **contemplation** stage, people still are not ready for structured exercise programs but they are thinking about exercising within the next 6 mo. They are probably aware of the benefits of exercise but even more aware of the barriers. At this stage it may be beneficial to encourage individuals to try a variety of activities to get a taste of what might be involved in regular exercise.

People in the **preparation** stage have taken some steps toward engaging in regular exercise. They may have contacted a physician, joined a gym, or bought a new pair of running shoes or the latest exercise gadget advertised on television. They are probably exercising irregularly but have plans of exercising 3 or more times per week beginning sometime in the next month. In this stage individuals can benefit from information about goal setting, suggestions for safe and enjoyable activities, and the recognition of obstacles to regular exercise.

Individuals in the **action** stage have modified their behaviour and have begun to exercise regularly but have been doing so for less than 6 mo. At this stage information about techniques for staying motivated, overcoming obstacles, and enhancing confidence can be useful.

The **maintenance** stage is achieved when there is little risk of returning to sedentary behaviours, usually after a period of 6 mo of regular exercise. Efforts, however, still need to be made to avoid relapse, or returning to an earlier stage of change. To prevent relapse individuals can focus on refining specific types of exercise behaviour, injury avoidance, rewarding themselves for the attainment of goals, and methods of reducing boredom.

Some researchers have suggested a sixth stage called **termination**. Exercisers in the termination stage have exercised regularly for at least 5 yr and are considered by some to have exited from the cycle of change. The argument is made that after 5 yr of regular exercise, relapse does not occur. Others argue that although long-term exercisers may be resistant to relapse, uncontrollable factors such as

long-term illness or injury may still result in these people becoming sedentary.

When interventions are matched to the relevant stages of change, intervention programs have a much higher chance of success. Success in some programs may be moving people from the precontemplation stage to the contemplation stage. Research has demonstrated the effectiveness of basing exercise interventions on the transtheoretical model with adolescents, college students, sedentary employees, and adults 65 yr and older.

EXERCISE ADDICTION

Some people have no trouble adhering to exercise (i.e., achieving the maintenance stage). In fact, some exercisers become addicted to exercise. Individuals can be defined as being addicted to exercise when physical or psychological withdrawal symptoms are experienced after 24 to 36 h without exercise. An **addiction** occurs when there is dependence on or commitment to a habit, practise, or habit-forming substance to the extent that its cessation causes trauma. Signs of exercise addiction include increased tolerance to exercise, withdrawal symptoms (e.g., mood disorders) when one cannot exercise, relief of withdrawal symptoms when exercise is resumed, rigid exercise schedules, and the compulsion to exercise.

Some believe that exercise can be a positive addiction because there are many physical and psychological benefits of exercise. If discomfort or other negative effects are experienced when an individual lets more than 1 to 2 d pass without exercising, the individual may be more likely to exercise on a regular basis. In this situation, when the individual has control over the exercise, it should be considered to be a positive health habit, not an addiction.

Addicts do not have control over their exercise. Although exercise is usually a positive and healthy habit, there are those who think that more is always better. These individuals can end up with overuse injuries and psychosocial problems. Although minimum exercise guidelines have been suggested for fitness benefits, there is little information about how much is too much. For those addicted to exercise, exercise becomes a detriment. They may exercise despite pain and injury. Similar to individuals addicted to gambling or drugs, the substance of the addiction becomes more important than anything else.

In some cases, life is determined by when and where exercise is available. Minor injuries can develop into serious conditions. But in addition to the probable physical damage that can be incurred by excessive exercise, psychosocial problems turn into major predicaments. Exercise addicts have lost jobs and families because exercise became more important than work or relationships. In exercise addicts, exercise controls the individual.

Encouraging participation in a wide range of exercise and recreational activities can be the first step in helping individuals who are addicted to exercise. If one part of the body is injured, a person can engage in some other form of exercise or recreational activity that can allow the affected body part to recover. In addition, alternative forms of recreational activity can replace compulsive exercise behaviour. Providing a range of activities increases the chance that individuals will interact with others, possibly becoming less self-absorbed. Also, making decisions about the activities in which to participate begins to give individuals power and control over exercise.

EFFECTS OF EXERCISE ON PSYCHOLOGICAL FACTORS

Up to this point, this chapter has concentrated on how psychological factors influence exercise participation and adherence. The remainder of this chapter focuses on how exercise can influence psychological factors. Traditionally when the positive effects of exercise are discussed, the majority of the benefits are seen as being physiological. For example, it is well established that regular physical exercise is associated with lower cholesterol, lower blood pressure, reduced weight, and a decreased percentage of body fat. There are, however, many psychological benefits of exercise as well. Exercise is associated with enhanced psychological well-being, reduced state anxiety, decreased depression, and improved cognitive performance.

EXERCISE AND PSYCHOLOGICAL WELL-BEING

Exercise has been shown to influence how people feel about themselves. When people exercise they tend to feel more positive and self-confident than when they are sedentary. This relationship between exercise and psychological well-being has been demonstrated both in terms of long-term exercise and single bouts of exercise. Although there is general agreement that there are positive psychological effects of exercise, there is disagreement about why this relationship exists. Some possible explanations follow.

Individuals may experience a sense of mastery or achievement through exercise. They feel they are successful when they are able to walk for a longer period of time, run faster, stretch further, or lift more weight than they could before. It may be this experience of achievement that makes people feel better about themselves. Through mastery in exercise, people may realise that they have the capacity for change. They realise that they were able to change their exercise habits, their fitness levels, or even their body shapes. This realisation that change is possible may transfer to other areas of life, giving individuals a greater feeling of control.

Some argue that exercise is psychologically beneficial because it provides a distraction from problems and frustrations. They believe that instead of exercise having a positive effect in and of itself, it really just allows people to take a time-out from the problems in their lives. This temporary respite from worry may be the main positive influence of exercise.

Another proposed explanation for the positive effect of exercise on psychological well-being is that exercise causes biochemical changes that in turn influence psychological factors. The most commonly suggested biochemical change is an increase in endorphin levels. Endorphins are naturally occurring substances with opiate-like qualities. Endorphins are important in regulating emotion and the perception of pain. If exercise increases the release of endorphins, it may be the endorphins that cause the enhanced psychological well-being. Although theoretically this argument makes sense, the relationship has not been proven. Endorphin levels vary across individuals exercising at the same relative intensity, making it difficult to prove that any changes in psychological well-being are attributable to exercise-induced changes in endorphin levels.

Although a definitive explanation of the positive effects of exercise on psychological well-being does not currently exist, the psychological benefits of exercise should not be overlooked. Just because we do not understand why something happens does not mean we should fail to take advantage of the phenomenon. Exercise has been shown to increase the quality of life in many populations. Studies investigating individuals with multiple sclerosis, breast cancer, kidney transplants, mild traumatic brain injury, human immunodeficiency virus, chronic respiratory diseases, end-stage renal disease, or Parkinson's disease have all demonstrated enhanced quality of life as a result of exercise. In addition to enhancing self-confidence, feelings of control, and general well-being, exercise appears to reduce negative moods.

EXERCISE AND NEGATIVE MOOD STATES

Exercise effectively reduces depression. Although exercise is valuable in reducing depression in mentally healthy individuals, greater decreases in depression through exercise occur in those requiring psychological care. Both long-duration, lower-intensity and short-duration, higher-intensity exercise training have been found to be effective in reducing depression. Whatever the type of exercise, the more sessions per week (within reason) or the more weeks in the exercise program, the greater the decrease in depression.

Exercise also has been shown to effectively decrease both state and trait anxiety. Individuals who are initially low in fitness or high in anxiety tend to achieve the greatest reductions in anxiety from exercise. Unlike depression, however, anxiety appears to decrease only with endurance exercise. Short-duration, high-intensity exercise such as weight training may increase anxiety.

We need to keep in mind that exercise will probably not have a positive effect on mood states and quality of life for everyone. Some individuals find exercise to be extremely aversive and take part only because of pressure from medical staff, family, and friends. If coerced into exercising, they may focus on the negative aspects of the experience, find the entire experience quite onerous, and actually increase their stress levels.

For most people, however, exercise has been so effective in decreasing anxiety and depression that clinical psychologists have used endurance-exercise training as a form of therapy. In some cases it is the sole psychotherapeutic tool, but more often it is used in conjunction with other modes of therapy. Running is the most common form of exercise chosen for therapy. Obviously, however, exercise is not a panacea for anxiety and depression. It is difficult to get the average person to exercise regularly; it is even more difficult to get depressed people to exercise regularly.

EXERCISE AND COGNITIVE PERFORMANCE

Exercise not only influences how individuals feel, it influences how well they think! Regular exercise is associated with improved cognitive performance. Studies have shown that regular exercisers perform better on reasoning tests, mathematics tests, memory tests, and intelligence quotient tests than individuals who do not exercise regularly. There is also some evidence that exercisers have better creativity and verbal ability than nonexercisers. Although the reasons for this relationship between exercise and cognitive functioning are not completely understood, it has been suggested that exercise in older individuals may slow neurological deterioration. For younger individuals, exercise may increase the vascular development of the brain as well as increase the number of synapses in the cerebellar cortex in the brain (of the type discussed in chapter 18). More research is needed to ascertain whether there is a cause–effect relationship.

SUMMARY

Motivation is made up of direction, intensity, and persistence. Exercise participation motivation is the direction component, referring to the initiation of exercise. Knowledge, attitudes, and beliefs about exercise influence motivation toward exercise. Of greater importance, however, is individuals' self-efficacy, or their belief in their abilities to exercise.

Exercise adherence is the persistence facet of motivation. Biological, psychological, sensory, and situational factors interact to influence exercise adherence. Setting effective goals with target dates and strategies can strengthen self-motivation, which, in turn, can enhance exercise adherence.

The transtheoretical model suggests that individuals progress through five stages of change: precontemplation, contemplation, preparation, action, and maintenance (with perhaps the sixth stage of termination). Matching intervention programs to the relevant stage of change increases the chances that the intervention will be successful in changing exercise behaviour.

Although a lot of energy has gone into research and programs designed to increase the percentage of the population that is exercising regularly, a small percentage of people are addicted to exercise. Exercise is not a positive health habit for these people; rather, it controls their lives, sometimes to the detriment of jobs and personal relationships.

Although there are many psychological factors that influence if and how much people exercise, the reverse is also true. Exercise affects a number of psychological factors. Exercise has a positive effect on psychological well-being and quality of life. This positive effect may be due to mastery experiences, distraction, biochemical changes, or other factors. In addition to enhancing self-confidence, feelings of control, and general well-being, exercise decreases depression and anxiety. Finally, exercisers may have a cognitive advantage over nonexercisers.

FURTHER READING

Berger, B.G., Pargman, D., & Weinberg, R.S. (2006). Motivational strategies to enhance exercise adherence. In *Foundations of exercise psychology* (2nd ed.) (pp. 245-262). Morgantown, WV: Fitness Information Technology.

Dishman, R.K., & Chambliss, H.O. (2010). Exercise psychology. In J.M. Williams (Ed.), *Applied sport psychology: Personal growth to peak performance* (6th ed.) (pp. 563-595). Boston: McGraw-Hill.

Roberts, G.C., & Kristiansen, E. (2010). Motivation and goal setting. In S.J. Hanrahan & M.B. Andersen (Eds.), *Routledge handbook of applied sport psychology: A comprehensive guide for students and practitioners* (pp. 490-499). London: Routledge.

Weinberg, R.S., & Gould, D. (2007). Exercise and psychological well-being. In *Foundations of sport and exercise psychology* (4th ed.) (pp. 397-413). Champaign, IL: Human Kinetics.

CHAPTER 21

PHYSICAL ACTIVITY AND PSYCHOLOGICAL FACTORS ACROSS THE LIFE SPAN

The major learning concepts in this chapter relate to

- considering whether participating in sport affects personalities or whether people with certain types of personalities are more likely to participate in sport,
- debating whether sport participation positively or negatively affects the psychosocial development of children,
- the influence of peers on exercise behaviour in adolescents,
- the effects of exercise on life satisfaction in the aged,
- ways in which older adults may be encouraged to engage in exercise and sport,
- the termination of athletic careers as a challenging transition, and
- preparing well in advance for the termination of athletic careers.

Like the other biophysical subdisciplines of human movement studies, sport and exercise psychology provides basic concepts that are of value in explaining changes in human movement across the life span and in response to training and practise. This chapter discusses some basic concepts of sport and exercise psychology and explores their application to human movement at different points in the life span.

CHANGES IN PERSONALITY

As noted in chapter 19, some personality differences have been found between athletes and nonathletes. Although there is some question about whether these differences are real or merely artifacts arising from problems with operational definitions and the validity of questionnaires, there is also considerable debate about what these differences might mean (assuming they are real). Some conclude that participation in sport causes personality changes; by participating in sport people become more independent and extroverted. If this is the case, if you want to help people become more independent and extroverted, having them participate in sport may be beneficial. Others, however, argue that participation in sport does not cause personalities to change. Instead, they suggest that individuals with certain types of personalities are more likely to participate in sport. Participants are more likely to have the characteristics of extroversion and independence before taking part in sport.

Which side is right? Does participation in sport lead to certain personality traits, or do certain personality traits lead to participation in sport? Although there is no definitive answer to this question, most of the evidence suggests that the latter is more accurate: Individuals with certain personality traits seem to gravitate toward sport. However, exceptions always exist. Although extroverted, independent individuals may be more likely to participate in sport, there are many individuals participating in sport who are either introverted or dependent on others. Additionally, there is some evidence indicating that participation does influence the personality development of young participants. Although involvement in sport as an adult probably has little if any effect on personality, involvement in sport when young may influence personality development as individuals are maturing.

PSYCHOSOCIAL DEVELOPMENT THROUGH SPORT PARTICIPATION

Experiences in youth sport may have an effect on factors in addition to personality development. Childhood experiences in sport can also influence values, attitudes, and beliefs. Through sport children may learn about cooperation, respect, leadership, assertiveness, discipline, and fair play. They may learn that hard work results in positive achievement and develop self-esteem and self-confidence as a result.

Not all effects of participation in sport are positive, however. For example, if children encounter numerous negative events in sporting situations, they are likely to avoid the sporting environment as adults. If fitness activities such as running and push-ups are used as punishment in youth sport, children may learn to associate fitness-related activities with penalties. If running is something we must do when we have made an error or misbehaved, why should we ever do it voluntarily? Also, if youth coaches believe that winning is the most important result of participation in sport, children may learn about aggression and cheating instead of assertiveness and fair play. Similarly, just as children can develop positive self-esteem and self-confidence through participation in sport, they can develop negative self-esteem and lose self-confidence if their experiences in sport are demeaning or humiliating. Sport itself is neither good nor bad. The psychosocial development of children through sport is largely dependent on the quality of the sport experience, and this, in turn, is often determined by the quality of the adult leadership provided.

DESIGN OF YOUTH SPORT

Adults need to be aware that although sport programs have the potential to benefit the growth and development of children, they also have the potential to be detrimental. Many adults mistakenly believe that any sport experience is better than none. Although all children should have the opportunity to participate in sport, these opportunities need to allow children to experiment, make mistakes, and succeed without fear or pressure. Many of the problems in youth sport occur because of the inappropriate behaviour of adults. Adults sometimes design youth sport programs using the same structure that is used in professional adult sports. It is not uncommon, unfortunately, for children as young as 7 yr of age to be cut from teams or permanently benched. Winning is frequently overemphasised, youngsters occasionally are harangued because they do not perform or think like adults, and parent behaviour on the sidelines sometimes keeps children from enjoying the sporting experience.

For sport to have a positive effect on the psychosocial development of children, adults need to remember that children are children, not miniature adults. When fun and development are emphasised instead of winning, many potential problems are avoided and the chances of sport being a beneficial experience are augmented. It needs to be recognised that growth and development include psychological and social considerations in addition to the traditional emphasis on the physical. When fun and development are stressed instead of winning, the negative experiences of fear, inadequacy, anxiety, and inferiority are more easily avoided.

Research has demonstrated that the quality or type of adult leadership in sport influences the attitudes of the children involved. Youngsters who have coaches who use encouragement and praise and provide technical instruction like their coaches more and think their coaches are better teachers compared with children who have coaches who are either negative or fail to provide information. These results are not surprising. Less obvious, however, is that the children with positive coaches

also like their teammates, enjoy being involved in the sport, and desire to continue participating to a much greater extent than children who play under less-positive coaches. Participating in sport can be a good experience for youngsters. Nevertheless, parents, teachers, and coaches need to realise that they have to actively structure the sporting environment to ensure that sport is perceived as a pleasure rather than as an ordeal. See "Having Coaches Who Focus on Effort and Personal Development Is More Important Than Winning" for a study investigating the relative influences of coaching behaviours and won–lost records on young athletes' sport enjoyment and evaluations of their coaches.

HAVING COACHES WHO FOCUS ON EFFORT AND PERSONAL DEVELOPMENT IS MORE IMPORTANT THAN WINNING

The concept of motivational climate is introduced in chapter 19. Coaches can create a mastery-involving motivational climate by emphasising skill development and effort rather than winning. When coaches focus on winning, punish athletes for making mistakes, or favour the most-talented players, they create an ego-involving motivational climate. Nevertheless, sport usually has outcomes that can be identified as winning or losing, both of which clearly influence the motivation and affect of young athletes. The purpose of the study by S.P. Cumming and colleagues was to examine the effect of motivational climate and won–lost percentage on 268 young (age 10-15 yr) basketball players' enjoyment of their team experiences and their evaluations of their coaches (e.g., coaches' knowledge of basketball and teaching ability, how much athletes want to play for the coach in the future).

At the end of the regular competitive season the players completed questionnaires designed specifically for young athletes. Won–lost percentage for the regular season (excluding playoffs) was calculated for each of the 50 teams by dividing the number of games won by the number of games played and multiplying by 100.

Mastery climate was significantly positively related to won–lost percentage, enjoyment, and coach evaluations. Ego climate was unrelated to won–lost percentage and negatively related to enjoy-

ment and coach evaluations. Won–lost percentage was unrelated to enjoyment and weakly related to coach evaluations. Overall, motivational climate was a much stronger predictor of enjoyment and athletes' evaluations of their coaches than was winning.

These results support the assertion that winning is not a prerequisite for enjoying participation in youth sport. Athletes who perceived that their coaches focused on effort and personal development liked playing for their coaches more, thought their coaches were more knowledgeable about basketball, thought their coaches were better at instructing basketball, had a greater desire to play for the same coach the following season, and enjoyed their basketball experiences more than players who perceived that their coaches focused on outcome. That motivational climate has a greater effect than winning is good news because coaches can control the motivational climates they create. In addition, even in a sport like basketball in which every game results in a winning team and a losing team, players on both teams can have positive experiences if the coaches provide mastery-oriented environments.

Source

Cumming, S.P., Smoll, F.L., Smith, R.E., & Grossbard, J.R. (2007). Is winning everything? The relative contributions of motivational climate and won–lost percentage in youth sports. *Journal of Applied Sport Psychology, 19,* 322-336.

ADOLESCENCE, PEERS, AND EXERCISE

In the adolescence stage of development (approximately 13-19 yr of age), peers usually are a more powerful influence than parents or coaches. Peer relationships can be thought of in terms of friendship and peer acceptance. Both of these factors influence affect and perceptions of physical self-worth, which in turn influence motivation and physical activity. When trying to achieve a sense of personal identity, adolescents often turn to those who are in the same developmental and chronological stage. Peers are the models who influence slang, style, values, and desires. Because it is difficult and unusual for adolescents to strive to be unlike their peers, their behaviour is often dependent on what their friends are doing. This modelling of behaviour applies to sport and exercise. If their friends are into in-line skating, then there is every chance that they will endeavour to do the same. Unfortunately, if their friends spend their recreational time playing video games rather than exercising, then they are likely to follow suit. In addition, low perceived peer acceptance predicts low levels of physical activity, and peer acceptance predicts physical self-worth.

EXERCISE IN THE AGED

Although participation in sport has a greater influence on the psychosocial development of children and adolescents than on that of adults, the effect of exercise on psychological health continues throughout life. Unfortunately, prejudicial and discriminatory views about ageing exist. This discrimination is

known as age stratification. Age stratification often results in lower self-expectancies of older individuals. It is commonly believed that older individuals should "act their age," which most people think means being less competitive and having poorer physical performances. Two myths contribute to age stratification: people need to exercise less as they age, and exercise is hazardous to the health of the elderly. In fact, as we have discussed, exercise becomes increasingly important for maintaining and improving quality of life as individuals get older.

EXERCISE AND LIFE SATISFACTION IN THE AGED

Research has revealed that exercise is related to life satisfaction in the aged. Older adults who exercise regularly report significantly enhanced health, increased stamina, less stress, increased work performance, and more positive attitudes toward work than those who do not exercise regularly. Exercise may also increase self-efficacy in the aged. For example, a study examining a 5 wk swimming program for adults over 60 yr of age found that the program had multiple positive effects. Not surprisingly, participants had greater belief in their swimming ability and were more confident in their swimming competence. More importantly, participants also had increased generalised feelings of self-efficacy and competence. For example, they felt they could do chores more easily and were more able to use public transportation, which greatly strengthened their independence. Figure 21.1 demonstrates the effects of a 6 mo exercise program for 60- to 75-yr-old adults who had previously been sedentary. Hap-

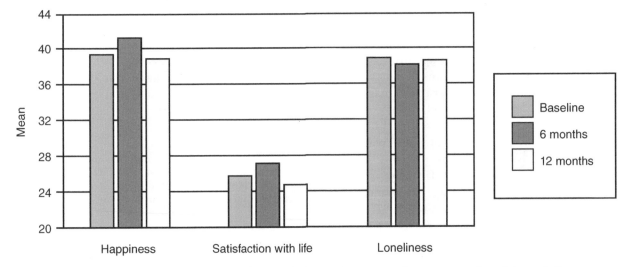

Figure 21.1 Changes in subjective well-being during a 6 mo exercise program for older adults.

piness and life satisfaction increased and loneliness decreased; however, when the exercise program ended, subjective well-being decreased.

As noted in chapter 20, exercise can improve self-concept and self-esteem. Exercise can also decrease stress, reduce muscle tension and anxiety, and lessen depression. These psychological benefits of exercise are advantageous to people of all ages, but in some respects may be particularly useful for the elderly. Depression is the major mental health problem of the elderly. Studies have shown that participation in a mild exercise program can significantly decrease depression in relatively independent adults (aged 60-80 yr) as well as clinically depressed residents of nursing homes (aged 50-98 yr). In a 6 mo randomised control study of 60- to 75-yr-old sedentary Brazilians, those who cycled 3 times per week (but not those in the control group) showed significant decreases in depression and anxiety and significant improvements in emotional and social functioning (see figure 21.2).

ENCOURAGING PARTICIPATION OF OLDER ADULTS IN EXERCISE AND SPORT

Exercise in older adults is associated with enhanced self-efficacy, life satisfaction, and happiness as well as decreased tension, anxiety, and depression. Nevertheless, with age stratification, participation moti-

vation and exercise adherence motivation are even worse in older adults than in the general population. If people are concerned that exercising is undignified or inappropriate for them because they are older, or that exercising will cause a heart attack or worsen existing medical conditions, they have additional barriers to overcome. Exercise programmers can help combat age stratification by including pictures and videos of older exercisers in their advertising. Some fitness centres are targeting the upper end of the age spectrum by having classes and social events specifically designed for older members. In addition to scheduling senior weights circuits or over-50 aerobics classes, a few programs now offer classes for those with arthritis and exercise classes based on ballroom dancing. Individuals who are expressly trained to assess older people with a variety of possible medical conditions and then prescribe suitable exercise for them can help potential exercisers overcome some of their fears and concerns.

Obviously, not all older adults are hesitant about exercising. The growing popularity of masters sport should help overcome some of the prejudicial and discriminatory views about the aged. The Seventh World Masters Games were held in Sydney, Australia, in 2009. Almost 29,000 competitors from 95 countries made these Games the largest participatory multisport competition in the world. Some older people definitely are active! Masters competitions are held every 4 yr at the world level and annually at the national level, and many countries hold ongoing local and regional competitions.

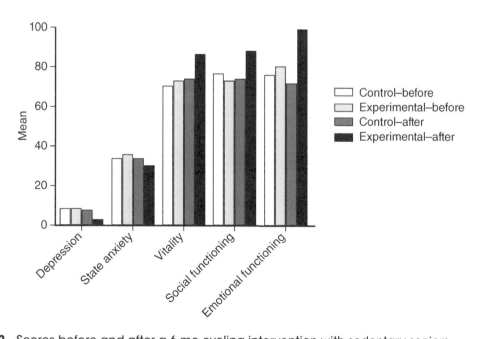

Figure 21.2 Scores before and after a 6 mo cycling intervention with sedentary seniors.

One of the more popular masters sports is swimming. Masters swim meets differ from traditional swimming competitions in that there are no heats. People are assigned to races in an event on the basis of previous best times. Each race therefore has people with similar performances, although a variety of ages may be represented. Results, however, are not based on any single race. Instead, results are based on the times in 5-yr age groups (e.g., ages 45-49 yr or ages 50-54 yr). Consequently, a fast 65-yr-old may finish fourth in a particular race yet still win the event for that age group because the first 3 finishers were in other age groups. One of the advantages of this system is that the emphasis is automatically placed on individual performances and participation.

Different masters sports tend to have different climates and structures regarding competition and training. A large percentage of masters swimming competitors, for example, train and compete on a regular basis yet focus on participation and improving (or maintaining) their times. In other sports, however, masters competitions are organised only at the national level. Although some of these competitors train regularly and compete locally in nonmasters events, a number of them begin exercising immediately before the masters competition in the hopes that the last-minute effort will get them through their events.

Greater availability of local masters events may encourage more regular involvement by more people. The growth of local and regional leagues and clubs for masters athletes in the United States suggests this notion is correct. For example, the Men's Senior Baseball League (MSBL)/Men's Adult Baseball League (MABL), a national organisation that began in 1986 with 60 members, now has 3,200 teams and 45,000 members. Organised masters sport leagues such as the MSBL/MABL may encourage some older people, who would not do so otherwise, to be physically active.

TERMINATION OF ATHLETIC CAREERS

For elite athletes, one of the major transitions in their lives is when their sporting careers end. Although sporting opportunities continue to exist for many of these athletes through participation in masters sports, the structure of their lives is often altered. No longer are they representing their country at the open level or being paid to compete by a professional sporting organisation. Although they still may be able to participate in other forms of competition, competitive performance is rarely the main focus of their lives.

VOLUNTARY VERSUS INVOLUNTARY TERMINATION

The effect of retirement from elite sport on athletes will depend, in part, on the circumstances surrounding the career termination. If athletes voluntarily retire because they no longer enjoy training or competing or because they have decided that other areas of their lives have priority, they may adapt quickly to the transition because they retain a sense of control. Athletes who are forced to end their careers because of injury or deselection may find the process to be difficult. Research has shown that individuals who voluntarily terminate their careers feel less emptiness and disappointment, and more relief and curiosity about what is to come, compared with those who involuntarily end their careers.

Retirement from elite sport can lead to depression. Some athletes have equated the experience to the ending of a close personal relationship. For some competitors, self-esteem decreases after career termination because they suddenly feel a loss of identity. (They may perceive a precipitous drop from "hero" to "zero".) Although having a strong self-identity as an athlete may have a positive effect on performance while competing in sport, it may lead to greater difficulties in adjusting to career termination. The stronger the athletic self-identity, the more time needed to adjust to career termination. The self-satisfaction and self-expression athletes gained through their participation in sport is no longer available to them, sometimes making them question their own identities.

To experience depression (or a feeling of flatness) after being removed from the environment in which one has spent a lot of time is normal. Schedules change. Goals that have been a source of energy are suddenly no longer there. If athletes finish their elite careers with a poor performance, they may feel that their previous devotion to their sport was for nothing. Finishing with a major success, however, does not guarantee a smooth transition.

ASSISTING THE TRANSITION PROCESS

When athletes retire from elite sport, either voluntarily or involuntarily, they can employ a number of strategies to ease the transition. Just as social support can help enhance exercise adherence (as mentioned in chapter 20), it can help in a period of major transition, such as career termination. Athletes can go to others for emotional support (reassurance, being loved in a time of stress), esteem support (agreement, positive comparison with others), instrumental support (direct assistance), informational support (advice, suggestions), and network support (feeling of belonging to a group of people who share a common interest). Table 21.1 provides some examples of these types of social support. Not all athletes will require all types of social support, but it is worth knowing that social support can be more than just a hug and an "Everything is going to be okay." Making others aware of the decision to retire from elite sport is the first step toward setting up social support. Other retired athletes can also be a valuable source of social support. Sharing experiences and discussing what has and has not been effective in dealing with the retirement process can be beneficial.

Retiring athletes also need to have alternative activities in which to redirect their energy. Ideally this process is initiated while the athlete is still participating in his or her sport. Some athletes may choose to become involved in other aspects of sport such as coaching, administration, or broadcasting.

Others may prefer to invest their energy into their families, further education, or entirely new careers. When participating in new activities it is important to avoid comparisons with previous sport competence. This aspect is particularly relevant to athletes who were national or world champions because it is improbable that they will be the best in the country or the world at their new careers. As illustrated in chapter 18, competence in any new activity is developed over a long period of time.

Athletes may find the transition to be easier if they apply skills that they learned through their participation in sport. For example, goal setting can be used not only in fitness enhancement or technique improvement but in any aspect of life that requires a change in behaviour or sustained effort over time. Many elite athletes also have become adept at controlling stress and anxiety. The skills that they employed to maintain calm and focus attention in a major championship or an international competition can be used to control stress in a new career. Similarly, if athletes have discovered that certain types of thinking are self-defeating in sporting performances, chances are good that comparable thinking would be harmful in work performances as well. Chapter 22 provides additional information on the transfer of skills from sport to work.

A percentage of retiring athletes may need to learn to take greater control of their lives. Depending on the sport, coaches, managers, and others may have made all the major decisions for the athletes in the past. Some coaches are dictatorial and some sports traditionally regiment the lives of the athletes.

Table 21.1 Types of Social Support That May Help Athletes Retiring From Sport

Type of support	Examples
Emotional	Loving them for who they are off the field, court, or track Supporting their decision to retire from sport
Esteem	Praising their nonsporting abilities (e.g., communication skills, organisational talents, artistic expertise)
Instrumental	Introducing them to potential future employers Teaching them basic computer skills (or other new, employable skills)
Informational	Suggesting that the skills they used to enhance their performances in sport can also enhance their performances in other careers Informing them of locally offered courses that they are qualified to attend
Network	Forming a group of individuals who either are currently retiring from sport or have already successfully retired from sport Joining a social club, church group, or volunteer organisation

If individuals were used to being told when to train, when to sleep, what to wear, what to eat, where to go, and how to get there, they might find it difficult to suddenly be on their own outside this decision-making process. Athletes may need to learn how to plan their own lives in terms of both new vocations and day-to-day responsibilities.

Whether it is gaining social support from significant others, finding new activities to pursue, learning to take greater control of one's life, applying skills previously learned in sport to new situations, or changing attitudes and priorities, adapting to retirement from elite sport takes time. The adaptation will be quicker for some than for others, but the process is easier when preparations are made before retirement. Athletes who during their sporting careers enjoy participating in other activities, change their routines in the off season, develop ideas for new careers, and create positive attitudes toward retirement are able to adjust quickly to the transition processes. Sporting organisations can help by educating athletes about the importance of having a balanced lifestyle, keeping in contact with athletes after they have left, and providing opportunities for athletes to maintain involvement in the organisation. Adaptation is quicker when retirement is perceived by athletes as presenting new opportunities rather than as the end of life as they know it.

SUMMARY

Involvement in sport as an adult probably has little if any effect on personality. However, involvement in youth sport may influence the development of children's personalities. Through participation in sport children can develop self-confidence, self-esteem, respect for others, leadership skills, and a sense of fair play. Nevertheless, if the sporting experience is poorly structured by coaches or parents who are more interested in winning than in children having fun or learning, children can develop negative self-esteem and lose self-confidence.

Although coaches and parents can have a huge effect on children, adolescents tend to be influenced more by peers than by adults. The behaviour of peers largely determines whether adolescents will engage in physical activity.

Although participation in sport and exercise may not play a large role in the psychosocial development of older adults, regular exercise has been shown to enhance self-efficacy and decrease depression and anxiety in the aged. Age stratification (prejudicial and discriminatory views about ageing), however, means that fitness centres and exercise professionals need to develop programs and marketing campaigns that specifically target older adults. Although the number of participants in the World Masters Games is steadily increasing, more effort needs to be made at the local level to increase the percentage of older adults involved in regular physical activity.

For elite athletes who have made sport the focus of their lives for many years, the termination of their athletic careers can be a challenging transition. Difficulties in career termination arise when athletes involuntarily retire from sport, have a strong self-identity as an athlete, access few sources of social support, refrain from participating in other activities, and have a poor attitude toward the transition process. Preparing for career termination well in advance can help many athletes adjust to the transition.

FURTHER READING

Chase, M.A. (2010). Children. In S.J. Hanrahan & M.B. Andersen (Eds.), *Routledge handbook of applied sport psychology: A comprehensive guide for students and practitioners* (pp. 377-386). London: Routledge.

Lavalee, D., Park, S., & Tod, D. (2010). Career termination. In S.J. Hanrahan & M.B. Andersen (Eds.), *Routledge handbook of applied sport psychology: A comprehensive guide for students and practitioners* (pp. 242-249). London: Routledge.

Medic, N. (2010). Masters athletes. In S.J. Hanrahan & M.B. Andersen (Eds.), *Routledge handbook of applied sport psychology: A comprehensive guide for students and practitioners* (pp. 387-395). London: Routledge.

Taylor, J., & Lavallee, D. (2010). Career transition among athletes: Is there life after sports? In J.M. Williams (Ed.), *Applied sport psychology: Personal growth to peak performance* (6th ed.) (pp. 542-562). Boston: McGraw-Hill.

Weinberg, R.S., & Gould, D. (2011). *Foundations of sport and exercise psychology* (5th ed.). Champaign, IL: Human Kinetics.

CHAPTER 22

PSYCHOLOGICAL ADAPTATIONS TO TRAINING

The major learning concepts in this chapter relate to

- the relationship between aerobic fitness and response to psychological stress,
- the limitations of research on the effects of sport participation on personality,
- the phases of training stress syndrome,
- measures of overtraining and burnout,
- strategies to overcome and prevent training stress syndrome,
- mental skills that may be acquired through participation in sport and applied to other situations in life, and
- the importance of systematic practice of mental skills.

As mentioned in chapter 20, both single bouts of exercise and long-term involvement in exercise have been shown to have a positive effect on mood. This chapter focuses on the psychological effects of prolonged participation in sport and exercise. The topics covered include the role of fitness in the response to psychological stress, personality changes as a result of participation in sport, overtraining and burnout, and the development and transfer of mental skills over time.

AEROBIC FITNESS AND THE RESPONSE TO PSYCHOLOGICAL STRESS

As discussed in chapter 20, exercise is effective in reducing depression. Depression can be related to stress. For example, failing in an achievement situation that is perceived to be important may be stressful. This failure may lead to lowered self-esteem,

INTERVAL EXERCISE REDUCES CHILDREN'S CARDIOVASCULAR REACTIVITY TO PSYCHOLOGICAL STRESS

Reactivity to psychological stress is associated with elevated resting blood pressure, hypertension, and atherosclerosis. Therefore, reducing stress reactivity in children may reduce susceptibility to cardiovascular disease later in life. Roemmich and colleagues completed two studies with children to investigate whether an acute bout of interval exercise reduces cardiovascular reactivity to stress and whether physical activity or aerobic fitness predict stress reactivity.

In their first study, children aged 8 to 12 yr either watched videos or cycled in intervals for 25 min before being given 10 min to prepare and 5 min to deliver a speech about why they are a good friend. The stress associated with the speech was increased by telling the children that their speeches would be videotaped and judged by their peers. Although heart rate did not differ between the groups during the speeches, the reactivity of diastolic blood pressure of the exercise group was 63% lower than that of the video group.

In their second study, a similar protocol was used (with a different sample), but this time all children completed both protocols (in a counterbalanced order on different days). In this study, diastolic and systolic blood pressure and heart rate were significantly lower during the speech stressor in the exercise condition than in the video condition. Fitness, but not physical activity, predicted heart rate, but not blood pressure, reactivity to psychological stress.

These studies were the first to investigate cardiovascular stress reactivity in children. They were also the first to use interval exercise rather than constant-load exercise as the manipulation. This last point is important because children do not naturally play at a constant moderate or high level of intensity. Instead, children tend to be active in short, intense bursts interspersed with periods of relatively low-intensity activity. The good news is that this natural style of play appears to have potential benefits in terms of cardiovascular reactivity.

Source

Roemmich, J.N., Lambiase, M., Salvy, S.J., & Horvath, P.J. (2009). Protective effect of interval exercise on psychophysiological stress reactivity in children. *Psychophysiology, 46,* 852-861.

which in turn may lead to depression. Involvement in regular exercise has been shown to both reduce existing depression and prevent the onset of depression. Fitter people report lower levels of depression after prolonged stress than do less-fit individuals.

Depression is not the only area in which fitness level plays a role. Aerobically fit individuals are able to deal with many forms of psychological stress more effectively than less-fit individuals. Probably the greatest indication of this difference is the varied levels of cardiovascular arousal after exposure to psychological stress. When comparing highly fit individuals with individuals with low levels of fitness, the fitter individuals have smaller increases in heart rate and blood pressure in response to any particular psychological stressor. Moderate-endurance training, such as participating in aerobic dance classes, has been shown to be more effective than relaxation training in reducing heart rates before, during, and after the experience of psychological stress. In one study, rowing on an ergometer for 40 minutes 4 times per week for 16 weeks increased the $\dot{V}O_2$max and power of reasonably fit firefighters and reduced their responses to fire-related stress. This positive effect of exercise on stress reactivity holds true for children as well as adults and for acute as well as long-term exercise (see "Interval Exercise Reduces Children's Cardiovascular Reactivity to Psychological Stress").

CHANGES IN PERSONALITY

Researchers have investigated changes in psychological factors in response to participation in sport as well as exercise. A number of studies have investigated the personality profiles of athletes of varying levels of skill. Studies that have compared athletes within a team have found no significant results; that is, there are no meaningful personality differences between successful and unsuccessful athletes or between starters and bench players within a team. Elite athletes (e.g., international-level competitors), however, can be distinguished from novice athletes on some personality characteristics. For example, there is some evidence that highly successful athletes are greater risk-takers than are novices; they are more likely to take chances that might result in injury or failure. Nevertheless, generalisations about expert–novice differences are difficult to draw because there have been many of the same inconsistencies in this research as in that on athlete–nonathlete personality differences.

First, a variety of personality inventories and questionnaires have been used, making interpretations across studies difficult. Also, some studies have focused on traits and some on states, and some have inappropriately merged the two together. The available studies are also difficult to compare because of the different sports investigated. Generally, athletes in team sports are more extroverted and dependent than athletes from individual sports. Therefore, comparing elite gymnasts with novice softball players can be puzzling. Just as there are difficulties in defining athletes and nonathletes (as mentioned in chapter 19), there are similar problems when trying to define elite and novice athletes. Is someone who competes nationally in swimming and recreationally in tennis a novice or an elite athlete? Finally, even if consistent differences were found between elite athletes and novice athletes, there is no indication that training, competing, and developing physical skills over a long period of time cause personalities to change. It may be that individuals with specific personality types are more likely to develop into elite athletes than are individuals with other personality types. Not enough evidence exists to support the use of personality tests for screening or selecting athletes.

Personality researchers have also tried to determine whether the type of sport in which individuals participate affects their personalities. Once again, there are few consistent findings, as well as problems with the methodology frequently used in such studies. Even logical-sounding assumptions have not been supported by the research. For example, there is no evidence that athletes in body-contact sports are any more aggressive than athletes in noncontact sports. One consistent finding is that participants in high-risk activities such as parachuting and auto racing usually have higher levels of sensation seeking than athletes in low-risk activities such as golf or bowling. Although some studies have controlled for factors such as age and education when comparing the participants of high- and low-risk sports, and there is general agreement in terms of whether the activities selected are high or low risk, the structure of the studies does not allow researchers to determine causality. Descriptive studies that do not follow participants longitudinally across time cannot determine whether people high in sensation seeking are attracted to high-risk sports or whether participation in high-risk sports causes individuals to be high in sensation seeking.

CHANGES IN MOTIVATION: STALENESS, OVERTRAINING, AND BURNOUT

As discussed earlier, training stress needs to be imposed physiologically to achieve training gains. Too much training stress, however, can lead to a negative psychophysiological reaction. Both psychological and physiological factors are involved in the negative adaptation to training. Although this section focuses on the psychological factors, it must be emphasised that psychological and physiological factors interact with each other. For example, being unable to maintain training loads physically can decrease enthusiasm for training. Similarly, having low levels of enthusiasm can decrease the ability to maintain training loads.

DEFINITIONS IN TRAINING STRESS

Studying the area of training stress can be confusing because different terms mean different things for different people. There is general agreement that there is a series of stages through which individuals progress when negatively adapting to training stress. Exercise physiologists often suggest that athletes first experience overtraining, then over-reaching, and then staleness. In the exercise physiology literature, staleness is usually perceived as being equivalent to burnout.

On the other hand, many sport psychologists refer to the negative adaptation to training as **training stress syndrome**. This syndrome is made up of three phases: staleness, overtraining, and burnout.

▶ **Staleness** is the initial failure of the person to cope with the psychological and physiological stress created by training. The body and mind attempt to adapt to the training demands, but the demands exceed the person's present capabilities. Staleness is characterised by an increased susceptibility to illness, flat or poor performance, physical fatigue, and a loss of enthusiasm.

▶ **Overtraining** is the repeated failure of the person to cope with chronic training stress. Symptoms of overtraining include an increased resting heart rate, chronic illness, mental and emotional exhaustion characterised by grouchiness and anger, and being overly bothered by minor stresses.

▶ **Burnout**, the final phase of the syndrome, is when the athlete is exhausted both physiologically and psychologically from frequent but usually ineffective efforts to meet excessive training and competition demands. When experiencing burnout, athletes lose interest in the sport, are extremely exhausted, and generally do not care about their training or performance. In some cases there is resentment toward the activity.

Monotony, repetition and boredom, too much stress, or too much training can cause staleness. Because one of the characteristics of staleness is a flat or poor performance, many coaches and athletes react to staleness by increasing the training load. This further increase in stress and training demands often leads to overtraining. In addition to too much stress or pressure, too much repetition, or too much training, overtraining can be caused by lack of proper sleep, a loss of confidence, and the feeling that one is never successful. If the training load is maintained or increased at this point, burnout may result.

Burnout is the result of an ongoing process. Athletes do not just awaken one morning and suddenly experience burnout. Burnout is usually caused by excessive devotion with little positive feedback over a long time, severe training conditions, unrealistic expectations, and insufficient recovery time from competitive stress. Nontraining stressors such as having a poor relationship with one's coach or not having enough time to spend on other important relationships are also related to staleness and burnout. Burnout typically results in dropout. It should be noted, however, that although burnout often leads to dropout, many people drop out for reasons other than burnout.

MEASURING OVERTRAINING AND BURNOUT

Traditionally, sport psychologists have measured overtraining by focusing on the relationship between overtraining and mood states. The **profile of mood states** (POMS) has been used in many studies and frequently demonstrates that athletes' moods worsen during states of overtraining. Because the paper-and-pencil POMS is less expensive to administer as a measurement of overtraining than are analyses of biochemical changes that also accompany overtraining, it is sometimes used with athletes as a screening device.

Some sport psychologists have argued that the POMS is inadequate because it does not measure the process of recovery. An alternative measure of overtraining is the **recovery stress questionnaire for athletes** (RESTQ-Sport) that measures the frequency of current stress and recovery-associated activities. Studies using both instruments have found that the POMS and RESTQ-Sport scales are correlated and therefore measure similar constructs. The argument put forth for using the RESTQ-Sport is that it provides useful information that is not available from the POMS. The POMS can reveal that mood states have worsened but cannot shed light on whether that is due to increases in stress or decreases in recovery.

Neither questionnaire attempts to measure burnout. The concept of burnout originated in the domain of human services (i.e., people working in helping professions). Burnout in human-service providers is characterised by emotional exhaustion, a reduced sense of accomplishment, and depersonalisation. **Depersonalisation** is characterised by a detached and callous attitude toward clients. Sport psychologists have suggested that depersonalisation, although relevant to human-service providers, is not a key feature of burnout for athletes. The sport equivalent of depersonalisation is **sport devaluation**, when athletes stop caring about their sport and their own performances. With this concept in mind, the athlete burnout questionnaire was created to measure three factors: emotional and physical exhaustion, reduced sense of accomplishment in sport, and sport devaluation. The validity of the athlete burnout questionnaire is strong: Those who score higher on the subscales of the athlete burnout questionnaire score lower on measures of social support, coping, enjoyment, commitment, and intrinsic motivation.

WHAT MIGHT HELP?

A number of strategies can be implemented if signs of staleness, overtraining, or burnout emerge. The simplest course of action is to introduce a time-out. Taking a break from training (and in some cases competition) can refresh and invigorate athletes. For athletes to be able to recharge their batteries, it is important that the time-out is a break from everything, not just physical training. An emotional vacation is needed. Some coaches grudgingly allow their athletes to take a break physically, believing that allowing a respite from physical work is all that is needed. Athletes who during a physical pause in training continue to observe training sessions, analyse game film, or otherwise mentally involve themselves in their sport gain little (if anything) from the physical respite.

Training stress syndrome is a psychophysiological process. Although reducing the intensity or volume of training may prevent the progression of the syndrome from staleness to overtraining or from overtraining to burnout, rarely is it enough to eradicate the syndrome entirely. The more significant the changes in day-to-day activities, the more effective the break will be. If a swimmer takes a week off of weight training, swim training, time trials, and team meetings yet spends the week living in an apartment with three roommates who are also teammates, the athlete remains in a swimming environment. With three roommates draping wet bathing suits around the apartment, leaving for and arriving from training, complaining about the session's format, and generally talking about swimming, the swimmer fails to experience a real time-out. Stress levels may even increase in that situation if the swimmer feels guilty or ashamed of the current inactivity.

Rather than focusing on the idea of overtraining, it may be more useful to consider the problem as one of under-recovery. Increases in stress are okay if there are accompanying increases in recovery. Similarly, a small amount of stress can be detrimental if the athlete has even less recovery (i.e., athletes need to be able to create a balance of stress and recovery). Recovery should be a planned component of athletes' schedules. It is fairly obvious that athletes need to have adequate nutrition and rest between training sessions, but having a mental break from the sport is just as important. Even the keenest athletes who are quite happy to eat and breathe their sports 24 h a day can benefit from a day off every once in awhile. Small breaks allow individuals some time to develop other aspects of their lives (important if they are to have any semblance of balance in their lives) and can increase the enthusiasm and energy levels of athletes once they return.

AVOIDING TRAINING STRESS SYNDROME

Coaches can help athletes avoid or overcome training stress syndrome by providing variety in training. Because monotony, repetition, and boredom can contribute to staleness and overtraining, activities that break the tedium and routine of training can be beneficial. Variety can be achieved in many ways. Some examples include introducing new drills, changing the training schedule, altering the training venue, encouraging fitness training through cross-training (i.e., participating in activities that are not specific to the sport), involving new faces (e.g., guest coaches, admired athletes), incorporating fun and games, and inviting and using athletes' input in the design of training.

A club swim team that was having problems with dropouts, staleness, and even behaviour dealt with the difficulties by introducing board games to Friday training sessions. Fridays were selected because attendance was worst at the end of the week. The process began when the coach brought in an old game of Monopoly. The Chance and Community Chest cards were replaced with Land Activity and Just for Fun cards that contained silly games or activities that were different from those used in traditional training sessions. Buying houses and hotels earned swimmers the right to specify the next activity for his or her lane. The game helped the swimmers feel in control of the session and created a sense of anticipation in the team. The game's popularity encouraged the coach and certain swimmers to adapt other pre-existing board games and create entirely new board games for use at the pool. Attendance records, punctuality during the week, and personal improvements in fitness and time trials were used to determine which swimmers got to select the games for each Friday session. Attendance, training productivity, and enthusiasm all greatly increased. There were fewer signs of staleness and fewer disciplinary problems. This specific example illustrates how effective providing variety in training can be.

The majority of the additional strategies that help athletes deal with staleness, overtraining, and burnout involve the development of a variety of mental skills. Goal setting can help athletes pursue realistic expectations. Training stress syndrome is exacerbated by athletes trying to achieve performances that are unrealistic and, as a consequence, expending a lot of energy with little, if any, return. Short-term goals give meaning to training sessions, help take minds off of how long the season may be, and increase positive self-concept because feelings of success are generated every time a short-term goal is achieved.

MANAGING STRESS

Because too much stress is one of the causes of training stress syndrome, skills that help athletes manage stress are extremely useful. Time-outs, variety in training, reduction in training, and goal setting can, as we've seen, help prevent training stress syndrome. Relaxation, positive self-talk, and learning to say no are additional strategies that can help athletes deal with negative training syndrome.

The most obvious skill in this regard is the ability to relax. The two most common physical relaxation strategies are progressive muscular relaxation and breathing exercises. Progressive muscular relaxation exercises involve contracting and then relaxing specific muscle groups systematically throughout the body. Many times if athletes are tense, telling them to relax is ineffective because they do not know how to relax. By increasing the tension in muscles and then relaxing, athletes become more aware and sensitive to what the presence and absence of muscle tension feels like and they learn that they can control the level of tension. Many individuals find it difficult to decrease the tension in a muscle voluntarily. By first increasing the tension and then relaxing, however, the level of tension in the muscle automatically drops below the initial level of tension. After regular practice of this exercise, athletes learn to relax the muscles without first having to tense them.

Breathing exercises are also useful for cultivating relaxation. When stressed, many individuals either hold their breath or breathe in a rapid and shallow manner. By learning to control breathing, stress reduction can be achieved. Controlled, deep, slow breathing is the aim of most breathing relaxation exercises. This type of breathing physically relaxes the body. The advantage of using breathing as a form of relaxation is that we are always breathing. With the possible exception of underwater hockey, athletes can always increase their awareness of their breathing even while being physically active.

Focusing on breathing not only enables athletes to control their breathing, and therefore aid physical relaxation, it also may help them mentally relax. Stress has both physical and mental repercussions. Mentally, athletes may begin to worry, create self-doubt, say negative things to themselves, and focus on factors that are irrelevant or counterproductive to performance. Concentrating on breathing is a useful way of refocusing attention and controlling self-talk. Practising functional cue words such as *relax* or *focus* in conjunction with exhaling helps athletes stop negative self-talk. Breathing then becomes a technique for relaxing the body both physically and mentally.

Another useful skill for individuals experiencing staleness, overtraining, or burnout is learning to say no. Although physical training is often a common source of stress, it is rarely the only source. Most athletes who experience training stress syndrome are hardworking, idealistic individuals who strive for high achievement. Because these people are highly motivated and often have a track record of getting things done, other people frequently ask them for favours and help. These athletes want to be successful and often agree to do too much. Any single act does not require too much effort or too much time, but when added together the demand is greater than the individual's resources. Every time the athlete agrees to appear at a public promotion for the sport or team, stay late at training to help another athlete, develop a textbook or training video, attend a charity fundraiser, or complete an extra assignment or project at work or at home, the total amount of stress is increased. When the person has the time, energy, and capabilities to attend to all the demands, this schedule is not problematic. But for people who are in a situation where the demands already exceed the capabilities, the inability to say no exacerbates the problem. Saying no sounds easy, but often if individuals know they are capable of doing what is being asked, they believe they may get positive recognition for doing the task (remember, a lack of positive feedback is a cause of burnout) and they do not want to take the chance of disappointing the individual requesting the favour. It often is easier in the short term to say yes than it is to say no.

There is no single intervention for treating athletes experiencing staleness, overtraining, or burnout. Research is currently being undertaken to determine which intervention or combination of interventions is most effective under specific circumstances. Severe cases of burnout may require professional counseling.

CHANGES IN MENTAL SKILLS

Participation in sport can lead to the evolution and improvement of an assortment of mental skills. These mental skills are largely believed to serve one function: the enhancement of performance. Although the acquisition of mental skills can help athletes enhance their performances, these skills also can increase the enjoyment of participation and provide individuals with an arsenal of skills that can positively affect the quality of life in areas outside of sport. Table 22.1 provides several examples of skills that athletes are likely to acquire in sport and of situations outside of sport in which those skills may be applicable.

Table 22.1 Mental Skills That May be Acquired Through Participation in Sport and Applied to Other Situations in Life

Skill	Possible nonsport applications
Arousal increase	Feeling tired and needing to make a presentation
Communication	Organising a major event
Concentration	Studying at home with loud roommates
Emotional control	Maintaining calm when the boss is rude
Goal setting	Changing health or work habits
Imagery	Rehearsing difficult confrontations
Injury rehab	Dealing with work-related injuries
Preparation	Being mentally ready for a job interview
Relaxation	Getting to sleep quickly
Self-confidence	Speaking in front of a large group
Self-talk	Remaining positive when things look bad
Team harmony	Working on a group project
Time management	Studying adequately for all exams
Travel	Avoiding jet lag

TRANSFERRING SKILLS FROM SPORT TO WORK

Interviews with elite athletes have revealed that these athletes have been able to transfer a number of skills and abilities from sport to work. Cognitive skills (e.g., controlling attention, making decisions), self-management skills (e.g., goal setting, stress management), interpersonal skills, communication skills, leadership skills, and the ability to handle negative events are just some of the abilities that individuals felt they had acquired through sport and then successfully applied to subsequent jobs. They also mentioned that personal characteristics such as assertiveness, confidence, ambitiousness, flexibility, conscientiousness, determination, persistence, and dependability were developed in sport but were beneficial in non-sport-related careers.

Being able to demonstrate these skills or personal characteristics in sport, however, does not mean that athletes will automatically be able to transfer them to new jobs. Skills are more likely to transfer when individuals are interested in their jobs, are self-confident, and are aware of the skills they have. Skills are less likely to transfer from sport when individuals have no career goals, little time to devote to work, or no perceived connection in their minds between sport and work. The structure of the work environment also influences skill transfer. Job control, coworker support, recog-

nition by supervisors, and job responsibility help skill transfer. Low job clarity, negative reactions of others, and excessive rules and restrictions hinder the transfer of skills.

ACQUIRING MENTAL SKILLS

Participation in sport does not automatically lead to the acquisition of multiple mental skills. Some successful athletes have developed mental skills through trial and error. Others have copied the skills of previously successful athletes, and some have obtained them through the guidance of good coaches or sport psychologists. Unfortunately, there are many individuals who never adequately develop these skills. A fairly large percentage of these people drop out from sport because they do not enjoy their participation in sport or they perceive their performances to be less than satisfactory. Providing a mental-skills training program for participants enhances the formation of mental skills that individuals can use throughout life and can increase the levels of both performance and enjoyment of participation. These two factors often go hand in hand. For example, if athletes always get extremely nervous before competitions to the point of being physically ill, chances are they will not perform their best immediately after being sick. If, however, they can learn to control their anxiety and therefore avoid being sick before competition, they will improve their performances and probably enjoy their participation much more.

For mental-training programs to be effective, they must be designed around the needs of the participants. No single technique will work for everyone. For example, although controlled, deep breathing can be an effective relaxation technique, a proportion of individuals will find that focusing attention on breathing actually increases anxiety because they no longer feel that breathing is natural and they develop fears of suffocation. Though this reaction is rare, it does exist in some people. Mental-training programs therefore need to be flexible and individualised. It is also imperative that individuals systematically practise their mental skills in the same way they practise their physical skills. Watching and understanding the principles in an instructional video about how to pole vault does not mean that we will be able to physically pole vault with correct technique. Practice would be needed. Similarly, listening to a lecture and understanding the principles of relaxation does not mean that we will be able to relax on command in extremely anxiety-provoking situations. Extensive, systematic practice is needed to bring about this positive adaptation.

SUMMARY

Training affects people psychologically as well as physically. Aerobically fit individuals are able to deal with psychological stress more effectively than less-fit individuals. Fitter individuals have smaller increases in blood pressure and heart rate than less-fit individuals in response to stress. Research also suggests that exercise results in lowered levels of negative affect and anxiety during stressful situations.

Research in the area of personality is not robust enough to indicate whether participation in particular sports or forms of exercise affects personalities. Although some differences have been found in the personalities of team-sport athletes versus individual-sport athletes and in participants in high-risk versus low-risk activities, the structure of the research cannot determine whether participation in the activities caused the personality differences. It may be that people with different personalities are attracted to different sports.

Although participation in sport and exercise can have many positive psychological and physiological effects, it can also result in staleness, overtraining, or burnout if participants allow themselves insufficient recovery. Burnout in sport is characterised by emotional or physical exhaustion, a reduced sense of accomplishment, and sport devaluation. Breaks from training, variety in training, goal setting, practising relaxation skills, learning to say no, and positive self-talk can help athletes avoid and deal with negative adaptations to training.

The mental skills that athletes acquire to avoid burnout and enhance performance can enhance their quality of life in areas outside of sport. Cognitive, self-management, interpersonal, and leadership skills can be transferred from sport to other careers. Nevertheless, instead of assuming that athletes will automatically develop these mental skills as an inherent part of participating in sport, extensive and systematic practice of mental skills should be incorporated into the structure of training.

FURTHER READING

Berger, B.G., Pargman, D., & Weinberg, R.S. (2006). Exercise as a stress management technique: Psychological and physiological effects. In *Foundations of exercise psychology* (2nd ed.) (pp. 151-170). Morgantown, WV: Fitness Information Technology.

Goodger, K., Lavallee, D., Gorely, T., & Harwood, C. (2010). Burnout in sport: Understanding the process—From early warning signs to individualized intervention. In J.M. Williams (Ed.), *Applied sport psychology: Personal growth to peak performance* (6th ed.) (pp. 492-511). Boston: McGraw-Hill.

Kellmann, M. (2010). Overtraining and recovery. In S.J. Hanrahan & M.B. Andersen (Eds.), *Routledge handbook of applied sport psychology: A comprehensive guide for students and practitioners* (pp. 292-302). London: Routledge.

Weinberg, R.S., & Williams, J.M. (2010). Integrating and implementing a psychological skills training program. In J.M. Williams (Ed.), *Applied sport psychology: Personal growth to peak performance* (6th ed.) (pp. 361-391). Boston: McGraw-Hill.

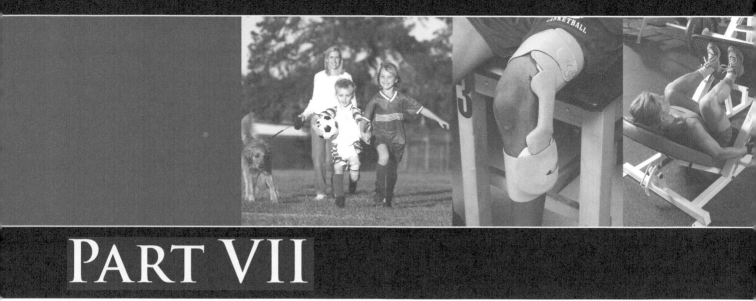

PART VII

MULTI- AND CROSS-DISCIPLINARY APPLICATIONS OF HUMAN MOVEMENT SCIENCE

In this final section of the text we first examine the issue of specialisation versus generalisation in the discipline of human movement studies and introduce an important distinction between multidisciplinary and cross-disciplinary research. We then examine some examples of multi- and cross-disciplinary research in practise, drawn from a range of areas of application.

SPECIALISATION VERSUS GENERALISATION

In describing the history of the discipline of human movement studies in chapter 2, we note that up until as late as the 1960s researchers interested in human movement typically had broad interests and backgrounds and frequently examined questions that would now be considered to belong to a number of different subdisciplinary fields. For example, the pioneer researcher Franklin Henry [see "Franklin Henry (1904-1993)" in chapter 2]

actively researched issues in both exercise physiology and motor control and educated a number of graduate students who played key roles in the establishment of the subdiscipline of sport and exercise psychology. As the discipline of human movement studies has matured and as specialised scientific research on human movement has proliferated, it has become increasingly difficult for individual researchers to remain expert in more than one of the subdisciplines of the field. The expansion of knowledge is such that it is now impossible for any one person to stay abreast of all the latest developments in each subdiscipline of human movement studies. Indeed, knowledge is expanding at such a rate and becoming so specialised that expertise is generally confined to select areas in a subdiscipline. Although it may have been realistic in the formative years of the discipline for an exercise physiologist, for example, to be knowledgeable in all the areas of human exercise physiology, the situation is now very different. Most researchers and academics possess limited knowledge of each of the major subdisciplines of the field and in-depth knowledge of only a few specialist areas (e.g., strength training, endurance training, muscle sore-ness, temperature regulation, immune function, or overtraining).

Although the increased specialisation of the dis-cipline has allowed greatly increased understanding of many aspects of human movement, the danger of such specialisation is the potential for fragmentation of the field, a concern discussed in part I. When fragmentation of a discipline occurs, researchers become so interested in their own specialist field that they fail to fully appreciate the significance of knowledge from other subdisciplines and the importance of integrating information from across the different subdisciplines in order to ultimately advance understanding. Even with increasing spe-cialisation it is important to retain a sound general knowledge and understanding of the entire field of human movement studies and an interest in, and an appreciation of, how the various systems in our bodies interact.

If we are truly interested in the key disciplinary question of how and why humans move, we need to consider not just functional anatomy, biome-chanics, exercise physiology, or sport and exercise psychology in isolation. Rather, we must integrate the various subdisciplines to produce a coherent, global view about human movement. Increasingly we should expect to see specialists from the different subdisciplines of human movement studies working together in research teams to understand complex problems from a number of perspectives.

Research involving more than one of the special-ist subdisciplines can be classified as either **multi-disciplinary** or **cross-disciplinary**, depending on the nature and extent of the integration that occurs (see figure 1.2). In multidisciplinary research, spe-cialists from the various subdisciplinary fields all investigate a common problem but do so only from the perspective of their own subdiscipline. What results is a number of perspectives on a particular research issue and no particular attempt to integrate these different perspectives into a consolidated viewpoint.

An example may help illustrate this approach. Suppose that a group of researchers, including a biomechanist, a sport psychologist, and an exer-cise physiologist, are interested in determining the features that characterise elite long-distance runners. The top five male and female 10,000 m runners at a national championship are selected as participants for a study and given a battery of tests. The exercise physiologist chooses to measure anaerobic threshold and $\dot{V}O_2$max; the biomechanist measures leg strength, anthropometry, and muscle activity patterns; and the psychologist conducts tests to ascertain mental imagery ability and pain tolerance. On the basis of their data, each then ascertains the physiological, biomechanical, and psychological predictors, respectively, of an athlete's 10,000 m running performance. No particular effort is made to ascertain the relative importance of the different measures or to determine whether there is a certain combination of physiological, psycho-logical, and biomechanical factors that best distin-guishes elite runners. Such an approach is termed multidisciplinary because it involves a number of the subdisciplines of human movement, although each specialist area essentially works independent of the others.

Cross-disciplinary research, on the other hand, relies on two or more specialists working together on a problem, taking their own perspective to the situation but integrating their views with the theories of others from different backgrounds. This approach is in many ways preferable to unidisci-plinary and multidisciplinary approaches because it has the potential to create a greater understand-ing of human movement than would be possible if the topic was investigated in a fragmented way. A human being moves in a certain way because of a host of interacting factors, and looking from only one perspective or level of analysis may lead to an incomplete, and perhaps flawed, understanding of the problem. In cross-disciplinary research a genuine attempt is made to cross the traditional subdisciplinary boundaries and for each subdisci-plinary perspective to attempt to understand the perspective of the others so that a new consolidated perspective may be developed. The biomechanist and motor control specialist who work together to decide the best type of kinematic feedback to assist learning, or the exercise physiologist and psychologist who work together to promote adher-ence to exercise with health benefits, are engaged in cross-disciplinary work.

EXAMPLES OF MULTIDISCIPLINARY AND CROSS-DISCIPLINARY APPROACHES

The three chapters in this section present examples that suggest how integration of the methods and knowledge of two or more subdisciplines can contribute to understanding of some central issues in the field of human movement studies. Chapters 23 and 24 provide examples of current integrative research that has applications to human health, chapter 23 in relation to chronic disease prevention and management and chapter 24 in relation to injury prevention and rehabilitation. Chapter 25 introduces examples of integrative research that have application to performance enhancement in sport and in the workplace.

CHAPTER 23

APPLICATIONS TO HEALTH IN CHRONIC-DISEASE PREVENTION AND MANAGEMENT

The major learning concepts in this chapter relate to

▸ the use of multi- and cross-disciplinary research in helping improve the prevention and management of chronic disease;

▸ the major causes of disease, death, and disability worldwide;

▸ the connection between disease and physical activity;

▸ the financial cost of physical inactivity;

▸ assessment of physical activity and sedentary behavior;

▸ current levels of physical activity; and

▸ how much physical activity is necessary for health benefits.

By far the most common rationalisation for encouraging people to be physically active is health. This is an outcome of the indisputable evidence supporting a decrease in the development of lifestyle-related disease in those who are physically active or aerobically fit. Governments, doctors, and health promoters now recommend regular physical activity for virtually all individuals, including children, persons who are elderly, and those with disease or disability.

Physical activity can be broadly defined as any muscular action that requires energy expenditure. This action can be spontaneous (e.g., fidgeting), unstructured (e.g., walking the dog), or structured (e.g., exercise or sport). Describing an individual as physically active generally means that individual does sufficient activity (defined by specific physical activity guidelines) that is of at least moderate intensity (e.g., brisk walking). An inactive individual suggests that individual is not performing sufficient moderate to vigorous activity (defined by specific physical activity guidelines). Sedentary behaviour, on the other hand, is waking behaviour that requires low levels of energy expenditure, such as sitting. Tasks that require more energy than sedentary behaviours but are not at least moderate intensity are usually described as low-intensity activities. Physical activity is thought to be an intricate combination of both biological function and behavior and it has been suggested that central neural processes may control individual levels of physical activity. Conversely, physical activity behavior is an observed course of action, differentiated through various external influences such as the social, cultural, or physical environments and our own psychological state.

Despite recognition of the health benefits of living a more active life, a large proportion of the child and adult populations living in middle- and high-income countries are insufficiently active. Substantial evidence suggests that the decline in active transport, minimal manual household tasks, and advances in electronic leisure pursuits have contributed to reductions in levels of activity. In turn, this has led to an excessive accumulation of energy and the escalation of overweight and obesity and related comorbidities. The worldwide pandemic of overweight and obesity and other lifestyle-related diseases has led to a renewed interest in physical activity for both preventative purposes and treatment.

This chapter begins by examining the major causes of disease, death, and disability worldwide and the extent to which these can be attributed to physical inactivity. We then discuss measuring physical activity, current levels of physical activity, and give consideration to the amount of physical activity necessary for better health, as well as the effect of sedentary behavior on health.

MAJOR CAUSES OF DISEASE AND DEATH GLOBALLY

The causes of disease and death have changed dramatically over the past 100 yr. Whereas infectious diseases such as tuberculosis were the major killers in the early part of the past century, the major causes of illness and death in most middle- and high-income countries are now heart disease, stroke, and cancer, which collectively account for approximately two-thirds of all deaths each year. Worldwide, cardiovascular diseases are the leading cause of death. Table 23.1 presents the four leading causes of death in 2004 by country income.

It is estimated that by 2030 ischemic heart disease will be the leading cause of death worldwide, even in low-income countries. The decrease in infectious diseases is attributable to improvements in hygiene (e.g., water and waste treatment), medical care (e.g., access to drugs), and public health (e.g., countrywide immunisation programs). However, the increase in ischemic heart disease, cerebrovascular disease, and cancers is believed to relate to lifestyle. Diseases of "comfort," such as cardiovascular diseases, respiratory diseases, diabetes, cancers, musculoskeletal disorders, and psychological disorders, have common risk factors that are directly related to a cluster of poor lifestyle choices, including lack of adequate physical activity and poor diet.

When the actual causes of death are computed, physical inactivity and poor diet combined were responsible for 15% of the total deaths in the United States. This percentage is marginally less than that for smoking—the number-one cause of death—which accounted for 18% of total deaths. It has been suggested that physical inactivity and poor diet will soon overtake tobacco as the leading cause of death in the United States. Physical inactivity and poor diet have already overtaken tobacco as the leading cause of death elsewhere in the world.

As mentioned earlier, cardiovascular diseases are predicted to become the leading cause of death in low-, middle-, and high-income countries by 2030. Ample evidence now links physical inactivity to cardiovascular disease, independent of obesity. Globally, it is estimated that 1.5 billion adults are overweight; more than half a million of these adults are obese. Being obese substantially increases the risk of developing cardiovascular diseases, and physical inactivity is a key determinant of obesity. The combined effect on cardiovascular health of a continued increase in levels of physical inactivity, sedentary behaviour, and obesity will be overwhelming.

CARDIOVASCULAR DISEASE

Since the early work of Jerry Morris in the United Kingdom and Ralph Paffenbarger in the United States—epidemiologists studying the link between

Table 23.1 Leading Causes of Death in Low-, Middle-, and High-Income Countries

Top 4 leading causes of death	Low income	Middle income	High income
1	Low respiratory infections	Cerebrovascular disease	Ischemic heart disease
2	Ischemic heart disease	Ischemic heart disease	Cerebrovascular disease
3	Diarrheal diseases	Chronic obstructive pulmonary disease	Lung cancer
4	Human immunodeficiency virus (HIV), acquired immune deficiency syndrome (AIDS)	Low respiratory infections	Low respiratory infections

Data from Mathers and Loncar, 2006.

lifestyle and cardiovascular disease—it has been clear that being inactive contributes to cardiovascular disease. Indeed, about 12% to 48% of all cardiovascular disease may be attributed to physical inactivity. Risk of cardiovascular disease increases with age in adults. Before menopause, women have a lower risk than do men, but the difference becomes much smaller after menopause. Family history of early cardiovascular disease in a close relation (e.g., siblings, parents, or grandparents) also increases disease risk. Risk of cardiovascular disease is increased in individuals with impaired glucose tolerance (the ability of the body to regulate blood glucose levels) or diabetes (disease involving impaired glucose tolerance). Elevated blood pressure and high cholesterol are also associated with the development of cardiovascular disease. Blood pressure, cholesterol, and glucose tolerance all relate to physical inactivity; therefore, being inactive considerably increases the risk of heart and vascular disease.

OBESITY

Population surveys indicate that more than half of adults in the United States and 30% to 50% of Europeans are overweight or obese. The greatest increase in the prevalence of obesity over the past decade has occurred in rapidly developing countries such as India and China. Obesity in Chinese children has increased by 180%, and it is estimated that nearly one quarter of China's total population will become overweight or obese if urbanisation continues at the present rate. An associated rise in cardiovascular disease is expected in countries where obesity rates are rapidly increasing.

Overweight and obesity are most commonly defined using the **Quetelet index** (see table 23.2) or body mass index (BMI). This requires dividing body mass (kg) by height (m^2). In Caucasians, overweight is defined as a BMI > 25 and obesity is defined as a BMI > 30.

Table 23.2 Definitions of Overweight and Obesity

Measure	Overweight	Obese
Quetelet index or body mass index	25-30	>30
Height–weight tables	10%-20% over recommended weight for height	>20% over recommended weight for height
Body composition	Adult male: 18%-20% fat Adult female: 27%-30% fat	Adult male: >20% fat Adult female: >30% fat
Waist-to-hip ratio	Adult male: >1.0 Adult female: >0.85	
Waist measurement	Adult male: >100 cm (39 in.) Adult female: >90 cm (35 in.)	

Obesity is usually caused by an imbalance between energy intake and energy expenditure. Physical inactivity is a major contributor to the deficits in energy expenditure. We now know that central obesity, or fat distributed around the upper body (chest and waist), carries a higher risk of a person developing cardiovascular diseases and diabetes than does fat deposited in the lower body. This upper-body fat distribution is known as the android shape, and lower-body fatness is known as the gynoid shape (see figure 23.1). Women tend to deposit fat primarily in the lower body (hips and thighs) before menopause and therefore have a lower risk of cardiovascular disease than do men.

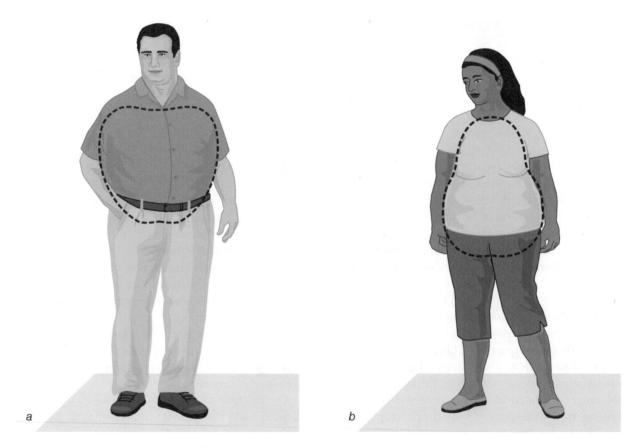

a

b

Figure 23.1 Pattern of fat distribution in the body. Left: android or male pattern (apple shape); right: gynoid or female pattern (pear shape).

Reprinted, by permission, from J.H. Wilmore and D.L. Costill, 2004, *Physiology of sport and exercise*, 3rd ed. (Champaign, IL: Human Kinetics), 679.

Waist circumference is a simple but effective way to determine the increased risk of cardiovascular disease from central adiposity. The threshold that determines whether waist circumference is a risk factor is 88 cm in adult women and 102 cm in adult men.

COST OF PHYSICAL INACTIVITY

About one third of all deaths in high-income countries result from diseases that are related to physical inactivity, and the health care cost is substantial. Physical inactivity is directly responsible for about 3% of total morbidity and mortality. This translates into health care bills ranging from $1.5 to 5 billion USD in the high-income countries. In contrast, it has been estimated that decreasing levels of physical inactivity by just 10% confers health care savings of at least $100 million USD. In a study of older adults, health care costs were halved in the active group compared with the inactive group. Simply put, increasing the number of active people could save tens of millions to hundreds of millions of dollars.

MEASURING PHYSICAL ACTIVITY AND SEDENTARY BEHAVIOUR

As mentioned at the beginning of the chapter, physical activity is a complex combination of biology and behavior. The assessment of physical activity is not a simple task, and difficulties magnify as age decreases. The challenges associated with assessing physical activity have received extensive attention, and both subjective (e.g., self-report) and objective (e.g., heart rate, accelerometers, pedometers) techniques are available. No one technique can be used to completely assess both the behavioral and biological components of physical activity. Because so many subjective and objective options are available, selection of the appropriate method requires careful consideration. Most important is consideration of the research question being addressed, the aspect of physical activity most pertinent to this question, and the population being assessed.

SELF-REPORT MEASURES

Self-report questionnaires are simple to use and therefore are often used to assess physical activity in large groups. Any self-report instrument is susceptible to report bias, and, in the context of physical-activity recall questionnaires, levels of physical activity are often overestimated. Although reliability is good for many instruments, validity—how self-report measures compare with objective methods of assessing physical activity—is generally low to moderate. These problems are compounded in children who find it even more challenging than adults to accurately recall physical activity. Interpretation of the information at the individual level is therefore not recommended because of the high degree of error.

Assessing sedentary behaviour can be even more problematic than assessing physical activity because sedentary behaviour tends to be generally unplanned and therefore much more difficult to accurately recall. Self-reported inventories of sitting time have been used successfully in adults and appear to be particularly useful in certain domains, such as the workplace. The validity of sitting-recall tools is moderate in adults, but, to date, validation of self-reported sedentary time for children has been inadequate. It is therefore recommended that sedentary behaviour in children should be assessed using objective methods.

OBJECTIVE SENSORS

Various objective tools are available for the assessment of both physical activity and sedentary behavior. The gold standard for free-living physical activity is doubly labeled water, which provides a precise measure of daily energy expenditure. However, given the expense of this technique and the need to couple it with other measures (e.g., resting energy expenditure and the thermal effect of food) to tease out the physical-activity portion of total daily energy expenditure, it is rarely used. Heart rate monitoring has been extensively used in the evaluation of physical activity, especially in children. Heart rate data provide an excellent measure of moderate and vigorous exercise, and, if calibrated against oxygen consumption, provide a good estimation of energy expenditure. However, heart rate is influenced by psychological stress, medication, and environmental factors, making measurement particularly problematic for low levels of physical activity.

Motion-sensing devices have become very popular. These small, lightweight devices offer a simple solution for good assessment of physical activity and sedentary behaviour. The pedometer, the simplest of movement sensors, assesses vertical movement, usually at the waist. The most commonly used output is the step or steps per day. Pedometers are inexpensive, small, and lightweight and are very useful in large-scale population studies. These days many pedometers have memories, negating the need for a person to remember to write down the steps taken per day. Because the steps output is tangible and easily understood, these devices can serve equally well as motivational prompts for increasing physical activity. The problem with the pedometer lies in its inability to detect changes in intensity and an inability to detect nonambulatory activities such as cycling. It is therefore of little use when one needs precise accounts of the amount of light, moderate, and vigorous physical activity undertaken. However, as a general marker of activity, the pedometer is very useful.

More complex motion sensors detect movement in more than one plane and record acceleration. The most sophisticated of these accelerometers detect motion in three axes: vertical (up–down), horizontal (forward–back), and oblique (side–side). These devices usually have a large data-storage capacity and high sampling rates and provide excellent assessment of movement across the full range of intensities, including body-posture data for the assessment of sitting. The drawback with accelerometers is that the data are not easily interpreted and the user must calculate intensity thresholds before use. Additionally, accelerometers do not provide accurate measures of some nonambulatory

movement such as cycling. The use of accelerometry in the measurement of physical activity is a good example of how a measurement tool initially developed primarily in the biomechanics field has become centrally important in making assessments in an area historically more aligned with the subdiscipline of exercise physiology.

An overview of the available methods for assessing physical activity, as well as the relative ease of using the device against the precision of the method, is provided in figure 23.2.

LEVELS OF PHYSICAL ACTIVITY IN ADULTS AND CHILDREN

When reviewing the available literature on levels of physical activity, it becomes apparent that a single marker of physical activity does not exist. Output from different objective devices is not interchangeable and therefore cannot be directly compared. Even when the same devices are used, no firm standards guide best practise, making comparisons of data sets challenging. Currently, the call for international standardisation of the objective assessment of physical activity has gathered considerable momentum, and it is hoped that best-practise guidelines for assessment of physical activity will emerge in the next 3 to 5 yr.

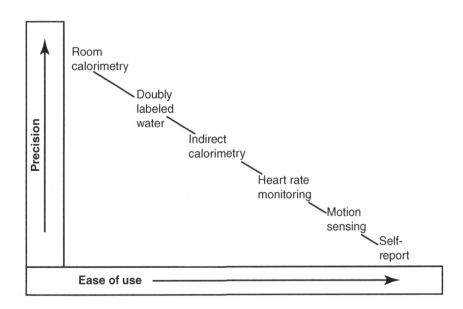

Figure 23.2 The precision of various techniques for the assessment of physical activity plotted against the ease of use.

When just pedometer steps are considered, healthy American adults normally take between 4,000 and 18,000 steps per day; the average number of steps is around 7,000 to 8,000 per day. Healthy children from the United States, New Zealand, and Australia take, on average, 12,000 to 15,000 steps per day. About one third of American adults are sedentary, defined as fewer than 5,000 steps per day, and only 16% are active or very active. A great deal of the normal waking day is spent seated, and the average child or adult spends very little time—usually less than 10 min—engaged in movement of vigorous intensity.

How much physical activity is sufficient for health? People interested in their health regularly ask this question. The answer to this question is somewhat complex because the optimal and minimal amounts of physical activity needed to achieve health benefits vary according to the individual's health goals, health status, fitness level, and age. This question also implies another question: "How sedentary do I need to be to damage my health?" A recent experimental study has provided initial evidence to help us to better understand the consequences of being sedentary. In this study, researchers used pedometers to record the average physical activity levels (6,000-10,000 steps per day) of healthy young adults. The participants were asked to radically reduce their daily step count to 1,500 steps per day. Over the course of only 3 wk, the researchers noted dramatic changes in the metabolic health of the participants. A 7% increase in central adiposity was accompanied by a decrease in insulin sensitivity, impairing the ability to regulate blood glucose, and an increase in plasma triglycerides, or fats. The findings from this study are striking and have helped focus attention on the serious health risks associated with sedentary behaviour.

RECOMMENDATIONS FOR PHYSICAL ACTIVITY

In recent years, recommendations by two influential organisations have provided guidance on the amount and type of physical activity needed to achieve health benefits. Many national and international organisations have followed with their own recommendations. Since 1978, American College of Sports Medicine (ACSM) has regularly issued position statements on the recommended amount of physical activity for healthy adults. The most recent recommendation published jointly by ACSM and American Heart Association (AHA) in 2007 recommended that "To promote and maintain health, all healthy adults aged 18 to 65 yr need moderate-intensity aerobic physical activity for a minimum of 30 min on 5 d each week or vigorous-intensity aerobic activity for a minimum of 20 min on 3 d each week." These recommendations are similar to previous recommendations, but two additions are worthy of note: the specific reference to vigorous-intensity exercise (see "A Little Pain for a Lot of Gain") and the inclusion of resistance training.

In 1996, the landmark U.S. Surgeon General's (USSG) report, *Physical Activity and Health*, clearly linked regular physical activity with a variety of health benefits. The USSG recommendations are outlined in "Summary of U.S. Surgeon General's Recommendations for Physical Activity" (U.S. Department of Health and Human Services, 1996). Despite some differences, the USSG and ACSM recommendations are more similar than dissimilar, and both sets provide a starting point for people developing their own physical-activity programs that will meet their specific needs.

UNDERSTANDING THE GUIDELINES

To explain what the ACSM and USSG guidelines mean, we need to consider the type, intensity, duration, and frequency of exercise recommended for health. The following section reviews how variations in our activities influence our ability to meet the physical activity guidelines.

TYPES OF RECOMMENDED EXERCISE

Beneficial adaptations to physical activity are relatively independent of exercise mode. That is, health benefits can be achieved through a variety of physical activities, and no single ideal exercise mode exists. The USSG suggests a range of modes of physical activity at work, during leisure time, or in activities of daily living (e.g., brisk walking, cycling, swimming, home repair, or yard work). Most adults prefer brisk walking, and the USSG notes the importance of preference.

The ACSM recommends whole-body exercise or activities that use large muscle groups, such as walking, jogging, swimming, and cycling. This type of exercise is of moderate to vigorous intensity, promotes loss of body fat and results in favorable changes in risk factors of cardiovascular disease.

INTENSITY OF EXERCISE

Both sets of guidelines recommend moderate-intensity physical activity. The USSG defines moderate-intensity activity as activity that can be sustained or accumulated over the day for 30 min. This report notes that more vigorous exercise leads to greater health benefits and should not be discouraged in those who are sufficiently healthy and willing to undertake more intense activity. The latest ACSM–AHA recommendations make specific mention of vigorous-intensity exercise and offer it as an alternative to moderate-intensity exercise. Evidence now suggests that vigorous-intensity activity can confer many health benefits. Reports suggest that adults may find short-duration activity easier to incorporate into day-to-day life. See "A Little Pain for a Lot of Gain" for an overview of the evidence that high-intensity activity is beneficial.

Neither the USSG nor the ACSM–AHA guidelines make specific mention of low-intensity activity (activity which requires more energy than sedentary behaviour but is lower than moderate to vigorous intensity). Yet we now know that low-intensity physical activity is important because when we consider the variation between individuals in energy expenditure during daily physical activity, differences are found in low-intensity movement such as ambulation and postural changes rather than in structured exercise. It is thought that this low-

A LITTLE PAIN FOR A LOT OF GAIN

The performance benefits of vigorous- or high-intensity training have been established for a number of decades in both adults and children. Despite the metabolic contrast between the maximal, all-out effort required during high-intensity exercise and the submaximal effort required by aerobic-endurance exercise, data suggest that the resulting oxidative adaptations to these two very different forms of exercise are remarkably similar. Recently, interest in the possible health benefits of high-intensity exercise has proliferated, and some intriguing findings have resulted.

Studies delivering high-intensity exercise generally favor cycle ergometers, largely because of the ease with which external work can be controlled. Initial studies used training protocols consisting of very exertive bouts of exercise that were short in duration interspersed with longer rest periods to provide high-intensity interval-training programs. Most notable has been the use of repeated Wingate anaerobic tests. The Wingate test requires a maximal effort, usually over 30 s. During this time, participants cycle against a fixed load, calculated on the basis of body mass. Training protocols have comprised a series of Wingate tests, such as 6 to 8 Wingate tests with longer rest periods of 4 to 5 min between. Training using this method provides very intense but short training sessions that provide substantial increments in oxidative metabolism. Additionally, changes in muscular markers of carbohydrate and lipid metabolism have been shown that are similar to changes that occur during aerobic endurance training. Further research started to reveal potential health benefits from this extremely demanding but time-efficient exercise. For example, after just 2 wk of high-intensity exercise lasting only 16 min, improvements were noted in insulin sensitivity in both lean and overweight sedentary adults.

Initial concerns about the safety of untrained individuals or those with known disorders engaging in this type of supramaximal activity led to some adaptations of the protocol. Alternative, more practical protocols for delivering high-intensity exercise have been developed. These protocols still require multiple intervals, but they are of longer duration (usually 1 min), and the recovery period is of similar duration (usually 1 min). Intensity is set at either maximal power (the highest power output elicited when completing a maximal oxygen uptake test on the cycle ergometer) or about 90% of maximum heart rate.

Again, the findings provide convincing evidence of health benefits, particularly for those with cardiovascular or metabolic disorders. Greater improvements in cardiac and vascular function have been identified following high-intensity exercise compared with conventional, moderate-intensity exercise, showing central cardiovascular benefit. Improvements in insulin sensitivity have been found, as well as increases in glucose-transport capacity of skeletal muscle and increases in content of skeletal muscle GLUT4 (an insulin-regulated glucose transporter protein), showing peripheral muscular benefit.

These studies have involved overweight, Type II diabetics and otherwise sedentary individuals.

Despite initial fears that intense exercise may be unsafe for these populations, it appears that high-intensity activity is well tolerated and safe. Perhaps more important, it was thought that high-intensity exercise would be less appealing than moderate-intensity exercise because it is so demanding, but recent evidence suggests otherwise. A group of recreationally active men showed greater enjoyment after participating in high-intensity exercise than after participating in conventional, continuous, moderate-intensity exercise. These findings need to be replicated in more at-risk populations, but they are very appealing in terms of exercise adherence. The extant evidence suggests that high-intensity or vigorous exercise requires minimum time but maximises health benefits and will probably appeal to those with time-deprived lives.

Sources

Bartlett, J.D., Close, G.L., MacLaren, D.P., Gregson, W., Drust, B., & Morton, J.P. (2011). High-intensity interval running is perceived to be more enjoyable than moderate-intensity continuous exercise: Implications for exercise adherence. *Journal of Sports Sciences, 29*, 547-553.

Gibala, M.J., & Little, J.P. (2010). Just HIT it! A time-efficient exercise strategy to improve muscle insulin sensitivity. *Journal of Physiology, 588*, 3341-3342.

Gibala, M.J., Little, J.P., MacDonald, M.J., & Hawley, J.A. (2012). Physiological adaptations to low-volume, high-intensity interval training in health and disease. *Journal of Physiology, 590*(Pt 5), 1077-1084.

Tjønna, A.E., Lee, S.L., Rognmo, O., Stølen, T., Bye, A., Haram, P.M., Loennechen, J.P., Al-Share, Q.Y., Skogvoll, E., Slørdahl, S.A., Kemi, O.J., Najjar, S.M., & Wisløff, U. (2008). Aerobic interval training vs. continuous moderate exercise as a treatment for the metabolic syndrome—"A pilot study." *Circulation, 118*, 346-354.

intensity, nonexercise activity is a key factor that distinguishes the lean from the obese; the obese sit for about 2 h/d more than the lean. In contrast, the lean stand or ambulate for about 2 h/d more than the obese. Lower-intensity movement is more closely associated with fat oxidation. Therefore, low-intensity activity may provide a very important means of enhancing fat loss by mobilising fatty acids from adipose tissue. Low- to moderate-intensity exercise is also less likely to cause musculoskeletal injury. This may be very important for obese individuals when they first embark on a physical activity program.

DURATION OF EXERCISE

The duration of physical activity is a function of intensity. Exercise at a lower intensity usually must be performed for a longer duration to provide the same health benefits as physical activity at a higher intensity.

The USSG and ACSM–AHA recommend at least 30 min of moderate-intensity activity each day, which the USSG states can be accumulated over the day in several shorter session (e.g., 3 sessions of 10 min) provided that the total time is at least 30 min. Longer-duration physical activity is also recommended for those who are interested and physically capable. The ACSM recommends between 20 and 60 min per session. The duration is dependent on the intensity—shorter duration for vigorous exercise and longer duration for moderate exercise.

FREQUENCY OF EXERCISE

For most people, exercise frequency is a matter of personal preference, accessibility, and convenience. The health benefits are very much related to the total amount of exercise over time (months to years). Lower-intensity physical activity may require more sessions, whereas more intense physical activity can be performed somewhat less frequently.

A minimum frequency of 3 d/wk is recommended for health benefits. The ACSM–AHA guidelines suggest that frequency of 3 to 5 sessions per week provides a good balance between optimal beneficial training effects and potential risk of injury. However, low-intensity, low-impact activities, such as walking, cycling, or swimming, are not associated with risk of injury; these activities may be undertaken daily. The USSG recommends taking part in moderate intensity physical activity such as brisk walking on most, preferably all, days. Physical activity is more likely to become an integral part of one's lifestyle if performed on a daily basis.

RESISTANCE TRAINING

The 2007 ACSM–AHA recommendations suggest that the normal recommendations for physical activity can be supplemented with resistance exercise. Although health benefits are most commonly associated with endurance-type exercise, resistance training also offers health benefits such as alterations in risk of heart disease, improvements in vascular function, favorable changes in body composition (e.g., reduced body fat and increased muscle mass), and counteraction of the age-related loss of muscular strength, bone density, and functional capacity. (Resistance training is discussed in more detail in chapter 14.)

As the name implies, in circuit training the participant moves around a circuit of exercise stations. Circuit training was designed for athletes as a way to develop all aspects of physical fitness, including muscular strength, power, endurance, agility, and cardiovascular endurance. Circuit training provides a time-efficient and versatile means of engaging in resistance exercise.

Circuit training can be modified for lower-intensity work to develop health in less-fit individuals by alternating weightlifting with aerobic-type activity. The participant usually spends 30 to 90 s at each station and takes short rest intervals between stations (see figure 23.3). A major advantage of circuit training is its versatility. Stations can be changed regularly to provide variety and to prevent overuse of a particular muscle group.

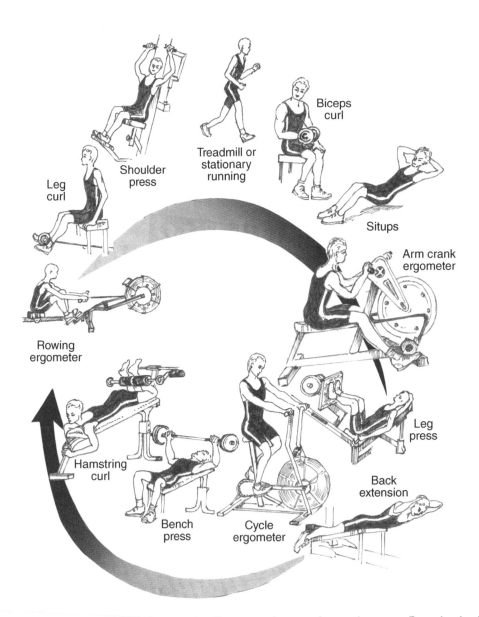

Figure 23.3 Circuit training for cardiovascular fitness and muscular endurance. Exercise includes alternating aerobic-type activities, such as running or cycling, and resistance work. The individual moves from station to station in a circuit, exercising for a given time, usually 30 to 90 s, and takes shorter rest intervals between stations.

SUMMARY

Evidence irrefutably shows that physical activity confers health benefits. The consequences of sedentary behaviour are also now being understood, and early data provide convincing evidence that sedentary behaviour results in dramatic increases in disease risk. The precise mechanisms for the health benefits of increased physical activity, as well as the mechanisms for the consequences of sedentary behaviour, have yet to be definitively understood. Increased availability of new imaging technologies and the adoption of cellular and molecular techniques are providing fresh insights into the health benefits of staying physically active and avoiding sedentary time. Developing effective means of bringing about increased uptake of physical activity among sedentary individuals requires a combination of methods and knowledge from both the biological and behavioral sciences.

FURTHER READING

Haskell, W.L., Lee, I.M., Pate, R.P., Powell, K.E., Blair, S.N., Franklin, B.A., Macera, C.A., Heath, G.W., Thompson, P.D., & Bauman, A. (2007). Physical activity and public health: Updated recommendation for adults from the American College of Sports Medicine and the American Heart Association. *Circulation, 116,* 1081-1093.

Lee, I.M., Matthews, C.E., & Blair, S.N. (2009). The legacy of Dr. Ralph Seal Paffenbarger, Jr.—Past, present, and future contributions to physical activity research. *President's Council on Physical Fitness and Sports Research Digest, 10,* 1-8.

Levine, J.A. (2007). Nonexercise activity thermogenesis—Liberating the life force. *Journal of Internal Medicine, 262,* 273-287.

Olsen, R.H., Krogh-Madsen, R., Thomsen, C., Booth, F.W., & Pedersen, B.K. (2008). Metabolic responses to reduced daily steps in healthy nonexercising men. *Journal of the American Medical Association, 299,* 1261-1263.

Paffenbarger, R.S. (2001). Jerry Morris pathfinder for health through an active and fit way of life. *British Journal of Sports Medicine, 34,* 217.

Sisson, S.B., Camhi, S.M., Tudor-Locke, C., Johnson, W.D., & Katzmarzyk, P.T. (2012). Characteristics of step-defined physical activity categories in U.S. adults. *American Journal of Health Promotion, 26,* 152-159.

U.S. Department of Health and Human Services. (1996). *Physical Activity and Health: A Report of the Surgeon General.* Atlanta, U.S: Department of Health and Human Services. Available from www.cdc.gov/nccdphp/sgr/sgr.htm.

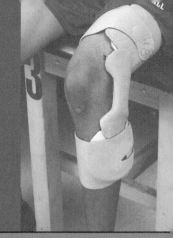

CHAPTER 24

APPLICATIONS TO HEALTH IN INJURY PREVENTION AND MANAGEMENT

The major learning concepts in this chapter relate to

▸ the use of multidisciplinary and cross-disciplinary research in helping improve injury prevention, management, and rehabilitation;

▸ the development of guidelines to assist in prevention of manual-handling injuries in the workplace;

▸ approaches to the prevention of overuse injuries in sport; and

▸ the prevention and management of injuries arising from osteoporosis, which are a major public health burden.

Prevention of, and rehabilitation from, injury is a central concern for many professions related to the discipline of human movement studies. In this chapter we examine a number of examples of integrative research on this topic. The first section describes a workplace setting in which manual-lifting tasks have been examined from a number of perspectives, including physiological, biomechanical, and psychological. The second section details the similar cross-disciplinary approaches that researchers have taken in attempts to prevent injuries associated with young people's participation in baseball and cricket. Finally, we

briefly examine injury prevention as it relates to osteoporosis, which is a chronic disease of growing public health significance.

PREVENTING MANUAL-LIFTING INJURIES IN THE WORKPLACE

Ergonomics, as noted earlier, is the field of study concerned with investigating the interface between humans and their working environment. It involves the disciplines of human movement studies and engineering as well as other areas of biological, medical, and technological sciences. A major area of concern in ergonomics is the prevention of work-related injuries. A disproportionately large number of injuries in the workplace are caused by overexertion while manually lifting and carrying objects. The back is the principal site for such injuries; more than one third of all workers' compensation claims in most industrialised Western countries are related to back injuries.

Guidelines related to the prevention of manual materials-handling injuries, especially those involving the back, have been developed in many countries. The guidelines for the United States were developed by the National Institute for Occupational Safety and Health (NIOSH). NIOSH considered information from a number of subdisciplines of human movement studies during the formulation of its guidelines. NIOSH defines limits for manual materials handling based on three criteria. These criteria are derived from the results of biomechanical, physiological, and psychophysical studies of lifting tasks.

Compression on the base of the lumbar spine is thought to be directly and causally related to some types of back injuries. A biomechanical criterion (based on a maximal acceptable lumbar compressive force) provides the limit to safety when lifts are performed infrequently. If only the biomechanical criterion existed, an apparently logical recommendation that might follow would be that small loads can be lifted frequently. This, however, is likely to lead to physiological fatigue, which, in turn, is a risk factor for back injuries. Therefore, for repetitive lifting tasks, it is more appropriate to use physiological criteria based on cardiovascular responses and metabolic fatigue to set limits for safe lifting. If only the physiological approach was used, a recommendation might be that large loads can be lifted occasionally, but this would clearly be inappropriate given the knowledge from functional anatomy and biomechanics that injury may result from the excessive forces generated during heavy lifts.

The psychophysical criterion is based on the lifter's perception of physical work. It is believed that each person internally monitors the responses of muscles, joints, and the cardiorespiratory system to estimate an acceptable workload and that this self-monitoring process appears to have a high level of accuracy. The psychophysical approach usually defines a limit that is a good compromise between the biomechanical and physiological results. The

psychophysical approach is applicable to most lifting tasks except for high-frequency lifting, when the object is lifted more than 6 times per minute for an extended period. In this situation, the physiological limit is recommended by NIOSH because it is lower than the psychophysical limit.

Many studies have compared the results from the psychophysical approach with those from either the biomechanical or physiological approaches. These studies have led to the recommendations provided above. The only true long-term measure of the validity of the NIOSH recommendations is their effectiveness in lowering the prevalence and severity of injury after their implementation. The important point in the current context is that this approach, aimed at reducing the incidence of work-related back injuries, relies on knowledge and criteria determined from a number of the biophysical subdisciplines of human movement studies.

PREVENTING AND MANAGING OVERUSE INJURIES IN SPORT

As noted in chapter 18, in order to become an expert performer one must usually perform literally millions of trials of practice. Repetition exposes the athlete, especially the growing athlete, to the danger of overuse injuries of the musculoskeletal system. In this section we examine how integrative research has been used in the sports of baseball and cricket in an attempt to address this issue.

INJURY PREVENTION IN BASEBALL: THE CASE OF LITTLE LEAGUE ELBOW

In the 1960s, epidemiological surveys highlighted a growing problem among child athletes involved in Little League baseball. These surveys revealed an alarmingly high incidence of conditions involving injury to the developing bony centres around the elbow. Because of the hypothesised relationship between the condition and the pitching action, the term *Little League elbow* was coined in 1960.

The first series of studies by Dr. Joel Adams involved radiological examination of 162 boys. Some were involved in Little League baseball (80 pitchers, 47 nonpitchers) and some were not (control group). This study revealed that the Little League players showed higher incidences of abnormal bone growth and development at the elbow than did the control group. The other major observation of this study was that a direct relationship existed between the magnitude of the changes at the joint and the amount of pitching or throwing. That is, the pitchers showed higher incidences and increased severity of injury to the medial aspect of the elbow than did the other players and the controls. Furthermore, symptoms are insidious in onset and typically reach their peak, and require medical attention for the first time, between 13 and 14 yr of age as the players move from Little League to Pony League.

Numerous follow-up studies have been conducted, including comprehensive radiological and clinical surveys of Little League baseball players in Eugene, Oregon, and Houston, Texas. These research programs identified early changes to the bone and cartilage of the elbows of many of the players involved in Little League baseball. Because the participants were 11 to 12 yr of age, very few had been playing long enough to develop pain in the elbow.

The rate of incidence of Little League elbow is highest at the onset of the adolescent growth spurt (see chapter 5), affecting boys who are 13 to 14 yr of age. During these years of rapid growth, the bones remodel in response to the applied loads. The stresses applied during the pitching action particularly, and throwing in general, often cause the medial epiphysis of the throwing arm to widen and develop abnormally.

Follow-up research after the initial clinical and epidemiological studies included biomechanical investigations of the pitching technique. High-speed film and video analysis identified various phases of the movement, including the wind-up, cocking, acceleration, and follow-through phases. It is generally believed that the end of the cocking phase and the start of the acceleration phase are the times at which the elbow is placed under the greatest stress. During these phases, the elbow is flexed and the humerus is laterally rotated, and the muscles on the medial side of the joint undergo eccentric activity before positively accelerating the forearm toward the target. During the follow-through, large loads are placed on the lateral side of the joint as the forearm undergoes **pronation** after ball release, leading to damage to the capitulum of the humerus and the head of the radius.

Another important finding of the biomechanical research is the relationship between style of throwing and injury rate. Children who pitch with sidearm motions are three times more likely to develop problems than are those who use a more overhand technique. Specific pitches, notably the curve ball, require the application of different loads at the elbow than do fast balls. In fact, the altered action required for curve balls places even higher loads on the medial aspect of the elbow, which in turn increases the prospect of injury to the medial humeral ossification centre.

Administrators of Little League baseball were extremely concerned about the high injury rates among their athletes. In response to the medical and biomechanical research findings, various controls have been put into place to try to reduce the number and severity of elbow injuries to pitchers. These measures have included general advice to coaches and parents regarding the number of pitches to be thrown during training sessions, typical warm-up and cool-down activities, suitable progressions and build-up in throwing during preseason, limits to the number of innings that pitchers can throw in a game, and limits to the number of games that can be pitched in a week. Many of these rule changes were instigated in 1972, and follow-up research has been conducted to support the positive effect of these changes. Although it is believed, in general, that the incidence rates of Little League elbow declined after the rule changes were implemented, the problem is still substantial. Recent surveys indicate that many questions relating injury of the elbow to training practises and throwing technique are still unanswered. The problem may have been reduced by the implementation of rule changes and altered coaching practises; however, the problem has not been eliminated and will continue to be a fertile area of future research.

INJURY PREVENTION IN CRICKET: THE CASE OF BACK INJURIES IN FAST BOWLERS

The history of Little League elbow has to some degree been emulated in cricket-playing countries, such as England, Australia, and India, where the incidence of vertebral stress fractures in young fast bowlers is a growing concern. A considerable understanding of back injuries in cricket fast bowlers now exists as a consequence of a systematic, cross-disciplinary research. One of the first tasks undertaken by a research team at the University of Western Australia, under the direction of the biomechanist Bruce Elliott, was to gather statistics on the number and types of injuries incurred by fast bowlers. This research revealed that approximately 45% to 55% of fast bowlers had abnormal radiological features in the lumbar spine. This incidence of lumbar spine abnormalities was 10 times higher than the rate that would have been expected given the players' age and sex.

After recognising the magnitude of the problem and that young fast bowlers were especially susceptible, the next step was to try to determine the causes of the problem. Through a series of studies that included physiological, biomechanical, and anthropometric measurements, a number of insights into the problem were gained. First, it was shown that poor physical preparation was not related to the onset or continued development of spinal abnormalities and low-back pain. Equal numbers of fast bowlers who were in excellent physical condition and in poor physical condition showed degenerative changes in the lumbar spinal vertebrae.

The biomechanical analyses were more revealing with respect to the source of the problem. Through high-speed film analysis and subsequent modelling, Elliott and colleagues were able to establish a link between bowling technique and the incidence of injury. They began by classifying a bowler's technique as side on, front on, or mixed. In both the side-on and front-on techniques the shoulders and pelvis are parallel. The bowlers who used a mixed technique that requires considerable trunk rotation about the vertical axis presented with a significantly higher incidence of injury than did the bowlers who used the other two styles. A relationship was therefore established between biomechanics and injury. Researchers at the University of Queensland, Australia, subsequently established that asymmetry of volume of the back muscles, determined by magnetic resonance imaging, is strongly correlated with the incidence of vertebral stress fractures at the level of the fourth lumbar vertebra.

Another aspect that was investigated was the notion of overuse. Logbook records indicated that athletes who bowled more had a higher incidence of stress fractures, disabling back injuries, or both. Bowlers who reported bowling in excess of 10 overs (approximately 60 balls) without a rest also had a higher incidence of back disability. These observations led to the formulation and implementation of recommendations about the number of balls that fast bowlers should bowl in matches and in practice in order to minimise the likelihood of injury. It has been recognised that overuse by itself does not explain the incidence and severity of injuries, so research involving technique analysis of young fast bowlers continues in an effort to prevent the repetition of potentially injurious techniques. These training recommendations and coaching guidelines are based on a linkage between aspects of exercise physiology, motor control, biomechanics, and injury.

In recent years it has also been recognised that injury prevention and especially rehabilitation has a large psychological component. Historically, relatively little attention has been devoted to the psychological effects of injury and how injury influences the mental well-being of an athlete. A Western Australian researcher, Sandy Gordon, has attempted to fill this void by describing and interpreting athletes' psychological responses to injury, to the rehabilitation process, and to their subsequent return to competition. This research is best thought of as an integral part of a holistic approach to understanding injury and subsequent rehabilitation. A case study published in 1990 that describes the psychological response to injury of an injured fast bowler is a valuable adjunct to the published work on injury mechanisms and their prevention.

PREVENTING INJURIES RELATED TO OSTEOPOROSIS

Osteoporosis is a bone disorder in which bone density decreases to a critically low level, making bones susceptible to fracture in a fall. Osteoporosis is a disease primarily of older Caucasian women of European descent, although elderly men may also be at risk. It has been estimated that 15% of women age 70 yr and more than half of women age 80 yr will suffer a bone fracture as a result of osteoporosis. Half of all older people suffering a bone fracture will become disabled, and 20% of women with hip fractures die within 1 yr. The annual cost of medical care and loss of independence due to osteoporotic bone fractures is estimated to be in the billions of dollars in North American and European countries. The costs to individuals are physical and psychosocial as well as financial. Obviously, effective strategies for reducing the anticipated costs to society are needed.

The causes of osteoporosis are multifaceted and include both genetic and lifestyle factors. As noted in chapters 3 and 5, bone is a dynamic tissue that is constantly remodelled. During ageing, the process of bone degradation exceeds the rate of synthesis of new bone, gradually reducing bone density and the loads that can be tolerated. The most commonly fractured sites are the vertebrae and hip (at the neck of the femur). Older, postmenopausal women are most at risk for osteoporosis because bone synthesis is reduced with low estrogen (female hormone) levels; bone loss is much slower in older men than in women of the same age. Medical treatment of osteoporosis often may include replacement of female hormones in postmenopausal women. Although a strong genetic predisposition to osteoporosis exists, lifestyle factors such as regular exercise and adequate nutrition, especially before adulthood, are important in the prevention of osteoporosis later in life.

Mechanical loading is necessary for maintenance of bone density. Bone loses mineral content, mass, and density in the absence of loading. For example, bone loss occurs in astronauts who spend long periods in space and in patients after prolonged bed rest. Weight-bearing activity and muscular contractions that increase mechanical loading of bone increase bone density. Bone density is highest in athletes who participate in weight-bearing sports, such as running, and in activities that induce muscular hypertrophy, such as weightlifting.

Because bone density can be influenced by load-bearing exercise, much attention has been focused on whether physical activity can prevent or counteract the loss of bone with ageing. Achieving a high bone mass early in life (before young adulthood) is of primary importance to preventing osteoporosis later in life. Thus, all children are encouraged to participate in various forms of physical activity. It is also recommended that young women especially aim to achieve and maintain optimal bone mass through a combination of adequate intake of calcium, a well-balanced diet, and regular weight-bearing exercise and avoidance of factors that reduce bone mass such as smoking, eating disorders, and menstrual disturbances. Smoking and excessive consumption of alcohol and caffeine are discouraged because of their direct effects on bone as well as their other adverse effects on health. Some research indicates that moderate exercise (<70% maximum heart rate) is as effective as strenuous exercise in maintaining bone density. Older individuals are strongly encouraged to continue cardiovascular and muscular conditioning to prevent or minimise bone loss in old age. Exercises to improve muscular strength and motor skills such as balance, coordination, and reaction time may also help prevent bone fractures by reducing the risk of falls in the elderly.

Evidence-based medicine is practised and strongly promoted in most countries with modern health facilities. Within this approach, the scientific method is used to determine whether medical interventions actually work and to compare the relative efficacy of different types of interventions. A number of strategies are being used to prevent and treat osteoporosis and its consequences. Therefore, an important research question is which of these approaches (or combination of approaches) is the most effective.

Regular exercise is appealing as an intervention because physical activity has many physical and psychological benefits. However, as noted in chapter 23, convincing the public about these benefits and then reaching a high degree of compliance so that most members of society participate in regular physical activity is not an easy task. Improving the public uptake of physical activity requires the combined skills of educators, psychologists, sociologists, and experts in marketing. Pharmaceutical interventions are attractive because of the appeal of a quick fix to the challenges posed by osteoporosis. Bisphosphonates, which reduce the rate of bone loss by inhibiting osteoclastic activity, have been recommended for older women in particular. Hormone-replacement therapy has also been recommended, but recent research has indicated that unacceptable side effects of long-term therapy may exist. Even more direct approaches involve the use of technology to address prevention of falls. Body-mounted accelerometers that detect potential balance loss and hip pads to protect the weak bone of the hip region from fracture by absorbing the impact energy of a fall have been suggested as part of a multidimensional approach to minimising the burdens associated with osteoporosis and fractures due to falls. Cross-disciplinary research and evaluation of the effectiveness of different interventions is needed to prevent and manage this chronic disease.

SUMMARY

Preventing and managing injury is a major issue for many areas of human endeavour. In the workplace, in the home, and in sport and recreation, finding effective means of both minimising the occurrence of injury and accelerating recuperation from injury is important for both personal and public health.

Because the risk factors for injury and the mechanisms underpinning injury are generally multifaceted, finding effective approaches to prevention and management necessitates the input and integration of knowledge from many disciplines. The most effective interventions are inevitably those with multidisciplinary and cross-disciplinary origins.

FURTHER READING

Manual-Lifting Guidelines

Dawson, A.P., McLennan, S.N., Schiller, S.D., Jull, G.A., Hodges, P.W., & Stewart, S. (2007). Interventions to prevent back pain and back injury in nurses: A systematic review. *Occupational Environmental Medicine, 64,* 642-650.

National Institute for Occupational Safety and Health (NIOSH): www.cdc.gov/NIOSH/

Nelson, N.A., & Hughes, R.E. (2009). Quantifying relationships between selected work-related risk factors and back pain: A systematic review of objective biomechanical measures and cost-related health outcomes. *International Journal of Industrial Ergonomics, 39,* 202-210.

Waters, T.R., Putz-Anderson, V., & Garg, A. (1994). *Applications manual for the revised NIOSH lifting equation.* Available: www.cdc.gov/niosh/docs/94-110/.

Waters, T.R., Putz-Anderson, V., Garg, A., & Fine, L.J. (1993). Revised NIOSH equation for the design and evaluation of manual lifting tasks. *Ergonomics, 36,* 749-776.

Pitching Injuries in Baseball

Adams, J.E. (1968). Bone injuries in very young athletes. *Clinical Orthopaedics, 58,* 129-140.

Anderson, M.W., & Alford, B.A. (2010). Overhead throwing injuries of the shoulder and elbow. *Radiology Clinics North America, 48,* 1137-1154.

Chen, F.S., Diaz, V.A., Loebenberg, M., & Rosen, J.E. (2005). Shoulder and elbow injuries in the skeletally immature athlete. *Journal of the American Academy of Orthopaedic Surgeons, 13,* 172-185.

Limpisvasti, O., ElAttrache, N.S., & Jobe, F.W. (2007). Understanding shoulder and elbow injuries in baseball. *Journal of the American Academy of Orthopaedic Surgeons, 15,* 139-147.

Sciascia, A., & Kibler, W.B. (2006). The pediatric overhead athlete: What is the real problem? *Clinical Journal of Sports Medicine, 16,* 471-477.

Wells, M.J., & Bell, G.W. (1995). Concerns on Little League elbow. *Journal of Athletic Training, 30*(3), 249-253.

Back Injuries in Cricket

Elliott, B. (2000). Back injuries and the fast bowler in cricket. *Journal of Sports Sciences, 18,* 983-991.

Elliott, B.C., Davis, J., Khangure, M., Hardcastle, P., & Foster, D. (1993). Disc degeneration and the young fast bowler in cricket. *Clinical Biomechanics, 8*(5), 227-234.

Elliott, B., Hardcastle, P., Burnett, A., & Foster, D. (1992). The influence of fast bowling and physical factors on radiologic features in high performance young fast bowlers. *Sports Medicine, Training, and Rehabilitation, 3*(2), 113-130.

Engstrom, C., & Walker, D.G. (2007). Pars interarticularis stress lesions in the lumbar spine of cricket fast bowlers. *Medicine and Science in Sports and Exercise, 39,* 28-33.

Engstrom, C., Walker, D.G., Kippers, V., & Mehnert, A.J. (2007). Quadratus lumborum asymmetry and L4 pars injury in fast bowlers: A prospective MR study. *Medicine and Science in Sports and Exercise, 39,* 910-917.

Gordon, S., & Lindgren, S. (1990). Psycho-physical rehabilitation from a serious sport injury: Case study of an elite fast bowler. *The Australian Journal of Science and Medicine in Sport, 22*(3), 71-76.

Ranson, C.A., Burnett, A.F., King, M., Patel, N., & O'Sullivan, P.B. (2008). The relationship between bowling action classification and three-dimensional lower trunk motion in fast bowlers in cricket. *Journal of Sports Sciences, 26,* 267-276.

Prevention of Injuries Related to Osteoporosis

Borer, K.T. (2005). Physical activity in the prevention and amelioration of osteoporosis in women: Interaction of mechanical, hormonal and dietary factors. *Sports Medicine, 35,* 779-830.

Gatti, J.C. (2002). Which interventions help to prevent falls in the elderly? *American Family Physician, 65*(11), 2259-2261.

Hellekson, K.L. (2002). NIH releases statement on osteoporosis prevention, diagnosis, and therapy. *American Family Physician, 66*(1), 161-162.

Howe, T.E., Shea, B., Dawson, L.J., Downie, F., Murray, A., Ross, C., Harbour, R.T., Caldwell, L.M., & Creed, G. (2011). Exercise for preventing and treating osteoporosis in postmenopausal women. *Cochrane Database of Systematic Reviews,* July 6(7), CD000333.

Khan, K., McKay, H., Kannus, P., Bennell, K., Wark, J., & Bailey, D. (2001). *Physical activity and bone health.* Champaign, IL: Human Kinetics.

Messinger-Rapport, B.J., & Thacker, H.L. (2002). A practical guide to prevention and treatment of osteoporosis. *Geriatrics, 57*(4), 16-23.

Winters-Stone, K. (2005). *ACSM action plan for osteoporosis.* Champaign, IL: Human Kinetics.

CHAPTER 25

APPLICATIONS TO PERFORMANCE ENHANCEMENT IN SPORT AND THE WORKPLACE

The major learning concepts in this chapter relate to

▸ the use of multidisciplinary and cross-disciplinary research in helping improve talent identification and performance optimisation,

▸ recognition of the multifactorial nature of talent and the challenges associated with early talent identification,

▸ the utilisation of stretch–shortening cycles to enhance muscular power,

▸ understanding and preventing overtraining syndrome in competitive athletes, and

▸ using biofeedback to augment control and regulation of key physiological parameters.

In this final chapter we explore some examples of how the consideration and integration of knowledge from across different biophysical subdisciplines of human movement can contribute to the enhancement of performance. In the first part of the chapter we explore an example of multidisciplinary research that seeks to understand the factors that contribute to talent in a particular sport.

The introduction of systematic approaches to talent identification clearly depends on a sound understanding of the relative contribution that different key factors may make to future successful performance. In the second part of the chapter we present some cross-disciplinary examples of attempts to optimise performance. The first example is related to the exploitation of the stretch–shortening cycles inherent in muscle, the second example is related to the psychophysiology of overtraining syndrome, and the third example is related to using biofeedback to enhance regulatory control over some of the physiological processes that are fundamental to successful performance.

TALENT IDENTIFICATION

For many years, managers in both industry and government have been concerned with selecting personnel to ensure the best possible match between the person and the task to be completed. Person–task matching is an important concern in ergonomics. A parallel problem in sport is talent identification—the identification of talented youngsters who have potential to become future champions in specific sports. The complex nature of elite sport performance necessitates that talent identification be broadly based on knowledge from a number of subdisciplines of human movement studies.

Structured systems for the development and nurturing of talent in specific sports exist in many countries around the world, although the evidence base for the variables used in athlete selection is frequently quite poor. Researchers at the Liverpool John Moores University in the United Kingdom have begun attempting to determine the best predictors of talent in soccer. Given the multifaceted nature of soccer, many potentially important predictors of future success in the sport exist: anthropometric variables (e.g., height, weight, body size, bone diameter, muscle girth, body fat, and somatotype), physiological variables (e.g., aerobic capacity, anaerobic power, and endurance), motor control variables (e.g., attention, anticipation, decision making, creative thinking, and technical skills), psychological variables (e.g., self-confidence, anxiety control, motivation, and concentration), and sociological variables (e.g., parental support, education, socioeconomic background, cultural background, and hours and resources available for practice) (figure 25.1).

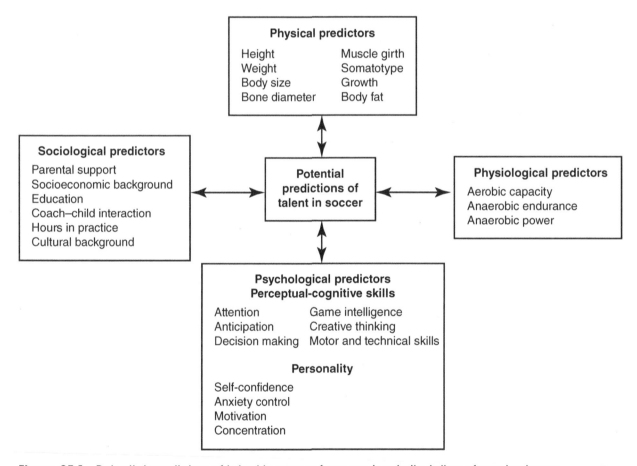

Figure 25.1 Potential predictors of talent in soccer from each subdiscipline of sport science.

Reprinted from A.M. Williams and A. Franks, 1998, "Talent identification in soccer," *Sport Exercise and Injury,* 4(4): 159-165. Reprinted with permission of Taylor & Francis, http://www.tandf.co.uk/journals.

As a step toward determining which variables are the best predictors of talent and future success, a team of researchers administered a battery of multidisciplinary tests to young players (16 yr of age) classified as either elite or subelite. The elite players were those who had signed for a professional club and had international experience playing youth soccer. A total of 28 tests were used, including items drawn from each of the biophysical subdisciplines of human movement studies as well as some soccer-specific practical-skills tests. Of the biophysical measures used, the best discriminators between the elite players and the subelite players were the physiological measures of agility and sprint time, the psychological measure of ego orientation, and the motor control measure of anticipation (see "Response Speed in Soccer Players" in chapter 18). The elite players were also significantly different from the subelite sample in that they were leaner, had greater aerobic power and superior fatigue tolerance, and performed better on the soccer-dribbling test. The key discriminators drawn from different subdisciplinary origins matched the multidimensional nature of soccer. This clearly highlights why it is imperative that multidisciplinary approaches, rather than unidisciplinary approaches, be used to examine research questions such as this. The future challenge for researchers is to establish baseline reference data for young players and to understand more fully how, when, and to what extent each of these key predictors of elite performance can be detected and developed. Existing approaches to early talent identification that lack a suitable evidence base have proven to be less-than-optimal vehicles for producing future champions.

PERFORMANCE OPTIMISATION

In contrast to talent identification, which relies largely on identifying individuals with appropriate genetic predispositions for success, in performance optimisation integrated knowledge from different subdisciplines of human movement studies is used to optimise performance within the limits of each individual's physical capabilities. We present in this section examples drawn from the stretch–shortening cycle of muscles, the monitoring of overtraining in athletes, and the use of psychophysiological feedback in recreational exercise and in competitive sport.

USING INTRINSIC STRETCH–SHORTENING CYCLES IN MUSCLE

The stretch–shortening cycle is the mechanism by which muscle generates an enhanced response when it is prestretched. In the field of strength and conditioning, activities that seek to exploit the stretch–shortening cycle have become known as *plyometrics*. Understanding the functional basis for plyometrics is an interdisciplinary problem.

A series of experiments conducted in Finland by Paavo Komi and colleagues in the 1970s and 1980s has led to an enhanced understanding of muscle function and techniques used in many human actions. Komi used the term *stretch–shortening cycles* to describe activities in which the muscles are stretched before concentrically contracting. It necessarily follows that the muscle needs to be acting eccentrically (i.e., contracting while increasing in length) for stretching to actually occur.

Beginning in the 1960s, Giovanni Cavagna, among others, noted that, compared with purely concentric contraction, prior eccentric contraction could develop greater force and result in higher efficiency. Komi established that the same was true for intact muscles and further showed that increased or augmented work and power were available during a concentric muscle contraction if it was preceded by an eccentric phase. His work over approximately 10 yr culminated in findings relating to efficiency of movement and the way in which this increased efficiency came about. Komi also showed that enhancement of performance does not occur when the muscle is fatigued.

Komi's early work is typified by an experiment using a specially designed sled to investigate concentric and eccentric muscular work of the lower limb. The sled was mounted on two parallel rails that were inclined approximately 40° above horizontal. A force plate, positioned at right angles to the rails at the bottom of the sled, was used to determine the forces applied to the ground. An oxygen-measuring system, placed nearby, allowed collection of expired gases from the participants and, hence, calculation of the physiological cost of work. A person could sit on this sled, be positioned at a fixed point up the rails, and be released or, alternatively, be required to push him- or herself up the sled to a certain height. The work done in either the eccentric condition (being released from a fixed point) or the concentric condition (pushing up to a certain height) was constant, but the power and physiological costs were found to be different.

The stretch–shortening cycle is usually associated with the effects of gravity on the lower-limb muscles during locomotion, but researchers from the Gulhane Military Medicine Academy in Turkey used an isokinetic machine to produce stretch–shortening cycles of upper-limb muscles. They showed that when an eccentric action precedes a concentric action of the medial rotators at the shoulder, performance is enhanced compared with that in other conditions. This has implications for throwing training.

The explanation of the augmented work (or increased efficiency) achieved during the stretch–shortening cycle has both biomechanical and neurophysiological aspects. In terms of biomechanics, energy is stored in elastic structures of the muscles and tendons during the eccentric phase and then used to do work during the concentric phase. Concurrently, during the stretching phase, muscle spindles and Golgi tendon organs (described in chapter 15) are stimulated, which results in increased potentiation of the muscles. Komi and colleagues were able to establish, through a series of cleverly designed experiments, that approximately two thirds of the increased work is due to elastic energy return and that one third is due to increased potentiation through reflex activity.

The implications of Komi's work are far reaching because virtually all motor activities use stretch–shortening cycles. At first glance it may appear that augmented work available through the stretch–shortening cycles would benefit only power athletes who require singular, maximal efforts. In such cases, techniques reflect the fact that muscles must be prestretched before contracting concentrically. However, closer examination reveals that the use of stretch–shortening cycles is equally important to endurance athletes, who are clearly interested in improving efficiency. For example, in running, stretching and then shortening takes place during the stance phase, when the calf muscles are stretched as the body rotates over the foot and the knee flexes. This energy can then be used to do work on the body during the concentric, push-off phase. Use of this elastic energy increases the economy of gait, possibly enhancing performance.

Komi's work, and that of others who have followed, is an excellent example of cross-disciplinary research in which physiology, biomechanics, biochemistry, and neurophysiology have led to an improved understanding of muscle-contraction dynamics. By using an activity that requires the large muscles of the lower limbs to perform sizeable amounts of work, the investigators could induce sufficiently large differences between the eccentric and concentric conditions and use a system that analyses expired gas to estimate oxygen metabolism. If only small limb movements had been chosen, the precision of the gas-analysis system would probably not have been sufficient to show significant differences. Through understanding the concept of mechanical work, Komi and colleagues recognised that the same amount of work could be done on the sled if participants accelerated the sled to a predetermined height or stopped the sled after it had been released from that height. The only difference was that the muscles had to perform concentrically in one situation and eccentrically in the other.

Knowledge of the anatomy of muscle, tendon, and neural pathways allowed these researchers to theorize on how these structures would influence the stretch–shortening cycle. If they had not known about muscle spindles, Golgi tendon organs, and the role of the Ia afferent neurons, they would not have been able to recognise that the increased work output after muscle stretch could be due to increased potentiation of the muscles through a reflex arc. Thus, to obtain a complete picture of the ways in which humans move, it is important to be eclectic and to investigate problems from multiple perspectives using a variety of techniques.

UNDERSTANDING OVERTRAINING SYNDROME IN ATHLETES

Overtraining syndrome (OTS) is a condition characterised by poor performance in competition, inability to maintain high training loads, persistent fatigue, changes in mood state, disrupted sleep patterns, and frequent illness. Overtraining syndrome reflects the body's inability to adapt to the cumulative fatigue resulting from daily intense training that is not balanced with adequate rest and recovery. Overtraining is of great importance to the high-performance athlete because persistent fatigue and lack of motivation interfere with the ability to continue to train and to optimally prepare for major events. Moreover, inconsistent or poor performance due to OTS may cause an athlete to prematurely retire from sport.

OTS has both physiological and psychological symptoms and outcomes, including decreased maximum heart rate and blood lactate levels, lower work rate at the lactate threshold, changes in hormone levels, increased fatigue, poor-quality sleep, and altered mood state and indicators of well-being. The relationship between physiological and psychological causes and consequences of OTS has been the subject of much speculation and research over the past 2 decades. It is unclear whether one precedes and possibly contributes to the other or whether both might result from a common cause. For example, psychological factors such as depression, anxiety, or the inability to concentrate can adversely affect the ability of athletes to maintain motivation to continue intense training and to compete effectively. Similarly, psychological factors such as mood disturbances can lead to disrupted sleep and eating patterns that, in turn, can negatively affect physiological functioning. On the other hand, physiological changes such as lower lactate threshold and inability to train intensely might cause athletes to become discouraged and anxious about their loss of form. Alternatively, OTS may involve changes in factors that affect both the physiological and psychological responses to intense training. Looking at OTS simplistically, negative moods could cause poor performances, poor performances could cause negative moods, or both negative moods and poor performances could be caused by a third over-riding factor. To better understand OTS, several research groups have studied simultaneous changes in physiological and psychological factors in athletes experiencing OTS.

OTS has been shown to reduce the lactate threshold, or exercise pace at which blood lactate starts to accumulate exponentially—a critical component of endurance-exercise capacity (see chapter 14). In one study (Urhausen et al., 1998), triathletes and cyclists were followed over 19 mo. In athletes showing symptoms of OTS, the ability to perform brief, maximal exercise (10-30 s) was not impaired, but endurance performance and lactate threshold declined by 27%. Maximal heart rate and blood lactate levels also declined in athletes showing symptoms of OTS. Athletes reported typical symptoms such as heavy muscles, fatigue, inability to concentrate, and disturbed sleep; mood state was also affected.

OTS has also been associated with changes in hormone levels. Hormones are messenger molecules that have far-reaching effects on the regulation of many physiological, psychological, and immuno-logical functions. Overtraining seems to particularly affect the production of **norepinephrine** (NEp; also known as noradrenaline), a hormone released in response to physical stress. For example, studies by Lehmann and colleagues (1991, 1992) reported progressive decreases in production of NEp in male distance runners over 4 wk of intensified running training. Changes in NEp production were also significantly correlated with self-reported symptoms of OTS on a complaints index; that is, lower NEp production was associated with more physical complaints (e.g., fatigue, heavy legs). In another study of elite swimmers (Mackinnon et al., 1997), NEp production was lower in swimmers showing symptoms of OTS compared with those who were able to adapt to higher training loads. Interestingly, low NEp levels appeared 2 wk before the onset of other symptoms, such as poor performance and high fatigue, suggesting that hormonal changes may precede and possibly contribute to later development of symptoms of OTS.

NEp is a hormone involved in regulation of heart rate and muscle metabolism, in particular the use of glycogen as a substrate to produce adenosine triphosphate (ATP). Low NEp production is consistent with data showing reduced maximum heart rate and ability of the muscles to use glycogen to produce ATP (as evidenced by the reduction in maximum blood lactate levels; see chapter 14). NEp also affects the immune system, and low NEp production might be related to the high frequency of minor illness such as colds and influenza reported among top athletes, especially during overtraining. Lower production of NEp suggests disturbance of the body's regulatory system, in particular stress hormones involved in arousal and the fight or flight response. In the fight or flight response, the body physiologically reacts to a perceived threat. In other words, physiological changes are dependent on psychological perceptions, reinforcing the connection between physiological and psychological factors. NEp level is related to a number of psychological factors. The main symptoms of depression are related to a lack of NEp in the brain, and many medications used in the treatment of depression increase the availability of NEp in the brain. NEp has also been associated with learning and memory. Blocking NEp receptors causes loss of memory. Therefore, if overtraining decreases the production of NEp, it is not surprising that overtrained athletes report an inability to concentrate and increased mood disturbance.

Coaches, athletes, and sports medicine and science practitioners are interested in monitoring athletes and possibly predicting which athletes might be susceptible to OTS. Effective monitoring could be used to modify training to prevent an athlete from training excessively during times of susceptibility. In one study (Hooper et al., 1995), 19 elite swimmers were followed for 6 mo as they prepared for national championships. A variety of physiological, biomechanical, and psychological variables were measured at five intervals, and a mathematical model was used to find which factors might predict the appearance of OTS symptoms. The best predictors were indicators of well-being, such as the athletes' daily ratings of fatigue, muscle soreness, perceived stress level, and quality of sleep. Changes in these variables appeared 4 wk before other symptoms of OTS, such as poor performance, appeared; these variables also predicted competitive performance in the championships. It was concluded that athletes' self-assessment on a daily basis may provide an easy, inexpensive, and effective means for monitoring how effectively athletes adapt to training and for preventing OTS.

The consistent finding that both psychological and physiological changes occur with OTS raises the question of whether a common cause of both types of symptoms exists. At present, the answer is not clear, but an interesting hypothesis has been put forward to try to explain the physiological, psychological, hormonal, and immunological changes that occur with OTS. It was suggested by Smith (2000) that OTS occurs in response to the persistent musculoskeletal stress arising from excessive training done without adequate recovery and this causes a chronic general inflammatory state in the body. That is, daily intense training without adequate recovery may cause microtrauma to the muscles, joints, and bones throughout the body and elicit release of inflammatory molecules, called **cytokines**. Cytokines are released during infection, inflammation, injury, trauma such as burns, and intense prolonged exercise.

Cytokines have far-reaching effects on virtually all body systems. For example, they cause fever, stimulate immune cells, influence energy metabolism, cause release of some hormones, and influence mood state and behaviour. Of special interest is that cytokines act on the central nervous system and are associated with sickness behaviours such as loss of interest, sleep disturbance, reduced appetite,

and depression, some of which also occur during OTS. These far-reaching effects of cytokines have probably evolved as a protective mechanism to force the body to take time to rest and recuperate from illness or injury. Cytokine production is normally highly regulated, and cytokines are usually released only in very small amounts for a limited time until health is restored.

Smith (2000) suggests that maintaining excessive training with inadequate recovery causes the body to adopt an abnormal and persistent state of inflammation that results in chronic elevation of cytokine levels. Cytokines may be the common mediator of both the physiological and psychological symptoms of OTS. Further experimental study is needed to determine whether this attractive hypothesis can explain the myriad symptoms of OTS and, importantly, whether such a relationship can be exploited to prevent OTS among athletes.

IMPROVING CONTROL AND PERFORMANCE THROUGH BIOFEEDBACK

Biofeedback is another area related to performance optimisation in which information from the subdisciplines of exercise physiology and sport and exercise psychology is integrated. Biofeedback instruments give individuals information about their biological processes. Through increasing awareness of biological processes such as heart rate, sweating, muscle tension, or brain activity, individuals can eventually learn to modify or control their physiological activation. Biofeedback provides a means by which we can learn to control bodily functions that traditionally have been thought to be beyond our control.

Researchers worldwide are interested in ascertaining whether biofeedback can be used productively to enhance performance. For example, a study in Spain (Moleiro & Cid, 2001) investigated the effectiveness of biofeedback in controlling the heart rate during low- and moderate-intensity exercise. The participants were 35 females between the ages of 17 and 25 yr who were randomly assigned to 1 of 4 experimental groups: medium exercise intensity (50% of maximal heart rate) with heart rate biofeedback, low exercise intensity (30% of maximal heart rate) with heart rate biofeedback, medium exercise intensity with verbal instructions to try to lower heart rates, and low exercise intensity with verbal instructions.

During the study the participants completed a submaximal test in the first session and then had four sessions on controlling heart rate. In these four sessions, participants received either biofeedback or verbal instructions. The participants who trained with biofeedback showed a greater attenuation of the increase in heart rate produced by exercise than did participants who trained with verbal instructions. The workload did not influence the voluntary control of heart rate. These results demonstrate that individuals can learn to voluntarily control physiological variables through the use of biofeedback techniques.

Biofeedback has also been included in intervention studies focusing upon enhancing sporting performance. Researchers from Arizona State University in the United States (Landers et al., 1991) directly tested the effectiveness of biofeedback in enhancing archery performance. Rather than focusing on heart rate, these researchers looked at electrocortical, or brain wave, biofeedback. Through use of an electroencephalograph (EEG), 24 archers were provided with correct EEG feedback, incorrect EEG feedback, or no feedback at all. The archers who received correct feedback significantly improved their performance, the archers who received incorrect feedback showed a significant performance decrement, and the control group showed no changes in performance. The connection of electrocortical activity to archery performance is based on research showing that shooters have reduced cortical activity in the left temporal area while shooting. In the archery study, the correct feedback reduced right temporal activation. One can conclude from the study that EEG biofeedback can affect performance and that the direction of the effect is determined by whether the feedback is incorrect or correct.

Since this early work, EEG-based biofeedback has been progressively refined in an attempt to both accelerate and enhance the learning effects that may be available. In recent times biofeedback has become of particular interest to therapists seeking to help patients recover previously automatic skills such as gait. Collectively, these various studies highlight the interaction of psychological, physiological, and neural factors and further illustrate the importance of cross-disciplinary research to knowledge advancement.

SUMMARY

Two different but complementary approaches to maximising performance on a particular movement task are possible. One approach is to ensure that the person selected to undertake the task has the most suitable attributes to perform on the task at a high level; this is essentially the talent-identification approach. A second approach is to optimise the learning and performance of the person undertaking the task by making best use of both the functional capabilities of the performer's own body as well as fully harnessing any external technologies, such as biofeedback, that may enable learning. In both approaches, best results are achieved when practice is guided by research that examines the situation from multiple and integrated perspectives.

FURTHER READING AND REFERENCES

Talent Identification in Sport

Baker, J., & Davids, K. (Eds.) (2007). Nature, nurture and sport performance. *International Journal of Sport Psychology, 38,* 1-143. (Special issue).

Carlson, R. (1993). The path to the national level in sports in Sweden. *Scandinavian Journal of Medicine and Science in Sports, 3,* 170-177.

Durand-Bush, N., & Salmela, J.H. (2001). The development of talent in sport. In R.N. Singer, H.A. Hausenblas, & C.M. Janelle (Eds.), *Handbook of sport psychology* (2nd ed.) (pp. 269-289). New York: Wiley.

Pearson, D.T., Naughton, G.A., & Torode, M. (2006). Predictability of physiological testing and the role of maturation in talent identification for adolescent team sports. *Journal of Science and Medicine in Sport, 9,* 277-287.

Reilly, T., Williams, A.M., Nevill, A., & Franks, A. (2000). A multidisciplinary approach to talent identification in soccer. *Journal of Sports Sciences, 18,* 695-702.

Vaeyens, R., Güllich, A., Warr, C.R., & Philippaerts, R. (2009). Talent identification and promotion programmes of Olympic athletes. *Journal of Sports Sciences, 27,* 1367-1380.

Vaeyens, R., Lenoir, M., Williams, A.M., & Philippaerts, R. (2008). Talent identification and development programmes in sport: Current models and future directions. *Sports Medicine, 38,* 703-714.

Williams, A.M., & Reilly, T. (2000). Talent identification and development in soccer. *Journal of Sports Sciences, 18,* 657-667.

Stretch–Shortening Cycle

Aydin, T., Yildiz, Y., Yildiz, C., & Kalyon, T.A. (2001). The stretch-shortening cycle of the internal rotator muscle group measured by isokinetic dynamometry. *Journal of Sports Medicine and Physical Fitness, 41*(3), 371-379.

Fleischmann, J., Gehring, D., Mornieux, G., & Gollhofer, A. (2010). Load-dependent movement regulation of lateral stretch shortening cycle jumps. *European Journal of Applied Physiology, 110,* 117-187.

Gerodimos, V., Zafeiridis, A., Perkos, S., Dipla, K., Manou, V., & Kellis, S. (2008). The contribution of stretch-shortening cycle and arm-swing to vertical jumping performance in children, adolescents, and adult basketball players. *Paediatric Exercise Science, 20,* 379-389.

Komi, P.V. (1984). The stretch-shortening cycle and human power output. In N.L. Jones, N. McCartney, & A.J. McComas (Eds.), *Human muscle power* (pp. 27-39). Champaign, IL: Human Kinetics.

Komi, P.V. (2000). Stretch-shortening cycle: A powerful model to study normal and fatigued muscle. *Journal of Biomechanics, 33*(10), 1197-1206.

Overtraining in Athletes

Hooper, S.L., Mackinnon, L.T., Howard, A., Gordon, R.D., & Bachmann, A.W. (1995). Markers for monitoring overtraining and recovery. *Medicine and Science in Sports and Exercise, 27,* 106-112.

Kellmann, M. (2010). Preventing overtraining in athletes in high-intensity sports and stress/recovery monitoring. *Scandinavian Journal of Medicine and Science in Sports, 20*(2), 95-102.

Lehmann, M., Dickhuth, H.H., Gendrisch, G., Lazar, W., Thum, M., Kaminski, R., Aramendi, J.F., Peterke, E., Wieland, W., & Keul, J. (1991). Training–overtraining. A prospective, experimental study with experienced middle- and long-distance runners. *International Journal of Sports Medicine, 12,* 444-452.

Lehmann, M., Schnee, W., Scheu, R., Stockhausen, W., & Bachl, N. (1992). Decreased nocturnal catecholamine excretion: Parameter for an overtraining syndrome in athletes? *International Journal of Sports Medicine, 13,* 236-242.

Mackinnon, L.T., Hooper, S.L., Jones, S., Gordon, R.D., & Bachmann, A.W. (1997). Hormonal, immunological, and haematological responses to intensified training in swimmers. *Medicine and Science in Sports and Exercise, 28,* 741-747.

Roose, J., de Vries, W.R., Schmikli, S.L., Backx, F.J., & van Doornen, L.J. (2009). Evaluation and opportunities in overtraining approaches. *Research Quarterly for Exercise and Sport, 80,* 756-764.

Smith, L.L. (2000). Cytokine hypothesis of overtraining: Physiological adaptation to excessive stress? *Medicine and Science in Sports and Exercise, 32,* 317-331.

Urhausen, A., Gabriel, H.H.W., Weiler, B., & Kindermann, W. (1998). Ergometric and psychological findings during overtraining: A long-term follow-up study in endurance athletes. *International Journal of Sports Medicine, 19,* 114-120.

Biofeedback Training

Hunt, M.A., Simic, M., Hinman, R.S., Bennell, K.L., & Wrigley, T.V. (2011). Feasibility of a gait retraining strategy for reducing knee joint loading: Increasing trunk lean guided by real-time biofeedback. *Journal of Biomechanics, 44,* 943-947.

Landers, D.M., Petruzzello, S.J., Salazar, W., Crews, D.J., Kubitz, K.A., Gannon, T.L., & Han, M. (1991). The influence of electrocortical biofeedback on performance in pre-elite archers. *Medicine and Science in Sports and Exercise, 23*(1), 123-129.

Moleiro, M.A., & Cid, F.V. (2001). Effects of biofeedback training on voluntary heart rate control during dynamic exercise. *Applied Psychophysiology and Biofeedback, 26*(4), 279-292.

Tate, J.J., & Milner, C.E. (2010). Real-time kinematics, temporospatial, and kinetic biofeedback during gait retraining in patients: A systematic review. *Physical Therapy, 90,* 1123-1134.

GLOSSARY

acceleration—The time rate of change of velocity.

accommodation—The process by which the curvature of the lens of the eye is adjusted for viewing objects at different distances; achieved through control of the ciliary muscles.

acetylcholine—A common neurotransmitter in the peripheral and central nervous systems.

achievement goal orientations—Individual definitions of success.

actin—Thin protein filaments of muscle.

action stage—The fourth stage of the transtheoretical model. In this stage, individuals have changed their behaviour, but have done so for less than 6 mo.

activation heat—Heat produced in the initial stage of a muscular contraction.

adaptation—The structural or functional adjustment of an organism to its environment that improves its chances of survival.

addiction—Dependence on or commitment to a habit, practice, or substance that is present to the extent that its cessation causes trauma.

adenosine triphosphate—High-energy phosphate molecule produced in all cells; cleavage of the terminal phosphate bond yields energy to fuel cellular work such as development of force in skeletal muscle.

aerobic (oxidative) energy system—Energy system of oxidative metabolism requiring oxygen to produce adenosine triphosphate; provides the major source of adenosine triphosphate for endurance exercise lasting longer than 3 min.

aerobic power—See $\dot{V}O_2max$.

afferent—Pertaining to sensory (or input) information transmitted from the sensory receptors to the central nervous system and brain.

age stratification—Hierarchical ranking of people into age groups that may promote prejudicial and discriminatory views about ageing.

alpha–gamma coactivation—The process involving simultaneous activation of both alpha motor neurons (to extrafusal muscle fibres) and gamma motor neurons (to intrafusal muscle fibres) that permits comparison of actual muscle length with intended muscle length.

alpha motor neurons—Large nerve cells that innervate skeletal (extrafusal) muscle fibres.

ambient vision—The aspect of vision, deriving from the whole visual field and the nerve pathway connecting the optic nerve to the superior colliculi in the midbrain, that is responsible for the location of objects and the perception of self-motion.

anaerobic capacity—The total amount of work, in kilojoules, accomplished during brief, high-intensity exercise such as the 30 s bicycle ergometer test; considered an indication of the capacity of the anaerobic glycolytic energy system.

anaerobic glycolytic energy system—Metabolic energy system that uses glycogen or glucose to produce adenosine triphosphate and produces lactic acid as a byproduct; the major source of adenosine triphosphate for maximal exercise lasting between 20 s and 3 min.

anaerobic power—Peak or maximal power, in watts, achieved during brief, high-intensity exercise such as the 10 s bicycle ergometer test; considered an indication of the capacity of the immediate energy system.

anaerobic threshold—See *lactate threshold*.

androgyny index—The use of pelvic and shoulder widths to distinguish between adult males and females.

angular acceleration—The rate of change of angular velocity. Usually expressed as two components: centripetal (or radial) and tangential acceleration.

angular displacement—The change in the angular position of a line during a time interval Δt.

angular velocity—The rate of change of angular displacement.

anterior—Front.

anthropometer—An instrument used to measure lengths of body segments.

anthropometry—Study of the size, proportions, and composition of the human body

anxiety—The subjective feeling of apprehension or worry often experienced when a situation is perceived to be threatening.

appositional growth—Addition or erosion, or both, of bone at the outer and inner surfaces of the shaft to cause changes in shaft diameter and thickness of the compact bone.

apraxia—The inability to carry out purposeful movements in the absence of paralysis or other sensory or motor impairments, generally following damage to the cerebral cortex.

arousal—General state of physiological alertness as controlled by activation of the reticular formation.

artery—A blood vessel that carries blood away from the heart.

arthritis—Inflammation of synovial joints.

arthrology—The study of the joints of the body.

articular cartilage—Cartilage forming the smooth bearing surface of a synovial joint.

articulation—See *joint*.

asceticism—The religious doctrine of the Middle Ages that demanded extreme physical self-denial in order to focus exclusively on spiritual matters.

associating—Attending to the body while exercising.

associative phase—The second phase in the learning of a new skill, in which movement patterns become more refined and consistent through practice.

ataxia—The breakdown or irregularity of muscular coordination, generally following damage to the cerebellum.

atrophic—Referring to the process whereby cells decrease in size (waste away) with disuse or disease.

attention-arousal set theory—A theory that states that imagery is effective because it primes the athlete for performance by helping the athlete reach the optimal level of arousal and focus attention on what is relevant.

autonomous phase—The final phase in the learning of a new skill, in which the control of movement appears to be automatic and free of the need for constant attention.

auxology—The study of growth.

axon—A single nerve fibre extending away from the cell body of a neuron and responsible for sending nerve impulses away from the cell body.

basal ganglia—Collection of nuclei, located in the cerebral hemispheres, that is intimately involved in movement control and coordination.

bending—A combination of tensile, compressive, and shear forces in a structure.

biarticular muscle—A skeletal muscle that crosses two synovial joints.

bicondylar callipers—An instrument used to measure widths of bones near their ends.

biofeedback—Immediate feedback of a biological phenomenon provided through use of electronic recording instruments or the use of instruments to obtain information about biological processes of which one is not normally aware. Through biofeedback training, individuals can learn to modify or control these biological processes.

biomechanics—Application of mechanics to the study of living systems.

blocked practice—A type of practice in which each skill component is practised repetitively as an independent block. Practice is fully completed on one skill component before practice is commenced on the next skill component.

body mass index—A simple estimate of body mass calculated as weight (kg) divided by height (m)^2.

brain stem—The section of the brain consisting of the medulla, pons, and midbrain, lying between the cerebrum and the spinal cord.

burnout—The third and final phase of training stress syndrome in which repeated failure to cope with training demands results in physiological and psychological exhaustion.

cadaver—A human body that has been embalmed after death for the purposes of anatomical dissection and instruction.

cadence—Frequency of a rhythmic motion, such as the gait cycle.

calcium—A mineral stored in bone and essential for muscle contraction.

callus—Unorganised meshwork of fibres or bone laid down after a fracture; eventually replaced by compact bone.

cancellous bone—See *spongy bone*.

cardiac muscle—The specialised striated muscle forming the walls of the heart.

cardiac output—The volume of blood pumped by the heart per minute.

cardiovascular disease—A general term that describes various diseases of the heart and blood vessels, including coronary heart disease, heart failure, high blood pressure (hypertension), stroke, and peripheral vascular disease (disease of the blood vessels in the limbs).

cartilage—A type of connective tissue with high water content.

cartilaginous joint—A type of joint in which the material between the bones is mainly composed of cartilage.

catastrophe model—When cognitive anxiety is low, the relationship between arousal and performance takes the form of the inverted U; when cognitive anxiety is high, any increase in arousal past the optimum point results in a rapid and dramatic decline in performance.

central nervous system—The nervous system consisting of the brain and the spinal cord.

central pattern generator—The capacity of the spinal cord (or networks of neurons in it) to generate rhythmic flexion–extension patterns that form the basis for locomotion.

central sulcus—A deep trench or vertical groove in the middle of each cerebral hemisphere that separates the frontal lobe from the parietal lobe. The motor cortex lies immediately forward of, and the sensory cortex immediately behind, the central sulcus.

centre of mass—The theoretical point in an object at which its entire mass appears to be concentrated; also known as centre of gravity.

centripetal acceleration—Component of acceleration of a point on a body given by the square of the angular velocity of the body multiplied by the radius of the circle on which the body is moving at the instant under consideration; directed toward the centre of the circle on which the body is moving at the instant under consideration.

cephalic index—Relationship between the breadth and length of the skull.

cephalocaudal principle—Principle stating that development occurs in a "head-down" or "head-to-tail" manner.

cerebellum—A subdivision of the brain, lying below the cerebral cortex and behind the brain stem, that plays a major role in movement coordination.

cerebrum—The largest region of the brain, consisting of the two cerebral hemispheres.

circuit training—Training in which the athlete moves around a circuit of exercise stations (exercise machines or activities).

clavicle—Collarbone.

climbing fibre—A type of afferent nerve fibre in the cerebellum.

closed-chain exercises—Exercises where an external force is applied to one or both feet during the exercise (e.g., the ground applies a force to both feet during a squatting exercise).

closed-loop control—A type of movement control where the movement is controlled continuously on the basis of sensory feedback arising during the movement itself.

cognate discipline—A related discipline.

cognitive anxiety—The *mental* (cognitive) component of the situation-specific experience of apprehension.

cognitive science—The hybrid field of study between experimental psychology and computer science concerned with understanding the computational processes of the brain and nervous system.

collagen—A fibrous protein that is an important constituent of connective tissues such as bone, ligament, tendon, and cartilage.

compact bone—Dense bone that does not appear porous.

comparison—The perceptual process responsible for determining the relative strengths of two stimuli.

compass gait—The simplest form of walking in which all joints, with the exception of the hip joints, are locked in the anatomical position.

complex joint—Synovial joint containing an intra-articular structure.

compliance—The ability of a material to store energy (as strain energy) and then return it to the object that initially possessed the energy.

compound joint—Synovial joint involving more than one pair of articulating surfaces.

compression—A type of force in which the two ends of a structure are squeezed together.

compressive strength—Strength in opposing breaking when a material is compacted.

computer-aided tomography—Production of images of sections through the body using an X-ray source; also referred to as computed tomography.

computer simulation—Solution of a set of mathematical (differential) equations describing the behaviour of a physical system. Complex simulations must be performed on a computer. Computer simulations of human movement are usually performed using time as the independent variable.

concentric—A dynamic muscle contraction in which the muscle shortens while developing tension.

conservation—A system in which the total mechanical energy remains unchanged.

contemplation stage—The second stage of the transtheoretical model. In this stage individuals are seriously thinking about changing their behaviour in the next 6 mo.

contralateral—On the opposite side of the body.

control parameters—Parameters (such as movement frequency) that, when manipulated to a critical value,

trigger a transition from one pattern of organisation to another.

coronal plane—A plane dividing the body into front and back sides.

corpus callosum—The thick band of neural tissue connecting the two cerebral hemispheres.

cortical bone—See *compact bone.*

critical fluctuations—A term from synergetics that describes the increasing variability in a particular pattern of organisation as the point of transition to a different pattern is approached.

critical period—A period during development when the organism (or a specific system or skill) is most readily influenced by both favourable and adverse environmental factors.

critical slowing down—A term from synergetics that describes the delayed response to unexpected perturbations that a particular pattern of organisation shows as it approaches the point of transition to a different pattern.

cross-bridge cycling—The process by which myofilaments are pulled toward each other to produce a muscle contraction.

cross-bridge hypothesis—Proposed explanation of the mechanics of striated muscle contraction.

cross-disciplinary—An approach in which a problem is examined using the methods of more than one subdiscipline and with some integration of the information from the different subdisciplines.

cross-sectional study—A study of a representative set of all the population at one specific point in time. For example, a cross-sectional study of motor development compares individuals of different ages all measured at about the same time.

crossed extensor reflex—A reflex that increases activation of the extensor muscles of the limb on the side opposite a limb undergoing flexion.

cutaneous receptors—Receptors that are located in the skin.

cytokines—Molecules that play a key role in regulating the intensity and duration of the response of the immune system to stress and illness.

degrees of freedom—The number of independent variables (e.g., motor units, muscles, joint angles) that must be simultaneously controlled in order to produce purposeful movement.

delayed-onset muscle soreness—Muscle soreness, characterised by tender and painful muscles, that usually appears in the few days after unaccustomed exercise; most commonly occurs after exercise with a large eccentric component, such as downhill running.

dendrites—The branches of a neuron that synapse with, and receive nerve impulses from, other neurons.

dense bone—See *compact bone.*

density—Mass per unit volume of a material.

depersonalisation—A characteristic of burnout in which an individual develops a detached and disinterested attitude toward others.

detection—The perceptual process responsible for determining the presence of particular stimuli.

displacement—The straight-line distance between the initial and final positions of a body.

dissection—Use of scalpels and other surgical instruments to reveal particular anatomical structures.

dissociating—Using attentional strategies to distract oneself from the fatigue-producing effects of exercise.

distance—Magnitude of the displacement of an object, along the path followed, between its initial and final positions.

dopamine—A neurotransmitter, the absence of which in parts of the basal ganglia gives rise to Parkinson's disease.

dorsal—Pertaining to the posterior side or back of the body.

dorsiflexion—Backward flexion or bending of a joint. Dorsiflexion of the foot draws the upper surface of the foot closer to the forward surface of the lower leg (i.e., the shin).

Down syndrome—A chromosomal abnormality associated with delays in both mental and motor development.

drive theory—A proposed linear relationship between arousal and performance.

dual X-ray absorptiometry—Determination of body composition using X-ray analysis.

dynamical models—Models that explain the control of movement largely on the basis of the physical properties of the musculoskeletal system.

dynamics—The study of objects in motion.

dynamometer—An instrument or machine used to measure muscular strength.

dyslipidemia—A condition in which blood lipid (fat) levels are outside the range recommended for good health.

eccentric—A dynamic muscle contraction in which the muscle lengthens while developing tension.

ectoderm—Primary germ layer that gives rise to the outer skin and nervous system.

ectomorphy—One component of a somatotype, based on a person's height and mass.

efferent—Pertaining to motor (or output) information transmitted from the brain and central nervous system out to the muscles.

efficient—Referring to mechanical work done divided by metabolic energy consumed.

ego orientation—The tendency to define success as being better than others.

electroencephalography—The recording of electrical activity from the brain.

electromyography—The recording of the electrical signal produced by skeletal muscle as it contracts.

electron microscope—A microscope in which an electron beam passes through a very thin section of tissue.

embryo—Developing human during the first quarter of intrauterine life.

endochondral ossification—Development of bone from a cartilaginous model.

endoderm—Primary germ layer that gives rise to the organs of the body.

endomorphy—One component of a somatotype, based on the thicknesses of the skinfolds.

endurance-exercise capacity—Performance measure such as the maximum time an individual can exercise at a given speed or total amount of work that can be accomplished in a given time.

epiphyseal plate—Cartilaginous growth plate in developing long bone.

equilibrium—The state in which every point on the body has the same velocity.

excess postexercise oxygen consumption—Oxygen consumption in excess of resting level during recovery after exercise; formerly known as oxygen debt.

excitation–contraction coupling—A series of steps between the excitation of a skeletal muscle by its nerve and the production of force.

exercise adherence motivation—The drive to maintain regular physical activity.

exercise and sport science—Term used to describe the field of study concerned with the application of the methods of science to understanding of exercise and sport.

exercise-induced asthma—Bronchoconstriction (narrowing of breathing tubes) causing reduced ventilation during exercise in persons who have asthma.

exercise-induced bronchospasm—See *exercise-induced asthma*.

exercise participation motivation—The drive to initiate regular physical activity.

exercise physiology—The subdiscipline of human movement studies concerned with understanding physiological responses to exercise.

exercise self-efficacy—The perception of one's ability to succeed at an exercise program.

explicit learning—Learning that occurs consciously and deliberately with the concurrent acquisition of verbalisable knowledge.

extension—Joint motion in the sagittal plane that increases the angle between the limb segments.

extensor thrust reflex—A reflex extension of the legs in response to stimulation of the soles of the feet; assists in supporting the body's weight against gravity.

external force—A force that does work to move a body; only external forces are included in free-body diagrams.

extrafusal muscle fibres—The characteristic skeletal muscle fibres, activated by alpha motor neurons. Their contraction causes voluntary movement.

extrapyramidal tract—All the descending motor pathways from the brain to alpha motor neurons other than the direct connections contained in the pyramidal tract.

extrovert—A person whose basic orientation is toward the external world.

fibrous joint—A type of joint in which the material between the bones contains, or resembles, fibres.

first law of thermodynamics—Law describing the relationship between energy production, heat liberation, and the rate of work performed on an external load.

flexion—Joint motion in the sagittal plane that decreases the angle between the limb segments.

flexion reflex—A reflex that produces flexion of the joints that draw an injured limb away from a painful stimulus.

focal vision—The aspect of vision, deriving from the central retina (fovea) and the nerve pathway connecting the optic nerve to the visual cortex, that is responsible for object recognition and the resolution of fine detail.

foetus—Developing human during the last three quarters of intrauterine life.

force—Measure of the effort applied.

fovea—The area near the centre of the retina in which cone cells are most concentrated and that therefore provides the most acute vision.

free-body diagram—Schematic diagram showing all the external forces acting on a system.

frontal plane—See *coronal plane*.

functional anatomy—The subdiscipline concerned with understanding the anatomical bases of human movement and the effects of physical activity on the musculoskeletal system.

gait—Term used interchangeably with walking in this text. One complete gait cycle is defined by the time between successive heel strikes of the same leg.

gamma motor neurons—Small nerve cells that innervate spindle (intrafusal) muscle fibres.

genotype—A genetically determined somatotype.

gerontology—The study of ageing.

glia—Non-neuronal cells in the brain and spinal cord that help regulate the extracellular environment of the central nervous system through the provision of metabolic and immunological support for the nerve cells.

glial cells—See *glia*.

glycogen—Storage form of glucose in the cells, comprising polymers of glucose molecules.

Golgi tendon organs—Specialised receptors located in tendons that respond to tendon tension.

goniometer—An instrument used to measure range of movement at a joint.

gross anatomy—The study of structures that can be seen with the unaided eyes.

ground reaction force—The pattern of force exerted on the ground during locomotion. Usually separated into three components: fore–aft, vertical, and medial–lateral.

gustatory information—Information derived through the sense of taste.

hamstrings muscles—The muscle group at the back of the thigh.

Haversian canals—The narrow system of blood vessels that supply oxygen and micronutrients to the bone cells (osteocytes).

Haversian system—The basic unit of compact bone, consisting of a canal containing blood vessels surrounded by multiple thin layers of bone.

Heath-Carter anthropometric somatotype—An example of a somatotyping technique based on anthropometric measurements.

hemiparesis—Muscular weakness or paralysis affecting one side of the body.

hemiplegia—Paralysis of one side of the body.

hippocampus—A portion of the brain believed to play a central role in memory and learning.

homeostasis—Maintenance of metabolic equilibrium in an organism.

human movement science—A term used to describe the field of study concerned with the application of the methods of science to the understanding of human movement.

human movement studies—The field of study concerned with understanding how and why people move and the factors that limit and enhance the capacity to move.

Huntington's disease—A degenerative, inherited disease that affects neurotransmission in the basal ganglia and is characterised by rapid, involuntary limb movements.

hyperplasia—Increase in the number of cells forming a tissue.

hypertrophic phase—In weight training, the second phase of increase in muscle size.

hypertrophy—An increase in the size of each cell forming a tissue.

imagery—A mental skill involving the use of all senses to create or recreate an experience in the mind.

immediate energy system—Metabolic energy system in which phosphocreatine is split to provide energy for adenosine triphosphate resynthesis; provides an immediate source of energy at the onset of exercise.

implicit learning—Learning that occurs without conscious awareness and without the concurrent acquisition of verbalisable knowledge.

inertia—The resistance of a body to a change in state of its motion.

information-processing model—A model that considers the human nervous system as a sophisticated processor of information, like a computer.

interaction framework—An approach that views behaviour as a function of the interaction of personality and environmental factors.

interdisciplinary—An approach in which a problem is examined using the methods of more than one subdiscipline and with tight integration of the information from the different subdisciplines.

internal force—A force that does no work to move a body. Internal forces are not included in free-body diagrams.

interneurons—Nerve cells that connect one nerve cell to another.

interval training—Form of exercise training of alternating work and rest intervals.

intrafusal muscle fibres—The modified skeletal muscle fibres found in the muscle spindle receptors and activated by gamma motor neurons.

introvert—A person who tends to be hesitant, reflective, withdrawn, and reserved.

inverted-U hypothesis—High performance occurs with an optimal level of arousal, and lesser performances occur with either low or very high arousal.

ipsilateral—On the same side of the body.

isoinertial—A situation involving isotonic muscle contraction.

isokinetic—Movement at constant speed (e.g., of a body joint).

isometric—Contraction of a muscle that is being held at constant length. At steady state, the muscle develops a constant force over time.

isotonic muscle contraction—Contraction occurring while a constant force is applied to a muscle.

joint—Union of two or more bones.

joint capsule—Thick connective tissue membrane forming the boundary of a synovial joint and enclosing the joint cavity.

joule—The Standard International unit of kinetic energy. It is equal to the work done when the point of application of a force of 1 newton is moved through a distance of 1 m in the direction of force

kinaesthetic—Pertaining to sensory information provided by the receptors located in the muscles, tendons, joints, and skin.

kinanthropometry—The scientific specialisation dealing with the measurement of humans in a variety of morphological perspectives, its application to movement, and the factors that influence movement.

kinematics—The branch of mechanics concerned with the description of motion.

kinesiology—The scientific study of movement. The term is often used more narrowly to refer to the subdisciplines concerned with understanding the anatomical and mechanical bases of human movement.

kinesthesis—Sensory information provided by the receptors located in the muscles, tendons, joints, and skin.

kinetic energy—Amount of mechanical energy a body possesses due to its motion.

Krause's end bulbs—Nerve ending receptors located in the skin.

lactate threshold—Point of inflection of the curve of blood lactate concentration versus exercise intensity, above which blood lactate concentration increases disproportionately with increasing exercise intensity. Considered to identify an exercise intensity that can be maintained at the upper limit of aerobic metabolic capacity without major input from the anaerobic glycolytic system.

lactic acid—Byproduct of anaerobic glycolytic breakdown of glucose or glycogen. Hydrogen ions dissociated from lactic acid increase acidity (decrease pH), causing fatigue in skeletal muscle.

lacuna—A small space in compact bone in which an osteocyte is situated.

lamella—A layer of bone material.

lamina—Plate-like structure forming part of the arch of a vertebra.

ligament—A dense regular connective tissue joining bone to bone.

light microscope—A microscope in which a beam of light passes through a thin, stained section of tissue.

line of action of the force—The direction in which a force is applied.

Little League elbow—Injury to the elbow joint in children caused by overuse from overarm throwing; so named because of its prevalence in junior and adolescent baseball pitchers.

locomotor reflexes—Reflexes present at birth or soon thereafter that are the primitive precursors to locomotion.

longitudinal study—A study in which the same individuals are measured over a period of years.

long-loop reflexes—A collective term for reflexes involving multiple synapses in which nerve impulses are passed from the local level of the spinal cord to higher levels of the cord and perhaps even to regions of the brain.

long-term potentiation—The prolonged increase in the efficiency of a synapse that occurs as a result of repeated stimulation; thought to be a fundamental neurophysiological mechanism for memory and learning.

macroscopic anatomy—See *gross anatomy*.

magnetic resonance imaging—Production of images of sections through the body using a very strong magnetic field.

magnetoencephalography—The recording of magnetic signals proportional to the electronencephalographic waves resulting from the electrical activity of the brain.

maintenance heat—Heat produced by a muscle during a steady-state contraction.

maintenance stage—The fifth stage of the transtheoretical model. In this stage, individuals have successfully changed their behaviour for at least 6 mo.

marrow—Tissue within the shafts of long bones and in the spaces within spongy bone.

maturation—Sequence of changes occurring between conception and maturity.

maximum heart rate—Highest possible heart rate, usually achieved during maximal exercise; estimated by the equation 220 minus age.

mechanics—Study of forces and their effects; often divided into statics and dynamics.

mechanomyography—A technique used to examine mechanical activity in muscle based around recording of muscle deformation.

medulla—Major anatomical component of the brain stem.

Meissner's corpuscles—Specialised receptors found primarily on the hairless surfaces of skin that respond to light pressure or touch.

menarche—Onset of menstruation.

meniscus—An intraarticular cartilaginous structure that is shaped like both a crescent and a wedge.

Merkel's discs (or corpuscles)—Specialised receptors found primarily on the hairless surfaces of skin that respond to light pressure or touch.

mesoderm—Primary germ layer that gives rise to the structures of the musculoskeletal system.

mesomorphy—One component of a somatotype, based on the development of the musculoskeletal system.

metabolic syndrome—A cluster of four conditions (obesity, type 2 diabetes, hypertension, and abnormal blood lipid profile) that greatly increase an individual's risk of cardiovascular disease.

metabolism—The chemical processes (liberating energy from food) that occur in living cells and sustain life and normal functioning.

microscopic anatomy—The study of structures that can be seen in thin sections of tissue through use of a microscope.

mind–body dualism—The philosophy that the mind and body are conceptually separate, distinct, and independent from each other.

minute ventilation—Volume of air inspired by the lungs per minute.

model—Set of mathematical equations used to describe the behaviour of a physical system. The equations are usually solved using a computer.

moment (of a force)—Measure of the turning effect of the force about that point.

moment (of inertia)—A body's resistance to rotation; mathematically, defined as the net moment applied to a body divided by the angular acceleration of the body.

monoarticular—Referring to a skeletal muscle that crosses a single synovial joint.

mossy fibre—A type of afferent nerve fibre in the cerebellum.

motivation—The drive, interest, and desire to undertake or achieve a particular goal; the directions and intensity of effort over time.

motor control—That subdiscipline of human movement studies concerned with understanding the processes that underlie the acquisition, performance, and retention of motor skills.

motor cortex—A strip of cerebral cortex immediately anterior of the central sulcus that is responsible for the relay of many of the motor commands from the brain to the muscles.

motor development—The field of study concerned with understanding the changes in motor control that occur throughout the lifespan.

motor equivalence—The capability of the motor system to produce the same movement outcome in a variety of ways.

motor learning—The field of study concerned with understanding the changes in motor control that occur in response to practice; also known as skill acquisition.

motor milestones—Identifiable stages in the development of specific fundamental motor skills.

motor program—A set of motor (efferent) commands set up in advance of a movement commencing that provide the potential for the movement to be completed without relying on sensory feedback during the movement; see also *open-loop control*.

motor skills—Skills that require movement of the whole body, a limb, or a muscle in order to be successfully performed.

motor unit—Structure consisting of a single α motor neuron and all the muscle fibres it innervates.

movement time—The time elapsed between the initiation and completion of a movement.

multidisciplinary—An approach in which a problem is examined using the methods of more than one subdiscipline but with little integration of the information from the different subdisciplines.

multiple sclerosis—A disease caused by damage to the myelin sheath surrounding nerves, resulting in hardening of the nerve tissue and consequent tremor, loss of coordination, and occasionally paralysis.

muscle fibre—Muscle cell. Humans have three types of skeletal muscle fibres that differ in their metabolic and physiological characteristics: type I (slow oxidative), type IIA (fast oxidative glycolytic), and type IIB (fast glycolytic).

muscle moment arm—The perpendicular distance between the line of action of a muscle and the axis of rotation of the joint.

muscle spindle—Specialised receptor located in the intrafusal fibres of skeletal muscle that responds to stretch.

muscle tone—The tension produced in skeletal muscles under resting conditions as a result of low-frequency discharge of the alpha motor neurons.

muscular endurance—Ability of a muscle to repeatedly develop and maintain submaximal force over time.

muscular power—Muscle force multiplied by contraction velocity. A muscle develops power when it undergoes a shortening contraction, and it absorbs power during a lengthening contraction.

muscular strength—The peak isometric force that a muscle can develop. Muscles usually develop peak isometric force at their resting length. The maximum strength of a muscle is directly proportional to its cross-sectional area.

musculoskeletal system—The system consisting of bones, joints, and muscles.

myelin—The fatty insulating material covering the axons of many neurons and responsible for increasing conduction velocity in those neurons.

myoblast—A muscle-forming cell.

myofibrils—Longitudinal bundles of thick and thin contractile filaments located in muscle fibres.

myology—The study of the muscular system.

myosin—The thick protein filaments of muscle.

myotatic reflex—The simple monosynaptic muscle-stretching reflex where the excitation of a muscle spindle receptor by the imposition of stretch to a muscle causes that muscle to contract.

neuromuscular training—Exercise program designed to enhance postural equilibrium and intermuscular control.

neuron—A nerve cell. The fundamental building block of the nervous system.

neurotransmitter—A chemical agent released by one nerve cell that acts on another nerve or muscle cell by altering its electrical activity or state.

neurotrophic phase—In weight training, the initial phase of motor learning.

newton—The Standard International unit for measurement of force that imparts an acceleration of 1 m/s on a mass of 1 kg.

newton metre—The Standard International unit for measurement of the moment of a force.

norepinephrine—A hormone released in response to physical stress that causes a host of physiological responses, including increased heart rate and blood pressure; can also act as a neurotransmitter.

olfactory information—Information derived through the sense of smell.

onset of blood lactic acid accumulation—See *lactate threshold*

open-loop control—A type of movement control where the efferent (motor) commands are planned before the movement is initiated and control is not dependent on any sensory feedback arising during the movement itself; see also *motor program*.

optimal level of arousal—The level of arousal at which performance is best; varies between sports and between athletes.

optimization—A technique used to determine the best way of doing a task.

osmolality—A measure of osmotic activity. Osmotic activity occurs when one solution, usually lower in mineral content, passes across a cell membrane to dilute the concentration of the same mineral content on the other side.

osteoarthrosis—The degeneration of articular cartilage.

osteoblast—A bone-forming cell.

osteoclast—A bone-eroding cell.

osteocyte—A bone cell.

osteology—The study of the skeletal system.

osteon—See *Haversian system*.

osteopenia—Reduced bone density.

osteoporosis—Reduced bone density below a certain level, likely to be associated with fractures.

otolith organs—Set of two specialised receptors (the utricle and the saccule) located in the inner ear in which the movement of calcium carbonate stones (the otoliths) against nerve endings signals horizontal and vertical linear acceleration of the head.

overtraining—The second phase of training stress syndrome in which there is repeated failure to cope with chronic training stress.

overtraining syndrome—See *training stress syndrome*.

ovum—Female sex cell containing half the normal number of chromosomes.

oxygen debt—See *excess postexercise oxygen consumption*.

oxygen deficit—Difference between the amount of energy required during exercise and that which can be supplied by oxidative metabolism. Energy requirements that cannot be met oxidatively must be supplied by the anaerobic energy systems (immediate and glycolytic systems).

Pacinian corpuscles—Specialised platelike nerve endings located deep in the skin that respond to pressure, deep compression, or high-frequency vibration.

palpation—Using the senses of touch and pressure to identify anatomical features on a living human.

palsy—A persisting movement disorder due to brain damage acquired around the time of birth.

Parkinson's disease—A disease characterised by tremor, rigidity, and delays in the initiation of movement caused by a deficiency in the production of the neurotransmitter dopamine in regions of the basal ganglia.

particle—An object that is infinitesimal in size but whose mass can be represented as being located at one point.

peak height velocity—Point of most rapid growth in height; a major landmark in pubertal growth.

perception—The set of processes through which a person interprets the current and future state of his or her internal and external environments.

perceptual–motor integration—The integration of perceptual and motor information in movement control in order to produce movements tightly tuned to environmental demands.

peripheral nervous system—The collection of all the nerve fibres that connect the receptors and the effectors to the brain and spinal cord.

peripheral neuropathy—Disturbances to the function or structure in the peripheral nervous system.

personality—The composite of the characteristic differences that make each person unique.

phenotype—A somatotype determined by both genetic and environmental factors.

phosphocreatine—High-energy phosphate molecule stored in muscle cells. Cleavage of the terminal phosphate bond yields energy used to rephosphorylate adenosine diphosphate to adenosine triphosphate at the onset of exercise.

phosphorus—A mineral stored in bone.

physical activity—According to the U.S. Surgeon General, bodily movement produced by contraction of skeletal muscle that increases energy expenditure above the basal level.

physical education—The profession concerned with education of people in, through, and about physical activity.

plasticity—Flexibility or adaptability of structure or function; often used specifically in reference to the ability of early embryonic cells to alter in structure or function to suit the surrounding environment.

polyarticular—Referring to a skeletal muscle that crosses more than two synovial joints.

pons—Major anatomical component of the brain stem.

positron emission tomography—Production of images of sections through the body using positively charged electrons as the energy source.

postnatal—After birth.

postural reflexes—The reflexes present in the first year of life that are responsible for keeping the head upright and the body correctly oriented with respect to gravity.

potential energy—Amount of mechanical energy a body possesses due to a change in its position; results from the gravitational acceleration experienced by the body as it moves from one position to another near the surface of the earth.

power—The time rate of doing work.

practice-dependent learning—Learning that is an immediate consequence of engagement in practice and typically estimated from changes in performance from the start of a practice session to the end of a practice session.

precontemplation stage—The first stage of the transtheoretical model. In this stage, individuals have no intention of changing behaviour.

premotor cortex—Area of the brain situated just forward of the primary motor cortex.

prenatal—Before birth.

preparation stage—The third stage of the transtheoretical model. In this stage, individuals have made some steps toward changing their behaviour.

primitive reflexes—Reflexes present at birth that function predominantly for protection and survival.

principle of work and energy—A fundamental law of mechanics that states that the work done on an object is equal to the change in its kinetic energy.

profile of mood states—A psychological inventory used to monitor individual differences in six mood-state components.

progressive muscular relaxation—A relaxation technique involving the systematic contraction and relaxation of specific muscle groups throughout the body.

pronation—Movement that brings the hand or foot into the prone (palm or sole facing down) position. Hand pronation is achieved by medial (inward) rotation of the forearm.

proprioception—The sense of the position of the body and the movement of the body and its limbs.

proximodistal principle—Principle stating that development of control over muscles close to the body (attached to the axial skeleton) occurs before development of control over muscles of the limbs (attached to the appendicular skeleton).

psychoneuromuscular theory—A theory that states that imagery aids skill learning because when a person imagines moving, the brain sends electrical signals to the muscles in the same order as in actual movement.

puberty—The period of rapid physical growth between childhood and adulthood.

Purkinje cells—Large branching neurons in the cerebellum.

pyramidal cells—Nerve cells within the pyramidal tract. So named because of their pyramid shape (i.e., broad at the top, with a large branching tree of dendrites funnelling down to a single, slender axon).

pyramidal tract—The major descending motor pathway consisting of nerve cells having their origins in the cerebral cortex and synapsing directly with motor neurons at the spinal cord level.

Quetelet index—Ratio of body mass (m^2) in kilograms to height. Used for recommending appropriate body mass; also known as body mass index.

radiographic—Approaches involving the making of film records (radiographs) of internal body structures using X-rays or gamma rays.

radiological—Pertaining to the use of X-rays to visualise structures in the body.

random practice—A type of practice in which, in contrast to blocked practice, component skills are practised together in an unstructured order. One trial of practice on any particular skill component can be followed, in random order, by a trial of practice on any other (or the same) skill component.

reaction time—The time elapsed between the presentation of a stimulus and the initiation of a response to that stimulus.

reciprocal inhibition—The inhibition of one motor neuron or nerve pathway by the simultaneous excitation of another having an opposing action (e.g., the inhibition of extensor motor neurons during the activation of flexor motor neurons).

recognition—The perceptual process responsible for the identification of stimulus patterns.

recovery-stress questionnaire for athletes—A sport-specific questionnaire that measures the frequency of current stress and recovery-associated activities.

reflex—An involuntary movement elicited by a specific stimulus and typically completed rapidly and without conscious thought.

reliability—The extent to which measurements can be repeated consistently.

repetition maximum (RM)—Maximum weight that can be lifted a specified number of times (e.g., 3RM = maximal weight that can be lifted only three times).

resistance training—Exercise program that builds muscular strength.

response time—The time elapsed between the presentation of a stimulus and the completion of a response to that stimulus (i.e., the sum of reaction time plus movement time).

resting heat—The heat released by a muscle during its resting state.

resultant force—Vector sum of all forces acting (on a body).

reticular formation—A loosely defined network of cells extending from the upper part of the spinal cord through the medulla and pons to the brain stem; responsible for the control and regulation of attention, alertness, and arousal.

rheumatoid arthritis—Inflammation of the synovial membrane.

righting reflexes—Reflexes designed to return the body to an upright position after any event causing a loss of balance.

rigid body—Idealized model of an object that does not deform or change its shape; that is, the distance between every pair of points on the body remains constant.

Ruffini corpuscles—Specialised platelike nerve endings located in the skin that respond to pressure or touch.

saccule—The otolith organ responsible for the detection of vertical linear acceleration of the head.

sagittal plane—A plane dividing the body into left and right sides.

sarcomere—The structural and functional unit of skeletal muscle, containing thick and thin filaments.

scalar—Any physical quantity that is fully described by a number (e.g., length, mass, moment of inertia).

scholasticism—The philosophical doctrine of the Middle Ages that placed emphasis on the development of the mind and scholarly pursuits to the exclusion of any concentration on the body or physical activity.

secular trend—Changes in physical dimensions between people of one generation and following generations.

selective attention—The perceptual process of attending to one stimulus or event in preference to all others.

self-efficacy—The confidence one has in one's ability to successfully perform a particular behaviour.

semicircular canals—Set of three specialised receptors, located in the inner ear and positioned at 90° to each other, that respond to angular acceleration of the body in three planes.

serial order—The challenge to the nervous system to structure efferent commands such that motor units are recruited, and movement elements are produced, in the correct sequence.

sexual dimorphism—Structural differences between males and females.

shear—Describing a type of force in which adjacent parts of a structure move parallel to each other in opposite directions.

shortening heat—The difference between heat liberation by a muscle when it shortens and the same muscle when it contracts isometrically.

simple joint—Synovial joint involving a single pair of articulating surfaces.

size principle—The principle by which motor units are recruited in order of their size, from smallest to largest.

skeletal muscle—Striated muscle forming the major muscles of the trunk and limbs.

skill acquisition—The process through which skills are learned.

skinfold calipers—An instrument used to measure the thickness of a skinfold.

sleep-dependent learning—Learning that occurs after the completion of a practice session that is a consequence of the memory consolidation during sleep; typically estimated from changes in performance from the end of one practice session to the start of the next practice session following sleep.

sliding-filament hypothesis—Proposed explanation of the mechanics of striated muscle contraction.

smooth muscle—Muscle tissue forming the walls of organs such as blood vessels.

social facilitation—The improvements in performance that occur in the presence of other people.

somatic anxiety—Physiological responses to perceived threat.

somatochart—A special diagram indicating the three components of a person's somatotype.

somatotype—A shorthand way of representing the shape and composition of a person.

spasticity—An increase in the tone (resting tension) of certain muscle groups involved in the maintenance of posture; thought to arise from damage to extrapyramidal motor fibres.

specific gravity—The density of a material compared with the density of water.

speed—Magnitude of velocity.

spermatozoan—Male sex cell containing half the normal number of chromosomes.

spinal nerve—One of the 31 pairs of nerves arising from the spinal cord.

spongy bone—Less-dense bone that appears to be porous.

sport and exercise psychology—The scientific study of human behaviour and cognition in sport and physical activity.

sport devaluation—A characteristic of burnout in sport in which athletes lose interest in their sport and their performance in it.

stadiometer—Instrument used to measure height.

staleness—The initial phase of stress syndrome in which the person fails to cope with the psychological and physiological stress of training.

state anxiety—The experience of apprehension at a particular point in time.

states—How individuals feel at a particular point in time; transitory feelings.

static equilibrium—The state in which a body has zero velocity and zero acceleration. A body is in equilibrium if and only if the sum of all forces and the sum of all moments acting on the body are zero.

statics—The study of the objects in equilibrium.

stature—Total body height.

step length—Distance between the left heel and the right heel when both legs are on the ground during walking.

step width—Pelvic width divided by ankle spread, where ankle spread is the distance between the two ankles when both legs are in contact with the ground.

stepwise multiple regression—A statistical procedure in which one dependent variable (e.g., body density) is determined from a number of predictor variables (e.g., skinfold thickness from different sites). The predictor variables are added in (or subtracted) one at a time (i.e., stepwise) to obtain the best possible prediction.

sternum—The breastbone.

stiffness—A change in force applied to a material divided by deformation of the material.

stimulus–response compatibility—A measure of the naturalness of the links between a particular signal (stimulus) and its associated (movement) response.

strain energy—Mechanical energy stored in the elastic tissues of muscle and tendon.

strength—The capacity to produce force against an external resistance.

stress—Force applied per unit area. Muscle stress, for example, is muscle force divided by muscle cross-sectional area.

stretch-shortening cycles—Cycles of activity in which muscles are first placed on stretch before contracting concentrically (i.e., shortening).

stroke volume—Volume of blood pumped by the heart with each beat.

subchondral bone—The thin layer of compact bone under the articular cartilage.

subcutaneous—Beneath the skin.

supplementary motor area—Area of the cerebrum, located forward of the motor cortex, involved in the production and control of skilled movement.

symbolic learning theory—A theory that states that imagery helps to develop a blueprint of a movement sequence without actually sending any messages to the muscles.

synapse—The junction between two neurons or nerve cells.

syndrome X—See *metabolic syndrome*.

synergetics—The study of complex pattern formation and transitions between patterns.

synovial fluid—A type of viscous fluid formed by the synovial membrane and contained in a synovial joint cavity.

synovial joint—A type of joint containing fluid in a cavity surrounded by a capsule.

synovial membrane—Thin membrane lining the inner surface of the joint capsule.

tangential acceleration—Component of acceleration of a point on a body given by the angular acceleration of the body multiplied by the radius of the circle on which the body is moving at the instant under consideration; directed at a tangent to the circle on which the body is moving at the instant under consideration.

task orientation—The tendency to have a self-referenced definition of success.

tendon—A dense regular connective tissue joining muscle to bone.

tensile strength—Strength in opposing breaking when a material is stretched.

tension—A type of force in which the two ends of a structure are pulled apart.

thalamus—A mass of grey matter near the base of the cerebrum.

thoracic index—Relationship between breadth and front-to-back diameter of the chest.

tonic reflexes—Reflexes that are the basis for posture; concerned with the maintenance of the position of one body part with respect to others.

torque—See *moment (of a force)*.

torsion—A combination of compressive, tensile, and shear forces that results in twisting of a structure.

total mechanical energy—The sum of the kinetic energy and gravitational potential energy of a rigid body.

trabeculae—Small, bony rods that form the framework of spongy bone.

trabecular bone—See *spongy bone*.

tract—A large bundle of nerve fibres in the central nervous system.

training stress syndrome—The psychological and physiological negative adaptation to training characterised by three phases: staleness, overtraining, and burnout.

trait anxiety—A predisposition to perceive situations as stressful or threatening.

traits—The stable and enduring predispositions each individual has to act in a certain way across a number of different situations.

transcranial magnetic stimulation—A technique for examining brain function in which a magnetic field is applied transiently to specific areas of the brain. Depending on the frequency and intensity of stimulation used, transcranial magnetic stimulation can either act to stimulate the neurons in a particular area or briefly disrupt processing at the site of stimulation.

transdisciplinary—See *interdisciplinary*.

transtheoretical model—A model suggesting that people go through a series of five stages when changing behaviour.

transverse plane—A plane through the body that is parallel to the floor.

turning effect—The rotation tendency arising from the application of a force.

type Ia afferents—Nerves providing sensory information from the noncontractile central portion of the muscle spindle receptors back to the spinal cord.

type Ib afferents—Nerves providing sensory information from the Golgi tendon receptors back to the spinal cord.

type II afferents—Nerves providing sensory information from the contractile end portion of the muscle spindle receptors back to the spinal cord.

type 2 diabetes—Mature-onset diabetes, characterised by insulin resistance and impaired glucose tolerance and related to obesity and inactivity.

U.S. Surgeon General's report—A landmark report, published in 1996, that summarised research showing a strong relationship between moderate physical activity and health and that called for further promotion of a physically active lifestyle for all Americans.

ultrasound scan—Production of images of sections through the body using an ultrasound probe. Moving pictures can also be recorded in real time.

utricle—The otolith organ responsible for the detection of horizontal linear acceleration of the head.

validity—The extent to which a measurement instrument actually measures the property it sets out to measure; generally established through correlation of the measure with an outside criterion or independent measure.

vector—A physical quantity that is described by both its magnitude and direction (e.g., displacement, velocity, acceleration, angular displacement, angular velocity, angular acceleration, force, moment).

velocity—The time rate of change of displacement.

ventral—Pertaining to the anterior or abdominal side of the body.

verbal-cognitive phase—The initial phase in the learning of a new skill where the emphasis is on conscious understanding of the task requirements.

vertebra—One of the bones forming the spine or vertebral column.

vertebrate—Animal with a spine or backbone.

vestibular apparatus—Sensory system in the inner ear consisting of the semicircular canals and otolith organs and responding to angular and linear accelerations on the head.

viscosity—A fluid's resistance to flow.

visual acuity—A measure of the capacity of the visual system to resolve fine detail in a viewed object.

visual cortex—Region at the back of the cerebrum involved in the processing of visual information.

$\dot{V}O_2$max—Maximal oxygen consumption or aerobic power, usually expressed as a volume of oxygen consumed per minute as measured in a progressive exercise test to volitional fatigue. High $\dot{V}O_2$max values are generally indicative of high capacity for endurance exercise.

waist-to-hip ratio—A simple measure of overweight and obesity; calculated as the measurement at the waist divided by measurement at the hips.

watt—The International System (SI) unit of measurement of power; equal to 1 joule/s.

Z disc—Boundary of sarcomere; site of attachment of thin filaments of skeletal muscle.

zygote—Single fertilised cell that is the result of union of a male and a female sex cell.

INDEX

Note: The italicized *f* and *t* following page numbers refer to figures and tables, respectively.

ABOUT THE AUTHORS

Bruce Abernethy, PhD, is professor of human movement science in the School of Human Movement Studies and deputy executive dean and associate dean (research) in the faculty of health sciences at the University of Queensland, Brisbane, Australia. He also holds a visiting professor appointment at the University of Hong Kong, where he was previously the inaugural chair professor and director of the Institute of Human Performance. He is also coeditor of *Creative Side of Experimentation.*

Abernethy earned his PhD from the University of Otago. He is an international fellow of the National Academy of Kinesiology (USA), a fellow of Sports Medicine Australia, and a fellow of Exercise and Sports Science Australia.

Stephanie J. Hanrahan, PhD, is a registered sport psychologist and an associate professor in the Schools of Human Movement Studies and Psychology at the University of Queensland, Brisbane, Australia. Hanrahan has over 20 years of experience in teaching human movement studies at the undergraduate level. She is a recipient of the University of Queensland's Excellence in Teaching Award. In addition to being part of the author team for the first two editions of *Biophysical Foundations of Human Movement,* Hanrahan has authored or edited nine other books.

Hanrahan is a fellow of the Australian Sports Medicine Federation and a fellow of the Association for Applied Sport Psychology, for which she is chair of the organization's International Relations Division. Hanrahan serves on the national executive committee of the College of Sport and Exercise Psychologists in the Australian Psychological Society.

Hanrahan earned her doctorate in sport psychology in 1990 from the University of Western Australia. She resides in Moorooka, Queensland, and enjoys traveling, Latin dancing, and kayaking.

Vaughan Kippers, PhD, is a senior lecturer in the School of Biomedical Sciences at the University of Queensland. He coordinates anatomy courses for students enrolled in medicine, physiotherapy, and occupational therapy programs. His major research involves the use of electromyography, in which the electrical signals produced by muscles as they contract are analyzed to determine muscular control of human movement.

Kippers is a fellow of the International Association of Medical Science Educators and is on the board of directors of that association. He is also secretary of the Australian and New Zealand Association of Clinical Anatomists.

Cycling and photography are Kippers' main interests. He commutes on a bicycle daily and regularly participates in long rides on weekends. He is a former president of Audax Queensland, an international long-distance cycling association.

Marcus G. Pandy, PhD, is a professor of mechanical and biomedical engineering in the department of mechanical engineering at the University of Melbourne, Parkville, Victoria, Australia. Pandy earned his PhD in mechanical engineering at Ohio State University in Columbus and then completed a postdoctoral fellowship in mechanical engineering at Stanford University. Before joining the University of Melbourne, he held the Joe J. King professorship in engineering at the University of Texas at Austin.

Pandy is an associate editor for the *Journal of Biomechanics* and a fellow of the Institute of Engineers Australia, the American Institute of Medical and Biological Engineering, and the American Society of Mechanical Engineers.

Ali McManus, PhD, is an associate professor and assistant director of the Institute of Human Performance at the University of Hong Kong. Her research focuses on the role exercise and free-living physical activity play in the health and well-being of children, the development of population measures of obesity and its associated health risks, and the provision of a more comprehensive understanding of the complex metabolic bases of exercise and physical activity in obese children.

McManus earned her PhD from the University of Exeter, UK. She lives in Clearwater Bay, Hong Kong, and enjoys going to the gym, horse riding, playing tennis, and spending time with her children, Tash and Bella, and husband, John.

Laurel T. Mackinnon, PhD, is a science writer and editor based in Brisbane, Queensland, Australia. She is also a former associate professor and now adjunct associate professor in the School of Human Movement Studies at the University of Queensland, Brisbane, Australia.

Mackinnon conducted research on the immune response to exercise in the 1980s and 1990s and is internationally recognized for her work on overtraining and immune function in athletes. She is the author of 6 books and 12 book chapters, including *Exercise and Immunology* (Human Kinetics, 1992), the first book to explore the intriguing relationship between exercise and immune response. She has published over 65 peer-reviewed articles in international journals.

Mackinnon has worked since 2000 as a science writer and editor. She is editing team manager for OnLine English, an Internet-based service that specializes in editing academic, research, and industry communications written by non-native speakers of English wishing to publish in English-language scientific journals.

Mackinnon is a fellow of the American College of Sports Medicine and a member of the Australasian Medical Writers Association. She is a former board member of the International Society of Exercise and Immunology (ISEI) and the Australian Association for Exercise and Sports Science. Mackinnon earned her PhD in exercise science from the University of Michigan.

She enjoys exercising, reading, and listening to classical and jazz music. Mackinnon resides in Brisbane, Queensland.